Neural Stem Cells for Brain and Spinal Cord Repair

Contemporary Neuroscience

Neural Stem Cells for Brain and Spinal Cord Repair

Edited by

Tanja Zigova, PhD

Center for Aging and Brain Repair, Departments of Neurosurgery, Anatomy and Pharmacology, University of South Florida College of Medicine, Tampa, FL

Evan Y. Snyder, MD,PhD

Department of Neurology, Harvard Medical School,
Harvard Institutes of Medicine & Beth Israel-Deaconess
Medical Center, Division of Newborn Medicine,
Children's Hospital-Boston, Boston, MA

Paul R. Sanberg, PhD, DSc

Center for Aging and Brain Repair, Departments of Neurosurgery, Neurology, Psychiatry and Pharmacology, University of South Florida College of Medicine, Tampa, FL

Humana Press
Totowa, New Jersey

For additional copies, pricing for bulk purchases, and/or information about other Humana titles, contact Humana at the above address or at any of the following numbers: Tel.: 973-256-1699; Fax: 973-256-8341; E-mail: humana@humanapr.com

Cover Illustration: Artist's conception of an in vitro-generated neurosphere clone, consisting of a clonogenic neural stem/progenitor cell and all of its progeny—neuronal and glial progenitor cells, and differentiating neurons and glia. Illustration by Thomas Reiniger and Bjorn Scheffler.

Cover design by Patricia F. Cleary.

Library of Congress Cataloging in Publication Data

Neural stem cells for brain and spinal cord repair / edited by Tanja Zigova, Evan Y. Snyder, and Paul R. Sanberg.
        p.;cm.—(Contemporary neuroscience)
    Includes bibliographical references and index.
    ISBN 1-58829-003-4 (alk. paper)
    1. Neurons. 2. Brain–Transplantation. 3. Stem cells. 4. Cell Transplantation. I. Zigova, Tanja. II. Snyder, Evan Y. III. Sanberg, Paul R. IV. Series.
    [DNLM: 1. Central Nervous System Diseases–therapy. 2. Cell Differentiation–physiology. 3. Nerve Regeneration. 4. Stem Cells–physiology. 5. Stem Cells–transplantation. WL 300 N493734 2002]
    RD124.N475 2002
    617.4'8–dc21   2002190227

Dedications

*To Professor Pasquale Graziadei for unveiling the mystery
of this exciting field to me
and
to Sasha for so many reasons...–Tanja*

*To my parents, Harry and Martha, my inspiration and
mentors–Evan*

To my loving mother, Molly Sanberg–Paul

Preface

The stem cell field has captured the imagination of the therapeutic community. Stem cells are defined as those cells that can give rise to a vast array of cell types ("pluripotency") as well as to other stem cells ("self-renewal"). The most primordial stem cell is that obtained from the inner cell mass of the blastocyst; these are termed "embryonic stem cells" and can theoretically give rise to all of the cells of the developing organism. By a series of developmental steps, these embryonic stem cells are presumed to give rise to cells with similar stem-like qualities, but that come to populate the various emerging organs of the embryo and fetus. These cells, termed "tissue-resident" or "tissue-derived stem cells, retain stem-like features, but somehow have "learned" their address within the body. They are postulated to exist not only in the embryo and fetus where they participate in specific organogenesis, but to persist into adulthood, possibly to maintain homeostasis or to mediate self-repair or regenerative processes. The tissue-resident stem cell that has been most extensively explored is the hematopoietic stem cell. It made sense that such a cell should exist because cells of the blood are constantly turning over. That stem cells could similarly reside in solid organs, however, was a revelation. Stem cells from the nervous system, neural stem cells (NSCs), were the first solid organ stem cell identified, isolated, expanded and grown in culture, characterized, transplanted, and employed in animal models. First "unveiled" by developmental biologists, the NSC has come to capture the imagination of the neuroregeneration community. Furthermore, because it was the first solid organ stem cell employed, and because there has grown such a rich knowledge of its behavior and therapeutic potential over the past 15 years, the NSC has come to serve as a prototype for understanding and exploiting stem cell biology within most other solid organs throughout the body. Because it was also the first solid organ stem cell to be abstracted from human material, it has shown the therapeutic community what potential might exist. Therefore, a book such as this will hold interest for all readers interested in the biology and therapeutic promise of the stem cell. Though the field is moving quickly, much work still needs to be done.

In the last few years, rapidly advancing NSC research has been providing encouraging evidence that such cells can serve as novel tools for repairing/restoring central nervous system anatomy and perhaps function damaged by acquired (e.g., from trauma) and neurodegenerative (e.g., genetically based) disorders. Scientists are exploring differences between different sources of embryonic, fetal, and adult stem cells and trying to understand their fundamental biological properties and functional characteristics that may determine the appropriateness of various cell sources for specific clinical applications. At present there has been a considerable amount of data gathered on multipotential NSCs as a source of transplantable progenitors for CNS repair. Their fundamental properties, such as proliferation and differentiation under specific culture conditions, have been described in a variety of mammalian species including humans. Their ability to join developmental programs within the recipient's brain as well as to participate in restoring missing enzymes or genetic information, or simply replacing lost/damaged cells in the diseased or injured brain, has become obvious. Here we have tried to outline the most important basic issues of NSC research with a primary focus on the translational aspects of the NSC field. Because the utilization of human embryonic and/or fetal tissue, though the "gold standard" for stem cell efficacy and safety, is alloyed with ethical and logistical considerations, researchers have begun the search for alternative yet equivalent sources of stem cells. The jury is still out as to whether they exist for the nervous system. (For example, it remains unclear whether stem cells derived from one tissue can "change addresses" and commitments to take on the identity of another.)

The promising area of stem cell research as it applies to nervous system dysfunction is still in its infancy and it may be quite premature to predict when it will generate cell therapies for patients.

Neural Stem Cells for Brain and Spinal Cord Repair is divided into three main sections. The first section "Stem Cells for and from the CNS: fundamental properties" summarizes potential sources of stem cells and their prospective advantages in cellular transplantation therapies, as well as their enormous plasticity to generate diverse cell types inside and outside of the original germ layer.

The second section: "In vitro and/or in vivo Manipulations of Stem/Progenitor Cells for the CNS" contains chapters illustrating the signaling pathways that regulate stem cell division and differentiation. Further chapters define methods of NSC expansion and propagation, neuromorphogenesis, factors determining cell fate both in vitro and *in situ*, and the induction of self-reparative processes within the brain.

While in the first two parts, the reader is given a detailed view of the fundamental biology of stem cells in the CNS; the third section, "Stem/progenitor Cells in Representative Therapeutic Paradigms for the CNS" gives the reader material on strategies that may lead to fruitful clinical therapeutic applications in the near future. It is intriguing to recognize that the same stem cell may play a therapeutic role by a number of mechanisms, at times simultaneously. These may include replacement of degenerated, dysfunctional, or maldeveloped cells, as well as the provision of factors (either produced intrinsically by the cells or induced to do so following genetic engineering) that may protect, correct, recruit, promote self-repair in, or mediate connectivity of host cells.

With deep gratitude we acknowledge the contributing authors who have given generously of their time and knowledge to make this information available to those who will read this book. Grateful thanks are also extended to Humana Press, without whose vision, patience, and support this book might never have been published.

*Tanja Zigova,*PhD
Evan Y. Snyder, MD,*PhD*
Paul R. Sanberg, PhD, DSc

Contents

Contributors

KAREN S. ABOODY • *Departments of Neurology, Pediatrics and Neurosurgery, Children's Hospital Harvard Medical School, Boston, MA, and Layton Bioscience, Sunnyvale, CA*

GUILLERMO ALEXANDER • *Department of Neurology, MCP-Hahnemann University, Philadelphia, PA*

S. AUSIM AZIZI • *Department of Neurology, Temple University, Philadelphia, PA*

JUNG H. BANG • *Brain Disease Research Center, Ajou University School of Medicine, Suwon, Korea*

CESARIO V. BORLONGAN • *Development and Plasticity Section, Cellular Neurobiology Branch, National Institute on Drug Abuse, NIH, Baltimore, MD*

MELISSA K. CARPENTER • *Geron Corporation, Menlo Park, CA*

ELENA CATTANEO • *Department of Pharmacological Sciences, University of Milano and Center of Excellence on Neurodegenerative Diseases, Milan, Italy*

HYUN B. CHOI • *Division of Neurology, Department of Medicine, University of British Columbia, Vancouver, BC, Canada*

LUCIANO CONTI • *Department of Pharmacological Sciences, University of Milano and Center of Excellence on Neurodegenerative Diseases, Milan, Italy, and Centre for Genome Research, University of Edinburgh, Edinburgh, UK*

MARCEL M. DAADI • *Layton Bioscience Inc., Sunnyvale, CA*

ROSSELLA GALLI • *Stem Cell Research Institute, DIBIT, H. San Raffaele, Milan, Italy*

STEVEN A. GOLDMAN • *Department of Neurology and Neuroscience, Cornell University Medical College, New York, NY*

ANGELA GRITTI • *Stem Cell Research Institute, DIBIT, H. San Raffaele, Milan, Italy*

KOZO HATORI • *Division of Neurology, Department of Medicine, University of British Columbia, Vancouver, BC, Canada*

JEAN KIM • *Brain Disease Research Center, Ajou University School of Medicine, Suwon, Korea*

SEUNG U. KIM • *Division of Neurology, Department of Medicine, University of British Columbia, Vancouver, BC, Canada, and Brain Disease Research Center, Ajou University School of Medicine, Suwon, Korea*

OSAMU KAKINOHANA • *Department of Anesthesiology, University of California, San Diego, La Jolla, CA*

BARBARA KRYNSKA • *Department of Neurology, Temple University, Philadelphia, PA*

VALERY G. KUKEKOV • *Departments of Neuroscience and Neurosurgery, The McKnight Brain Institute, and Stem Cell Program, University of Florida, Gainesville, FL*

MAHESH LACHYANKAR • *Departments of Neurology, Pediatrics and Neurosurgery, Children's Hospital Harvard Medical School, Boston, MA, and Layton Bioscience Inc., Sunnyvale, CA*

ERIN LAVIK • *Department of Materials Science and Engineering, Massachusetts Institute of Technology, Cambridge, MA*

ERIC D. LAYWELL • *Departments of Neuroscience and Neurosurgery, The McKnight Brain Institute, and Stem Cell Program, University of Florida, Gainesville, FL*

MIN C. LEE • *Division of Neurology, Department of Medicine, University of British Columbia, Vancouver, BC, Canada*

MYUNG A. LEE • *Brain Disease Research Center, Ajou University School of Medicine, Suwon, Korea*

SHAOXIANG LIU • *Departments of Neurology, Pediatrics and Neurosurgery, Children's Hospital, Harvard Medical School, Boston, MA*

PAUL LU • *Department of Neurosciences, University of California–San Diego, La Jolla, CA, and Veterans Administration Medical Center, San Diego, CA*

JEFFREY D. MACKLIS • *Division of Neuroscience, Children's Hospital, and Department of Neurology and Program in Neuroscience, Harvard Medical School, Boston, MA*

SANJAY S. P. MAGAVI • *Division of Neuroscience, Children's Hospital, and Department of Neurology and Program in Neuroscience, Harvard Medical School, Boston, MA*

MARTIN MARSALA • *Department of Anesthesiology, University of California, San Diego, La Jolla, CA*

MARK MATTSON • *LNS, National Institute of Aging, Baltimore, MD*

KATHERINE A. MORTATI • *Department of Neurology, MCP-Hahnemann University, Philadelphia, PA*

ATSUSHI NAGAI • *Division of Neurology, Department of Medicine, University of British Columbia, Vancouver, BC, Canada*

EIJI NAKAGAWA • *Division of Neurology, Department of Medicine, University of British Columbia, Vancouver, BC, Canada*

MARTA NUNES • *Department of Neurology and Neuroscience, Cornell University Medical College, New York, NY*

JITKA OUREDNIK • *Departments of Neurology, Pediatrics and Neurosurgery, Children's Hospital Harvard Medical School, Boston, MA, and Layton Bioscience Inc., Sunnyvale, CA*

VACLAV OUREDNIK • *Departments of Neurology, Pediatrics and Neurosurgery, Children's Hospital Harvard Medical School, Boston, MA, and Layton Bioscience Inc., Sunnyvale, CA*

AKIHIKO OZAKI • *Division of Neurology, Department of Medicine, University of British Columbia, Vancouver, BC, Canada*

JAMES J. PALACINO • *Departments of Neurology, Pediatrics and Neurosurgery, Children's Hospital Harvard Medical School, Boston, MA*

KOOK IN PARK • *Departments of Neurology, Pediatrics and Neurosurgery, Children's Hospital Harvard Medical School, Boston, MA; Department of Pediatrics, Yonsei University College of Medicine, Seoul, South Korea; and Department of Neurology, Beth Israel-Deaconess Medical Center, Harvard Institute of Medicine, Harvard Medical School, Boston, MA*

DANIEL A. PETERSON • *Neural Repair and Neurogenesis Laboratory, Department of Neuroscience, The Chicago Medical School, North Chicago, IL*

DARWIN PROCKOP • *Center for Gene Therapy, Tulane University, New Orleans, LA*

MAHENDRA S. RAO • *LNS, GRC, National Institute of Aging, Baltimore, MD*

JASODHARA RAY • *Laboratory of Genetics, The Salk Institute, La Jolla, CA*

THOMAS REINIGER • *Departments of Neuroscience and Neurosurgery, The McKnight Brain Institute, and Stem Cell Program, University of Florida, Gainesville, FL*

NEETA ROY • *Department of Neurology and Neuroscience, Cornell University Medical College, New York, NY*

JAE K. RYU • *Brain Disease Research Center, Ajou University School of Medicine, Suwon, Korea*

PAUL R. SANBERG • *Center for Aging and Brain Repair, Departments of Neurosurgery, Neurology, Psychiatry and Pharmacology, University of South Florida College of Medicine, Tampa, FL*

JUAN SANCHEZ-RAMOS • *Center for Aging and Brain Repair, Departments of Neurology, Neurosurgery, Psychiatry, and Pharmacology, University of South Florida, Tampa, FL, and James Haley VA Medical Center, FL*

BJORN SCHEFFLER • *Departments of Neuroscience and Neurosurgery, The McKnight Brain Institute, and Stem Cell Program, University of Florida, Gainesville, FL*

EMILY J. SCHWARZ • *Center for Gene Therapy, Tulane University, New Orleans, LA*

EVAN Y. SNYDER • *Department of Neurology, Harvard Medical School, Harvard Institutes of Medicine & Beth Israel-Deaconess Medical Center, Division of Newborn Medicine, Children's Hospital-Boston, Boston, MA*

SHIJIE SONG • *Center for Aging and Brain Repair, Department of Neurology, University of South Florida, Tampa, FL, and James Haley VA Medical Center, FL*

DENNIS A. STEINDLER • *Departments of Neuroscience and Neurosurgery, The McKnight Brain Institute, and Stem Cell Program, University of Florida, Gainesville, FL*

OLEG N. SUSLOV • *Departments of Neuroscience and Neurosurgery, The McKnight Brain Institute, and Stem Cell Program, University of Florida, Gainesville, FL*

BARBARA A. TATE • *Departments of Neurology, Pediatrics and Neurosurgery, Children's Hospital Harvard Medical School, Boston, MA, and Layton Bioscience Inc., Sunnyvale, CA*

ROSEANNE TAYLOR • *Department of Animal Science, Faculty of Veterinary Science, University of Sydney, Australia*

YANG D. TENG • *Departments of Neurology, Pediatrics and Neurosurgery, Children's Hospital, Harvard Medical School, Boston, MA, and Department of Neurosurgery, Brigham & Women's Hospital, Harvard Medical School, Boston, MA*

JOHO TOKUMINE • *Department of Anesthesiology, University of California, San Diego, La Jolla, CA*

MARK H. TUSZYNSKI • *Department of Neurosciences, University of California–San Diego, La Jolla, CA, and Veterans Administration Medical Center, San Diego, CA*

ANGELO VESCOVI • *Stem Cell Research Institute, DIBIT, H. San Raffaele, Milan, Italy*

MARY KATHERINE C. WHITE • *Departments of Neurology, Pediatrics and Neurosurgery, Children's Hospital, Harvard Medical School, Boston, MA, and Department of Neurology, Beth Israel-Deaconess Medical Center, Harvard Medical School, Boston, MA*

Martha Windrem • *Department of Neurology and Neuroscience, Cornell University Medical College, New York, NY*

Tony L. Yaksh • *Department of Anesthesiology, University of California, San Diego, La Jolla, CA*

Tong Zheng • *Departments of Neuroscience and Neurosurgery, The McKnight Brain Institute, and Stem Cell Program, University of Florida, Gainesville, FL*

Tanja Zigova • *Center for Aging and Brain Repair, Departments of Neurosurgery, Anatomy and Pharmacology, University of South Florida College of Medicine, Tampa, FL*

I
STEM CELLS FOR AND FROM THE CNS
THEIR SOURCE AND FUNDAMENTAL PROPERTIES

Sources of Cells for CNS Therapy

Melissa K. Carpenter, Mark Mattson, and Mahendra S. Rao

TRANSPLANT THERAPY

The goal in transplant therapy is to provide adequate numbers of cells to appropriate sites for useful cellular replacement. In principle, three types of therapy can be considered. One straightforward therapeutic approach is to use dissociated cells for replacement. In this approach, either purified populations of cells or mixed populations of cells can be used. A second therapeutic approach is to develop devices that are a combination of synthetic material and cells. A third possibility is to use stem cells and precursor cells to generate organs in culture. These synthetic organs can then be transplanted. Each of these approaches has been tried with varied amounts of success. Generating organs has been possible for some structures such as skin, blood vessels, lens, and bone (*see* ref. *1* and references therein), but the inherent complexity of even the simplest neural structures currently renders this an impractical approach for the nervous system. A combination of synthetic material and cells have been used in retinal implants, cochlear replacements, nerve electrode junctions, synthetic neural networks or guidance channels for nerve regrowth. These devices are still in developmental stages and a detailed discussion of such devices is beyond the scope of this chapter. In this chapter, we have limited our discussion to the relative advantages and disadvantages of the various human cell types that are available for dissociated cell replacement therapy.

As with any cell therapy in the brain, the following issues need to be considered: the cell source, the purity of the cells, the quantity of cells available, the ability of the cells to migrate, their ability to differentiate appropriately or inappropriately, and whether cues exist at the site of implantation to direct differentiation in vivo. An additional issue that needs to be considered when using dissociated cells is which cell type is appropriate for a particular neurological disorder. The potential role of transplanted

From: *Neural Stem Cells for Brain and Spinal Cord Repair*
Edited by: T. Zigova, E. Y. Snyder, and P. R. Sanberg © Humana Press Inc., Totowa, NJ

Table 1
Functional Uses of Transplanted Cells

Cellular replacement (neurons, astrocytes, oligodendrocytes)
Providing trophic support
Maintenance of synapses and axons
Myelination
Mobilization of endogenous stem cells
Introduction of foreign genes/proteins
Reduction in scar formation
Promotion of revascularization
Alteration of the immune response

Note: Different neurological disorders will require different subsets of the listed replacement functions. No single cell can adequately perform all of these functions.

cells in promoting functional recovery is summarized in Table 1. Functions such as promoting revascularization or modulating the immune response are likely as critical as replacing cells lost to damage. It is clear, however, that the functionally diverse roles expected of cellular replacement cannot be fulfilled by the transplantation of any single cell type; rather, specialized cells or titered mixtures of cells may be important in replacement strategies. Which cell types will work and how effective they will be is under active study and the variety of cells that have been used is quite astounding. These include macrophages and activated T-cells, multiple classes of glial cells, neuronal precursors, multipotent central nervous system (CNS) stem cells, pluripotent embryonic stem cells and their derivatives, and differentiated postmitotic cells (Table 2). In subsequent sections, we discuss the different sources of cells and their relative merits.

MULTIPLE SOURCES OF NEURAL DERIVATIVES EXIST

Conceptually, we have divided the sources of cells used for replacement into the following categories (summarized in Table 2): multipotent stem cells that can generate most differentiated cells in the nervous system, more restricted precursor cells that generate a subtype of differentiated cells, and differentiated cell populations. Each of these cell types can be further classified based on the time and source of isolation and the inherent ability to differentiate in a regionally specific manner. In addition to obtaining differentiated cells from multipotent or lineage-restricted precursors, the process of transdifferentiation has been described in which cells destined to contribute to other germ layer tissue have been shown to be able to con-

Table 2
Cells Used for Transplantation

Neural cells
  Multipotent
  Restricted precursors
  Differentiated
  Transdifferentiated
Non-neural cells
  Macrophages
  Transdifferentiated non-neural stem cells
Xenobiotic cells
  Neural, non-neural, and ES
Cell lines
  Immortalized
  Transformed cells
Pluripotent stem cells
  Embryonic germ cells
  Blastocyst-derived ES cells

Note: This is a list of the different kinds of cells used for therapy. Cells have been used as purified populations or mixtures of cells.

tribute to neural structures. Similarly, astrocytes have been shown to possess unique properties that might allow them to differentiate into other neural derivatives. Xenografts have been discussed as an alternative to allografts and have been shown to survive and make connections after transplantation. Furthermore, human embryonic stem (ES) and embryonic germ (EG) cells have been described and shown to differentiate into reasonable numbers of neural derivatives. An alternative to precursor or ES cells is genetically modified immortalized cell lines from different stages of development. Finally, it has been suggested that manipulating the immune system may allow the endogenous regeneration ability of the nervous system to be expressed and thus may be an alternative to generating neural derivatives. Thus, at least in principle, multiple sources of neural cells exist that could be used for cellular therapy (Table 2). Each of these cell sources needs to be evaluated for their efficacy. Some of the issues that we have considered in evaluating cells in subsequent sections are listed in Table 3. Currently, our operating assumption is that postmitotic cells do poorly in most transplant paradigms and that dividing immature cells that likely will respond to environmental signals to differentiate into region-specific phenotypes will be more valuable. In this chapter, we discuss recent advances in our

Table 3
Issues for Cell Transplant

Reliable and renewable source of cells
Ability to isolate a desired subpopulation
Ability to obtain adequate numbers of specific cells
Ability of cells to respond to cues in the adult and damaged environment
Ability of the cells to migrate and integrate in a site-specific fashion
Ability of the cells to modulate the immune response
Possible tumor formation from grafted cells

Note: Issues that need to be considered in evaluating cells for transplant therapy are listed. Although we have made some progress in identifying sources of cells for transplant therapy, substantial additional data characterizing these cells in vitro and in vivo is required before patient therapy can be considered.

understanding of these cells in the developing nervous system, the relative advantages and disadvantages of different classes of cells, their potential therapeutic use, and evidence from transplant studies.

MULTIPOTENT PRECURSOR CELLS

One approach to cellular replacement is to transplant multipotential cell populations (which can generate any required cell population) regardless of the specific need. The assumption in this approach is that appropriate cues will exist within the environment to direct tissue- and site-specific differentiation. With this view, several groups have worked on isolating multipotent stem cells and success has been reported by several groups (Table 4). Both fetal and adult stem cells have been isolated, have been maintained in culture for several years, and have been shown to retain their ability to differentiate even after prolonged passage in culture *(2–7)*. Two kinds of culture system have been used to grow multipotent stem cells. Suspension neurosphere-type cultures *(8)* have been used by many groups because these cultures offer several advantages (discussed below). Other groups have used adherent culture systems *(7)* to grow fetal neural stem cells. Figure 1 shows examples of stem cells grown under these culture conditions.

The potential advantages and disadvantages of using multipotent stem cells for transplantation are summarized in Table 5. Perhaps the single biggest advantage of multipotent stem cells is the ability of these cells to generate large numbers of cells required for transplantation. Furthermore, several investigators have established stem cell cultures in which cells can be

Table 4
Multipotent Stem Cells

Species	Age	Tissue	Clonal?	Differentiation potential	Reference
Human	Adult	Temporal lobe	N	Colonies= neurons only glia only neurons + glia neurons + non-neuronal cells	Kirschenbaum et al. (*161*)
Human	Fetal	Forebrain	N	Neurons, astrocytes,	Chalmers-Redman et al. (*2*)
Human	Fetal 7–10 wk	Cerebrum	N	Neurons, astrocytes, oligodendrocytes	Brustle et al. (*158*)
Human	Fetal 15 wk	Telencephalon	Y	Neurons, astrocytes	Flax et al. (*93*)
Human	Fetal 13 wk	Brain	Y	Neurons, astrocytes, oligodendrocytes	Sah et al. (*85*)
Human	Fetal	Cortex	N	Neurons, astrocytes	Svendsen et al. (*9*)
Human	Fetal 10.5 wk	Diencephalon	Y	Neurons, astrocytes, oligodendrocytes	Vescovi et al. (*3*)
Human	Fetal 6–12 wk	Forebrain	N	Neurons, astrocytes, oligodendrocytes	Carpenter et al. (*4*)
Human	Fetal 12–18 wk	Neural tube	N	Neurons, astrocytes	Piper et al. (*7*)

Note: Some examples of human multipotent neural stem cells are listed. Both fetal and adult tissue has been used to isolate multipotent stem cell populations. Most cells obtained have not been analyzed as clonal cultures and clonal HNSC (human neural stem cell lines) lines have not been obtained.

Adherent Cultures   **Suspension Cultures**

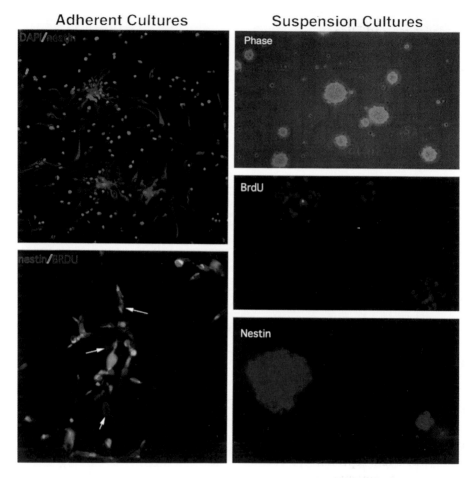

Fig. 1. Multipotent stem cell cultures. Two methods of culturing multipotent stem cells are shown. Cells have either been cultured in suspension cultures or as adherent cultures. In both cases cells express nestin, incorporate BRDU, and can be maintained in an undifferentiated state over multiple passages and under appropriate differentiation, stimuli can generate neurons, astrocytes, or oligodendrocytes in vitro and in vivo. A list of the different reports of isolation of human neural stem cells is provided in Table 4. [Suspension culture photos reproduced from Carpenter et al. *(4)*.]

grown as floating spheres over long time periods *(3,4,9,10)*. This culturing technique provides an economical way to generate large cell numbers with the ability to passage cells and the potential to adapt cells to bioreactor culture vessels. Another advantage of multipotent stem cells is that, in

Table 5
Advantages and Disadvantages of Multipotent Stem Cells

Advantages	Disadvantages
Can be passaged and large numbers of cells obtained	Loss of properties after passaging
Neurosphere cultures may be easily adapted to bioreactor cultures	Limited differentiation in vivo
Foreign gene introduction possible	Limited ability to bias differentiation
May respond to environmental cues	Unknown ability to respond to cues in the adult and damaged environment
All major neural types can be obtained	Possibility of heterotopias
Fetal stem cells may not be regionally specified	Unknown immunological compatibility

Note: The potential advantages and disadvantages of multipotent stem cells for transplant therapy are listed. Note that many of the disadvantages of using multipotent stem cells may apply to other cell types also.

principle, cells are spontaneously immortal. Most cells divide for only limited time periods in culture and then undergo senescence. However, multipotent stem cells appear to divide for much longer periods than other cells and express higher levels of telomerase than differentiated cells (*11*; our unpublished results). Multipotent stem cells therefore offer a realistic possibility to generate the vast numbers of cells required for transplantation in a reliable and reproducible manner.

There are, however, also potential shortcomings of using multipotent cells. One of the major problems is the efficiency of differentiation and, consequently, the efficiency of a transplant. Surprisingly little is known about the quantitative aspect of cellular differentiation following transplantation. For example, in many transplant experiments, the number of undifferentiated cells was large and only a small subset of cells differentiated appropriately. In other transplant experiments in which cells were transplanted into the brain parenchyma, only limited migration was seen (see, e.g., ref. *12*). In such cases, the area that could be reached by the transplant might be too small to confer a functional recovery (*13*).

Another possible "shortcoming" of using multipotential cells rather than other kinds of cells could be the influence of the host environment, the very same tool that is required to determine the fate of the transplanted populations. It has been shown that the adult brain harbors stem cells and these stem cells seem to be quiescent and not able to respond adequately or appropriately to damage (reviewed in ref. *14*). It is therefore unclear whether providing additional stem cells will necessarily be sufficient to adequately repair tissue in many brain regions. In the case of stem cells, it has been well established that the embryonic environment provides different cues than an adult environment and allows for different developmental processes to take place. It is unclear if appropriate environmental signals exist that will allow fetal stem cells to go through the series of lineage restrictions that are a normal part of the developmental process, before they become competent to differentiate into mature cell types. For example, transplant experiments in which multipotent stem cells have been transplanted into the spinal cord and hippocampus show that the adult spinal cord does not contain neuronal differentiation signals. Spinal cord stem cells will readily generate neurons in the hippocampus but not in the spinal cord *(15–17)*. Thus, undifferentiated multipotent neural stem cells cannot be used for neuronal replacement therapy in the adult intact or lesioned spinal cord.

A third potential problem is the stability of the cells in culture and the retention of their ability to differentiate in a site-specific and injury-specific fashion. Prolonged passage alters the differentiation potential of cells either increasing the repertoire of differentiation (see, e.g., refs. *18* and *19*) or decreasing the ability to differentiate into specific lineages *(20)*. Quinn and colleagues, for example, have shown that multipotent cells lose their ability to differentiate into neurons after prolonged culture.

The use of neural stem cells (NSC) as a reservoir of uncommitted cells that can be manipulated in vitro to yield more committed precursor cells may represent an alternative and equally valid use of NSCs. We have shown that more restricted precursors can be isolated from multipotent cells. Using cell surface markers, we have isolated neuronal restricted and glial restricted precursor cells from the embryonic spinal cord stem cells and shown that these cells will survive and integrate after transplantation *(21)*. Selecting the desired cell population from a pool of uncommitted multipotent stem cells bypasses many of the problems associated with transplanting undifferenti-ated multipotent stem cells. The possibility of tumors or unrestricted growth is reduced and one can deliver precise mixtures of selected cells that are appropriate for therapy without having to rely on the environment to provide appropriate cues. Overall multipotent stem cells represent a viable source of cells for therapy. Many issues remain unresolved, such as the following: How best to maintain cells? How to transplant? Should cells be preselected before transplantation? Where should the cells be transplanted? How should transplants be monitored?

Neural Crest Derivatives

The neural crest develops later in the evolution of vertebrates and is of importance because of the variety of cell types that are derived from this specialized population of cells *(22)*. Crest cells segregate from the developing neural tube at or around the time of neural tube closure. The exact process of delamination is species dependent *(23,24)*. Crest stem and precursor cells have been isolated and their multipotency and lineage restrictions demonstrated (reviewed in ref. *25*). These cells could, in principle, serve as an alternative source of multipotent stem cells, although several problems exist. In human tissue, crest migration occurs quite early (wk 5–6 of gestation) and fetal tissue at this early stage is generally unavailable. Thus, in humans at least, crest cells do not appear to be a viable source for therapy. Recently, in rodents, neural crest stem cells have been isolated at later stages in development *(26,27)* raising the possibility that it may be possible to isolate such populations in humans as well. However, even if this is possible, it has been difficult to grow neural crest cells for prolonged periods in culture; thus, the overall number of cells available would be unsuitable for any therapy.

Derivatives of neural crest however may be of therapeutic use. In particular, Schwann cells, the peripheral glia cell, may be of use in demyelinating lesions. Human Schwann cells have been isolated and have been maintained in culture for prolonged periods *(28–30)* and recent reports have suggested that rodent Schwann cells may not undergo senescence, thus serving as an inexhaustible source of cells without genetic manipulation *(31)*. Schwann cells can be obtained from adult tissue, and given their ready expansion, a small peripheral nerve biopsy can provide sufficient numbers of cells for autologous transplants. Based on these results, Schwann cells have been considered a viable alternative for therapy. An additional advantage of Schwann cells that is of specific interest in CNS repair is based on observations that Schwann cells may overcome the inhibitory effect of astrocytes on axonal regeneration. For example, Guenard and colleagues *(32)* reported that when purified astrocyte populations were seeded onto semipermeable guidance channels and grafted across an adult transected sciatic nerve, they effectively inhibited nerve regeneration. If, however, a mixture of Schwann cells/astrocytes was seeded, the inhibition could be overcome providing the number of Schwann cells was large enough *(32)*. Subsequently, Harvey and Plant *(33)* provide some evidence that Schwann cells might be useful for promoting axonal regeneration in the visual system and Guest et al. *(34)* showed, using human Schwann cells, that this population seems to be capable of promoting neurite outgrowth in the transected spinal cord of nude rats.

However, there is a potential disadvantage to be considered. Astrocytes prevent Schwann cell migration in vivo and restrict the extent of Schwann cell myelination seen (reviewed in ref. *35*). The complex interaction of astrocytes and Schwann cells that is required for successful remyelination places serious limitations on their potential in therapeutic applications. It should be noted, however, that all of the studies so far have used adult-derived or in vitro-expanded cell populations. It remains to be seen whether embryo-derived cells follow the same rules as their adult counterparts. Overall, in our opinion, neural crest cells themselves may not be a useful source of multipotent stem cells until alternative sources of crest cells can be identified. Neural crest derivatives such as Schwann cells, smooth muscle, and so forth, may be useful and need to be further evaluated in transplant paradigms.

Placodal Precursor Cells

Another set of precursor cells that contribute to the developing nervous system are placodal precursor cells. Several placodes develop in the cranial region at the junction between the neural tube and the ectoderm. Like neural crest, placodal precursors delaminate/involute from the overlying ectoderm in discrete locations. These cells then differentiate in a stereotypic fashion to form discrete placodes. Placodal cells undergo sequential steps of differentiation to generate neural cells in the cranial ganglia as well as non-neural derivatives such as part of the cranial mesenchyme, cells of the lens, and hair cells (reviewed in refs. *36–38*). Placodal cells differ from neural crest, however, in the markers they express, their timing of differentiation, and their differentiation capability *(37,39)*.

Placodal cells have received relatively limited attention from stem cell biologists although regeneration of hair cells and replacement of lens are important therapeutic targets. Regeneration of olfactory neurons is well known and receptor cells project back into the olfactory bulb and form synapses during normal replacement or when subjected to injury. The basal cells serve as neuronal stem cells *(40,41)*. Several groups have isolated human olfactory epithelial stem cells *(42,43)* and have shown that these cells can be grown as neurospheres and, when transplanted into the brain, can differentiate into neurons, astrocytes, and oligodendrocytes *(44,45)*.

Few detailed comparisons have been made between the ability of olfactory epithelial-derived stem cells and stem cells from other brain regions. Such comparisons would be useful in determining if these cells could serve as a source of precursors for autologous CNS therapy.

It is likely that multipotent stem cells (or precursor cells) persist in other placodes as well *(46)* and likely are derived from a self-renewing stem cell, although these placodal precursor cell have not been characterized in detail. This stem cell could also serve as a potential source of autologous precursors for therapy. However, to our knowledge, clonal stem cells have not been derived from placodes other than the olfactory placode.

A second placodal population that may be functionally important are olfactory ensheathing cells. The use of olfactory ensheathing cells (OECs) in wound cavities or spinal cord injuries had been discussed as early as 1995 *(47)*. Implanted ensheathing cells derived from the fetal rat olfactory bulb were shown to be able to survive in nonolfactory CNS regions and could support the growth of nonolfactory axons *(48)*, and cells derived from the adult olfactory bulb and transplanted into various lesion models allow for the regeneration of injured axons into the adult CNS *(49–51)*. The ability of OECs to form myelin and support axon regeneration gives these cells the same range of properties that had been documented for Schwann cells, but these cells may be superior to Schwann cells because of their ability to interact and intermingle with astrocytes. Schwann cells and olfactory placode cells offer the advantages of being readily amplified in culture and there is a possibility of obtaining an autologous source. Thus, they may be more useful than astrocytes and oligodendrocytes in some indications.

Overall, placodal precursor cells represent a potential alternative source of autologous cells. Both multipotent and more restricted precursors may be useful and need to be further evaluated. Data on human placodal cells are limited, but given the possibility of obtaining stem cells from cadaveric tissue as well as from adults examining the potential of these cells should be actively pursued.

RESTRICTED PRECURSORS

In addition to multipotent stem cells for therapy, it is possible to use more restricted precursor cells. Restricted precursors differ from multipotent cells in that they have a more limited repertoire of differentiation. Our laboratory has identified two such restricted precursors: a neuron-restricted precursor (NRP) and a glial-restricted precursor (GRP). Several other such restricted precursors have been identified (discussed below, see also Table 6) and

Table 6
Restricted Precursor Cells

Neuron-restricted precursors
   N-CAM immunoreactive HNRP cells; e.g., Piper et al. *(54)*
   Alpha-I tubulin expressing fetal HNRP cells; e.g., Wang et al. *(162)*
   Alpha-I tubulin expressing adult HNRP cells; e.g., Roy et al. *(111)*
Glial-restricted precursors
   Oligodendrocyte precursors fetal; e.g., Zhang et al. *(70)*
   Oligodendrocyte precursors adult; e.g., Scolding *(160)*, Roy et al. *(52)*
   Bipotential oligodendrocyte-astrocyte precursors fetal; e.g., Gregori
     et al. *(163)*

Note: Some examples of human restricted precursor cells are listed. Both fetal and adult tissue has been used to isolate restricted precursor cell populations. Most human cell populations obtained have not been analyzed as clonal cultures and non immortalized clonal restricted precursor cell lines have not been obtained.

human homologs of NRPs *(52–54)* and GRPs *(55;* Mayer-Proschel, personal communication) have been described. Only a limited amount of transplant data exists (Mayer-Proschel, personal communication), and to our knowledge, no human restricted precursors have been examined in a transplant paradigm. Nevertheless, these cells offer the possibility of replacing selective populations of cells or in titered mixtures replacing precise combinations of cells. The relative advantages and disadvantages of NRP and GRP cells is discussed next.

Neuronal Precursors

Historically therapeutic neuronal replacement has not been a widely considered option in most cases of injury. Although neuronal replacement has been considered imperative in degenerative therapy, neuronal transplants have not been widely tested. One important reason for this has been the observation that the adult CNS is inhibitory to axonal outgrowth; thus, transplanted neurons are unlikely to be able to generate long axons or make appropriate connections. Recent studies have suggested, however, that this inhibitory environment can be overcome (reviewed in ref. *56*) and that regions distal to the injury site may not be as inhibitory as previously thought *(57,58)*. Further important studies have suggested that fetal neurons may not respond to extrinsic inhibitory influences *(59–62)*. Such findings indicate that transplants of neurons for enhancing connectivity after spinal cord injury may be a viable option and several recent studies have been undertaken to test this hypothesis.

Neuron-restricted precursors (NRPs) may have multiple uses in therapy. NRPs may be a source of replacement neurons or may provide an intermediate synaptic target to maintain projection neurons, or transplanted neurons may serve to bridge connections between axons upstream of the damaged site and targets present downstream. Integrated transplanted precursor cells may allow for the formation of novel connections that could subserve the function of lost long-projection axons. Neurons may also provide trophic support for denervated axons as well as growth-promoting and regulatory molecules to regulate glial reaction and remyelination. The availability of NRP cells offers the potential of replacing solely neurons that are lost in a neurological disorder without the concomitant addition of potentially unwanted glial cells.

Some potential disadvantages of restricted cells also exist. In general, intermediate precursor cells often do not grow as extensively as more primitive stem cells. Thus, the degree of amplification achieved by culturing multipotent stem cells cannot be expected for NRP cells. An important recent observation that suggests that not all neuronal precursors are created equal comes from studies analyzing neuronal precursors from different brain regions. Progenitors from the hippocampus, but not from the cerebellum or midbrain, produce hippocampal pyramidal neurons *(63)*, suggesting a bias in differentiation potential that is maintained in culture. Likewise, Ling et al. *(64)*, analyzing rodent mesencephalic neural precursors, showed that, under appropriate conditions, as many as 50% of the neurofilament immunoreactive neurons appear dopaminergic. This frequency is much higher than that obtained from any other precursor cell population (reviewed in ref. *10*). Similarly, Luskin and colleagues noted that transplanting Svz(a) neurons into the striatum resulted in predominantly GABA-ergic rather than dopaminergic differentiation, raising the possibility that the restriction in developmental potential cannot be reversed even in vivo (reviewed in ref. *65*).

Apart from a bias in differentiation potential, differences in migration ability have been noted as well. Alvarez-Buylla and colleagues compared the migration ability of medial ganglion eminence (MGE) precursors and lateral ganglion eminence (LGE) precursors, and they noted that under identical transplant conditions, LGE neurons migrated extensively, whereas MGE neurons appeared to be restricted in their ability to migrate to the cortex *(66)*. A similar difference between Svz(a) precursor cells migration and spinal cord precursor migration was observed *(21)*. Svz(a) precursors were restricted to migrating along the rostral migratory stream, whereas spinal cord precursors migrated much more extensively but did not appear to migrate into the hippocampus. It is important to note that the cells in these experiments were isolated from E12–13 rodent embryos, a stage at which neurons are still dividing and have neither migrated nor received innervation. These data indicate that lineage commitment has occurred early in development, and although neuronal precursors offer several advantages, it will be important to isolate precursors from appropriate regions for optimal results. This represents an additional level of complexity and further restricts the number of neuronal precursor cells of the appropriate type available for therapy.

Glial Precursors

Glial cells may play a role in both promoting axonal growth and inhibiting regeneration. Oligodendrocyte precursors are critical for generating the myelin-producing oligodendrocytes and for remyelinating surviving axons. Astrocytes may serve a critical role in promoting oligodendrocyte remyelination, regulating the blood-brain barrier and immune response, modulating local glutamate metabolism and providing trophic support for surviving neurons (reviewed in ref. *66*). On the other hand, it has been suggested that myelin proteins and myelin-associated proteins produced by oligodendrocytes contain growth-inhibitory molecules and that reactive astrocytes are responsible for scar formation that may form a barrier to axonal regrowth (reviewed in ref. *66*). Transplanting cells for therapy therefore requires that the timing of replacement and the numbers and kinds of glial cells supplied be critically evaluated. Reconstitution of an appropriate glial environment will promote axonal regrowth and recovery, whereas inappropriate cells may well inhibit recovery. Multiple studies are underway to determine which of the many glial precursors identified (Table 6) will be useful for therapy.

Several different types of glial precursor have been identified in the developing and adult rodent brains (reviewed in ref. *67*). These include CNS myelinating cells such as O2A and GRP cells *(67)*, peripheral nervous system (PNS) myelinating Schwann cells (see above), and specialized aldynoglia *(68)*, such as olfactory ensheathing cells (see above). Multiple transplant studies in which each of these cell types was transplanted into demyelinating lesions have shown moderate to extensive remyelination, with measurable improvement in conduction velocity and frequency–response properties of the remyelinated axons (see, e.g., ref. *69*). Human homologs

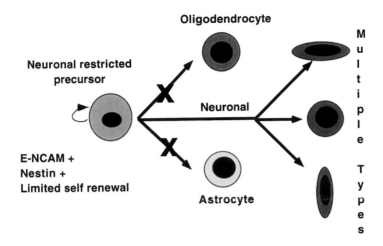

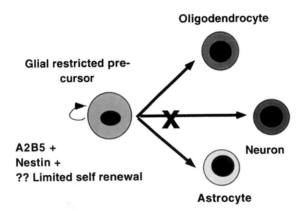

Fig. 2. Lineage-restricted precursors. An example of neuronal and glial restricted precursors is shown. Although many subtypes of neuronal and glial precursors have been described (*see* Table 6), all of these cell types express characteristic markers, proliferate to a limited extent in vivo, and are more restricted in their differentiation potential than multipotent neural stem cells.

of these precursor cells have been identified (Table 2). These include oligodendrocyte–astrocyte precursors, oligodendrocyte precursors, and astrocyte precursors (Fig. 2 and Table 6). These cells may have a greater potential to develop into astrocytes or oligodendrocytes as compared to multipotent stem cells and may be an alternative choice for transplant therapy in regions where glial replacement is required. Although each of these cell types can promote myelination, it is not clear which is the best cell for such purposes. In some cases, providing additional precursors similar to resident precursors may be therapeutically of greater benefit. In other cases, providing a novel cell type that may be more resistant to ongoing damage or host immune system may be a better choice.

Zhang et al. *(70)* have used rat NSCs from which they were able to isolate and expand a cell population they named oligospheres. According to their results, this population is able to generate an impressive amount of myelin-producing cells when retransplanted into the adult spinal cord. Although the authors did not compare myelination properties of stem cells and stem-derived precursor cells directly, a comparison with other published data does suggest that oligospheres appeared to myelinate more efficiently. Several other groups have also identified myelinating precursor cells (Table 6). Some of these cells can differentiate into astrocytes after transplantation in vivo (Dr. Mayer-Proschel, University of Rochester, personal communication) and may be better able to promote myelination. Data on the transplantation of other human myelinating precursors are limited, but based on the culture data, we would predict that these other glial precursors may be equally viable options for therapy.

Astrocyte precursors have been relatively less well studied in transplant paradigms, although the data available from rodent experiments clearly illustrate the importance of this class of cell. Blakemore and colleagues have shown that type-1 astrocytes facilitate the repair of demyelinating lesions by activating host oligodendrocytes/precursors *(71)*. Goldstein and colleagues *(72,73)* have shown that fetal astrocytes transplanted into a hemisected spinal cord will significantly reduce glial scar formation. Houle and colleagues *(74)* have shown that even in a chronic lesion with dense scar formation, a reduction in glial scar volume or a substantial reduction in new scar formation was seen with fetal transplants (presumably astrocytes). Other investigators have shown that in addition to a reduction in scar formation type-1 astrocytes also migrate along blood vessels and may help reconstitute the blood-brain barrier and reduce the inflammatory response. Bernstein and Goldberg *(75,76)* have noted that astrocytes can provide trophic support for denervated neurons. In general, the data indicate that fetal astrocytes are better than adult astrocytes as cell replacement therapy. The availability of astrocyte-specific precursor cells from fetal tissue (see, e.g., refs. *77* and *78*) and adult tissue (see, e.g., ref. *79*) will allow a better evaluation of the role of astrocytes in promoting repair. Another important property of astrocytes is their ability to migrate extensively and to integrate seamlessly into the host parenchyma. Fetal astrocytes or their precursors will, therefore, be important candidates for drug and gene delivery and may be even better than

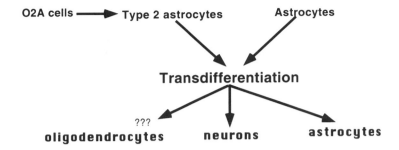

Fig. 3. Transdifferentiation of glial cells. Recent reports have described glial differentiation into neurons and oligodendrocytes. Either oligodendrocyte precursors (O2A) cells have been first differentiated into type-2 astrocytes that acquire the ability to differentiate into neurons in culture or astrocytes (presumably type-1) have been harvested directly from the brain and are shown to retain the ability to generate neurospheres and subsequently differentiate into neurons and oligodendrocytes.

oligodendrocyte precursors, which tend to be more difficult to infect/transfect (our unpublished results).

Human astrocytes have been isolated and cultured *(77–79)* and are even commercially available (e.g., Clonetics and Cloneexpress). Limited data on the use of human astrocytes are available, but based on the data with rodent astrocytes human cells remain a viable option for cell transplant therapy. It remains to be determined if human astrocytes are spontaneously immortal and can, under appropriate conditions, dedifferentiate into stem cells that can generate neurons astrocytes and oligodendrocytes (see above). If human astrocytes exhibit these properties, their therapeutic range will be considerably expanded.

IMMORTALIZED CELL LINES

Obtaining primary tissue from fetal or adult sources remains a problem and maintaining stem and precursor cells for prolonged periods in culture remains difficult and may have unintended consequences on the properties of these cells (see above). More restricted precursors have been maintained for only short periods of time in culture and it may be impossible to maintain these more restricted precursors for prolonged periods without genetic manipulation. An approach to bypass this problem is to immortalize precursor cells using oncogenes or telomerase or to utilize spontaneously transformed precursor cell populations. The immortalization of stem cells offers the attractive possibility of generating clonally identical cells for transplantation. Cell lines once established could yield highly reproducible results *(80–87)*.

Several precursor cell populations have been immortalized or spontaneously generated tumor cell lines have been subcloned and characterized. Some examples of such cell lines are described below. A subclone NT2/D1 derived from a human teratocarcinoma line has been shown to differentiate primarily into neurons (hNT). The hNT cells have been transplanted in several models and has shown some therapeutic benefit *(88–91)*. A recent trial has been initiated in which these cells have been transplanted into stroke patients *(92)* and some therapeutic benefit was reported.

Snyder et al. have transplanted multipotent neural stem cells that have been immortalized by the introduction of the oncogene v-*myc*. They have subsequently demonstrated that the fate of the transplanted cell depends on the environment with a bias toward the missing cellular phenotype *(93,94)*. Another potential advantage of these multipotent stem cells appears to be their ability to move great distances in the intact brain *(94,95)*. Therefore, cells can be transplanted into the ventricles and will reach regions that are inaccessible to surgery. As more is learned about the properties of multipotent stem cells, one can reasonably expect that the therapeutic potential of NSCs will widen further. Taken together, these observations suggest that multipotent cells could be a good choice for any transplantation approach. These cells are readily available, have a reasonable ability to divide and survive in the host environment, and can differentiate into all desired phenotypes in a site-specific manner.

A human PNET cell line "Dev" has been developed by Derrington and colleagues *(96)*. This cell line was derived from a primitive neuroectodermal tumors (PNETs), which are thought to derive from the malignant transformation of pluripotent CNS precursors. Clonal lines are multipotential and can be maintained as undifferentiated cells or stimulated to differentiate. The majority of differentiated cells express neuronal phenotypes, but distinct subpopulations express oligodendrocytic and astrocytic markers. HNSC100 is a stem cell line that has been obtained by immortalizing human stem cells using an avian *myc* oncogene *(97)*. These cells have been maintained in culture and shown to differentiate after transplantation *(98)*. Human neuronal precursor cells have been immortalized from the hippocampus and the spinal cord using v-*myc (85,86)* and multiple human neuroblastoma cell lines (peripheral neuronal cell lines) have also been isolated. One such cell line is the SHSY clone, which has been extensively used to study cell signaling, although not for transplants.

Glial tumor subclones have been established from oligodendrogliomas and astrocytomas *(99–101)*. It has been noted that growth in serum will alter their differentiation characteristics, but growth in a serum-free medium allows the maintenance of undifferentiated cells that are capable of differentiating into oligodendrocytes and astrocytes *(102)*. Glial precursors have been immortalized, but few published results have described the characteristics of the immortalized precursors or their behavior after transplantation.

Although immortalized cells represent one large source of cells, several potential problems exist for their use in the clinic. A serious problem is the potential of immortalized cells to form malignant or benign tumors. Stem cells are dividing cells, and rapid proliferation, in general, inhibits differentiation. Thus, genetically altered cells may not respond appropriately to differentiation signals and may form an undifferentiated mass of cells. A second potential problem with immortalized cells is their stability in culture and how well they mimic the properties of the nontransformed precursor cell. Immortalizing oncogenes genes not only affect the cell life-span but also alter their differentiation potential (see, e.g., ref. *103*). Likewise, whereas primary O-2A progenitor cells seem to rarely give rise to astrocytes *(104)*, the same cell population immortalized with an inducible SV40-large T-antigen will yield a cell line that gives rise predominately to astrocytes upon transplantation *(71,105)*. A third problem with using immortalized cells or any long-term passaged cells is the change in properties after prolonged culture. We have noted, for example, that p75 expressing immortalized crest cells *(83)* lose expression of p75 and appear biased to smooth-muscle differentiation.

Although several concerns have been voiced about the use of immortalized cells, other investigators have argued that these may be safe to use. Somewhat surprisingly, it has been noted that immortalized cells do not form tumors after transplantation and seem to be able to respond to appropriate cues to differentiate *(89,91)*. In the absence of other clearly superior alternatives, investigators have begun using cell lines for therapy (see, e.g., ref. *92*). As additional results accumulate, better comparisons can be made between immortalized/transformed cell sources and fetal or adult stem cells maintained in culture by epigenetic means.

DEDIFFERENTIATED/TRANSDIFFERENTIATED CNS CELLS

A general operating assumption in the development of the nervous system is the idea of a progressive restriction in developmental potential. More differentiated cells have a more limited repertoire of fate choices and fully differentiated cells do not have any alternative fates. This view has been challenged by a series of recent results. Kondo and Raff *(18)* reported recently that oligodendrocyte precursors, which normally only make oligodendrocytes and type-2 astrocytes in culture and only oligodendrocytes after transplantation *(106)*, can be induced to become multipotent stem cells (Fig. 3). The authors show that exposure to serum to induce type-2 astrocyte differentiation followed by prolonged exposure to FGF, a known mitogen for stem cells, was sufficient to induce the proliferation of a cell population that could differentiate into neurons, astrocytes (type-1 astrocytes), and oligodendrocytes in culture. Steindler and colleagues *(19)* have independently demonstrated the ability of type-1 rodent astrocytes to generate neurons after a prolonged period in culture. The authors noted that the ability of astrocytes to generate neurospheres and neurons was limited to the early postnatal period and this ability was lost in most astrocytes after the second week of development. The ability of astrocytes to generate neurons in the adult, however, was retained in a specialized astrocyte present in the subependymal zone (SEZ). A third glial population, the radial glial cell, has also been shown to retain the ability to generate neurons. Malesta and colleagues *(107)*, using FACS sorting to follow the differentiation of radial glial cells showed that radial glia could generate both neurons and astrocytes.

These results suggest an alternative source of cells for therapy. Glial cells are more abundant than stem cells and can be harvested from adult tissue. Astrocytes divide in vivo and can be maintained for prolonged periods in culture. The frequency of neuronal differentiation from dedifferentiating astrocytes is quite remarkable and is better than that reported for any passaged stem cell line. Furthermore, two recent reports *(31,108)* have suggested that glial cells in the nervous system may be spontaneously immortal and will not undergo senescence, unlike most somatic cells. This is different from even tissue-specific stem cells that have been suggested to undergo senescence (reviewed in ref. *14*). Thus, dedifferentiating cells may not only be an alternative to stem cells but may represent a better solution to the problem of obtaining sufficient numbers of cells for therapy.

Although the data clearly suggest that dedifferentiated astrocytes represent a viable alternative to stem cells for transplant therapy several questions remain. One question that remains unanswered is whether neurons derived from dedifferentiated astrocytes will integrate and behave appropriately after transplantation. Such transplant experiments have not been performed. Another question that remains unanswered is whether one can direct endogenous cells to transdifferentiate in vivo. Astrocytes, for example, proliferate in vivo after injury, but no newborn neurons are seen at sites of injury. Likewise, when rodent glial-restricted precursors are transplanted in vivo, these cells do not generate neurons (see, e.g., ref. *21*). A third question that remains to be answered is how stable the process of transdifferentiation is. Do the neurons derived from glial precursors revert to glia or maintain their neuronal identity. Little data are available for transdifferentiated cells derived from astrocytes, but the process of transdifferentiation from mesenchymal cells appears unstable, with cells reverting after removal of the DMSO *(109)*.

The process of dedifferentiation may not be limited to glial cells, Brewer and colleagues *(110)* have shown that postmitotic neurons can be induced to re-enter the cell cycle. They isolated hippocampal neurons and maintained them in high concentrations of growth factor (FGF) in defined medium. The authors showed that a significant number of the cells re-entered mitosis and generated new neurons. The authors did not examine if the newly dividing cells were capable of generating astrocytes as well, but their data clearly demonstrated the ability of postmitotic neurons to dedifferentiate into a dividing neuronal precursor. Temporal lobe biopsies have been performed and stem cells and precursor cells isolated *(52,55,111)*. It is possible that neurons could likewise be harvested and dedifferentiated to obtain significant numbers of neurons for therapy. It is important to note that transdifferentiation from human astrocytes or glial precursors has not been demonstrated; nor has it been demonstrated that human glial cells fail to undergo senescense after repeated passaging. In the absence of these results and the potential issues that remain to be addressed even in rodent transdifferentiation experiments, we would suggest that transdifferentiation remains an exciting but unproven alternative for cellular therapy.

NEURAL DIFFERENTIATION OF NON-NEURONAL PRECURSOR CELLS

Another recently described source of neuron astrocytes and oligodendrocytes is transdifferentiation of non-neural stem cells (summarized in Fig. 4). Several recent reports have described the generation of neurons from skin, skeletal muscle, hematopoietic stem cells (HSCs) and mesenchymal stem cells (see below). HSCs are of particular interest as they can be readily harvested, persist in the adult, and are already used in a variety of clinical therapies. Bone marrow cells taken from either the patient or an immunologically similar relative can be administered to the patient without concerns of immune attack on the donor cells. Use of bone marrow cells also eliminates ethical concerns with the use of cells (embryonic stem cells or neural progenitor cells) obtained from human embryos *(112)*.

Multiple studies have suggested that HSCs can generate nonhematopoietic lineages (reviewed in ref. *113*). Recent studies have shown that the HSC can differentiate into muscle *(114)*, endothelial cells *(115)*, liver *(116)*, and brain

Fig. 4. Transdifferentiation of non-neural cells. The ability of non-neural tissue to differentiate into neurons, astrocytes and oligodendrocytes is diagrammed. In many experiments it is not clear whether the tissue specific stem cell transdifferentiates or whether a subpopulation persists that has the appropriate neural differentiation ability. Note that predominantly neuronal and astrocytic differentiation has been described. Oligodendrocyte differentiation has not been accomplished in most such assays.

cells *(117)*. Differentiation into neural derivatives has been demonstrated both in vitro and in vivo. When bone marrow cells from GFP transgenic mice were transplanted into irradiated mice, some transplanted cells expressed neuron-specific genes *(118)*. Many of the transplanted cells that expressed neuronal markers were located in the olfactory bulb, a site where neurons turnover extensively throughout life. It is not known whether the marrow-derived neural cells were functional, as most of the cells that expressed neuronal markers elaborated only short neuritelike processes. HSCs can also acquire neural cell phenotypes in culture. Thus, bone marrow cells maintained in culture can express nestin, a marker of neural progenitor cells, when maintained in the presence of platelet-derived growth factor *(119)*. Additional studies have provided evidence that HSCs and stromal cells can give rise to oligodendrocyte precursor cells after long-term culture *(120)*.

Similar studies using stromal cell cultures have suggested that stromal cells and mesenchymal stem cells also possess the ability to differentiate into neural derivatives (*109,121*; reviewed in ref. *122*). In one study, stromal cells were purified by negative selection. When exposed to retinoic acid and brain-derived neurotrophic factor, a small percentage of the stromal cells expressed the neuronal marker NeuN *(121)*. Woodbury et al. *(109)* showed that a much larger percentage of cells differentiate into neural derivatives, although the differentiation was not sustained. At this early point in time the possibility that bone marrow cells can be used as a source of progenitor cells to replace damaged neurons or oligodendrocytes in the central nervous system is very speculative. If efficacy of bone marrow cells in replacing damaged neural cells and improving functional outcome is established in future studies of animal models of neurological disease, then stromal cells offer a number of advantages as a source of neural progenitor cells over any other source yet discussed. One disadvantage to these bone-marrow-derived cells is that the population capable of transdifferentiation appears to be a very rare population. Therefore, the isolation of these cell for therapies may not be practical. Furthermore, there may be many differences between different human donors, and the characterization of these cells from individual donors before transplantation may not be feasible.

It is important to note that the transdifferentiation studies have only demonstrated that cells from one germ layer can express markers/proteins of cells from another germ layer. It is too early to know if this marker expression represents functional differentiation. The cells that have been transdifferentiated are typically maintained for extended periods in vitro, and karyotypes have not been evaluated to demonstrate that these cells are not transformed. To date, functional data have not been described in any of the transdifferentiation experiments.

XENOBIOTIC SOURCES OF NEURAL CELLS

An important issue that remains unresolved for human cell replacement therapy is localizing a substantial and reliable source of cells. Fetal tissue is limiting. Multipotent stem cells maintained in culture appear to change with time, human ES cells (see below) are still being evaluated and the numbers of intermediate precursors or transdifferentiated cells, although promising alternatives, are currently not viable sources of clinical therapy. Several investigators have explored the use of animal tissue for transplants (reviewed in ref. *123*). However, xenografted neural tissue itself generates a number of practical, ethical, safety, and immunological issues that have to be addressed prior to any clinical xenotransplant program *(91)*.

The main hurdles include immunological reaction of the recipient against the transplant, the physiological differences, biogenetically disparate recipients, and the possibility of transferring infectious organisms from the graft into the recipient (summarized in Table 7). Of various species considered, pigs and swine have been considered the best source of organ transplant and have been studied quite extensively (reviewed in ref. *124*). In principle, these animals could also be used for dissociated cell therapy, and a limited number of experiments have suggested that this is indeed possible in rodents *(125–129)* and human *(130–132)*.

Perhaps the single biggest unsolved problem is the immune response. The brain is not an immunologically privileged site. Grafts placed in the brain survive longer than elsewhere, but are ultimately rejected (reviewed in ref. *133*). The immune response to xenotransplants is more acute and prolonged than with allografts and some progress has been made with reducing the

Table 7
Problems with Xenotransplant

Ethical
Public policy
Immunological
    Hyperacute
    Acute
    Chronic
Physiological
    Hormones, enzymes, receptors, adhesion molecules, etc.
Zoonosis

Note: Issues that need to be resolved before xenotransplants can be used for transplant therapy are listed. Several problems remain but rapid progress has been made toward resolving at least some of these problems. In the absence of any clearly superior alternative (see, however, discussion on human ES cells), animal tissue remains a possible source of cells for therapy.

hyperacute and acute phases of transplant rejection using a variety of different strategies *(124)*. However, despite the limited success, significant problems remain and represent a barrier to the use of xenotransplants for therapy.

In addition to the problem of rejection, several other problems exist *(124)*. A problem that appears difficult to address is the issue of physiological differences between otherwise identical tissue/cells (reviewed in ref. *134*). Of importance for dissociated cell xenotransplants are known differences among adhesion molecules, enzymes, cytokine receptors, and response to hormones (reviewed in ref. *134*). It remains to be determined how many physiological differences can be tolerated in long-term transplant therapy. A third issue of concern is the possibility that zoonotic diseases will be transferred into the transplant recipient to the human population at large *(135,136)*. These fears have been amplified by outbreaks of ebola and hantavirus, the isolation of porcine endogenous retrovirus (PERV), and the demonstration that PERVs can infect human cells in culture *(124)*. Several strategies have been undertaken to solve these problems. These include the generation of transgenic pigs, induction of tolerance, maintenance of pathogen-free swine herds, and attempts to generate porcine ES cell lines for genetic manipulation. Many of these strategies hold promise and are being extensively pursued. Overall however, in our opinion xenotransplants do not currently represent a viable alternative for a neural transplant program (see ref. *91* for a detailed discussion). However, the significant progress made in recent years on prolonging the survival of xenotransplanted tissue and the current absence of a significantly better alternative suggest that xenotransplants must continue to be evaluated with all other cell types as a potential source of cells for future therapy. Indeed, in a recent trial initiated by Diacrin, porcine fetal tissue grafts have been used to treat patients with Parkinson's disease.

MACROPHAGE CELLS FOR THERAPY

In addition to serving as an alternative source of neural cells, hematopoietic cells may be therapeutically useful as a source of macrophages for therapy. Bone-marrow-transplantation studies in which the bone marrow of a recipient animal is replaced with the bone marrow of a donor animal have provided strong evidence that many, if not all, of the microglia in the brain arise from the bone marrow. Two lines of evidence strongly support this conclusion. First, when bone marrow cells from transgenic mice expressing green fluorescent protein (GFP) are transplanted into irradiated wild-type mice, the vast majority of wild-type microglia in the brain are replaced with GFP-positive microglia over a period of several months *(137,138)*. Second, bone marrow transplants can essentially cure mice with lipid storage disorders in which abnormal microglia play a central role in the neurological phenotype *(139)*. The apparent ability of hematopoeitic stem cells to differentiate into microglia can readily cross the blood-brain barrier has suggested their use as delivery agents.

Existing data support the potential of bone marrow cells to promote the regeneration of damaged nerves. Implantation of damaged nerves into the region of a subsequent sciatic nerve crush in rats enhances regeneration of the sensory nerves, and this regenerative "conditioning" effect is associated with macrophage activation *(140)*. In another study, macrophages that had been exposed to peripheral nerve segments in vitro were transplanted into the transected rat spinal cord *(141)*. The macrophages promoted tissue repair and partial recovery of motor function, which were associated with apparent regrowth of descending inputs to motor neurons. A group in Israel has enrolled patients for autologous macrophage therapy in eight spinal cord injury patients. Although results are still unavailable, the data from the rat transplant studies suggest optimism.

PLURIPOTENT STEM CELLS AS A SOURCE
OF NEURAL DERIVATIVES

An alternative to somatic stem cells, xenotransplants, and transdifferentiated or immortalized cells is the use of embryonic stem (ES) cells. ES cells were initially isolated from mice *(142,143)* and have been extensively

Table 8
Comparison of the Properties of ES, EG, and Transformed EC Cells

	Mouse ES	Rhesus ES [Thomson et al. *(164)*]	Human ES [Thomson et al. *(145)*]	Human EC [Andrews et al. *(152)*, Kannagi et al. *(148)*]	Human EG [Shamblott, et al. *(147)*]
SSEA-1	+	–	–	–	+
SSEA-3	–	+	+	+	+
SSEA-4	–	+	+	+	+
Tra-1-60	–	+	+	+	+
Tra-1-81	–	+	+	+	+
Alkaline phosphatase	+	+	+	+	+
Oct-4	+	?	+	?	?
Karyotype	Diploid	Diploid	Diploid	Aneuploid	Diploid
LIF responsive	+	–	–	?	+

used in the generation of genetically modified mice (reviewed in ref. *144*). ES cells are unique because of their ability to contribute to all embryonic derivatives, including cells of the germ line. In addition, in vitro, ES cells have been shown to differentiate into ectoderm, endoderm, and mesoderm. Another pluripotent cell is the embryonic germ cell (EG cell), which is isolated from primordial germ cells (PGCs) migrating along the genital ridge. EG cells, like ES cells, have been shown to contribute to multiple tissue derivatives and also are capable of generating transgenic animals. ES and EG cells could, therefore, serve as a source of cells for neural transplantation.

Several pluripotent human cell lines have been isolated (summarized in Table 8). Recently, ES cells have been isolated from human blastocysts and could be maintained in culture without differentiation for prolonged periods and express telomerase activity and cell surface markers that are characteristic of human EC and nonhuman primate ES cells. The expression pattern of stage-specific embryonic antigens SSEA-3 and SSEA-4 was similar to human EC cells, but did not correspond to mouse ES cells, indicating species differences (Table 9). Upon injection into SCID mice, the human ES cells formed teratomas, which contained derivatives of all three embryonic germ layers *(145,146)*.

Table 9
Comparison of the Properties of ES Cells and Other Somatic Cells

Somatic stem cell	Embryonic stem cell

1. Limited expansion
  (i) Repeat derivations required
  (ii) Different donors required =
   repeated QC
	1. Unlimited quantity ≥ 250 population
	 doublings/year
	  (i) Cell banking and bulk
	   manufacturing possible
	  (ii) Stable long-term cultures with
	   opportunity for genetic
	   engineering

2. Rare populations, limited lineages	2. Clonal line possible
3. Restricted lineage differentiation capacity	3. Multilineage differentiation capacity
4. Loss of phenotype over prolonged culture	4. Stable differentiation potential
5. Autologous transplantation	5. Allogenic transplant
6. Ability to respond to cues in the adult and damaged environment	6. Ability to respond to cues in the adult and damaged environment (mouse cells)
7. Ability to migrate and integrate in a site-specific fashion	7. Ability to migrate and integrate in a site-specific fashion (mouse)

Note: Embryonic stem cells offer many potential advantages over any other cell type. A concern unique to ES cells is the potential of forming teratomas after transplant. Whether this should be a significant concern is discussed in the text.

Another population of pluripotent cells that have been derived recently are human embryonic germ cells. These EG cells also showed some of the characteristics associated with pluripotent cells *(147)*. These cells were isolated from the genital ridges of 5- to 9-wk-old fetuses and formed embryoid bodies, containing derivatives of the three embryonic germ layers. These cells have exhibited normal and stable karyotypes over 10 passages and been continuously maintained over 20 passages in culture. Surprisingly, the cells exhibit a stage-specific cell surface marker SSEA-1 that is normally seen in undifferentiated mouse ES cells *(148)* and differentiated human ES cells. It is yet to be demonstrated that these cell lines differentiate in vivo. In general, EG cells have not been as widely used as ES cells in developmental studies and genetic modification.

One of the basic differences between ES cell lines and EG cells is the source material. Human and mouse ES cells are derived from the inner cell mass of blastocysts. In contrast, the EG cells are isolated from primordial germ cells undergoing migration to the genital ridges. The third human pluripotent cell type can be isolated from teratocarcinomas and is termed embryonal carcinoma (EC) cells. Each of these cell types is maintained in different culture conditions. Human ES cells are maintained on mouse feeder cells and do not appear to respond to leukemia inhibitory factor (LIF) *(145,146)*. In contrast, the human EG cells are maintained in a medium containing serum, bFGF, LIF, and forskolin *(147)*. Some human EC lines are feeder dependent *(149,150)*, but the growth conditions for individual lines varies.

Of the three pluripotent cells described (ES, EC, and EG cells), human ES cell lines appear to be the best candidates for cell therapy. Human ES cells have demonstrated remarkable stability in vitro. They can undergo over 250 population doublings and remain karyotypically and phenotypically stable *(151)*. Further, it has been shown that clonal ES cells have a normal karyotype, and surface marker expression and can generate complex teratomas *(151)*, indicating that the human ES cells are, indeed, pluripotent. Teratomas from human EC lines such as the NTERA2 line do not show the same sort of complexity. Rather, simple tubular structures and neural rosettes are reported that resemble primitive gut neuroepithelium *(152)*. To date, teratoma formation from human EG cells has not been reported.

All three of these types of human pluripotent stem cells have the ability to generate neurons in vitro. The NTERA2 human EC cells described by Andrews *(152)* can be serially expanded and then induced to differentiate in vitro by exposure to retinoic acid, and, as previously discussed, hNTERA2 cells have been transplanted in several models and have shown some therapeutic benefit *(88–91)*. Although experiments have shown remarkably high levels of neuronal formation (>95%) *(153)*, these carcinoma cells are aneuploid and it is possible that this may lead to instability of the cell phenotype. The human EG cells have been shown to differentiate in vitro by the spontaneous formation of embryoid bodies (EBs). Evaluation of the EBs indicated positive immunoreactivity against markers for all three germ layers. Positive neurofilament immunoreactivity was found in EBs formed in four of six human EG lines, indicating that these cells have the capacity to form neurons *(147)*.

Human ES cells have the ability to generate neurons in vitro and in vivo. Teratoma formation by these lines indicated that the cells had the capacity to generate neural derivatives *(145,146)*. Reubinoff et al. *(146)* demonstrated that when the human ES cells are allowed to become overconfluent, somatic differentiation is seen. In these cases, they were able to manually subdissect

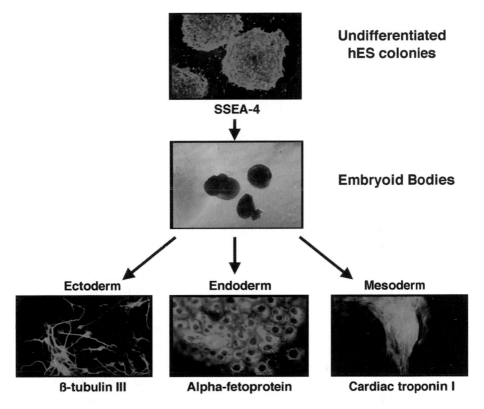

Fig. 5. Human ES cell cultures. The growth and differentiation of human ES cells is illustrated. Cells are maintained on feeder layer and then allowed to differentiate to form embryoid bodies. Embryoid bodies are then replated to obtain ectodermal, endodermal, and mesodermal differentiation. Differentiation can be biased by modifying culture conditions and purified populations can be isolated by cell sorting.

regions from the overconfluent cultures, which subsequently gave rise to differentiated cells that expressed mature neuronal markers *(146).*

Our lab has also demonstrated neuronal formation in vitro (Fig. 5). In these experiments, we used human ES cells maintained in the absence of feeders *(154,155),* which had undergone between 125 and 150 population doublings. EBs were generated and exposed to retinoic acid before plating into a mitogen cocktail to allow neural progenitors to proliferate. In these cultures, 70–90% of the cells express PS-NCAM or A2B5, markers for neural progenitor cells. We have demonstrated that the NCAM and A2B5 cells in these populations have the capacity to proliferate and to differentiate into neurons. These cultures were then transferred into differentiation conditions in which the mitogens were exchanged for neurotrophins. After 2 wk in these conditions, 30% of the cells express the mature neuronal marker MAP-2. Subpopulations of the neurons in these cultures also expressed neurofilament, TH, glutamate, GABA, glycine, and synaptophysin. In addition, the NCAM expressing neurons in these cultures were assessed for functional activity. Ninety-four percent of these neurons responded to depolarization and 100% of the neurons tested generated action potentials *(155)*. In addition to neurons, we also identified GFAP and GalC immunoreactive cells in these cultures.

Few transplant experiments using cells derived from differentiating human ES cells have been performed. Experiments done with rodent ES cell cultures and with human EC cells have shown that human ES cells may be useful for therapy. McDonald and colleagues have presented suggestive evidence that rodent-ES-cell-derived cells may be useful in spinal cord injury. When embryoid bodies were transplanted in a spinal cord injury model, significant improvement and remyelination were seen *(156)*. Although the behavioral improvement was only evaluated for 30 d, the remyelination was clearly superior to that seen in other transplants. In another report, mouse differentiated ES cells transplanted into quinolinic acid-lesioned rats retained a neuronal phenotype after transplantation *(157)*. Similarly, transplantation of differentiated mouse ES cells into embryonic mice demonstrated that these cells can integrate into the CNS and form neurons, oligodendrocytes, and astrocytes *(158)*.

Of importance is the recent demonstration that melanocytes and potential neural crest precursors can be directly isolated from murine ES cells *(159)*. Melanocytes have been generated from ES cells by coculturing them with a bone-marrow-derived stromal cell line along with dexamethasone and steel factor. All of the above results taken together suggest that it may be possible to obtain both CNS and PNS derivatives from differentiating human ES cells and that these cells may be useful for transplant therapy.

Cells derived from ES cell cultures offer several potential advantages over somatic stem cells (summarized in Table 9). One potential advantage for neurons, glia, or precursor cells derived from ES cells is that they may not be regionally specified and thus have greater capacity for site-specific integration. Further, because it is possible to obtain regionally specific phenotypes by manipulating ES cell culture conditions, it may be possible to obtain appropriately specified phenotypes. Thus, ES-cell-derived precursors may be preferable to fetal or adult precursor cells. Perhaps the greatest advantage of the human ES cells is the ability to generate large numbers of cells. The human ES cells have demonstrated remarkable stability in vitro. After 250 population doublings, the cells still express telomerase and ES-associated surface and molecular markers and they retain the capacity to generate derivatives of all three germ layers in vitro and vivo. In addition, we have demonstrated that these cells have the capacity to generate mature neurons after as many as 150 population doublings *(155)*. In considering cell therapies in clinical situations, it is important to generate enough cells to make master and working cells banks and perform quality control assays on the cell banks. The remarkable stability of the human ES cells over long-term expansion will allow the generation of these banks and assays.

Given the many advantages of ES cells, it is likely that these cells will become the cell of choice to obtain appropriate derivatives for therapy. Nevertheless, multiple transplant experiments need to be performed to validate the potential of ES cells and to demonstrate their efficacy in a transplant paradigm.

CONCLUSION

Multiple cell types are being evaluated for cellular replacement. It is our belief that appropriate cell derivatives from a cultured source of cells will be preferable for therapeutic applications. Of the many potential sources of cells, ES cells have the highest potential, but currently no alternative cell source can be ruled out. Direct side-by-side comparisons are necessary to evaluate the efficacy of the cells to determine the optimum cell type and optimum mode of therapy.

ACKNOWLEDGMENTS

We gratefully acknowledge the input of all members of our laboratory provided through discussions and constructive criticisms. MR is supported by NIDA, NINDS, NIA, and MDA. MKC is supported by Geron Corporation. MM was supported by the NIA.

REFERENCES

1. Kamb, A. and Rao, M. S. (2000) Stem cells and gene discovery, in *Stem Cells and CNS Development* (Rao, M. S., ed.), Humana, Totowa, NJ.
2. Chalmers-Redman, R. M., Priestley, T., Kemp, J. A., and Fine, A. (1997) In vitro propagation and inducible differentiation of multipotential progenitor cells from human fetal brain. *Neuroscience* **76,** 1121–1128.
3. Vescovi, A. L., Gritti, A., Galli, R., and Parati, E. A. (1999) Isolation and intracerebral grafting of nontransformed multipotential embryonic human CNS stem cells. *J. Neurotrauma* **16,** 689–693.

4. Carpenter, M. K., Cui, X., Hu, Z. Y., Jackson, J., Sherman, S., Seiger, A., et al. (1999) In vitro expansion of a multipotent population of human neural progenitor cells. *Exp. Neurol.* **158,** 265–278.
5. Milward, E. A., Lundberg, C. G., Ge, B., Lipsitz, D., Zhao, M., and Duncan, I. D. (1997) Isolation and transplantation of multipotential populations of epidermal growth factor-responsive, neural progenitor cells from the canine brain. *J. Neurosci. Res.* **50,** 862–871.
6. Gritti, A., Frolichsthal-Schoeller, P., Galli, R., Parati, E. A., Cova, L., Paganao, S. F., Bjornson, C. R., and Vescovi, A. L. (1999) Epidermal and fibroblast growth factors behave as mitogenic regulators for a single multipoint stem cell-like population from the subventricular region of the adult mouse forebrain. *J. Neurosci.* **19(9),** 3287–3297.
7. Piper, D. R., Mujtaba, T., Rao, M. S., and Lucero, M. T. (2000) Immunocyto-chemical and physiological characterization of a population of cultured human neural precursors. *J. Neurophysiol.* **84,** 534–548.
8. Reynolds, B. A. and Weiss, S. (1992) Generation of neurons and astrocytes from isolated cells of the adult mammalian central nervous system. *Science* **255,** 1707–1710.
9. Svendsen, C. N. Caldwell, M. A., and Ostenfeld, T. (1999) Human neural stem cells: isolation, expansion and transplantation. *Brain Pathol.* **9(3),** 499–513.
10. Svendsen, C. N. and Rosser, A. E. (1995) Neurones from stem cells? *Trends Neurosci.* **18,** 465–467.
11. Ostenfeld, T., Caldwell, M. A., Prowse, K. R., Linskens, M. H., Jauniaux, E., and Svendsen, C. N. (2000) Human neural precursor cells express low levels of telomerase in vitro and show diminishing cell proliferation with extensive axonal outgrowth following transplantation. *Exp. Neurol.* **164,** 215–226.
12. Fricker, R. A., Carpenter, M. K., Winkler, C., Greco, C., Gates, M. A., and Bjorklund, A. (1999) Site-specific migration and neuronal differentiation of human neural progenitor cells after transplantation in the adult rat brain. *J. Neurosci.* **19,** 5990–6005.
13. Learish, R. D., Brustle, O., Zhang, S. C., and Duncan, I. D. (1999) Intraven-tricular transplantation of oligodendrocyte progenitors into a fetal myelin mutant results in widespread formation of myelin. *Ann. Neurol.* **46(5),** 716–722.
14. Cai J. and Rao, M. S. (2002) Aging and stem cells, in *Stem Cells: A Cellular Fountain of Youth* (Mattson, M. P. and Van Zant, G., eds.), Elsevier, Amsterdam, pp. 97–116.
15. Horner, P. J., Power, A. E., Kempermann, G., Kuhn, H. G., Palmer, T. D., Winkler, J., et al. (2000) Proliferation and differentiation of progenitor cells throughout the intact adult rat spinal cord. *J. Neurosci.* **20,** 2218–2228.
16. Schuldiner, M., Yanuka, O., Itskovitz-Eldor, J., Melton, D. A., and Benvenisty, N. (2000) From the cover: effects of eight growth factors on the differentiation of cells derived from human embryonic stem cells. *Proc. Natl. Acad. Sci. USA* **97,** 11,307–11,312.
17. Cao, Q. L., Zhang, Y. P., Howard, R. M., Walters, W. M., Tsoulfas, P., and Whittemore, S. R. (2001) Pluripotent stem cells engrafted into the normal or

lesioned adult rat spinal cord are restricted to a glial lineage. *Exp. Neurol.* **167,** 48–58.

18. Kondo, T. and Raff, M. (2000) Oligodendrocyte precursor cells reprogrammed to become multipotential CNS stem cells. *Science* **289,** 1754–1757.
19. Laywell, E. D., Rakic, P., Kukekov, V. G., Holland, E. C., and Steindler, D. A. (2000) Identification of a multipotent astrocytic stem cell in the immature and adult mouse brain. *Proc. Natl. Acad. Sci. USA* **97,** 13,883–13,888.
20. Quinn, S. M., Walters, W. M., Vescovi, A. L., and Whittemore, S. R. (1999) Lineage restriction of neuroepithelial precursor cells from fetal human spinal cord. *J. Neurosci. Res.* **57,** 590–602.
21. Yang, H., Mujtaba, T., Venkatraman, G., Wu, Y. Y., Rao, M. S., and Luskin, M. B. (2000) Region-specific differentiation of neural tube-derived neuronal restricted progenitor cells after heterotopic transplantation. *Proc. Natl. Acad. Sci. USA* **97,** 13,366–13,371.
22. Northcutt, R. G. and Gans, C. (1983) The genesis of neural crest and epidermal placodes: a reinterpretation of vertebrate origins. *Q. Rev. Biol.* **58,** 1 28.
23. Nieto, M. A., Bradley, L. C., Hunt, P., Das Gupta, R., Krumlauf, R., and Wilkinson, D. G. (1992) Molecular mechanisms of pattern formation in the vertebrate hindbrain. *Ciba Found. Symp.* **165,** 92–102; discussion 102–107.
24. Scherson, T., Serbedzija, G., Fraser, S., and Bronner-Fraser, M. (1993) Regulative capacity of the cranial neural tube to form neural crest. *Development* **118,** 1049–1062.
25. Anderson, D. J. (2000) Genes, lineages and the neural crest: a speculative review. *Philo. Trans. R. Soc. (Lond) B* **355,** 953–964.
26. Morrison, S. J., White, P. M., Zock, C., and Anderson, D. J. (1999) Prospective identification, isolation by flow cytometry, and in vivo self-renewal of multipotent mammalian neural crest stem cells. *Cell* **96,** 737–749.
27. Hagedorn, L., Floris, J., Suter, U., and Sommer, L. (2000) Autonomic neurogenesis and apoptosis are alternative fates of progenitor cell communities induced by TGFbeta. *Dev. Biol.* **228,** 57–72.
28. Askanas, V., Engel, W. K., Dalakas, M. C., Lawrence, J. V., and Carter, L. S. (1980) Human schwann cells in tissue culture: histochemical and ultrastructural studies. *Arch. Neurol.* **37,** 329–337.
29. Scarpini, E., Kreider, B. Q., Lisak, R. P., Meola, G., Velicogna, M. E., Baron, P., et al. (1988) Cultures of human Schwann cells isolated from fetal nerves. *Brain Res.* **440,** 261–266.
30. Scarpini, E., Meola, G., Baron, P. L., Beretta, S., Velicogna, M., Moggio, M., et al. (1987) Human Schwann cells: cytochemical, ultrastructural and immunological studies in vivo and in vitro. *Basic Appl. Histochem.* **31,** 33–42.
31. Mathon, N. F., Malcolm, D. S., Harrisingh, M. C., Cheng, L., and Lloyd, A. C. (2001) Lack of replicative senescence in normal rodent glia. *Science* **291,** 872–875.
32. Guenard, V., Gwynn, L. A., and Wood, P. M. (1994) Astrocytes inhibit Schwann cell proliferation and myelination of dorsal root ganglion neurons in vitro. *J. Neurosci.* **14,** 2980–2992.

33. Harvey, A. R. and Plant, G. W. (1995) Schwann cells and fetal tectal tissue cografted to the midbrain of newborn rats: fate of Schwann cells and their influence on host retinal innervation of grafts. *Exp. Neurol.* **134,** 179–191.

34. Guest, J. D., Rao, A., Olson, L., Bunge, M. B., and Bunge, R. P. (1997) The ability of human Schwann cell grafts to promote regeneration in the transected nude rat spinal cord. *Exp. Neurol.* **148,** 502–522.

35. Blakemore, W. F. and Franklin, R. J. (2000) Transplantation options for therapeutic central nervous system remyelination. *Cell Transplant.* **9,** 289–294.

36. Le Douarin, N. M., Dupin, E., Baroffio, A., and Dulac, C. (1992) New insights into the development of neural crest derivatives. *Int. Rev. Cytol.* **138,** 269–314.

37. Northcutt, R. G. (1990) Ontogeny and phylogeny: a re-evaluation of conceptual relationships and some applications. *Brain Behav. Evol.* **36,** 116–140.

38. Schlosser, G. and Northcutt, R. G. (2000) Development of neurogenic placodes in *Xenopus laevis*. *J. Comp. Neurol.* **418,** 121–146.

39. LaBonne, C. and Bronner-Fraser, M. (1999) Molecular mechanisms of neural crest formation. *Annu. Rev. Cell Dev. Biol.* **15,** 81–112.

40. Costanzo, R. M. (1991) Regeneration of olfactory receptor cells. *Ciba Found. Symp.* **160,** 233–242; discussion 243–248.

41. Calof, A. L., Rim, P. C., Askins, K. J., Mumm, J. S., Gordon, M. K., Iannuzzelli, P., et al. (1998) Factors regulating neurogenesis and programmed cell death in mouse olfactory epithelium. *Ann. NY Acad. Sci.* **855,** 226–229.

42. Pagano, S. F., Impagnatiello, F., Girelli, M., Cova, L., Grioni, E., Onofri, M., et al. (2000) Isolation and characterization of neural stem cells from the adult human olfactory bulb. *Stem Cells* **18,** 295–300.

43. Roisen, F. J., Klueber, K. M., Lu, C. L., Hatcher, L. M., Dozier, A., Shields, C. B., et al. (2001) Adult human olfactory stem cells. *Brain Res.* **890,** 11–22.

44. Goldstein, B. J., Fang, H., Youngentob, S. L., and Schwob, J. E. (1998) Transplantation of multipotent progenitors from the adult olfactory epithelium. *Neuroreport* **9,** 1611–1617.

45. Huard, J. M., Youngentob, S. L., Goldstein, B. J., Luskin, M. B., and Schwob, J. E. (1998) Adult olfactory epithelium contains multipotent progenitors that give rise to neurons and non-neural cells. *J. Comp. Neurol.* **400,** 469–486.

46. Farbman, A. I. (1997) Injury-stimulated neurogenesis in sensory systems. *Adv. Neurol.* **72,** 157–161.

47. Doucette, R. (1995) Olfactory ensheathing cells: potential for glial cell transplantation into areas of CNS injury. *Histol. Histopathol.* **10,** 503–507.

48. Smale, K. A., Doucette, R., and Kawaja, M. D. (1996) Implantation of olfactory ensheathing cells in the adult rat brain following fimbria-fornix transection. *Exp. Neurol.* **137,** 225–233.

49. Li, Y., Field, P. M., and Raisman, G. (1998) Regeneration of adult rat corticospinal axons induced by transplanted olfactory ensheathing cells. *J. Neurosci.* **18,** 10,514–10,524.

50. Kato, T., Yokouchi, K., Fukushima, N., et al. (2001) Related continual replacement of newly-generated olfactory neurons in adult rats. *Neurosci. Lett.* **307(1),** 17–20.

51. Ramon-Cueto, A., Cordero, M. I., Santos-Benito, F. F., and Avila, J. (2000) Functional recovery of paraplegic rats and motor axon regeneration in their spinal cords by olfactory ensheathing glia. *Neuron* **25,** 425–435.
52. Roy, N. S., Wang, S., Harrison-Restelli, C., Benraiss, A., Fraser, R. A., Gravel, M., et al. (1999) Identification, isolation, and promoter-defined separation of mitotic oligodendrocyte progenitor cells from the adult human subcortical white matter. *J. Neurosci.* **19,** 9986–9995.
53. Roy, N. S., Wang, S., Jiang, L., Kang, J., Benraiss, A., Harrison-Restelli, C., et al. (2000) In vitro neurogenesis by progenitor cells isolated from the adult human hippocampus. *Nat. Med.* **6,** 271–277.
54. Piper D. R., Mujtaba, T., Keyoung H., et al. (2001) Identification and characterization of neuronal precursors and their progeny from human fetal tissue. *J. Neurosci. Res.* **66,** 356–368.
55. Wang, S., Roy, N. S., Benraiss, A., and Goldman, S. A. (2000) Promoter-based isolation and fluorescence-activated sorting of mitotic neuronal progenitor cells from the adult mammalian ependymal/subependymal zone. *Dev. Neurosci.* **22,** 167–176.
56. Huber, A. B. and Schwab, M. E. (2000) Nogo-A, a potent inhibitor of neurite outgrowth and regeneration. *Biol. Chem.* **381,** 407–419.
57. Davies, S. J., Fitch, M. T., Memberg, S. P., Hall, A. K., Raisman, G., and Silver, J. (1997) Regeneration of adult axons in white matter tracts of the central nervous system. *Nature* **390,** 680–683.
59. Reier, P. J., Anderson, D. K., Thompson, F. J., and Stokes, B. T. (1992) Neural tissue transplantation and CNS trauma: anatomical and functional repair of the injured spinal cord. *J. Neurotrauma* **9(Suppl 1),** S223–S248.
60. Reier, P. J., Stokes, B. T., Thompson, F. J., and Anderson, D. K. (1992) Fetal cell grafts into resection and contusion/compression injuries of the rat and cat spinal cord. *Exp. Neurol.* **115,** 177–188.
61. Clowry, G., Sieradzan, K., and Vrbova, G. (1991) Transplants of embryonic motoneurones to adult spinal cord: survival and innervation abilities. *Trends Neurosci.* **14,** 355–357.
62. Tessler, A. (1991) Intraspinal transplants. *Ann. Neurol.* **29,** 115–123.
63. Shetty, A. K. and Turner, D. A. (1998) In vitro survival and differentiation of neurons derived from epidermal growth factor-responsive postnatal hippocampal stem cells: inducing effects of brain-derived neurotrophic factor. *J. Neurobiol.* **35,** 395–425.
64. Ling, Z. D., Potter, E. D., Lipton, J. W., and Carvey, P. M. (1998) Differentiation of mesencephalic progenitor cells into dopaminergic neurons by cytokines. *Exp. Neurol.* **149,** 411–423.
65. Rao, M. S. (1999) Multipotent and restricted precursors in the central nervous system. *Anat. Rec.* **257,** 137–148.
66. Wichterle, H., Garcia-Verdugo, J. M., Herrera, D. G., and Alvarez-Buylla, A. (1999) Young neurons from medial ganglionic eminence disperse in adult and embryonic brain. *Nat. Neurosci.* **2,** 461–466.
67. Lee, J. C., Mayer-Proschel, M., and Rao, M. S. (2000) Gliogenesis in the central nervous system. *Glia* **30,** 105–121.

68. Gudino-Cabrera, G. and Nieto-Sampedro, M. (2000) Schwann-like macroglia in adult rat brain. *Glia* **30,** 49–63.

69. Jeffery, N. D., Crang, A. J., O'Leary M, T., Hodge, S. J., and Blakemore, W. F. (1999) Behavioural consequences of oligodendrocyte progenitor cell transplantation into experimental demyelinating lesions in the rat spinal cord. *Eur. J. Neurosci.* **11,** 1508–1514.

70. Zhang, S. C., Ge, B., and Duncan, I. D. (2000) Tracing human oligodendroglial development in vitro. *J. Neurosci. Res.* **59,** 421–429.

71. Franklin, R. J., Crang, A. J., and Blakemore, W. F. (1993) The reconstruction of an astrocytic environment in glia-deficient areas of white matter. *J. Neurocytol.* **22,** 382–396.

72. Goldberg, W. J. and Bernstein, J. J. (1988) Fetal cortical astrocytes migrate from cortical homografts throughout the host brain and over the glia limitans. *J. Neurosci. Res.* **20,** 38–45.

73. Goldberg, W. J. and Bernstein, J. J. (1988) Migration of cultured fetal spinal cord astrocytes into adult host cervical cord and medulla following transplantation into thoracic spinal cord. *J. Neurosci. Res.* **19,** 34–42.

74. Houle, J. (1992) The structural integrity of glial scar tissue associated with a chronic spinal cord lesion can be altered by transplanted fetal spinal cord tissue. *J. Neurosci. Res.* **31,** 120–130.

75. Bernstein, J. J. and Goldberg, W. J. (1989) Maintenance of host medullary nucleus gracilis neurons after C3 homografting of fetal spinal cord into host fasciculus gracilis. *Brain Res.* **488,** 180–185.

76. Bernstein, J. J. and Goldberg, W. J. (1989) Rapid migration of grafted cortical astrocytes from suspension grafts placed in host thoracic spinal cord. *Brain Res.* **491,** 205–211.

77. Oster-Granite, M. L. and Herndon, R. M. (1978) Studies of cultured human and simian fetal brain cells. I. Characterization of the cell types. *Neuropathol. Appl. Neurobiol.* **4,** 429–442.

78. Bressler, J. P., Cole, R., and de Vellis, J. (1980) Cell culture systems to study glial transformation. *Dev. Toxicol. Environ. Sci.* **8,** 187–192.

79. Whittemore, S. R., Sanon, H. R., and Wood, P. M. (1993) Concurrent isolation and characterization of oligodendrocytes, microglia and astrocytes from adult human spinal cord. *Int. J. Dev. Neurosci.* **11,** 755–764.

80. Ryder, E. F., Snyder, E. Y., and Cepko, C. L. (1990) Establishment and characterization of multipotent neural cell lines using retrovirus vector-mediated oncogene transfer. *J. Neurobiol.* **21,** 356–375.

81. Renfranz, P. J., Cunningham, M. G., and McKay, R. D. (1991) Region-specific differentiation of the hippocampal stem cell line HiB5 upon implantation into the developing mammalian brain. *Cell* **66,** 713–729.

82. Nakafuku, M. and Nakamura, S. (1995) Establishment and characterization of a multipotential neural cell line that can conditionally generate neurons, astrocytes, and oligodendrocytes in vitro. *J. Neurosci. Res.* **41,** 153–168.

83. Rao, M. S. and Anderson, D. J. (1997) Immortalization and controlled in vitro differentiation of murine multipotent neural crest stem cells. *J. Neurobiol.* **32,** 722–746.
84. Martinez-Serrano, A. and Bjorklund, A. (1997) Immortalized neural progenitor cells for CNS gene transfer and repair. *Trends Neurosci.* **20,** 530–538.
85. Sah, D. W., Ray, J., and Gage, F. H. (1997) Bipotent progenitor cell lines from the human CNS. *Nat. Biotechnol.* **15,** 574–580.
86. Li, R., Thode, S., Zhou, J., Richard, N., Pardinas, J., Rao, M. S., et al. (2000) Motoneuron differentiation of immortalized human spinal cord cell lines. *J. Neurosci. Res.* **59,** 342–352.
87. Raymon, H. K., Thode, S., Zhou, J., Friedman, G. C., Pardinas, J. R., Barrere, C., et al. (1999) Immortalized human dorsal root ganglion cells differentiate into neurons with nociceptive properties. *J. Neurosci.* **19,** 5420–5428.
88. Saporta, S., Borlongan, C. V., and Sanberg, P. R. (1999) Neural transplantation of human neuroteratocarcinoma (hNT) neurons into ischemic rats. A quantitative dose-response analysis of cell survival and behavioral recovery. *Neuroscience* **91,** 519–525.
89. Muir, J. K., Raghupathi, R., Saatman, K. E., Wilson, C. A., Lee, V. M., Trojanowski, J. Q., et al. (1999) Terminally differentiated human neurons survive and integrate following transplantation into the traumatically injured rat brain. *J. Neurotrauma* **16,** 403–414.
90. Hurlbert, M. S., Gianani, R. I., Hutt, C., Freed, C. R., and Kaddis, F. G. (1999) Neural transplantation of hNT neurons for Huntington's disease. *Cell Transplant.* **8,** 143–151.
91. Barker, R. A., Kendall, A. L., and Widner, H. (2000) Neural tissue xenotransplantation: what is needed prior to clinical trials in Parkinson's disease? Neural Tissue Xenographic Project. *Cell Transplant* **9(2),** 235–246.
92. Kondziolka, D., Wechsler, L., Goldstein, S., Meltzer, C., Thulborn, K. R., Gebel, J., et al. (2000) Transplantation of cultured human neuronal cells for patients with stroke. *Neurology* **55,** 565–569.
93. Flax, J. D., Aurora, S., Yang, C., Simonin, C., Wills, A. M., Billinghurst, L. L., et al. (1998) Engraftable human neural stem cells respond to developmental cues, replace neurons, and express foreign genes. *Natl. Biotechnol.* **16,** 1033–1039.
94. Yandava, B. D., Billinghurst, L. L., and Snyder, E. Y. (1999) "Global" cell replacement is feasible via neural stem cell transplantation: evidence from the dysmyelinated shiverer mouse brain. *Proc. Natl. Acad. Sci. USA* **96,** 7029–7034.
95. Aboody, K. S., Brown, A., Rainov, N. G., Bower, K. A., Liu, S., Yang, W., et al. (2000) From the cover: neural stem cells display extensive tropism for pathology in adult brain: evidence from intracranial gliomas. *Proc. Natl. Acad. Sci. USA* **97,** 12,846–12,851.
96. Derrington, E. A., Dufay, N., Rudkin, B. B., and Belin, M. F. (1998) Human primitive neuroectodermal tumor cells behave as multipotent neural precursors in response to FGF2. *Oncogene* **17,** 1663–1672.

97. Villa, A., Snyder, E. Y., Vescovi, A., and Martinez-Serrano, A. (2000) Establishment and properties of a growth factor-dependent, perpetual neural stem cell line from the human CNS. *Exp. Neurol.* **161,** 67–84.

98. Rubio, F. J., Bueno, C., Villa, A., Navarro, B., and Martinez-Serrano, A. (2000) Genetically perpetuated human neural stem cells engraft and differentiate into the adult mammalian brain. *Mol. Cell Neurosci.* **16,** 1–13.

99. Bocchini, V., Casalone, R., Collini, P., Rebel, G., and Lo Curto, F. (1991) Changes in glial fibrillary acidic protein and karyotype during culturing of two cell lines established from human glioblastoma multiforme. *Cell Tissue Res.* **265,** 73–81.

100. Izumi, I., Mineura, K., Watanabe, K., and Kowada, M. (1994) Establishment of the two glioma cell lines: YH and AM. *Hum. Cell* **7,** 101–105.

101. Andres-Barquin, P. J., Hernandez, M. C., Hayes, T. E., McKay, R. D., and Israel, M. A. (1997) Id genes encoding inhibitors of transcription are expressed during in vitro astrocyte differentiation and in cell lines derived from astrocytic tumors. *Cancer Res.* **57,** 215–220.

102. Mao, X., Barfoot, R., Hamoudi, R. A., and Noble, M. (1998) Alleletyping of an oligodendrocyte-type-2 astrocyte lineage derive from a human glioblastoma multiforme. *J. Neurooncol.* **40,** 243–250.

103. Gao, W. Q. and Hatten, M. E. (1994) Immortalizing oncogenes subvert the establishment of granule cell identity in developing cerebellum. *Development* **120,** 1059–1070.

104. Espinosa de los Monteros, A., Zhang, M., and De Vellis, J. (1993) O2A progenitor cells transplanted into the neonatal rat brain develop into oligodendrocytes but not astrocytes. *Proc. Natl. Acad. Sci. USA* **90,** 50–54.

105. O'Leary, M. T. and Blakemore, W. F. (1997) Oligodendrocyte precursors survive poorly and do not migrate following transplantation into the normal adult central nervous system. *J. Neurosci. Res.* **48,** 159–167.

106. Espinosa de los Monteros, A., Zhao, P., Huang, C., Pan, T., Chang, R., Nazarian, R., et al. (1997) Transplantation of CG4 oligodendrocyte progenitor cells in the myelin-deficient rat brain results in myelination of axons and enhanced oligodendroglial markers. *J. Neurosci. Res.* **50,** 872–887.

107. Malatesta, P., Hartfuss, E., and Gotz, M. (2000) Isolation of radial glial cells by fluorescent-activated cell sorting reveals a neuronal lineage. *Development* **127,** 5253–5263.

108. Tang, D. G., Tokumoto, Y. M., Apperly, J. A., Lloyd, A. C., and Raff, M. C. (2001) Lack of replicative senescence in cultured rat oligodendrocyte precursor cells. *Science* **291,** 868–871.

109. Woodbury, D., Schwarz, E. J., Prockop, D. J., and Black, I. B. (2000) Adult rat and human bone marrow stromal cells differentiate into neurons. *J. Neurosci. Res.* **61,** 364–370.

110. Brewer, G. J. (1999) Regeneration and proliferation of embryonic and adult rat hippocampal neurons in culture. *Exp. Neurol.* **159,** 237–247.

111. Roy, N. S., Benraiss, A., Wang, S., Fraser, R. A., Goodman, R., Couldwell, W. T., et al. (2000) Promoter-targeted selection and isolation of neural

progenitor cells from the adult human ventricular zone. *J. Neurosci. Res.* **59,** 321–331.

112. Kaji, E. H. and Leiden, J. M. (2001) Gene and stem cell therapies. *JAMA* **285,** 545–550.

113. Weissman, I. L. (2000) Translating stem and progenitor cell biology to the clinic: barriers and opportunities. *Science* **287,** 1442–1446.

114. Ferrari, G., Cusella-De Angelis, G., Coletta, M., Paolucci, E., Stornaiuolo, A., Cossu, G., et al. (1998) Muscle regeneration by bone marrow-derived myogenic progenitors. *Science* **279,** 1528–1530.

115. Gunsilius, E., Duba, H. C., Petzer, A. L., Kahler, C. M., Grunewald, K., Stock-hammer, G., et al. (2000) Evidence from a leukaemia model for maintenance of vascular endothelium by bone-marrow-derived endothelial cells. *Lancet* **355,** 1688–1691.

116. Mitaka, T. (2001) Hepatic stem cells: from bone marrow cells to hepatocytes. *Biochem. Biophys. Res. Commun.* **281,** 1–5.

117. Bjornson, C. R., Rietze, R. L., Reynolds, B. A., Magli, M. C., and Vescovi, A. L. (1999) Turning brain into blood: a hematopoietic fate adopted by adult neural stem cells in vivo. *Science* **283,** 534–537.

118. Brazelton, T. R., Rossi, F. M., Keshet, G. I., and Blau, H. M. (2000) From marrow to brain: expression of neuronal phenotypes in adult mice. *Science* **290,** 1775–1779.

119. Mezey, E., Chandross, K. J., Harta, G., Maki, R. A., and McKercher, S. R. (2000) Turning blood into brain: cells bearing neuronal antigens generated in vivo from bone marrow. *Science* **290,** 1779–1782.

120. Chandross, K. and Mezey, E. (2002) Plasticity of adult bone marrow stem cells, in *Stem Cells: A Cellular Fountain of Youth* (Mattson, M. P. and Van Zant, G., eds.), Elsevier, Amsterdam, pp. 74–96.

121. Sanchez-Ramos, J., Song, S., Cardozo-Pelaez, F., Hazzi, C., Stedeford, T., Willing, A., et al. (2000) Adult bone marrow stromal cells differentiate into neural cells in vitro. *Exp. Neurol.* **164,** 247–256.

122. Prockop, D. J. (1997) Marrow stromal cells as stem cells for nonhematopoietic tissues. *Science* **276,** 71–74.

123. Brevig, T., Holgersson, J., and Widner, H. (2000) Xenotransplantation for CNS repair: immunological barriers and strategies to overcome them. *Trends Neurosci.* **23,** 337–344.

124. Fishman, J. A. (1998) Infection and xenotransplantation. Developing strategies to minimize risk. *Ann. NY Acad. Sci.* **862,** 52–66.

125. Bjorklund, A. (1992) Dopaminergic transplants in experimental parkinson-ism: cellular mechanisms of graft-induced functional recovery. *Curr. Opin. Neurobiol.* **2,** 683–689.

126. Wictorin, K., Brundin, P., Sauer, H., Lindvall, O., and Bjorklund, A. (1992) Long distance directed axonal growth from human dopaminergic mesence-phalic neuroblasts implanted along the nigrostriatal pathway in 6-hydroxydo-pamine lesioned adult rats. *J. Comp. Neurol.* **323,** 475–494.

127. Larsson, L. C., Czech, K. A., Brundin, P., and Widner, H. (2000) Intrastriatal ventral mesencephalic xenografts of porcine tissue in rats: immune responses and functional effects. *Cell Transplant.* **9,** 261–272.
128. Larsson, L. C. and Widner, H. (2000) Neural tissue xenografting. *Scand. J. Immunol.* **52,** 249–256.
129. Lambrigts, D., Sachs, D. H., and Cooper, D. K. (1998) Discordant organ xenotransplantation in primates: world experience and current status. *Transplantation* **66,** 547–561.
130. Deacon, T., Schumacher, J., Dinsmore, J., Thomas, C., Palmer, P., Kott, S., et al. (1997) Histological evidence of fetal pig neural cell survival after transplantation into a patient with Parkinson's disease. *Nat. Med.* **3,** 350–353.
131. Hauser, R. A., Freeman, T. B., Snow, B. J., Nauert, M., Gauger, L., Kordower, J. H., et al. (1999) Long-term evaluation of bilateral fetal nigral transplantation in Parkinson disease. *Arch. Neurol.* **56,** 179–187.
132. Lindvall, O. (1997) Neural transplantation: a hope for patients with Parkinson's disease. *Neuroreport* **8,** 3–10.
133. Brevig, T., Pedersen, E. B., and Finsen, B. (2000) Molecular and cellular mechanisms in immune rejection of intracerebral neural transplants. *Novartis Found. Symp.* **231,** 166–177; discussion 177–183, 302–306.
134. Hammer, C. (1998) Physiological obstacles after xenotransplantation. *Ann. NY Acad. Sci.* **862,** 19–27.
135. Fishman, J. A. (1995) Pneumocystis carinii and parasitic infections in transplantation. *Infect. Dis. Clin. North Am.* **9,** 1005–1044.
136. Fishman, J. A. (1997) Xenosis and xenotransplantation: addressing the infectious risks posed by an emerging technology. *Kidney Int.* **58(Suppl.),** S41–S45.
137. Ono, K., Takii, T., Onozaki, K., Ikawa, M., Okabe, M., and Sawada, M. (1999) Migration of exogenous immature hematopoietic cells into adult mouse brain parenchyma under GFP-expressing bone marrow chimera. *Biochem. Biophys. Res. Commun.* **262,** 610–614.
138. Wu, Y. P., McMahon, E., Kraine, M. R., Tisch, R., Meyers, A., Frelinger, J., et al. (2000) Distribution and characterization of GFP(+) donor hematogenous cells in Twitcher mice after bone marrow transplantation. *Am. J. Pathol.* **156,** 1849–1854.
139. Platt, F. M. and Butters, T. D. (1998) New therapeutic prospects for the glycosphingolipid lysosomal storage diseases. *Biochem. Pharmacol.* **56,** 421–430.
140. Miyauchi, A., Kanje, M., Danielsen, N., and Dahlin, L. B. (1997) Role of macrophages in the stimulation and regeneration of sensory nerves by transposed granulation tissue and temporal aspects of the response. *Scand. J. Plast. Reconstr. Surg. Hand. Surg.* **31,** 17–23.
141. Rapalino, O., Lazarov-Spiegler, O., Agranov, E., Velan, G. J., Yoles, E., Fraidakis, M., et al. (1998) Implantation of stimulated homologous macrophages results in partial recovery of paraplegic rats. *Nat. Med.* **4,** 814–821.
142. Evans M. J. and Kaufman, M. H. (1981) Establishment in culture of pluripotential cells from mouse embryos. *Nature* **292(5819),** 154–156.

143. Martin, G. R. (1981) Isolation of a pluripotent cell line from early mouse embryos cultured in medium conditioned by teratocarcinoma stem cells. *Proc. Natl. Acad. Sci. USA* **78(12)**, 7634–7638.

144. Torres, M. (1998) The use of embryonic stem cells for the genetic manipulation of the mouse. *Curr. Top. Dev. Biol.* **36**, 99–114.

145. Thomson, J. A., Itskovitz-Eldor, J., Shapiro, S. S., Waknitz, M. A., Swiergiel, J. J., Marshall, V. S., et al. (1998) Embryonic stem cell lines derived from human blastocysts. *Science* **282**, 1145–1147.

146. Reubinoff, B. E., Pera, M. F., Fong, C. Y., Trounson, A., and Bongso, A. (2000) Embryonic stem cell lines from human blastocysts: somatic differentiation in vitro. *Nat. Biotechnol.* **18**, 399–404.

147. Shamblott, M. J., Axelman, J., Wang, S., Bugg, E. M., Littlefield, J. W., Donovan, P. J., et al. (1998) Derivation of pluripotent stem cells from cultured human primordial germ cells. *Proc. Natl. Acad. Sci. USA* **95**, 13,726–13,731.

148. Kannagi, R., Nudelman, E., Levery, S. B., and Hakomori, S. (1982) A series of human erythrocyte glycosphingolipids reacting to the monoclonal antibody directed to a developmentally regulated antigen SSEA-1. *J. Biol. Chem.* **257(24)**, 14,865–14,874.

149. Roach, S., Cooper, S., Bennett, W., and Pera, M. F. (1993) Cultured cell lines from human teratomas: windows into tumour growth and differentiation and early human development. *Eur. Urol.* **23**, 82–87; discussion 87–88.

150. Pera, M. F., Cooper, S., Mills, J., and Parrington, J. M. (1989) Isolation and characterization of a multipotent clone of human embryonal carcinoma cells. *Differentiation* **42**, 10–23.

151. Amit, M., Carpenter, M. K., Inokuma, M. S., Chiu, C. P., Harris, C. P., Waknitz, M. A., et al. (2000) Clonally derived human embryonic stem cell lines maintain pluripotency and proliferative potential for prolonged periods of culture. *Dev. Biol.* **227**, 271–278.

152. Andrews, P. W. (1984) Retinoic acid induces neuronal differentiation of a cloned human embryonal carcinoma cell line in vitro. *Dev. Biol.* **103**, 285–293.

153. Pleasure, S. J., Page, C., and Lee, V. M. (1992) Pure, postmitotic, polarized human neurons derived from NTera 2 cells provide a system for expressing exogenous proteins in terminally differentiated neurons. *J. Neurosci.* **12(5)**, 1802–1815.

154. Xu, C., Inokuma, M. S., Denham, J., Golds, K., Kundu, P., Gold, J. D., Carpenter, M. K. (2001) Feeder-free growth of undifferentiated human embryonic stem cells. *Nat. Biotechnol.* **19(10)**, 971–974.

155. Carpenter M. K., Inokuma M. S., Denham J., Mujtaba, T., Chiu, C.-P., and Rao, M. S. (2001) Enrichment of neurons and neural precursors from human embryonic stem cells. *Exp. Neurol.* **172(2)**, 383–397.

156. McDonald, J. W., Liu, X. Z., Qu, Y., Liu, S., Mickey, S. K., Turetsky, D., et al. (1999) Transplanted embryonic stem cells survive, differentiate and promote recovery in injured rat spinal cord. *Nat. Med.* **5**, 1410–1412.

157. Dinsmore, J., Ratliff, J., Deacon, T., Pakzaban, P., Jacoby, D., Galpern, W., and Isacson, O. (1996) Embryonic stem cells differentiated in vitro as a novel source of cells for transplantation. *Cell Transplant.* Mar-Apr; **5(2),** 131–143.
158. Brustle, O., Spiro, A. C., Karram, K., Choudhary, K., Okaabe, S., and McKay, R. D. (1997) In vitro-generated neural precursors participate in mammalian brain development. *Proc. Natl. Acad. Sci. USA* Dec. 23: **94(26),** 14,809–14,814.
159. Yamane, T., Hayashi, S., Mizoguchi, M., Yamazaki, H., and Kunisada, T. (1999) Derivation of melanocytes from embryonic stem cells in culture. *Dev. Dyn.* **216,** 450–458.
160. Scolding, N. J. and Franklin, R. J. (1997) Remyelination in demyelinating disease. *Baillieres Clin. Neurol.* **6,** 525–548.
161. Kirschenbaum, B., Nedergaard, M., Preuss, A., Barami, K., Fraser, R. A., Goldman, S. A. (1994) In vitro neuronal production and differentiation by precursor cells derived from the adult human forebrain. *Cereb. Cortex* **4(6),** 576–589.
162. Wang, S., Wu, H., Jiang, J., Delohery, T. M., Isdell, F., Goldman, S. A. (1998) Isolation of neuronal precursors by sorting embryonic forebrain transfected with GFP regulated by the T alpha 1 tubulin promoter. *Nature Biotechnology* **16(2),** 196–201.
163. Gregori, N., Proschel, C., Noble, M., Mayer-Proschel, M. (2002) The tripotential glial-restricted precursor (GRP) cell and glial development in the spinal cord: generation of bipotential oligodendrocyte-type-2 astrocyte progenitor cells and dorsal-ventral differences in GRP cell function. *J. Neuroscience* **22(1),** 248–256.
164. Thomson, J. A., Kalishman, J., Golos, T. G., Durning, M., Harris, C. P., Becker, R. A., Hearn, J. P. (1995) Isolation of a primate embryonic stem cell line. *PNAS* **92(17),** 7844–7854.

2

Neural Stem Cells and Their Plasticity

Angela Gritti, Angelo Vescovi, and Rossella Galli

INTRODUCTION

Stem cells are functional units in both development and tissue homeostasis and can be found in a variety of embryonic and adult mammalian tissues. These cells are thought to arise from totipotent embryonic stem (ES) cells of the inner cell mass of the blastocyst from which distinct groups of precursors segregate into the three main germ layers (ectoderm, mesoderm, and endoderm) at around the time of gastrulation. Gradually, these cells will mature into fate-restricted organ- and tissue-specific somatic stem cells (SCs) *(1)*, which are responsible for the growth of tissues during development. The number of SCs declines when the tissues approach maturity and remains rather constant throughout life.

Throughout adulthood, SCs are responsible for tissue maintenance and repair, although the latter function may be carried out at very different rates in various organs. For instance, although many tissues like the hemopoietic system, the epidermis and the intestinal epithelium are known to undergo continuous, extensive cell replacement, the mature mammalian central nervous system (CNS) has long been considered incapable of significant cell turnover. This view has begun to change in the last few decades and, lately, the existence of *de novo* neurogenesis in the adult brain and the presence of stem cells in the mammalian CNS have emerged. Currently, a restricted area of the adult forebrain, the subventricular zone lining the forebrain ventricles (a remnant of the embryonic subventricular zone) is thought to be the largest stem cell compartment of the adult brain. Adult neural stem cells (ANSCs) have been isolated from this region and have been propagated in vitro (reviewed in refs. *2–5*). Although it was expected that ANSCs may display a certain degree of plasticity both in terms of growth and expansion rate and differentiation ability, it was also generally held that ANSC fate was restricted to generating exclusively the three major brain cell types, namely neurons, astrocytes, and oligodendrocytes.

From: *Neural Stem Cells for Brain and Spinal Cord Repair*
Edited by: T. Zigova, E. Y. Snyder, and P. R. Sanberg © Humana Press Inc., Totowa, NJ

In this chapter, we focus on a series of recent findings that document an unexpected degree of plasticity of SCs. Emphasis is put on discussing the capacity of ANSCs to generate non-neural cell lineages. A preliminary examination of some basic notions on the functional properties and on the emerging role of the microenvironment in regulating ANSC behavior is preparatory to the discussion that follows on the extreme plasticity of these cells.

GENERAL CONCEPTS ON ADULT STEM CELLS

During development, stem cells proliferate and their progeny undergo a process of progressive lineage restriction and, eventually, generate the terminally differentiated cells that form the mature tissues. Although diversification of distinct mature cell types is complete at or soon after birth, many tissues in the adult organism undergo continuous physiological cell turnover and repair and must therefore embody a population of rather pliable SCs. These are often relatively quiescent or slow proliferating cells, but they retain a significant ability to increase their activity to replace dead and/or injured cells, and very often this occurs through the generation of an intermediate, fast-proliferating transit-amplifying cell population *(6,7)*.

Despite the significant effort that has gone into defining the specific molecular and/or antigenic markers for the various types of adult stem cell, their identification in many tissues is still carried out on an operational basis and, essentially, relies on the retrospective assessment of critical functional characteristics. A notable exception is represented by the hemopoietic stem cells, for which a wide array of such markers is, indeed, available.

According to the most widely accepted operational definition *(6,7)*, stem cells are undifferentiated cells (i.e., lacking antigenic markers typical of mature cells), displaying an extensive proliferation potential that is inextricably linked to their extensive (possibly throughout life) self-renewal capacity. SCs are also believed to be multipotent, in the sense that they can give rise to a wide array of mature progeny of the tissue in which they reside and should possess the ability to regenerate their tissue of origin, even following significant damage.

Self-renewal is defined as the capacity of a cell to perpetuate itself and, at least in invertebrates, it can be achieved at the single-cell level by a deterministic type of division in which one cell is identical to its mother and another more differentiated cell is invariably generated at each cell cycle. In vertebrates, self-renewal is rather viewed as the property of a cell population as a whole and is, thus, interpreted as the capacity to maintain the number of stem cells in a given cell compartment at a steady level. Yet, under particular circumstances, the stem cell population can be expanded or reduced in size, if necessary. This is made possible by the fact that a fixed stem cell population is physiologically maintained by an even balance between the number of symmetric divisions that generate two stem cells or, alternatively, two differentiated cells at each cycle. Shifting this equilibrium in favor of the first or second type of division will either determine an increase or, alternatively, a decrease in the number of stem cells within the population. This mechanism is likely to provide a system by which the size of the stem cell population and the number of differentiated progeny generated can be varied in response to changes in the extracellular environment or in intercellular communication, brought about by various injuries or pathological situations.

Multipotency is the ability of a single cell to generate many different types of mature progeny. In principle, multipotent stem cells should be able to give rise to all the cell types that constitute their tissue. In vivo, this criterion may be difficult to assess, and stem cells may appear to generate only a subset of the differentiated cell types of a given tissue. However, the global fate potential of SCs may be unraveled by in vitro assays *(8–13)*, in which a candidate cell is challenged under environmental conditions that may not be as readily available in vivo.

A notable feature of SCs is that they are generally located in specific restricted tissue regions, within which a cytoarchitectural and/or biochemical confinement may create a specific niche. Within the niche, conditions are maintained so that SCs can retain their peculiar attributes and, particularly, their life-long ability to self-renew and generate mature progeny. Although in some tissues like the liver and the hemopoietic system the anatomical location of the stem cell niche and its relationship to the tissue-specific stem cells are not perfectly clear, in other systems like the epidermis in nonhaired skin, the hair follicle and the small intestine, the niche is neatly, spatially defined and the stem cells residing therein can be identified by their morphology and relative position within the niche itself *(14,15)*.

THE ADULT NEURAL STEM CELL
AND ITS *INTRA-GERM-LAYER* PLASTICITY

The mammalian brain with its complex network of connections develops from a much simpler embryonic, neuroepithelial structure called the neural tube, contains uncommitted proliferating neural precursors that, initially, reside in the luminal cell layer or ventricular zone (VZ). As development proceeds, a new germinal layer appears beneath the original VZ, which is called the subventricular zone (SVZ). The thickness of the VZ then gradually decreases and the latter is eventually reduced to a continuous cell monolayer (ependyma) lining the ventricular cavities of the adult brain. A remnant of the primitive SVZ persists in the forebrain throughout adulthood as an actively mitotic layer. A large body of evidence has been accumulating that suggests the persistence of intense neurogeneic activity within this region. This has raised questions as to the existence of neural stem cells in the adult brain and of specific niches within which the appropriate set of neurogeneic signals is preserved throughout life to continuously support neurogenesis (reviewed in refs. *16–18*). Recently, *bona fide* neural stem cells have been isolated from the adult SVZ *(8,19)* and numerous reports from different groups have suggested that a multipotent stem cell compartment resides within the periventricular region of the adult mammalian forebrain *(9,20)*. Furthermore, putative or "potential" stem cells have subsequently been isolated from non-neurogeneic regions of the adult brain *(10,21,22)*. This reinforces the idea that cells endowed with different degrees of stemness may reside throughout the adult CNS and that an appropriate neurogeneic microenvironment, which is likely found only within specific locations in the mature brain, is necessary for these stem cells to express their neurogeneic capacity in vivo.

The current view identifies the periventricular region of the forebrain as the main stem cell compartment of the adult mammalian CNS. From here, large numbers of cells are born in the neonatal *(23)* and adult *(11)* rodent brain that migrate along a restricted pathway to the olfactory bulb (OB), where they differentiate into interneurons. These stem cells can be isolated from the ependymal layer and/or the SVZ and, when propagated in long-term cultures, they retain extensive self-renewal, multipotency and stable functional features over time (Fig. 1). Culturing of these stem cells takes place in the presence of epidermal growth factor (EGF) and/or basic fibroblast growth factor (FGF2) *(8,9,19,20)* in serum-free medium, so that a steady expansion of the stem cell population is obtained. Notably, the actual rate of ANSCs expansion strictly depends on the pattern of mitogenic signals to which they are exposed. An intensive expansion of the stem cell number takes place in the concomitant presence of EGF and FGF2 *(12)*, indicating that symmetric divisions in which a stem cell gives rise to two daughter stem cells must occur with a high frequency under these conditions *(24)*. Yet, when only one of the two factors is used as a mitogen the very same ANSCs revert to a significantly slower rate of expansion, showing that symmetric divisions yielding two differentiated cells at each cycle are significantly increased in the presence of a single growth factor *(12)*. These findings describe an important aspect of the ANSCs' physiology, that is, their

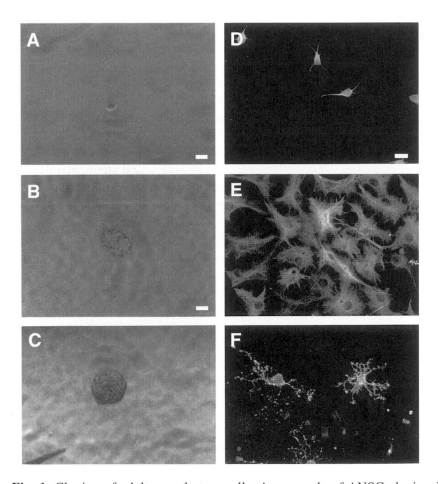

Fig. 1. Cloning of adult neural stem cells. An example of ANSC cloning is shown. (**A–C**) A single adult neural stem cell is shown after plating in isolation in a single well. After 7 d (**B**) and 14 d (**C**), this cell formed a cluster, which was serially subcultured every 6 d to establish a continuous culture. Growing cells expressed the neuroepithelial antigen, nestin (data not shown). (**D–F**) To induce differentiation, a fraction of these cells were plated in 1% fetal calf serum (FCS), in the absence of growth factors. The detection of neurons (**D**; MAP2; $16.3 \pm 0.55\%$ of total cell number [TCN]; $n = 5$, $\pm$ SEM), astroglia (**E**; GFAP; $72.6 \pm 5.9\%$ TCN; $n = 5$) and oligodendroglia (**F**; GalC; $2.5 \pm 0.9\%$ TCN; $n = 5$) among the progeny of the cell in (A) indicate its multipotentiality *(3,4)*. Secondary clones of the cell displayed in (A) produced an average of 38 ± 5.8 ($n = 6$) cells capable of producing tertiary, multipotential clones, thereby demonstrating self-maintenance. Scale bars: A, 15 µm; B and C, 40 µm; D and E, 15 µm.

capacity to "interpret" the exposure to various combinations of epigenetic signals in order to vary their self-renewal activity in relation to changes in the extracellular environment.

That the fate of ANSCs is under tight environmental control is also shown by the observation that, upon removal of growth factors (GFs) from the culture medium, spontaneous differentiation of the ANSCs progeny into a mixed neuronal/glial cell population rapidly ensues (Fig. 1). Remarkably, the timing of appearance of the mature CNS lineages is similar to that observed within most embryonic, neurogeneic regions in vivo: neurons are generated first, followed by astroglial cells and, eventually, by the onset of oligodendrocyte production. Furthermore, survival of the various differentiated stem cell progeny, as well as their maturation (indicated by the expression of defined antigenic properties), is dependent on extracellular signals. For example, FGF2 prevents the differentiation/maturation of glial and neuronal/glial precursors in serum-free medium, and the addition of low concentration of serum is a necessary step to improve neuronal maturation and the expression of glial antigens *(25)*. Also, neurotrophins play a distinctive role in affecting both the survival and differentiation of neurons in culture *(26,27)*, whereas several other molecules (bone morphogenetic proteins [BMPs], platelet-derived growth factor [PDGF], FGFs, leukemia inhibitory factor [LIF], ciliary neurotrophic factor [CNTF]) regulate the proliferation, the lineage selection, and the differentiation/maturation of ANSC-derived astrodendroglial and oligodendroglial progenitors *(13,28–31)*.

An intriguing aspect of ANSCs' plasticity is the ability to alter their developmental fate in response to extrinsic cues that may act in an instructive fashion. In fact, it has been shown that once differentiation has begun, the ratio of neuronal versus glial differentiation can be altered by exposing the stem cell progeny to various combinations of epigenetic signals. For example, the final outcome of the differentiation process in murine ANSCs can be biased by PDGF in favor of the acquisition of a neuronal fate, so that almost half of the differentiated progeny turns into neurons *(32)*. A similar phenomenon is observed in fetal human NSC cultures in the presence of LIF *(33)*. In striking contrast, CNTF appears to act instructively on murine ANSCs to generate a "larger than normal" proportion of glial cells *(32)*.

It is worth noting that the neuronal progeny of the neural stem cells is truly functionally active, as demonstrated by the ability of neurons derived from long-term passaged stem cells to elicit action potentials *(9)*. With respect to the neurotransmitter phenotypes that this progeny appears to acquire in vitro, it is clear that the large majority of the cells display a GABA-ergic or glutamatergic phenotype when differentiation is induced by the simple removal of GFs in serum-free culture medium. However, a certain degree of plasticity has been bestowed upon ANSCs also with respect to the choice of the neurotransmitter phenotype. In fact, there is evidence that specific neurotransmitter phenotypes such as the catecholaminergic one can be induced by exposing the ANSC progeny to various differentiation conditions *(34–36)*.

It should be emphasized that the epigenetic modulation of ANSC activity described above for ex vivo situations can also occur in vivo. This is shown by the dramatic increase in the number of newborn cells that is observed in the adult brain following intraventricular administration of EGF, FGF-2 *(36,37)*, or brain-derived neurotrophic factor (BDNF) *(38,39)*. Interestingly, whereas FGF-2 and BDNF induce an increase in the number of neurons, EGF enhances differentiation into the glial lineage *(36,37)*. Moreover, EGF has an inhibitory effect on the progression of SVZ-derived cells through their usual migratory routes and induces their displacement from the SVZ into the adjacent brain parenchyma *(36)*. This strongly suggests that different cell functions, including cell proliferation, survival, commitment, and migration, can be brought about under different environmental conditions in a rather complex fashion, in vivo. A recent confirmation of these phenomena has come from the recent work of Fallon et al., showing that the infusion of transforming growth factor (TGF)-α in the striatum of 6-hydroxy-dopamine-lesioned rats results in the proliferation of SVZ cells and in their migration toward the injection site, followed by differentiation of the newly born cells into intrastriatal, TH-immunoreactive neurons *(40)*.

Altogether, these observations lead to the conclusion that, in ANSCs, epigenetic signals work in concert with the cell autonomous genetic program to regulate a series of critical neurogenetic steps during which growth, fate determination, differentiation, and maturation are finely and timely regulated. The above-discussed findings underline how the rate and extent by which the enormous number of different cell types that make up the mammalian CNS are generated is subjected to flexible regulation. These phenomena can be regarded as an example of intratissue or intra-germ-layer plasticity, because they concern the flexible generation of mature neural cell lineages, so that all the progeny that are produced by ANSCs are derivatives of the same embryonic germ layer (i.e., the ectoderm).

ADULT NEURAL STEM CELLS AND THEIR *TRUE* IDENTITY

In the last 3 yr, a plethora of studies have contributed to unravel the functional characteristic of the ANSCs. However, the true identity of these cells is still the object of heated debate. The cellular composition of the

SVZ region has been previously described, and four cell types have been characterized therein: ependymal cells, type A cells (neuroblasts), type B cells (astrocytes), and type C cells (immature precursors) *(41)*. Separate studies have shown that the ependymal cells that line the luminal surface of the adult ventricular wall *(42)* and the astrocytes that resides in the adjacent SVZ (type B cells) *(11)* are the source of the multipotent ANSCs. To provide direct evidence that type B cells can give rise to neurons, Doetsch et al. *(11)* used a transgenic mice expressing the avian leukosis viral (ALV) receptor driven by the promoter/enhancer of the gene encoding the glial-specific protein GFAP (glial fibrillary acidic protein). When these mice were infected with the ALV encoding alkaline phosphatase (AP), they generated AP-expressing SVZ astrocytes that gave rise to a population of rapid dividing cells (type C cells) that, in turn, gave rise to neuronal cells (type A cells) migrating to the OB. By using a similar experimental approach, Seri et al. *(43)* recently suggested that the granule neurons of the hippocampus, the other major neurogenetic region of the adult brain, are generated by astrocyte-like stem cells through a transit-amplifying cell population (type D cells).

A recent report has proposed that adult SVZ-derived astrocytes grown as a monolayer can resume neural stem cells features when replated in suspension cultures in the presence of FGF2 and EGF *(44)*. Moreover, SVZ-derived multipotent progenitors have been demonstrated to express radial glial markers *(45,46)*, indicating that monolayer astrocytes retain a sort of immature "radial glia" phenotype that could be functionally related to the type B astrocytes found in the SVZ. These results strongly suggest that multipotent stem cells of the adult brain may, indeed, represent a small subset of astrocyte-like V cells in the SVZ. However, the possibility that, in certain conditions, ependymal cells themselves may act as stem cells, directly or through the generation of SVZ, remains open.

NEURAL STEM CELL, THE NEUROGENETIC PROCESS, AND THE BRAIN MICROENVIRONMENT

The molecular signals that are required for the maintenance of a neuro-genetic or, better, neuronogenetic capacity in the adult brain are poorly understood. Nevertheless, it appears that one can borrow the notion of stem cell niches found in many mature tissues and infer the existence of neurogenetic domains within the adult brain. For instance, adult rat hippocampus-derived progenitor cells grafted into the neurogenic (the SVZ) or non-neurogenic (the cerebellum) region of adult hosts specifically give rise to neurons only in the former *(47)*. Furthermore, SVZ cells transplanted in the SVZ of a recipient animal generate large numbers of new neurons *(11,48)*, whereas the same cells transplanted to the non-neurogenetic brain region (cortex and striatum) produce astrocytes almost exclusively *(49)*. Similarly, transplantation of in vitro expanded spinal-cord-derived ANSCs *(22)* into the spinal cord of adult rats resulted in the production of glial cells only. However, after their heterotopic transplantation into the dentate gyrus of the hippocampus, a known neurogenetic region in the adult brain *(50)*, the same cells integrated in the granular layer and differentiated into neurons, whereas engraftment into other hippocampal regions resulted in the production of glial cells only *(51)*.

Many efforts have gone into unraveling the molecular signals that drive neurogenesis and in trying to determine their underlying mechanism of action. The scenario that has emerged so far proposes that many different types of molecule (secreted factors, membrane proteins, extracellular matrix components) as well as different cell types act and interact to contribute to the specification of particular domains in the neural stem cell's microenvironment.

Studies on the SVZ in vitro indicate that EGF and FGF2 may be essential components of the stem cell niche. As discussed previously, these mitogens can maintain the proliferation and self-renewal of ANSCs isolated from the SVZ. Infusion of EGF or FGF2 into the forebrain ventricles causes the expansion of the SVZ cell population *(52,53)*, and the importance of EGF signaling for adult neurogenesis in vivo is suggested by the reduced dorsolateral SVZ cell proliferation found in the TGF-α knockout mice *(54)*. Other factors have recently been implicated as potential regulators of stem cell activity in the SVZ. Among them are ephrins *(55)* and noggin, a polypeptide that binds bone morphogenetic proteins (BMPs), preventing their activation of BMP receptors *(56)*. Lim et al. *(57)* proposed that noggin produced by ependymal cells antagonizes BMP autocrine signaling of type B cells (which normally blocks the neurogenic pathway), creating a neurogenetic environment in the adjacent SVZ and driving SVZ cells toward the acquisition of a neuronal identity.

The permissive environment provided by extracellular matrix, cell surface molecules, and special supporting cells (i.e., radial glia) allows the displacement of neuronal precursors from the site of genesis to the site of their full differentiation during development. This is somehow reproduced within those regions of the adult brain that are endowed with structural plasticity and/or the capacity for active neurogenesis. For example, the polysialylated "embryonic" isoform of the neural cell adhesion molecule N-CAM (PSA-NCAM), which plays an important role in cell migration and cell shaping

(58–60), is selectively expressed on the membrane of newly generated cells in the SVZ *(61)* and in the hippocampus *(62)*.

The role of integrins in maintaining stem cells and progenitors in the proper position within a stem cell niche, as well as in activating signal transduction pathways essential for stem cell proliferation and survival, has been documented for extra-CNS stem cells (epidermis and intestinal stem cells) (reviewed in ref. *14*). The contribution of these molecules to the stem cell microenvironment in the adult CNS is currently unknown, but the interactions between integrins and their various ligands has been implicated in neuronal migration *(63)*, in the regulation of neurite outgrowth *(64)*, and in the enhancement of myelin membrane formation by oligodendrocytes *(65)*. Moreover, these molecules have been suggested to play both signaling and structural functions in adult synapses during plasticity *(66)*.

The formation of a specific microenvironment in the niche may also entail the presence of specific cell types that may provide mechanical and/or trophic sustenance to the stem cell. In a way somewhat similar to that observed with stromal and hemopoietic stem cells in the bone marrow, in the developing CNS, radial glial cells provide support for neuronal migration and supply instructive and neurotrophic signals that are required for the survival, proliferation, and differentiation (reviewed in ref. *67*). Similarly, in the adult brain, short radial glialike cells are present in the hippocampal dentate gyrus *(68)*, whereas SVZ astrocytes (type B cells) that surround the cells migrating toward the OB express embryonic cytoskeletal proteins such as vimentin and nestin *(17,69)*.

Once the appropriate set(s) of epigenetic cues is established/maintained within the niche, self-renewal of the neural stem cell compartment ultimately depends on the modulation imposed by these signals upon cell intrinsic regulatory mechanisms. The identity of these intrinsic molecules and the signaling cascades activated upon interaction with epigenetic factors is complex and still poorly understood. Nuclear factors controlling gene expression in stem and progenitor cells, molecules involved in the control of asymmetric divisions of stem cells, and clock mechanisms that set the number of division rounds within the population may function as cell intrinsic regulator factors (reviewed in ref. *70*). Interestingly, some of the transcription factors and their downstream effectors involved in the regulation of cell cycle and proliferation in the stem cell niches of the epidermis and intestinal epithelium (reviewed in ref. *70*) have also been found in the developing and adult CNS as well as in neural-derived primary cultures and cell lines *(71–74)*. This suggests that an evolutionarily conserved set of intrinsic mechanisms may be acting in different types of somatic stem cell with the ultimate role of maintaining the appropriate size of stem cell compartment.

DEVELOPMENTAL REPERTOIRES OF ADULT SOMATIC STEM CELLS: *EXTRA-GERM-LAYER* PLASTICITY OF NEURAL STEM CELLS

As discussed, SCs from many different adult mammalian tissues, including the bone marrow, muscle, skin, gut, and central nervous system, have been isolated and characterized. In each tissue, these cells are responsible for the cellular turnover elicited by physiological cell depletion or by pathological situations or injury. Therefore, it is almost tautological to view the differentiation potential of SCs as being restricted to the sole production of mature cells that belong to the same tissue in which the SCs reside. However, several lines of evidence have recently challenged this dogmatic notion.

The initial finding of the intragerm layer transdifferentiation capacity of an adult stem cell—that is, the production of a mature progeny normally found in a tissue that is different from that in which the stem cell resides, but sharing a common embryonic germ layer origin—was reported in 1998, when Ferrari et al. *(75)* showed that mesodermal bone marrow precursors can differentiate into skeletal muscle cells following chemically induced damage of the adult tibialis anterioris in the mouse. This was followed by the report from Bjornson et al. demonstrating that, when injected into sublethally irradiated adult mice, ANSCs—that are ectodermal derivatives—could give rise to hematopoietic cells, which are mesodermal in origin *(76)*. This provided the first demonstration that somatic stem cells that were derived from a given germ layer could undergo *trans-germ-layer* differentiation and could generate cells of a distinct embryonic origin.

Later, this striking developmental flexibility began to emerge as a more general feature of other types of adult somatic stem cell. Thus, bone marrow precursors were soon proved to contribute to the regeneration of extramesodermal organs such as the liver and to be capable of converting into neural cell lineages *(77–80)*. Similarly, mesodermal derivatives such as mesenchymal stem cells were shown to give rise to astrocytes and, possibly, neurons, both in vivo *(81)* and in vitro *(82,83)*.

The unexpected plasticity of adult CNS stem cells was quite astounding, particularly because the nervous system has always been depicted as the most "quiescent" of the adult tissues, at least with reference to cell turnover in adult life. Yet, documentation of such extraordinary capability has been extended by the work of Clarke et al. that demonstrated that, upon injection into the mouse blastocyst, ANSCs can integrate into many different tissues derived from the three main germ layers *(84)*. In this work, the contribution of ANSCs to two major mesodermal lineages such as blood and the skeletal muscle was not observed, whereas our group demonstrated soon thereafter that ANSCs or human fetal neural stem cells do differentiate into skeletal muscle, in vivo and in vitro *(85)* (Fig. 2). This discrepancy is likely justified by the different cell systems and assays used. In the work of Clarke et al., ependymal-derived ANSCs were injected into an embryonic environment whereas, in our experiments, adult SVZ stem cells underwent trans-germ-layer differentiation in the context of an adult organism. Thus, whereas in a regenerating adult tissue the "inducing" cues direct transdifferentiation toward the generation of the local, tissue-specific lineages *(85)*, distinct signals are likely to act upon exogenous cells injected into a blastocyst. These signals will compete in trying to direct the fate of the transplanted cells toward many alternative developmental pathways. In the latter situation, the final fate acquired by the implanted cells will be the result of their exposure to numerous, yet unpredictable factors. To compound the problem, the overall pattern of composition of these factors will depend on the precise stage of blastula development and on a series of unpredictable positional cues that relate to the initial site of integration of the transplanted cells. This situation is likely to cause a significant degree of variability in the outcome of the blastula experiments, which will require the analysis of a significant number of grafts to allow for a categorical conclusion to be reached as to the actual overall potential of the donor cells. Recent data from Pipia et al. seem to confirm this view *(86)*. In contrast with the work of Clarke et al., when these authors injected neural stem cells into blastocysts, the very first lineage to be colonized by the neural cells in their assay was, indeed, the hemopoietic one *(86)*.

The many reports describing CNS stem cell plasticity underline the need for specific, rather peculiar, conditions in order for SCs to be able to express their latent, generalized developmental potential(s). Two main conditions have been identified to be necessary for CNS stem cell transdifferentiation to occur both in vivo and in vitro. First, the stem cells ought to be in a highly undifferentiated state. In fact, we reported that the proliferation state of ANSCs represents an important element that can influence the rate at which ANSCs convert into non-CNS cells. Thus, ANSCs need to be in a state of active proliferation in order to undergo efficient conversion into non-neural cells either in vivo or in vitro *(76,84,85)*. As described by Galli et al. *(85)*, when ANSC-derived terminally differentiated progeny (neurons and glia) were exposed to cues that induce the myogenic phenotype, almost no

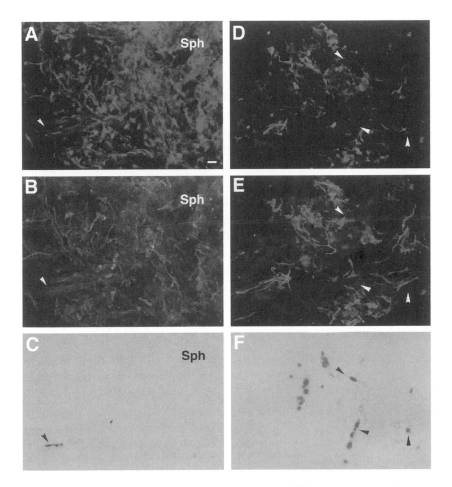

Fig. 2. Inhibition of myogenic conversion in ANSCs by neural cell contact. Equal numbers of ANSCs derived from MLC3F animals that carry the nuclear *lacZ* transgene under the control of the heavy-chain myosin promoter were plated as clustered cells (neurospheres; **A–C**) or following dissociation to single cells (**D–F**) onto C2C12 myogenic cells to induce conversion into muscle. (**A–C**) Myogenic conversion never took place within neurosphere cores (Sph), where densely packed neurons (**A**, microtubule-associated protein 2, MAP2) and astrocytes (**B**, glial fibrillary acidic protein, GFAP) were present. ANSC-derived myocytes were detected by X-gal reaction only at the outer margins of the same sphere (**C**, arrowhead). When dissociated cells from NSMLC-derived neurospheres were cocultured with C2C12 (**D–F**), a significantly greater number of myocytes (arrowheads in **D–F**) developed: These were uniformly dispersed among neurons (**D**, MAP2) and astrocytes (**E**, GFAP). Expression of neural antigens in MLC3F-derived myogenic cells was never observed (arrowheads, **A–F**). Scale bar = 8 µm.

conversion to skeletal muscle was observed, thus emphasizing the need for ANSCs to be in a high immature "naive" state, in order to successfully deal with a cohort of new and "unfamiliar" signals.

Second, it can be speculated that, to undergo *trans-germ-layer* differentiation, ANSCs ought to become exposed to specific microenvironments rich in instructive cues. These cues, apparently, become available during the regenerative phase that follows an injury in an adult target tissue or during embryonic development, as CNS stem cells have been shown to repopulate both the early developing gastrula *(84)* as well as the regenerating hemopoietic system *(76)* and muscle *(85)*.

The fact that only ANSCs but not their differentiated neuronal or glial progeny are susceptible to *trans-germ-layer* differentiation *(85)* highlights an important difference between this phenomenon and transdifferentiation as conceived in its classical form. In fact, transdifferentiation implies the ability of differentiated cells to acquire a new identity by turning off one set of lineage-specific genes and activating genes found in another differentiated cell type, whereas *trans-germ-layer* differentiation seems to reflect the *de novo* expression of a broader developmental potential of ANSCs that may become evident only under peculiar conditions. In the context of this observation it is worth noting that no molecular and biochemical overlapping of specific neural and non-neural markers could ever be observed between ANSCs and their extra-CNS progeny. In fact, markers of hematopoietic precursors were never detected in ANSCs before injection into damaged animals *(76)* and expression of muscle determination factors or muscle differentiation markers (MyoD, Myf5, myogenin and myosin) was never observed in ANSCs prior to their exposure to "myogenic" environments *(85)*. In addition, the total absence of expression of mesodermal- and endodermal-specific antigens (i.e., TROMA-1) was demonstrated in ANSCs before injection into the developing blastula *(84)*, yet these cells can give rise to derivatives from both layers. In all these examples, ANSCs expressed appropriate neural stem cell antigens such as nestin in their undifferentiated state and neuronal and glial markers upon differentiation.

An important issue ensues from the unexpected *trans-germ-layer* plasticity of somatic stem cells and concerns the mechanisms that govern this striking phenomenon and the identification of the signals released by the "inductive" environments. In spite of the relative abundance of examples of *trans-germ-layer* differentiation, very little is known as to the identity of the factors responsible for this transition. Nevertheless, based on our recent work showing the generation of skeletal muscle from ANSCs, it is still possible to propose some hypothesis. We have established the initial in vitro model by which neuro-myogenic conversion can be elicited and studied under controlled conditions. Using this system, it appeared immediately clear that neuro-myogenic conversion could be observed only when ANSCs were cocultured with C2C12 myogenic cells or with primary myoblasts, and never with nonmyogenic cells. Interestingly, conversion required direct cell–cell contact and did not take place when neural and myogenic cells were physically separated by a porous membrane, nor was it observed when ANSCs were exposed to myogenic cell-derived extracellular matrix or to medium in which muscle cells had been grown. Furthermore, when ANSC underwent myogenic induction as undissociated clonal colonies, the proportion of cells that underwent conversion to skeletal muscle dropped by almost 80% when compared to the same number of ANSCs that were dissociated before coculturing. This happened despite the fact that ANSCs clusters rapidly spread onto myoblasts—so that the cells in the sphere were in contact with the C2C12 cells. The occasional myogenic conversion in these cocultures occurred exclusively at the exterior of the cluster of neural cells and never within its core, for the latter contained only neurons and glia. Thus, when neural cells are clustered together, neural-to-neural signals override the myoblast-derived myogenic cues, blocking neuro-myogenic conversion of ANSCs. Therefore, in analogy to what is observed during the dedifferentiation of retinal pigmented epithelial into lentoids *(87)*, loss of cell-to-cell communication among ANSCs seems to emerge as one of the major determinants influencing the onset of neural-to-muscle conversion.

Based on these findings, we can infer that, at least in the specific case of neuroectodermal–mesodermal conversion, cell-to-cell contact between the inducing (muscle) and the induced (neural) cells must occur. Because cell fusion was not a prerequisite for neuro-myogenic conversion in our system, this kind of direct interaction underlines the necessity of a direct exchange of information between the inducer and the induced cells. This may imply the interaction of cell surface receptors and ligands, the formation of gap junction structures, as well as the involvement of short-range-acting molecules. However, it should be emphasized that, based on these observations, it cannot be ruled out that also secreted factors may be implicated in neuro-myogenic conversion. Yet, it can be argued that diffusible molecules alone cannot enforce such an extreme change of cellular identity. This finds confirmation in the observation that culturing ANSCs in the sole presence of hematopoietic growth factors and cytokines—that, alone, would allow for the growth and maturation of hemopoietic precursors—fails in eliciting conversion of neural cells into blood. In fact, the latter can only be achieved through transplantation into irradiated animals, following their integration into the bone marrow *(76)*. Finally, it emerges that a fine interplay between antagonistic cues takes place in the neuro-myogenic conversion phenomenon. In this system, the induction of the muscle fate elicited upon ANSCs is simultaneously counteracted by a "neuralizing" kind of signaling that takes place between the neural cells when they are in direct contact with each other. This phenomenon may be viewed as a classical "community effect." Hence, a cohort of instructive signals rather than a single effector is likely to direct the change from a brain-specific fate to a mesodermal one, as observed in ANSCs. In vivo, these signals can be found either in the extracellular microenvironment that has been perturbed by a lesion or are elicited through a direct cell-to-cell interaction between the host and donor cells (Fig. 3).

Along with the influence of the external milieu in controlling stem cell flexibility, cell-autonomous modifications may be involved in the transdifferentiation process. Therefore, it appears of fundamental importance to identify the genetic determinants that can activate the cascade of events that lead to cell lineage interconversion. Initial data are now becoming available in this area of investigation. For instance, the gene called *Pax7* has been involved in controlling the capacity of adult muscle satellite cells to transdifferentiate into hemopoietic cells *(88)*. In fact, gene deletion by homologous recombination provided adult muscle stem cells with a markedly increased potential for hematopoietic differentiation. Furthermore, similarities between the gene expression pattern observed during embryonic development and during transdifferentiation have been documented for the cornea–lens transdetermination phenomenon *(89)*. This implies that the different processes of embryogenesis, regeneration and transdifferentiation are likely highly interrelated at the molecular level. In a similar fashion, PTF-1 and PDX-1 expression in pancreatic acinar cells has been associated to the neogeneration of endocrine islet cells, and exocrine cells can transdifferentiate and acquire characteristics typical of precursors active during β-cell neogenesis *(90)*.

CONCLUSIONS

The different studies demonstrating the *trans-germ-layer* developmental potential of adult stem cells lead to a quite clear conclusion; that is, that many SCs possess the capacity to reactivate an apparently dormant set of developmental programs when challenged under peculiar environmental conditions. The reactivation of these programs does not require genetic manipulation or nuclear transplantation and is driven, at least in part, by a cohort of extracellular cues that exist in the developing organism as well as within regenerating adult tissues. This observation seems to reinforce the

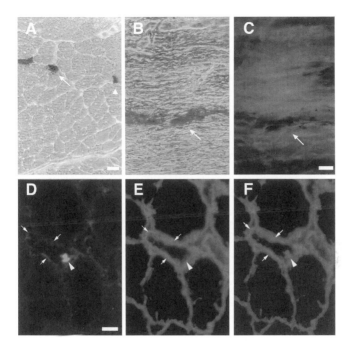

Fig. 3. Clonally derived mouse adult ROSA-26 and human embryonic neural stem cells differentiate in vivo to muscle fiber. 5×10^5 ANSCs that were clonally derived from ROSA26 animals (**A–C**) that constitutively express the *lacZ* transgene or human fetal neural stem cells (**D–F**) were injected into regenerating tibialis anterior (TA) of *scid/bg* mice. (**A**) After 3 wk, both ROSA26-derived single cells (arrowhead) and small regenerating fibers (arrow) expressing cytoplasmic β-gal were detected at the injection site. (**B–C**) Longitudinal 20-μm-thick sections show β-gal-positive fibers (arrow in **B**) within a group of fibers expressing a sarcomeric myosin heavy chain (arrow in **C**). The fluorescence of the β-gal-positive fibers in (**C**) is quenched by the X-gal reaction product. Scale bar in **A**: 100 μm; **B–C**: 50 μm. (**D–F**) In **D**, two human nuclei are detected deep inside a dystrophin-positive fiber (**E**, dystrophin) (**F**, merged), as expected in regenerating muscle fibers. Scale bar: 25 μm. (Courtesy of U. Borello; see also ref. *65.*)

idea that the cellular and molecular mechanisms that are responsible for the appropriate development of the mammalian body become partly reactivated during tissue regeneration. This underlines the importance and the impact that studies on mammalian development and on basic stem cells physiology bear on the establishment of novel therapeutic strategies for many human diseases that are untreatable with conventional therapies.

The existence of the *trans-germ-layer* differentiation phenomenon opens new therapeutic vistas that were unimaginable until a few years ago.

One may envision a scenario in which cells from a healthy tissue can be "transengineered" and used for the therapy of an ill part of the body in the same individual. The development of such a refined kind of autologous intervention hinges on the understanding of the basic cellular and molecular mechanisms that underline the conversion of cells derived from different germ layers into another. It is clear that this area of investigation will likely become a most explosive one in the years to come.

REFERENCES

1. Faust, C. and Magnuson, T. (1993) Genetic control of gastrulation in the mouse. *Curr. Opin. Genet. Dev.* **3,** 491–498.
2. Cameron, H. A. and McKay, R. (1998) Stem cells and neurogenesis in the adult brain. *Curr. Opin. Neurobiol.* **8,** 677–680.
3. Temple, S. and Alvarez-Buylla, A. (1999) Stem cells in the adult mammalian central nervous system. *Curr. Opin. Neurobiol.* **9,** 135–141.
4. Kuhn, H. G. and Svendsen, C. N. (1999) Origin, functions, and potential of adult neural stem cells. *BioEssays* **21,** 625–630.
5. Gage, F. H. (2000) Mammalian neural stem cells. *Science* **287,** 1433–1438.
6. Potten, C. S. and Loeffler, M. (1990) Stem cells: attributes, cycles, spirals, pitfalls and uncertainties. Lesson for and from the crypt. *Development* **110,** 1001–1020.
7. Loeffler, M. and Potten, C. S. (1997) Stem cells and cellular pedigrees—a conceptual introduction, in *Stem Cells* (Potten, C. S., ed.), Academic, London, pp. 1–27.
8. Reynolds, B. A. and Weiss, S. (1992) Generation of neurons and astrocytes from isolated cells of the adult mammalian central nervous system. *Science* **255,** 1707–1710.
9. Gritti, A., Parati, E. A., Cova, L., Frolichsthal, P., Galli, R., Wanke, E, et al. (1996) Multipotential stem cells from the adult mouse brain proliferate and self renew in response to basic fibroblast growth factor. *J. Neurosci.* **16,** 1091–1100.
10. Palmer, T. D., Markakis, E. A., Willhoite, A. R., Safar, F., and Gage, F. H. (1999) Fibroblast growth factor-2 activates a latent neurogenic program in neural stem cells from diverse regions of the adult CNS. *J. Neurosci.* **198,** 487–497.
11. Doetsch, F., Caille, I., Lim, D. A., Garcia-Verdugo, J. M., and Alvarez-Buylla, A. (1999) Subventricular zone astrocytes are neural stem cells in the adult mammalian brain. *Cell* **97,** 703–716.
12. Gritti, A., Frolichsthal, P., Galli, R., Parati, E. A., Cova, L., Pagano, S. F., et al. (1999) Epidermal and fibroblast growth factors behave as mitogenic regulators for a single multipotent stem cell-like population from the subventricular region of the adult mouse forebrain. *J. Neurosci.* **19,** 3287–3297.
13. Mi, H. and Barres, B. A. (1999) Purification and characterization of astrocyte precursor cells in the developing rat optic nerve. *J. Neurosci.* **19(3),** 1049–1061.

14. Fuchs, E. and Segre, J. A. (2000) Stem cells: a new lease on life. *Cell* **100,** 143–155.
15. Weissmann, I. L. (2000) Stem cells: units of regeneration and units in evolution. *Cell* **100,** 157–168.
16. Gage, F. H., Ray, J., and Fisher, L. J. (1995) Isolation, characterization, and use of stem cells from the CNS. *Annu. Rev. Neurosci.* **18,** 159–192.
17. Peretto, P., Merighi, A., Fasolo, A., and Bonfanti, L. (1999) The subependymal layer in rodents: a site of structural plasticity and cell migration in the adult mammalian brain. *Brain Res. Bull.* **49,** 221–243.
18. Temple, S. (1999) The obscure origins of adult stem cells. *Curr. Biol.* **9,** 397–399.
19. Richards, K. J., Kilpatrick, T. J., and Bartlett, P. F. (1992) De novo generation of neuronal cells from the adult mouse brain. *Proc. Natl. Acad. Sci. USA* **9,** 8591–8595.
20. Morshead, C. M., Reynolds, B. A., Craig, C. G., McBurney, M. W., Staines, W. A., Morassutti, D., et al. (1994) Neural stem cells in the adult mammalian forebrain: a relatively quiescent subpopulation of subependymal cells. *Neuron* **13,** 1071–1082.
21. Weiss, S., Dunne, C., Hewson, J., Wohl, C., Wheatley, M., Peterson, A. C., et al. (1996) Multipotent CNS stem cells are present in the adult mammalian spinal cord and ventricular neuroaxis. *J. Neurosci.* **16,** 7599–7609.
22. Shihabuddin, L. S., Ray, J., and Gage, F. H. (1997) FGF-2 is sufficient to isolate progenitors found in the adult mammalian spinal cord. *Exp. Neurol.* **148,** 577–586.
23. Luskin, M. B. (1993) Restricted proliferation and migration of postnatally generated neurons derived from the forebrain subventricular zone. *Neuron* **11,** 173–189.
24. Morrison, S. J., Shah, N. M., and Anderson, D. J. (1997) Regulatory mechanisms in stem cell biology. *Cell* **88,** 287–298.
25. Vescovi, A. L., Reynolds, B. A., Fraser, D. D., and Weiss, S. (1993) bFGF regulates the proliferative fate of unipotent (neuronal) and bipotent (neuronal/ astroglial) EGF-generated CNS progenitor cells. *Neuron* **11(5),** 951–966.
26. Vicario-Abejon, C, Johe, K. K., Hazel, T. G., Collazo, D., and McKay, R. D. (1995) Functions of basic fibroblast growth factor and neurotrophins in the differentiation of hippocampal neurons. *Neuron* **15,** 105–114.
27. Bibel, M. and Barde, Y. A. (2000) Neurotrophins: key regulators of cell fate and cell shape in the vertebrate nervous system. *Genes Dev.* **14,** 2919–2937.
28. Barres, B. A., Raff, M. C., Gaese, F., Bartke, I., Dechant, G., and Barde, Y. A. (1994) A crucial role for neurotrophin-3 in oligodendrocyte development. *Nature* **367(6461),** 371–375.
29. Raff, M. C., Lillien, L. E., Richardson, W. D., Burne, J. F., and Noble, M. D. (1988) Platelet-derived growth factor from astrocytes drives the clock that times oligodendrocyte development in culture. *Nature* **333(6173),** 562–565.
30. Mabie, P. C., Mehler, M. F., Marmur, R., Papavasiliou, A., Song, Q., and Kessler, J. A. (1997) Bone morphogenetic proteins induce astroglial differentiation of oligodendroglial–astroglial progenitor cells. *J. Neurosci.* **17(11),** 4112–4120.

31. Gross, R. E., Mehler, M. F., Mabie, P. C., Zan, Z., Santschi, L., and Kessler, J. A. (1996) Bone morphogenetic proteins promote astroglial lineage commitment by mammalian subventricular zone progenitor cells. *Neuron* **17(4),** 595–606.

32. Johe, K. K., Hazel, T. G., Muller, T., Dugich-Djordjevic M. M., and McKay, R. D. (1996) Single factors direct the differentiation of stem cells from the fetal and adult central nervous system. *Genes Dev.* **10,** 3129–3140.

33. Galli, R., Pagano, S. F., Gritti, A., and Vescovi, A. L. (2000) Regulation of neuronal differentiation in human CNS stem cell progeny by leukemia inhibitory factor. *Dev. Neurosci.* **22,** 86–95.

34. Daadi, M. M. and Weiss, S. (1999) Generation of tyrosine hydroxylase-producing neurons from precursors of the embryonic and adult forebrain. *J. Neurosci.* **19,** 4484–4497.

35. Yan, J., Studer, L., and McKay, R. D. (2001) Ascorbic acid increases the yield of dopaminergic neurons derived from basic fibroblast growth factor expanded mesencephalic precursors. *J. Neurochem.* **76,** 307–311.

36. Wagner, J., Akerud, P., Castro, D. S., Holm, P. C., Canals, J. M., Snyder, E. Y., et al. (1999) Induction of a midbrain dopaminergic phenotype in Nurr1-overexpressing neural stem cells by type 1 astrocytes. *Nat. Biotechnol.* **17,** 653–659.

37. Craig, C. G., Tropepe, V., Morshead, C. M., Reynolds, B. A., Weiss, S., and van der Kooy, D. (1996) In vivo growth factor expansion of endogenous subependymal neural precursor cell populations in the adult mouse brain. *J. Neurosci.* **16,** 2649–2658.

38. Kuhn, H. G., Winkler, J., Kempermann, G., Thal, L. J., and Gage, F. H. (1997) Epidermal growth factor and fibroblast growth factor-2 have different effects of neural progenitors in the adult rat brain. *J. Neurosci.* **17,** 5820–5829.

39. Zigova, T., Pencea, V., Wiegand, S. J., and Luskin, M. B. (1998) Intraventricular administration of BDNF increases the number of newly generated neurons in the adult olfactory bulb. *Mol. Cell. Neurosci.* **11,** 234–245.

40. Fallon, J., Reid, S., Kinyamu, R., Opole, I., Opole, R., Baratta, J., et al. (2000) In vivo induction of massive proliferation, directed migration, and differentiation of neural cells in the adult mammalian brain. *Proc. Natl. Acad. Sci. USA* **97(26),** 14,686–14,691.

41. Lim, D. A. and Alvarez-Buylla, A. (1999) Interaction between astrocytes and adult subventricular zone precursors stimulates neurogenesis. *Proc. Natl. Acad. Sci. USA* **96,** 7526–7531.

42. Johansson, C. B., Momma S., Clarke, D. L., Risling, M., Lendahl, U., and Frisén, J. (1999) Identification of a neural stem cell in the adult mammalian central nervous system. *Cell* **96,** 25–34.

43. Seri, B., Garcia-Verdugo, J. M., McEwen, B. S., Alvarez-Buylla, A. (2001) Astrocytes give rise to new neurons in the adult mammalian hippocampus. *J. Neurosci.* **21,** 7153–7160.

44. Laywell, E. D., Rakic, P., Kukekov, V. G., Holland, E. C., and Steindler, D. A. (2000) Identification of a multipotent astrocytic stem cell in the immature and adult brain. *Proc. Natl. Acad. Sci. USA* **97,** 13,883–13,888.

45. Hartfuss, E., Galli, R., Heins, N., and Gotz, M. (2001) Characterization of CNS precursor subtypes and radial glia. *Dev. Biol.* **229,** 15–30.
46. Noctor, S. C., Flint, A. C., Weissman, T. A., Dammerman, R. S., and Kriegstein, A. R. (2001) Neurons derived from radial glial cells establish radial units in neocortex. *Nature* **409(6821),** 714–720.
47. Suhonen, J. O., Peterson, D. A., Ray, J., and Gage, F. H. (1996) Differentiation of adult hippocampus-derived progenitors into olfactory neurons in vivo. *Nature* **383,** 624–627.
48. Lois, C. and Alvarez-Buylla, A. (1994) Long-distance neuronal migration in the adult mammalian brain. *Science* **264,** 1145–1148.
49. Herrera, D. G., Garcia-Verdugo, J. M, and Alvarez-Buylla, A. (1999) Adult-derived neural precursors transplanted into multiple regions of the adult brain. *Ann. Neurol.* **46,** 867–877.
50. Gage, F. H. (1998) Neurogenesis in the adult human hippocampus. *Nat. Med.* **4,** 1313–1317.
51. Shihabuddin, L. S., Horner, P. J., Ray, J., and Gage, F. H. (2000) Adult spinal cord stem cells generate neurons after transplantation in the adult dentate gyrus. *J. Neurosci.* **20,** 8727–8735.
52. Craig, C. G., Tropepe, V., Morshead, C. M., Reynolds, B. A., Weiss, S., and van der Kooy, D. (1996) In vivo growth factor expansion of endogenous subependymal neural precursor cell populations in the adult mouse brain. *J. Neurosci.* **16,** 2649–2658.
53. Kuhn, H. G., Dickinson-Anson, H., and Gage, F. H. (1996) Neurogenesis in the dentate gyrus of the adult rat: age-related decrease of neuronal progenitor proliferation. *J. Neurosci.* **16,** 2027–2033.
54. Tropepe, V., Craig, C. G., Morshead, C. M., and van der Kooy, D. (1997) Transforming growth factor-alpha null and senescent mice show decreased neural progenitor cell proliferation in the forebrain subependyma. *J. Neurosci.* **17,** 7850–7859.
55. Conover, J. C., Doetsch, F., Garcia-Verdugo, J. M., Gale, N. W., Yancopoulos, G. D., and Alvarez-Buylla, A. (2000) Disruption of Eph/ephrin signaling affects migration and proliferation in the adult subventricular zone. *Nat. Neurosci.* **3,** 1091–1097.
56. Wilson, P. A. and Hemmati-Brivanlou, A. (1997) Vertebrate neural induction: inducers, inhibitors, and a new synthesis. *Neuron* **18,** 699–710.
57. Lim, D. A., Tramontin, A. D., Trevejo, J. M., Herrera, D. G., Garcia-Verdugo, J. M., and Alvarez-Buylla, A. (2000) Noggin antagonizes BMP signalling to create a niche for adult neurogenesis. *Neuron* **26,** 713–726.
58. Bonfanti, L., Olive, S., Poulain, D. A., and Theodosis, D. T. (1992) Mapping of the distribution of polysialylated neural cell adhesion molecule throughout the central nervous system of the adult rat: an immunohistochemical study. *Neuroscience* **49,** 419–436.
59. Chazal, G., Durbec, P., Jankovsky, A., Rougon, G., and Cremer, H. (2000) Consequences of neural cell adhesion molecule deficiency on cell migration in the rostral migratory stream of the mouse. *J. Neurosci.* **20,** 1446–1457.

60. Hu, H., Tomasiewicz, H., Magnuson, T., and Rutishauser, U. (1996) The role of polysialic acid in migration of olfactory bulb interneuron precursors in the subventricular zone. *Neuron* **16,** 735–743.

61. Bonfanti, L. and Theodosis, D. T. (1994) Expression of polysialylated neural cell adhesion molecule by proliferating cells in the subependymal layer of the adult rat, in its rostral extension and in the olfactory bulb. *Neuroscience* **62,** 291–305.

62. Seki, T. and Arai, Y. (1993) Highly polysialylated neural cell adhesion molecule (NCAM-H) is expressed by newly-generated granule cells in the dentate gyrus of the adult rat. *J. Neurosci.* **13,** 2351–2358.

63. Dulabo, L., Olson E. C., Taglienti M. G., Eisenhuth S., McGrath B., Walsh C. A., et al. (2000) Reelin binds alpha3beta1 integrin and inhibits neuronal migration. *Neuron* **27,** 33–44.

64. Ivins, J. K., Yurchenco, P. D., and Lander, A. D. (2000) Regulation of neurite outgrowth by integrin activation. *J. Neurosci.* **20,** 6551–6560.

65. Buttery, P. C. and ffrech-Constant, C. (1999) Laminin-2/integrin interactions enhance myelin membrane formation by oligodendrocytes. *Mol. Cell. Neurosci.* **14,** 199–212.

66. Murase, S. and Schuman, E. M. (1999) The role of cell adhesion molecules in synaptic plasticity and memory. *Curr. Opin. Cell. Biol.* **11,** 549–553.

67. Chanas-Sacre, G., Rogister, B., Moonen, G., and Leprince, P. (2000) Radial glia phenotype: origin, regulation, and transdifferentiation. *J. Neurosci. Res.* **61,** 357–363.

68. Cameron, H. A., Woolley, C. S., McEwen, B. S., and Gould, E. (1993) Differentiation of newly born neurons and glia in the dentate gyrus of the adult rat. *Neuroscience* **56,** 337–344.

69. Jankovski, A. and Sotelo, C. (1996) Subventricular zone-olfactory bulb migratory pathway in the adult mouse: cellular composition and specificity as determined by heterochronic and heterotopic transplantation. *J. Comp. Neurol.* **371,** 376–396.

70. Watt, F. M. and Hogan, B. L. M. (2000) Out of eden: stem cells and their niches. *Science* **287,** 1427–1430.

71. Hughson, E., Dowler, S., Geall, K., Johnson, G., and Rumsby, M. (1998) Rat oligodendrocyte O-2A precursor cells and the CG-4 oligodendrocyte precursor cell line express cadherins, beta-catenin and the neural cell adhesion molecule, NCAM. *Neurosci. Lett.* **251,** 157–160.

72. Cho, E. A. and Dressler, G. R. (1998) TCF-4 binds beta-catenin and is expressed in distinct regions of the embryonic brain and limbs. *Mech. Dev.* **77,** 9–18.

73. Galceran, J., Miyashita-Lin, E. M., Devaney, E., Rubenstein, J. L., and Grosschedl, R. (2000) Hippocampus development and generation of dentate gyrus granule cells is regulated by LEF1. *Development* **127,** 469–482.

74. Satho, J. and Kuroda, Y. (2000) Beta-catenin expression in human neural cell lines following exposure to cytokines and growth factors. *Neuropathology* **20,** 113–123.

75. Ferrari, G., Cusella-De Angelis, G., Coletta, M., Paolucci, E., Stornaiuolo, A., Cossu, G., and Mavilio, F. (1998) Muscle regeneration by bone marrow-derived myogenic progenitors. *Science* **279,** 1528–1530.
76. Bjornson, C. R. R., Rietze, R. L., Reynolds, B. A., Magli, M. C., and Vescovi, A. L. (1999) Turning brain into blood: A hematopoietic fate adopted by adult neural stem cells in vivo. *Science* **283,** 534–537.
77. Petersen, B. E., Bowen, W. C., Patrene, K. D., Mars, W. M., Sullivan, A. K., Murase, N., et al. (1999) Bone marrow as a potential source of hepatic oval cells. *Science* **284,** 1168–1171.
78. Theise, N. D., Badve, S., Saxena, R., Henegariu, O., Sell, S., Crawford, J. M., et al. (2000) Derivation of hepatocytes from bone marrow cells in mice after radiation-induced myeloablation. *Hepatology* **31,** 235–240.
79. Mezey, E., Chandross, K.J., Harta, G., Maki, R. A., and McKercher, S. R. (2000) Turning blood into brain: cells bearing neuronal antigens generated in vivo from bone marrow. *Science* **290,** 1779–1782.
80. Brazelton, T. R., Rossi, F. M., Keshet, G. I., and Blau, H. M. (2000) From marrow to brain: expression of neuronal phenotypes in adult mice. *Science* **290,** 1775–1779.
81. Kopen, G. C., Prockop, D. J., and Phinney, D. G. (1999) Marrow stromal cells migrate throughout forebrain and cerebellum, and they differentiate into astrocytes after injection into neonatal mouse brains. *Proc. Natl. Acad. Sci. USA* **96,** 10,711–10,716.
82. Sanchez-Ramos, J., Song, S., Cardozo-Pelaez, F., Hazzi, C., Stedeford, T., Willing, A., et al. (2000) Adult bone marrow stromal cells differentiate into neural cells in vitro. *Exp. Neurol.* **164,** 247–256.
83. Woodbury, D., Schwarz, E. J., Prockop, D. J., and Black, I. B. (2000) Adult rat and human bone marrow stromal cells differentiate into neurons. *J. Neurosci. Res.* **61,** 364–370.
84. Clark, D. L., Johansson, C. B., Wilbertz, J., Veress, B., Nilsson, E, Karlstrom, H., et al. (2000) Generalized potential of adult neural stem cells. *Science* **288,** 1660–1663.
85. Galli, R., Borello, U., Gritti, A., Minasi, M. G., Bjornson, C., Coletta, M., et al. (2000) Skeletal myogenic potential of human and mouse neural stem cells. *Nat. Neurosci.* **3,** 986–991.
86. Pipia, G. G., Low H. P, Turner, H. P, McAuliffe C., Salmonsen, R., Quesenberry, P. J., et al. (2000) Transdifferentiation of neural precursor cells after blastocyst implantation. Society for Neuroscience, 30th Annual Meeting, New Orleans, LA, 2000, Vol. 1, p. 1101.
87. Kodama, R. and Eguchi, G. (1994) Gene regulation and differentiation in vertebrate ocular tissues. *Curr. Opin. Genet. Dev.* **4,** 703–708.
88. Seale, P., Sabourin, L. A., Girgis-Gabardo, A., Mansouri, A., Gruss, P., and Rudnicki, M. A. (2000) Pax7 is required for the specification of myogenic satellite cells. *Cell* **102,** 777–786.

89. Schaefer, J. J., Oliver, G., and Henry, J. J. (1999) Conservation of gene expression during embryonic lens formation and cornea-lens transdifferentiation in *Xenopus laevis. Dev. Dyn.* **215(4),** 308–318.
90. Rooman, I., Heremans, Y., Heimberg, H., and Bouwens, L. (2000) Modulation of rat pancreatic acinoductal transdifferentiation and expression of PDX-1 in vitro. *Diabetologia* **43,** 907–914.

3

Identification, Selection, and Use of Adult Human Oligodendrocyte Progenitor Cells

Martha Windrem, Neeta Roy, Marta Nunes, and Steven A. Goldman

INTRODUCTION

The adult mammalian forebrain continues to harbor several distinct populations of both multipotential stem and phenotypically biased progenitor cells. First demonstrated in the rodent olfactory bulb and hippocampus *(1–3)* as well as in songbird vocal control centers *(4,5)*, persistent neuronal precursor cells have now been identified in all higher vertebrates studied (reviewed in ref. *6*), including monkeys *(7)* and humans *(8,9)*. Both of these neuronal progenitors and the multipotential stem cells from which they derive appear largely restricted to the ventricular/subventricular zone (SVZ) *(4,10)*, the adult vestige of the fetal neuroepithelium.

In contrast to the restricted persistence of neuronal progenitor cells and ventricular zone stem cells, oligodendrocyte progenitor cells are widespread in the adult mammalian brain. The ventricular subependyma appears to be the predominant source of oligodendrocyte progenitor cells (OPCs) in early postnatal development *(11,12)*. However, OPCs appear to then disperse widely throughout the postnatal rodent brain parenchyma, to invade both the nascent cortex and subcortex. A number of groups have now reported the persistence of these glial progenitor populations in the parenchyma of the adult rodent brain *(13,14)*. Although these cells actively cycle and divide as undifferentiated cells in normal brain, their progeny may divert to a differentiated phenotype upon local injury, with predominant oligodendrocytic expansion *(15)*. Notably, retroviral lineage analysis revealed that single white-matter glial precursors could give rise to either or both astrocytes and oligodendrocytic progeny; as such, parenchymal glial progenitors may include both lineage-restricted and bipotential cells. The latter may correspond to the adult homolog of the O2A progenitor cell, which may give

From: *Neural Stem Cells for Brain and Spinal Cord Repair*
Edited by: T. Zigova, E. Y. Snyder, and P. R. Sanberg © Humana Press Inc., Totowa, NJ

rise to oligodendrocytes and type-2 fibrous astrocytes. Although initially described in optic nerve culture, these cells appear to be ubiquitous in the subcortical white matter and exist in the adult as well as perinatal rodent brain *(16,17)*.

OLIGODENDROCYTE PRECURSORS OF THE ADULT HUMAN BRAIN

The demonstration and identification of subcortical progenitor cells has been far more problematic in humans than in rodents. First, mitotic oligodendrocyte progenitors of the adult human white matter are relatively uncommon compared to their counterparts in the rodent brain. Whereas oligodendrocytes and preoligodendrocytes derived from the adult rat brain are actively mitotic in vitro *(18)*, antigenically homologous cells derived from the mature human brain are by-and-large postmitotic *(19,20)*. Although dividing cells giving rise to oligodendrocytes were noted in dissociates of the adult human white matter *(21)*, they initially appeared to be rare and were only identifiable post hoc in fixed cultures. Second, no methods or markers appropriate for the specific identification, much less the selective enrichment, of these mitotically competent oligodendrocyte progenitor cells had hitherto been reported. Third, the responses of human and rat oligodendrocyte precursors to exogenous mitogens and oligotrophic agents are not always analogous. As a result, even though standard protocols were developed long ago for the expansion of rodent oligodendrocyte progenitor cells, no such mitogenic conditions had hitherto been defined for stimulating the expansion of human oligodendrocyte progenitors *(22)*. Thus, not only have mitotic adult human OPCs not been preparable in the numbers or purity required for engraftment studies, but the very existence of a mitotically competent white-matter progenitor cell has been controversial *(19,20,23,24)*.

As a result of these considerations, live mitotic oligodendrocyte progenitors had never been isolated as such from human brain tissue until recently. Indeed, the very existence of a distinct pool of mitotic oligodendrocyte progenitor cells in the adult human subcortical white matter had remained uncertain and the isolation of these cells had been elusive. To address definitively the incidence and distribution of these cells and to circumvent the practical limitations of their apparent scarcity, we designed a strategy for the isolation and enrichment of native oligodendrocyte precursors from both postnatal and adult brain.

OLIGODENDROCYTE PROGENITOR CELLS
MAY BE SPECIFICALLY EXTRACTED
FROM ADULT HUMAN WHITE MATTER

To identify and purify oligodendrocyte progenitors from adult human subcortical white matter, we used a promoter-based selection and sorting strategy *(25,26)*. To this end, we used the early promoter for an early oligodendrocyte protein, 2',3'-cyclic nucleotide 3'-phosphodiesterase (CNP) *(27,28)* to drive green fluorescent protein (GFP) expression in young oligodendrocytes and their progenitors, followed by fluorescence-activated cell sorting (FACS) to separate these GFP-expressing oligodendrocyte progenitor cells. The CNP promoter was used for this purpose because CNP protein is the earliest known myelin-associated protein to be expressed in developing oligodendrocytes. It is expressed by newly generated cells of oligodendrocytic lineage even within the subventricular zone and appears to be expressed by their mitotic precursors as well *(27,29)*. Although CNP protein is expressed ubiquitously by oligodendrocytes at all ontogenetic stages, the 5' regulatory region of the CNP gene includes two distinct promoters, P2 and P1, that are sequentially activated during development. The more upstream P2 promoter (P/CNP2) directs transcription selectively to young oligodendrocytes and their progenitors, becoming inactive with maturation *(30,31)*. Thus, the CNP2 promoter (P/CNP2) was chosen for these experiments for its ability to target transgene expression to oligodendrocyte progenitors and their immature progeny.

When adult human subcortical dissociates were transfected with P/CNP2:GFP, GFP expression was observed to be restricted to a population of bipolar, largely A2B5+ precursor cells *(32)*. These cells incorporated the mitotic marker BrdU from the culture media and developed oligodendrocytic O4 expression in vitro. These data suggested that the P/CNP2:GFP-defined cells were mitotic oligodendrocyte progenitors. On this basis, we were then able to extract oligodendrocyte progenitors directly from adult white matter, using FACS to isolate the P/CNP2:hGFP+ cell pool from surgically resected temporal subcortex *(32)*. We found that an average of 0.6 ± 0.1% of all white-matter cells directed P/CNP2:hGFP expression. Given a transfection efficiency of 13.5% in these cultures, just over 4.4% of adult subcortical cells might then be estimated to reside as P/CNP2-defined progenitors. Immediately after FACS, these P/CNP2:hGFP-separated cells initially expressed the early oligodendrocytic marker A2B5, but failed to express the more differentiated markers O4, O1, or galactocerebroside. Some expressed astrocytic GFAP, but none expressed neuronal markers upon their initial identification. These cells were initially bipolar and most incorporated BrdU, indicating their mitogenesis in vitro; cell division persisted in serum-deficient base media containing fibroblast growth factor-2 (FGF2), platelet-derived growth factor (PDGF), and NT-3, but not in factor-deficient controls. When followed up to a month in culture, most (>90%) of the PCNP2:hGFP+ cells become oligodendrocytes, progressing through a stereotypic sequence of A2B5, O4, O1, and galactocerebroside expression, as in development (*see* ref. *16*). Thus, this strategy allowed us not only to establish the existence of oligodendrocyte progenitors in adult human white matter but to estimate their prevalence, while separating them in high yield in a form appropriate for engraftment and further analysis.

ANTIGENIC RECOGNITION OF P/CNP2:HGFP-DEFINED OLIGODENDROCYTE PRECURSOR CELLS

Based on their recognition of OPCs in adult rodents, a number of markers have been claimed to identify these cells in adult human tissues. The O4 antigen *(19)*, NG2 chondroitin sulfate *(33)*, and the PDGFα receptor *(22)* have each been reported to recognize oligodendrocyte progenitor cells in histological sections of the human brain. In each case, however, this has been claimed without independent verification of oligodendrocytic fate in human cells sorted on the basis of these markers. As a result, their cell-type specificity and utility remain unclear. O4, for instance, recognizes only postmitotic oligodendroglia in humans and does not appear to be expressed by mitotically competent oligodendrocyte progenitors *(20)*. NG2 recognizes a mitotic parenchymal cell type that has been reported to be oligodendrocytic in lineage *(33)*, but in humans, NG2-immunoreactive cells may also include microglia *(34)*, which are mitotically competent and morphologically similar to tissue progenitor cells (Nunes, Roy, and Goldman, unpublished observations). Within adult brain tissue, the PDGFα receptor may specifically recognize oligodendrocyte progenitors, and Scolding and colleagues have reported an incidence of PDGFαR-immunoreactive cells of 1–3% in adult human white matter *(22)*, a range similar to those predicted on the basis of P/CNP2:hGFP-based extraction. Nonetheless, whether the PDGFαR+ white-matter pool is entirely homologous to the P/CNP2:hGFP-defined population remains unclear.

We have noted that the A2B5-targeted GQ ganglioside is expressed by virtually all P/CNP2:hGFP-defined OPCs *(35)*. Although A2B5 is expressed by both immature neurons and glia during development, its expression in the adult brain appears largely restricted to parendymal progenitor cells

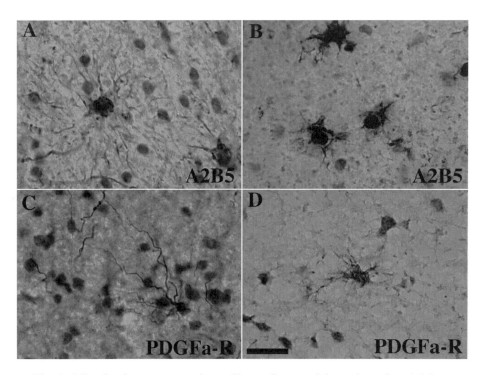

Fig. 1. Oligodendrocyte progenitor cells are dispersed throughout the adult human white matter. **(A,B)** Sections of the subcortical white matter immunoperoxidase stained with Mab A2B5 for its G_Q ganglioside epitope, which is characteristic of white-matter progenitor cells and serves as a surrogate for the P/CNP2:hGFP-defined progenitor cell population. **(C,D)** Sections stained for the PDGFα receptor, another marker for the adult oligodendrocyte progenitor cell population. Whether this cell type is phenotypically homologous to the A2B5-defined pool remains unclear, as is its relationship to the P/CNP2:hGFP-isolated pool. **(A)** and **(C)**: Tissue sampled from a 41-yr-old female; **(B)** and **(D)**: a 62-yr-old male. Scale bar = 20 μm.

(Nunes, Roy, Windrem, and Goldman, unpublished data). Thus, adult human white-matter progenitors may be sorted on the basis of A2B5-targeted immunolabeling. If A2B5 proves a specific surface antigenic surrogate for P/CNP2:hGFP defined oligodendrocyte progenitor cells, then its use as a target for FACS will be a significant practical advance, in that it will allow these cells to be extracted by surface-based sorting (see Fig. 3 below). This, in turn, will effectively circumvent the relative inefficiency, occasional liposomal toxicity, and extended periods of time in culture attending plasmid transfection-based sorting.

P/CNP2:hGFP⁺ PROGENITORS MAY GIVE RISE TO NON-OLIGODENDROCYTIC PHENOTYPES

The overwhelming majority of P/CNP2-purified progenitor cells developed into oligodendrocytes and did so regardless of ambient serum concentration, over a range of 0–10% platelet-depleted fetal bovine serum (FBS). However, almost a tenth developed GFAP expression and lost oligodendrocytic markers *(35)*. Interestingly, a small number, just over 1%, developed neuronal phenotype, as defined by TuJ1, NeuN, or Hu coexpression. The presence of sporadic P/CNP2:hGFP⁺ neurons and astrocytes suggested that the P/CNP2-separated white-matter progenitors might retain a degree of multilineage potential. This was indeed observed after FACS of the P/CNP2:hGFP⁺-defined cell pool *(35)*. By 3 wk after FACS, 74.1 ± 7.7% of the P/CNP2:hGFP⁺ sorted cells expressed oligodendrocytic CNP protein; 66.3 ± 6.8% were O4⁺. Of these, most matured as galactocerebroside⁺ oligodendrocytes under our culture conditions. Yet, some development of nonoligodendrocytic phenotypes was also noted: By 4 d after FACS, 6.5 ± 5.4% of the sorted cells expressed GFAP, and 11 ± 5% were GFAP⁺ by 3 wk in vitro. These were not simply false-positive contaminants, in that most expressed PCNP2:hGFP fluorescence. No neurons, as defined by Hu and/or TuJ1/βIII-tubulin-IR, were observed immediately prior to FACS. Remarkably though, 7.7 ± 4.4% of P/CNP2:hGFP-sorted cells were noted to mature into TuJ1⁺ neurons in the week thereafter. These data suggest that at least some progenitor cells of the adult human white matter retain the capacity to develop into neurons.

Perhaps the multilineage competence of human white-matter progenitor cells should not come as a surprise: In fetal animals, progenitor cells capable of giving rise to multiple lineages, including oligodendrocytes as well as neurons, have been derived from the cortical and subcortical parenchyma as well as from the ventricular zone *(36–38)*. Early postnatal rat cortex appears to retain similar multipotential progenitors *(39)*, and apparent O2A glial progenitor cells of the perinatal optic nerve have been reported to be "reprogrammable" to generate neurons *(40)*. Thus, the apparent lineage commitment of these tissue progenitors might depend on epigenetic factors, with the cells retaining far more lineage plasticity and competence than traditionally appreciated. Importantly, this may hold true for adult as well as perinatal brain progenitor cells: Under appropriate culture conditions, including constant mitogenic stimulation with FGF2 in serum-free media, single cells obtained from the nonventricular forebrain parenchyma of adult rats may generate both neurons and glia, including oligodendrocytes *(41,42)*. Taken together, these data suggest that progenitor cells of the adult subcortical parenchyma might be competent to generate multiple cell types, yet be restricted to oligodendrocytic lineage by virtue of their environment. As such, their predominant generation of oligodendrocytes in vitro might reflect an epigenetic bias imparted by their environment before harvest.

MODELS OF EXPERIMENTAL DEMYELINATION AND REMYELINATION

Together, these observations have suggested that progenitor cells capable of local cell genesis might persist throughout the subcortical white matter of the adult brain, where they might constitute a potential substrate for cellular replacement and local repair. In rats, these cycling precursors appear capable both of remyelination *in situ* and also upon implantation to areas of experimental injury *(43,44)*. Several models of induced demyelination are currently available for evaluating the in vivo growth and remyelination potential of adult oligodendrocyte progenitors. These include both murine and porcine models of experimental allergic encephalomyelitis (EAE), systemic cuprizone-induced widespread demyelination, radiation and ethidium bromide-targeted radiation injury, and local chemical demyelination by lysolecithin.

1. Whereas EAE is perhaps the most accepted experimental model of multiple sclerosis *(45)*, its relapsing and remitting course is attended by substantial fluctuations with time, often including transient recovery from acute inflammatory demyelinating episodes. In addition to this constantly wavering structural and functional baseline, neither the location nor the frequency of individual lesions can be predicted in EAE, so it is not a good model for attempting focal remyelination by introduced progenitors.
2. Systemic demyelinating agents, such as cuprizone, inflict widespread damage that may not be limited to oligodendrocytes and that is accompanied by a prominent inflammatory component *(46)*. As such, cuprizone poisoning may prove a useful model in studying the toxic and posthypoxic leukoencephalopathies, but it is not optimal for attempting focal remyelination of discrete stable lesions.
3. Radiation injury is highly inflammatory and includes all cellular elements, being particularly toxic for endothelial cells; although it has proven useful in assessing remyelination strategies *(43)*, it does not model clinical demyelination except for that suffered in radiotherapy-associated white-matter necrosis and perhaps that suffered in the inflammatory and immune vasculitides.
4. Ethidium bromide is a nonspecific DNA-intercalating transcriptional inhibitor and is used to kill cells whose cell bodies lie within the targeted field. It has the advantage of not damaging axons in passage, though, and its local injection yields lesions characterized both by local demyelination and frank oligodendrocytic cell loss *(47)*.

5. Lysolecithin lesioning has similarly developed as a useful agent for achieving predictable, focal lesions of the white matter *(15)*. It results in local demyelination with a mild inflammatory response. It is associated with some local oligodendrocytic loss, but less so than to ethidium, and with relative preservation of astrocytic and endothelial elements. Lysolecithin may thus be a more realistic model of inflammatory demyelination than ethidium or ethidium-radiation combinations. However, spontaneous remyelination is typical following circumscribed lysolecithin lesioning and follows a time-course that is dependent on both the type of lysolecithin, its volume, and concentration, as well as on the age of the animals.

EXPERIMENTAL REMYELINATION VIA IMPLANTATION OF CNS PROGENITOR CELLS

A variety of strategies have been used to assess the capacity of implanted neural precursor cells and oligodendrocyte progenitors to integrate and remyelinate within the experimentally demyelinated brain *(44)*. Using rat forebrain donor tissue, Blakemore and colleagues first reported that implanted homografts of oligodendrocyte precursors can remyelinate ethidium bromide-demyelinated rat brain and spinal cord *(43,48)*. Duncan et al. *(49)* similarly investigated the engraftability of FACS-enriched oligodendrocytes into demyelinated foci and observed efficient myelination by the engrafted cells. In a series of elegant studies, Duncan and colleagues subsequently established that both fetal and adult-derived neural progenitor cells could give rise to myelinogenic oligodendrocytes, which would indeed remyelinate deficient foci in vivo *(50–53)*. Using a more nominally com-

---

Fig. 2. *(see opposite page)* Oligodendrocyte progenitor cells may be specifically targeted and isolated from the white matter. **(A,B)** Representative sort of a human white matter sample, derived from the frontal lobe of a 42-yr-old woman during repair of an intracranial aneurysm. This plot shows 5×10^4 cells (sorting events) with their GFP fluorescence intensity plotted against forward scatter (a measure of cell size). **(A)** indicates the sort obtained from a nonfluorescent P/hCNP2:*lacZ*-transfected control; **(B)** indicates the corresponding result from a matched culture transfected with P/hCNP2:hGFP. **(C,D)** A bipolar A2B5$^+$/BrdU+ cell, 48 h after FACS. **(E,F)** By 3 wks post-FACS, P/CNP2:hGFP-sorted cells developed multipolar morphologies and expressed oligodendrocytic O4 (red). These cells often incorporated BrdU, indicating their in vitro origin from replicating A2B5$^+$ cells. **(G–I)** Matched phase **(G,I)** and immunofluorescent **(H,J)** images of maturing oligodendrocytes, 4 wk after P/CNP2:hGFP-based FACS. These cells expressed both CNP protein **(H)** and galactocerebroside **(J)**, indicating their maturation as oligodendrocytes. Scale bar = 20 µm. (From ref. *25*, with permission.)

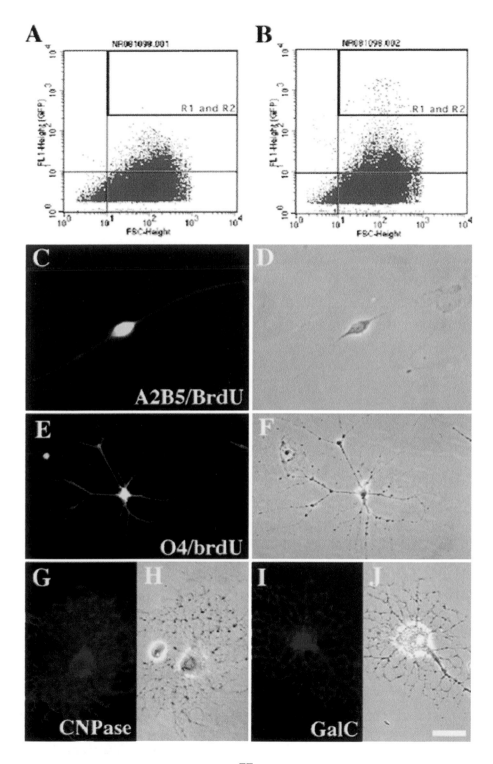

mitted progenitor phenotype, Noble et al. *(17)* similarly found that rat perinatal O2A progenitor cells are capable of maturing into myelinating oligodendrocytes after homograft to the adult rat white matter *(16)*.

Human fetal ventricular zone has also been assessed as a potential source of oligodendrocyte progenitor cells. A variety of sources have been used to this end, which have included tissue fragments and dissociates *(54–56)*, propagated oligospheres of both fetal *(57)* and adult *(58)* derivation, and FACS-sorted oligodendrocytes *(59)*. However, research with human donor cells in demyelinating models lags that of rodent-derived progenitors; despite the significant differences in oligodendrocyte progenitor biology between humans and rodents, to our knowledge no group has yet demonstrated that human oligodendrocyte progenitor cells may engraft and myelinate in any model of experimental demyelination. This omission may simply represent the operational difficulties implicit in the acquisition and extraction of human oligodendrocyte progenitor cells, which, by virtue of the delayed nature of oligodendrocyte development, are optimally derived at relatively late gestational ages *(59)*. As a result, the generation and use of human oligospheres, propagated and expanded in culture from ventricular zone progenitor cells under conditions biased toward oligodendrocytic differentiation *(57)*, may represent an especially promising approach to generating the scalable quantities of phenotypically homogeneous oligodendrocyte progenitor cells that will be needed for clinical implantation.

CONGENITAL DYSMYELINATION MAY ALSO BE A TARGET OF PROGENITOR-BASED THERAPY

Both rodent- and human-derived progenitors have also been assessed in several perinatal models of congenital dysmyelination. The myelinogenic potential of implanted fetal human brain cells was first noted in this regard in the shiverer mouse *(56,60)*, a mutant mouse deficient in myelin basic protein, homozygotes of which fail to develop central compact myelin. Using rat donor tissue, Warrington and Pfeiffer then established that the A2B5-defined oligodendrocyte progenitor phenotype was more efficient at migration and myelinogenesis in neonatal shiverers than the more mature O4-defined oligodendrocyte *(61)*. Yandava et al. *(62)* have since similarly noted that immortalized progenitors may give rise to a myelinogenic phenotype in shiverer, with consequent functional improvement in homozygous shiverer pups.

In parallel studies in other dysmyelinating models, Duncan and colleagues have found that oligodendrocytes derived from postnatal spinal cord can myelinate regions of congenitally hypomyelinated brain in the shaking pup, a canine mutant that exhibits dysmyelination resulting from a missense mutation in the proteolipid protein gene *(63)*. These workers found that fetal oligodendrocytes were able to engraft widespread regions of demyelinated central nervous system (CNS), with graft survival of over 6 mo. Although neonatal recipients fared best, adult recipients also exhibited graft oligodendrocyte survival and stable myelination. The same group subsequently demonstrated that oligosphere-derived cells raised from the neonatal rodent subventricular zone could engraft another dysmyelinated mutant, the myelin-deficient rat, upon perinatal intraventricular administration *(51)*. Together, these studies have suggested that congenital dysmyelination, like adult demyelination, may be an appropriate target for CNS progenitor cell-based therapy, in particular when directed at newborn recipients. As such, congenital diseases as diverse as the hereditary leukodystrophies, including Krabbe's, Canavan's and Tay-Sach's among others, as well as perinatal germinal matrix hemorrhages and the cerebral palsies, may all prove viable targets for cell-based therapeutic remyelination *(64)*.

EXPERIMENTAL REMYELINATION VIA IMPLANTATION OF NON-CNS PROGENITOR CELL TYPES

A wide range of other potentially myelinogenic cell types have also been implanted into experimental models of demyelination and dysmyelination, with varying degrees and claims of success.

Two myelinating phenotypes of the peripheral nervous system, Schwann cells *(65–68)* and olfactory ensheathing cells *(69,70)*, have each been used to remyelinate the demyelinated brain and spinal cord. Both phenotypes have been found capable of developing myelin upon implantation, but each made peripheral myelin, as defined by P0 protein expression *(69)*. Despite histological and ultrastructural evidence that these cells indeed myelinate central axons, whether the normative functional and conductive properties of central axons are regained after peripheral myelin-based ensheathment is unclear, as is the long-term fate of the remyelinated units. In fact, one must be circumspect in assessing the remyelination potential of these peripheral phenotypes, in that the complex interaction of axon and oligodendrocyte is now supplanted by that between the axon and fundamentally ectopic Schwann and ensheathing cells. Whether these cells are capable of the contact-dependent and humoral support of neuronal function normally exercised by central oligodendrocytes or, conversely, whether they are in turn supported by the axons with which they interact *(71,72)* remains unknown.

In addition to progenitors committed to neural lineage, embryonic stem cells have also been reported to myelinate demyelinated foci *(73,74)*.

Although tremendously promising, the use of such primitive phenotypes as vectors for remyelination is still limited by our inability to fully instruct all cells in the undifferentiated population to the desired phenotype. This, in turn, may lead to undesired phenotypes being generated within the transplant bed. Particularly concerning is the persistence of uncommitted progenitors within the implanted population, whose latent capacity for undifferentiated expansion, if not frank tumorigenesis, remains worrisome. These same concerns also pertain to the use of oncogenically immortalized stem cell lines, myelination by which has been demonstrated in shiverer mice, as already noted *(62)*. Even if rendered free of potentially neoplastic convertants, such embryonic, stem- (ES) cell-derived and immortalized oligodendrocytes would be subject to the same concerns regarding their ability to functionally remyelinate brain as the peripheral nervous system (PNS)-derived progenitor phenotypes already discussed.

Despite these concerns, though, both murine and human embryonic stem cells, as well as a variety of immortalized neuroectodermal lines, are each the subject of intense current study. In addition, primitive mesenchymal and marrow-derived stem cells, which may be capable of transdifferentiation or ectopic differentiation to neuroectodermal lineage *(75,76)*, are also the focus of significant interest as potential myelinogenic vectors. Together, these active and accelerating lines of investigation anticipate significant advances in our understanding of the biology and use of both ES cells and other preneural stem cell phenotypes.

OLIGODENDROCYTE PROGENITOR CELLS MAY ACT AS SUCH BY VIRTUE OF THE WHITE MATTER ENVIRONMENT

A number of studies have suggested that neural progenitor cells not yet committed to a terminally committed phenotype may respond to transplantation into a dysmyelinated environment by differentiating as oligodendrocytes. For example, a higher proportion of EGF-responsive murine neural stem cells differentiate as oligodendrocytes upon xenograft to the myelin-deficient rat than upon terminal differentiation in vitro *(50)*. Similarly, v-*myc*-transformed neural stem cells have been reported to differentiate preferentially as oligodendrocytes upon transplantation to perinatal shiverer mice *(62)*. In addition, premitotic progenitor cells may adjust to a more mature host environment by precociously exiting their cell cycles and terminally differentiating as oligodendrocytes. For instance, early human fetal CNS grafts generated myelin in murine recipients long before they would have done so in the donor human brains from which they were derived *(56)*. These observations suggest that implanted progenitors may respond to

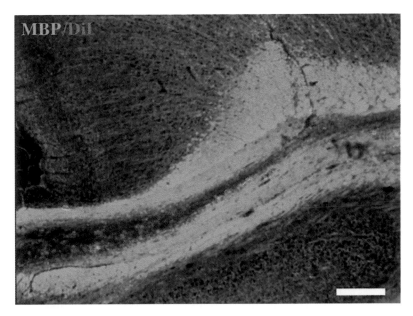

Fig. 3. Oligodendrocyte progenitor cells engraft upon implantation into demyelinated foci. A2B5-sorted human oligodendrocyte progenitor cells were transplanted into lysolecithin-induced demyelinated lesions in the corpus callosa of adult rats. In this section, myelin basic protein-immunoreactivity (green) delimits the dark, demyelinated lesion bed. The cannula track indicates the site of cell injection into the demyelinated lesion, which was induced 3 d before 10^5 sorted human cells were introduced as a 2-μL injection. The animal was sacrificed a week after implantation, by which point the DiI-tagged implanted human cells (red) had migrated widely throughout the demyelinated lesion bed. Of note, the transplanted cells actively migrated throughout the extent of the demyelinated plaque, but not beyond its borders, suggesting that normal myelin may typically impede the migration of these cells. Scale bar = 200 μm.

local cues within the host white-matter environment with oligodendrocytic differentiation, maturation and myelinogenesis.

OLIGODENDROCYTE PROGENITOR IMPLANTATION HAS THUS FAR BEEN LIMITED TO ANIMAL MODELS

Together, these reports have provided a firm conceptual foundation for the use of oligodendroglia and their precursors in transplantation for demyelinating disease. However, despite the overt differences in the biology of oligodendrocyte progenitors derived from rat and human CNS *(24)*, no prior studies have utilized adult primate or human donors or primate recipients. Furthermore, none of these studies has yet taken advantage of the specific advantages of adult-derived white-matter progenitor cells, which may very likely act as the actual remyelinating cell type of the damaged adult nervous system. As a result, it is fair to say that, at present, no unique combination of centrally derived oligodendrocyte progenitor cell and experimental lesion has generated sufficiently convincing and complete data to justify the transition from preclinical to clinical therapeutic modeling.

HUMAN WHITE-MATTER PROGENITOR CELLS SURVIVE AND INTEGRATE FOLLOWING XENOGRAFT TO DEMYELINATED FOCI OF THE ADULT RAT BRAIN

To establish whether adult white-matter-derived progenitor cells could survive xenograft to adult brain parenchyma, we implanted human white-matter progenitor cells into lysolecithin-lesioned rat donor sites. To this end, we first confirmed prior studies of lysolecithin-induced central demyelination *(15)*, which revealed that local lysolecithin infusion was associated with a discrete lesion of transcallosal myelin. When assessed 1 and 3 wk after 1-µL injections of 2% lysolecithin-V, these lesions exhibited a mild degree of reactive astrocytosis within, but not beyond, the demyelinated focus, the vascular architecture of which appeared intact. Oligodendrocytes were markedly diminished in number by 1 wk after injection, and no myelin could be visualized by staining for myelin basic protein along a 2-mm track within the corpus callosum.

Against this backdrop, we implanted A2B5-sorted adult oligodendrocyte progenitor cells stereotaxically into both normal and lysolecithin-lesioned adult rat brain. A2B5-directed immunomagnetic sorting was used to isolate the progenitors, based upon the expression of the A2B5 epitope by P/CNP2:hGFP-sorted cells *(25)*, and the relative restriction of A2B5 to this cell type in the adult human white matter *(78)*. The rats received bilateral lesions and were then implanted 3 d later with 10^5 sorted cells/2 µL, delivered unilaterally. Some donor cells were prelabeled with the lipophilic tracking dye PKH26 to allow their detection after implantation *(77)*. Other donor cells were instead localized using human-specific donor cell antigens. At both 1 and 2 wk later, the recipient brains were fixed and prepared for histology. We found that the engraftment sites harbored substantial populations of viable cells; by 2 wk after injection, most of these cells expressed oligodendrocytic CNP. Thus, oligodendrocyte progenitor cells extracted from the adult human white matter were able to successfully engraft lysolecithin-lesioned brain parenchyma.

These latter observations indicated that adult human white-matter progenitor cells can invade demyelinated white matter quickly and efficiently and that they are relatively restricted to that environment, survive in great numbers, and differentiate largely as oligodendrocytes. Whether these adult progenitors are capable of functionally significant remyelination remains to be established, as does their survivability after allograft and the fidelity of their interactions with the host axonal network. Even if these cells prove capable of mediating remyelination and structural repair, whether they are actually more effective in doing so than other myelinogenic cell types, remains a critical issue for future comparison.

CONCLUSION

A distinct population of white-matter progenitor cells, able but not necessarily committed to generate oligodendrocytes, remains ubiquitous in the adult human subcortical white matter. These cells are present in both sexes and into senescence and may constitute as many as 4% of the cells of adult human capsular white matter. Transduction of adult human white matter dissociates with plasmids bearing early oligodendrocytic promoters driving fluorescent reporters permits the separation of these cells in high yield and purity. They survive xenograft to demyelinated brain, and as such they may be appropriate vectors for cell-based remyelination strategies.

REFERENCES

1. Altman, J. and Das, G. D. (1966) Autoradiographic and histological studies of postnatal neurogenesis. I. A longitudinal investigation of the kinetics, migration and transformation of cells incorporating tritiated thymidine in neonate rats, with special reference to postnatal neurogenesis in some brain regions. *J. Comp. Neurol.* **126,** 337–390.
2. Kaplan, M. S. and Hinds, J. W. (1977) Neurogenesis in the adult rat: electron microscopic analysis of light radioautographs. *Science* **197,** 1092–1094.
3. Kaplan, M. S., McNelly, N. A., and Hinds, J. W. (1985) Population dynamics of adult-formed granule neurons of the rat olfactory bulb. *J. Comp. Neurol.* **239,** 117–125.
4. Goldman, S. A. and Nottebohm, F. (1983) Neuronal production, migration, and differentiation in a vocal control nucleus of the adult female canary brain. *Proc. Natl. Acad. Sci. USA* **80,** 2390–2394.
5. Nottebohm, F. (1985) Neuronal replacement in adulthood. *Ann. NY Acad. Sci.* **457,** 143–161.
6. Goldman, S. A. and Luskin, M. B. (1998) Strategies utilized by migrating neurons of the postnatal vertebrate forebrain. *Trends Neurosci.* **21,** 107–114.
7. Gould, E., Tanapat, P., McEwen, B., Flugge, G., and Fuchs, E. (1998) Proliferation of granule cell precursors in the dentate gyrus of adult monkeys is diminished by stress. *Proc. Natl. Acad. Sci. USA* **95,** 3168–3171.

8. Kirschenbaum, B., Nedergaard, M., Preuss, A., Barami, K., Fraser, R. A., and Goldman, S. A. (1994) In vitro neuronal production and differentiation by precursor cells derived from the adult human forebrain. *Cerebr. Cortex* **4,** 576–589.

9. Pincus, D. W., Keyoung, H. M., Harrison-Restelli, C., Goodman, R. R., Fraser, R. A., Edgar, M., et al. (1998) Fibroblast growth factor-2/brain-derived neurotrophic factor-associated maturation of new neurons generated from adult human subependymal cells. *Ann. Neurol.* **43,** 576–585.

10. Lois, C. and Alvarez-Buylla, A. (1993) Proliferating subventricular zone cells in the adult mammalian forebrain can differentiate into neurons and glia. *Proc. Natl. Acad. Sci. USA* **90,** 2074–2077.

11. Levison, S. W. and Goldman, J. E. (1993) Both oligodendrocytes and astrocytes develop from progenitors in the subventricular zone of postnatal rat forebrain. *Neuron* **10,** 201–212.

12. Luskin, M. B., Parnavelas, J. G., and Barfield, J. A. (1993) Neurons, astrocytes, and oligodendrocytes of the rat cerebral cortex originate from separate progenitor cells: an ultrastructural analysis of clonally related cells. *J. Neurosci.* **13,** 1730–1750.

13. Gensert, J. M. and Goldman, J. E. (1996) In vivo characterization of endogenous proliferating cells in adult rat subcortical white matter. *Glia* **17,** 39–51.

14. Reynolds, R. and Hardy, R. (1997) Oligodendroglial progenitors labeled with the O4 antibody persist in the adult rat cerebral cortex in vivo. *J. Neurosci. Res.* **47,** 455–470.

15. Gensert, J. M. and Goldman, J. E. (1997) Endogenous progenitors remyelinate demyelinated axons in the adult CNS. *Neuron* **19,** 197–203.

16. Noble, M. (1997) The oligodendrocyte-type 2 astrocyte lineage: in vitro and in vivo studies on development, tissue repair and neoplasia, in *Isolation, Characterization and Utilization of CNS Stem Cells* (Gage, F. and Christen, Y., eds.), Springer-Verlag, Berlin, pp. 101–128.

17. Noble, M., Wren, D., and Wolswijk, G. (1992) The O-2A(adult) progenitor cell: a glial stem cell of the adult central nervous system. *Semin. Cell Biol.* **3,** 413–422.

18. Engel, U. and Wolswijk, G. (1996) Oligodendrocyte-type-2 astrocyte (O-2A) progenitor cells derived from adult rat spinal cord: in vitro characteristics and response to PDGF, bFGF and NT-3. *Glia* **16,** 16–26.

19. Armstrong, R. C., Dorn, H. H., Kufta, C. V., Friedman, E., and Dubois-Dalcq, M. E. (1992) Preoligodendrocytes from adult human CNS. *J. Neurosci.* **12,** 1538–1547.

20. Gogate, N., Verma, L., Zhou, J. M., Milward, E., Rusten, R., O'Connor, M., et al. (1994) Plasticity in the adult human oligodendrocyte lineage. *J. Neurosci.* **14,** 4571–4587.

21. Scolding, N. J., Rayner, P. J., Sussman, J., Shaw, C., and Compston, D. A. (1995) A proliferative adult human oligodendrocyte progenitor. *Neuroreport* **6,** 441–445.

22. Scolding, N., Franklin, R., Stevens, S., Heldin, C., Compston, A., and Newcombe, J. (1998) Oligodendrocyte progenitors are present in the normal adult human CNS and in the lesions of multiple sclerosis. *Brain* **121,** 2221–2228.

23. Imaizumi, T., Lankford, K. L., and Kocsis, J. D. (2000) Transplantation of olfactory ensheathing cells or Schwann cells restores rapid and secure conduction across the transected spinal cord. *Brain Res.* **854,** 70–78.

24. Scolding, N. (1998) Glial precursor cells in the adult human brain. *Neuroscientist* **4,** 264–272.

25. Roy, N., Wang, S., Harrison-Restelli, C., Benraiss, A., Fraser, R., Gravel, P., Braun, P., and Goldman, S. A. (1999) Identification, isolation and enrichment of oligodendrocyte progenitor cells from the adult human subcortical white matter. *J. Neurosci.* **19,** 9986–9995.

26. Wang, S., Wu, H., Jiang, J., Delohery, T. M., Isdell, F., and Goldman, S. A. (1998) Isolation of neuronal precursors by sorting embryonic forebrain transfected with GFP regulated by the T alpha 1 tubulin promoter. *Nat. Biotechnol.* **16,** 196–201.

27. Scherer, S., Braun, P., Grinspan, J., Collarini, E., Wang, D., and Kamholz, J. (1994) Differential regulation of the 2',3'-cyclic nucleotide 3'-phosphodiesterase gene during oligodendrocyte development. *Neuron* **12,** 1363–1375.

28. Vogel, U., Reynolds, R., Thompson, R., and Wilkin, G. (1988) Expression of the CNPase gene and immunoreactive protein in oligodendrocytes as revealed by in situ hybridization and immunofluorescence. *Glia* **1,** 184–190.

29. Yu, W.-P., Collarini, E., Pringle, N., and Richardson, W. (1994) Embryonic expression of myelin genes: evidence for a focal source of oligodendrocyte precursors in the ventricular zone of the neural tube. *Neuron* **12,** 1353–1362.

30. Gravel, M., DiPolo, A., Valera, P., and Braun, P. (1998) Four-kilobase sequence of the mouse CNP gene directs spatial and temporal expression of lacZ in transgenic mice. *J. Neurosci. Res.* **53,** 393–404.

31. O'Neill, R., Minuk, J., Cox, M., Braun, P., and Gravel, M. (1997) CNP2 mRNA directs synthesis of both CNP1 and CNP2 polypeptides. *J. Neurosci. Res.* **50,** 248–257.

32. Roy, N., Wang, S., Benraiss, A., Harrison, C., Gravel, M., Braun, P., et al. (1998) Identification and high-yield enrichment of oligodendrocyte progenitor cells from the adult human subcortical white matter. *Soc. Neurosci. Abstr.* **24,** 506.4.

33. Chang, A., Nishiyama, A., Peterson, J., Prineas, J., and Trapp, B. D. (2000) NG2-Positive oligodendrocyte progenitor cells in adult human brain and multiple sclerosis lesions. *J. Neurosci.* **20,** 6404–6412.

34. Pouly, S., Becher, B., Blain, M., and Antel, J. (1999) Expression of a homologue of rat NG2 on human microglia. *Glia* **27,** 259–268.

35. Roy, N. S., Wang, S., Harrison-Restelli, C., Benraiss, A., Fraser, R. A., Gravel, M., et al. (1999) Identification, isolation, and promoter-defined separation of mitotic oligodendrocyte progenitor cells from the adult human subcortical white matter. *J. Neurosci.* **19,** 9986–9995.

36. Davis, A. A. and Temple, S. (1994) A self-renewing multipotential stem cell in embryonic rat cerebral cortex. *Nature* **372,** 263–266.
37. Qian, X., Davis, A. A., Goderie, S. K., and Temple, S. (1997) FGF2 concentration regulates the generation of neurons and glia from multipotent cortical stem cells. *Neuron* **18,** 81–93.
38. Williams, B. P., Read, J., and Price, J. (1991) The generation of neurons and oligodendrocytes from a common precursor cell. *Neuron* **7,** 685–693.
39. Marmur, R., Mabie, P., Gokhan, S., Song, Q., Kessler, J., and Mehler, M. (1998) Isolation and developmental characterization of cerebral cortical multipotent progenitors. *Dev. Biol.* **204,** 577–591.
40. Kondo, T. and Raff, M. (2000) Oligodendrocyte precursor cells reprogrammed to become multipotential CNS stem cells. *Science* **289,** 1754–1757.
41. Palmer, T. D., Ray, J., and Gage, F. H. (1995) FGF-2-responsive neuronal progenitors reside in proliferative and quiescent regions of the adult rodent brain. *Mol. Cell. Neurosci.* **6,** 474–486.
42. Richards, L. J., Kilpatrick, T. J., and Bartlett, P. F. (1992) De novo generation of neuronal cells from the adult mouse brain. *Proc. Natl. Acad. Sci. USA* **89,** 8591–8595.
43. Blakemore, W., Franklin, R., and Noble, M. (1996) Glial cell transplantation and the repair of demyelinating lesions, in *Glial Cell Development: Basic Principles and Clinical Relevance* (Jessen, K. and Richardson, W., eds.), BIOS Scientific, Oxford, pp. 209–220.
44. Franklin, R. J. and Blakemore, W. F. (1997) Transplanting oligodendrocyte progenitors into the adult CNS. *J. Anat.* **190,** 23–33.
45. Raine, C. S., Cannella, B., Hauser, S. L., and Genain, C. P. (1999) Demyelination in primate autoimmune encephalomyelitis and acute multiple sclerosis lesions: a case for antigen-specific antibody mediation. *Ann. Neurol.* **46,** 144–160.
46. Morell, P., Barrett, C. V., Mason, J. L., Toews, A. D., Hostettler, J. D., Knapp, G. W., et al. (1998) Gene expression in brain during cuprizone-induced demyelination and remyelination. *Mol. Cell. Neurosci.* **12,** 220–227.
47. Woodruff, R. H. and Franklin, R. J. (1999) Demyelination and remyelination of the caudal cerebellar peduncle of adult rats following stereotaxic injections of lysolecithin, ethidium bromide, and complement/anti-galactocerebroside: a comparative study. *Glia* **25,** 216–228.
48. Franklin, R. J. and Blakemore, W. F. (1995). Glial-cell transplantation and plasticity in the O-2A lineage—implications for CNS repair. *Trends Neurosci.* **18,** 151–156.
49. Duncan, I., Paino, C., Archer, D., and Wood, P. (1992) Functional capacities of transplanted cell-sorted adult oligodendrocytes. *Dev. Neurosci.* **14,** 114–122.
50. Hammang, J. P., Archer, D. R., and Duncan, I. D. (1997) Myelination following transplantation of EGF-responsive neural stem cells into a myelin-deficient environment. *Exp. Neurol.* **147,** 84–95.
51. Learish, R. D., Brustle, O., Zhang, S. C., and Duncan, I. D. (1999) Intraventricular transplantation of oligodendrocyte progenitors into a fetal myelin mutant results in widespread formation of myelin. *Ann. Neurol.* **46,** 716–722.

52. Zhang, S. C. and Duncan, I. D. (2000) Remyelination and restoration of axonal function by glial cell transplantation. *Prog. Brain Res.* **127,** 515–533.
53. Zhang, S. C., Ge, B., and Duncan, I. D. (1999) Adult brain retains the potential to generate oligodendroglial progenitors with extensive myelination capacity. *Proc. Natl. Acad. Sci. USA* **96,** 4089–4094.
54. Gansmuller, A., Lachapelle, F., Baron-Van Evercooren, A., Hauw, J. J., Baumann, N., and Gumpel, M. (1986) Transplantations of newborn CNS fragments into the brain of shiverer mutant mice: extensive myelination by transplanted oligodendrocytes. II. Electron microscopic study. *Dev. Neurosci.* **8,** 197–207.
55. Gumpel, M., Gout, O., Lubetzki, C., Gansmuller, A., and Baumann, N. (1989) Myelination and remyelination in the central nervous system by transplanted oligodendrocytes using the shiverer model. Discussion on the remyelinating cell population in adult mammals. *Dev. Neurosci.* **11,** 132–139.
56. Gumpel, M., Lachapelle, F., Gansmuller, A., Baulac, M., Baron van Evercooren, A., and Baumann, N. (1987) Transplantation of human embryonic oligodendrocytes into shiverer brain. *Ann. NY Acad. Sci.* **495,** 71–85.
57. Murray, K. and Dubois-Dalcq, M. (1997) Emergence of oligodendrocytes from human neural spheres. *J. Neurosci. Res.* **50,** 146–156.
58. Akiyama, Y., Honmou, O., Kato, T., Uede, T., Hashi, K., and Kocsis, J. D. (2001) Transplantation of clonal neural precursor cells derived from adult human brain establishes functional peripheral myelin in the rat spinal cord. *Exp. Neurol.* **167,** 27–39.
59. Grever, W. E., Zhang, S., Ge, B., and Duncan, I. D. (1999) Fractionation and enrichment of oligodendrocytes from developing human brain. *J. Neurosci. Res.* **57,** 304–314.
60. Lachapelle, F., Gumpel, M., Baulac, C., and Jacque, C. (1983) Transplantation of fragments of CNS into the brains of shiverer mutant mice: extensive myelination by transplanted oligodendrocytes. *Dev. Neurosci.* **6,** 326–334.
61. Warrington, A. E., Barbarese, E., and Pfeiffer, S. E. (1993) Differential myelinogenic capacity of specific developmental stages of the oligodendrocyte lineage upon transplantation into hypomyelinating hosts. *J. Neurosci. Res.* **34,** 1–13.
62. Yandava, B. D., Billinghurst, L. L., and Snyder, E. Y. (1999) "Global" cell replacement is feasible via neural stem cell transplantation: evidence from the dysmyelinated shiverer mouse brain. *Proc. Natl. Acad. Sci. USA* **96,** 7029–7034.
63. Archer, D., Cuddon, P., Lipsitz, D., and Duncan, I. (1997) Myelination of the canine central nervous system by glial cell transplantation: a model for repair of human myelin disease. *Nat. Med.* **3,** 54–59.
64. Tate, B., Bower, K., and Snyder, E. (2001) Transplant therapy, in *Stem Cells and CNS Development* (Rao M., ed.), Humana, Totowa, NJ, pp. 291–306.
65. Baron-Van Evercooren, A., Avellana-Adalid, V., Lachapelle, F., and Liblau, R. (1997) Schwann cell transplantation and myelin repair of the CNS. *Multiple Scler.* **3,** 157–161.

66. Baron-Van Evercooren, A., Clerin-Duhamel, E., Lapie, P., Gansmuller, A., Lachapelle, F., and Gumpel, M. (1992) The fate of Schwann cells transplanted in the brain during development. *Dev. Neurosci.* **14,** 73–84.

67. Baron-Van Evercooren, A., Gansmuller, A., Duhamel, E., Pascal, F., and Gumpel, M. (1992) Repair of a myelin lesion by Schwann cells transplanted in the adult mouse spinal cord. *J. Neuroimmunol.* **40,** 235–242.

68. Blakemore, W. F. and Crang, A. J. (1985) The use of cultured autologous Schwann cells to remyelinate areas of persistent demyelination in the central nervous system. *J. Neurol. Sci.* **70,** 207–223.

69. Franklin, R. J., Gilson, J. M., Franceschini, I. A., and Barnett, S. C. (1996) Schwann cell-like myelination following transplantation of an olfactory bulb-ensheathing cell line into areas of demyelination in the adult CNS. *Glia* **17,** 217–224.

70. Imaizumi, T., Lankford, K. L., Waxman, S. G., Greer, C. A., and Kocsis, J. D. (1998) Transplanted olfactory ensheathing cells remyelinate and enhance axonal conduction in the demyelinated dorsal columns of the rat spinal cord. *J. Neurosci.* **18,** 6176–6185.

71. Fernandez, P. A., Tang, D. G., Cheng, L., Prochiantz, A., Mudge, A. W., and Raff, M. C. (2000) Evidence that axon-derived neuregulin promotes oligodendrocyte survival in the developing rat optic nerve. *Neuron* **28,** 81–90.

72. Vartanian, T., Goodearl, A., Viehover, A., and Fischbach, G. (1997) Axonal neuregulin signals cells of the oligodendrocytic lineage through activation of HER4 and Schwann cells through HER2 and HER3. *J. Cell Biol.* **137,** 211–220.

73. Brustle, O., Jones, K., Learish, R., Karram, K., Choudhary, K., Wiestler, O., et al. (1999) Embryonic stem cell derived glial precursors: a source of myelinating transplants. *Science* **285,** 754–756.

74. Liu, S., Qu, Y., Stewart, T. J., Howard, M. J., Chakrabortty, S., Holekamp, T. F., et al. (2000) Embryonic stem cells differentiate into oligodendrocytes and myelinate in culture and after spinal cord transplantation. *Proc. Natl. Acad. Sci. USA* **97,** 6126–6131.

75. Mezey, E., Chandross, K., Harta, G., Maki, R., and McKercher, S. (2000) Turning blood into brain: cells bearing neuronal antigens generated in vivo from bone marrow. *Science* **290,** 1779–1782.

76. Sanchez-Ramos, J., Song, S., Cardozo-Pelaez, F., Hazzi, C., Willing, A., Freeman, T., et al. (2000) Adult bone marrow stromal cells differentiate into neural cells in vitro. *Exp. Neurol.* **164,** 247–256.

77. Horan, P. and Slezak, S. (1989) Stable cell membrane labeling. *Nature* **340,** 167–168.

78. Windrem, M., Roy, N., Wang, Z., Nunes, M., Benraiss, A., Fraser, R., Goodman, R., McKhann, G., Goldman, S. A. (2002) Progenitor cells derived from the human subcortical white matter disperse and differentiate as oligodendrocytes within demyelinated lesions of the adult rat brain. *J. Neuroscience Res.,* in press.

4

Generation, Characterization, and Transplantation of Immortalized Human Neural Crest Stem Cells

Eiji Nakagawa, Kozo Hatori, Atsushi Nagai,
Hyun B. Choi, Myung A. Lee, Jung H. Bang,
Jean Kim, Jae K. Ryu, Akihiko Ozaki,
Min C. Lee, Evan Y. Snyder, and Seung U. Kim

INTRODUCTION

Recent advances in stem cell biology have aroused wide and intense attention by investigators as well as the general public because of the stem cells' broad applications in basic biomedical research and transplantation therapies. Stem cells could be used in tissue transplantation therapies for clinical targets, including neurodegenerative diseases, spinal cord injuries, and strokes, among others. To realize the full potential of human multipotent stem cells, however, further research lies ahead of us.

The existence of immature multipotent stem cells has been identified in the embryonic and adult human brain (1–6) and several groups have generated stable, perpetual neural stem cell lines that were utilized for cell replacement or gene transfer therapies in animal models of human neurological disorders (7–13). To this end, we have previously generated permanent cell lines of human neural stem cells with the ability to proliferate, exhibit self-renewal, generate a large number of clonally related progeny, retain their multilineage potential over time, and produce new cells in response to injury or disease (3). To assess the potential of these human neural stem cells as a vehicle for cell replacement therapies, we implanted the human neural stem cells into the lateral ventricle of neonatal mice, where they integrated into the subventricular layer, and from this region, they migrated extensively to various anatomical regions and became neurons, astrocytes, and oligodendrocytes.

From: *Neural Stem Cells for Brain and Spinal Cord Repair*
Edited by: T. Zigova, E. Y. Snyder, and P. R. Sanberg © Humana Press Inc., Totowa, NJ

Having stable human neural stem cell lines established, we then proceeded to generate human neural crest stem cells, multipotent precursor cells for peripheral and autonomous nervous system and other tissues, derived from human embryonic dorsal root ganglia tissues using the safe and efficacious genetic technique previously used for generating human neural stem cell lines. Neural crest cells migrate from the dorsal aspect of the neural tube and differentiate into a variety of cell types in different locations. These cell types include sensory and sympathetic neurons and Schwann cells, adrenal chromaffin cells, melanocytes, endocrine cells, smooth muscle, skeletal muscle, and bone *(14)*. In vivo lineage tracing and in vitro clonal analyses in avian embryos have indicated that many neural crest cells are multipotent *(15,16)* and transplantation and culture studies suggest that the fate of multipotent neural crest cells can be determined by the environment *(17–19)*.

In the present study, we describe the generation of new cell lines of human neural crest stem cells, HNC10, isolated from primary cell culture of human embryonic dorsal root ganglia tissue. HNC10 cell lines are genetically modified human neural crest stem cells, maintained as a stable cell line in serum-free medium supplemented with basic fibroblast growth factor (bFGF), remain uncommited, undifferentiated, and multipotent and express phenotypes specific for neural crest stem cells such as nestin and low-affinity nerve growth factor receptor (LNGFR, p75). When HNC10 human neural crest stem cells were grown in serum containing medium, they differentiated into cell types known as progeny of neural crest stem cells such as neurons, Schwann cells, adrenal chromaffin cells, and skeletal muscle cells. When these cells were implanted into the brain of neonatal CD1 mice, externally introduced human neural crest stem cells migrated to other anatomical sites from the original grafted site, and they differentiated into neurons and glial cells.

GENERATION OF IMMORTALIZED HUMAN NEURAL CREST STEM CELL LINES

Twenty pairs of spinal dorsal root ganglia (DRG) were isolated from a 15-wk gestational embryo. (The permission to use embryonic tissue was granted by the Clinical Research Screening Committee involving Human Subjects of the University of British Columbia, and the embryonic tissues were obtained from the Anatomical Pathology Department of Vancouver General Hospital.) DRGs were dissociated into single cells by incubating for 30 min at 37°C in phosphate-buffered saline (PBS) containing 0.25% collagenase and 40 µg/mL DNase type 1 following the procedure previously described *(20,21)*. A suspension of dissociated cells (5×10^5 cells/mL) in serum-free culture medium was plated on polylysine-coated six-well plates (Falcon). Serum-free culture medium consisted of Dulbecco's modified Eagle's medium (DMEM) containing UBC1 supplements (containing human insulin, human transferrin, sodium selenite, progesterone, triiodothyronin, and other nutrients and antioxidants) *(22)*, 5 mg/mL glucose, 20 µg/mL gentamicin, 2.5 µg/mL amphotericin B, and 10 ng/mL bFGF (Peprotech). Culture medium was changed twice a week. DRG cultures grown for 1–4 wk consisted of DRG neurons, Schwann cells, and fibroblasts and were used for gene transfer experiments. Dissociated cells prepared from human embryonic DRG were initially grown in serum-free medium supplemented with bFGF. The cultures grown for 2–4 wk consisted of small (10 µm in diameter) or larger (15 µm in diameter) nerve cells in singles or clusters, more numerous spindle-shaped (15–20 µm in length) Schwann cells, and flat polygonal fibroblasts (Fig. 1A).

An amphotropic replication-incompetent retroviral vector encoding v-*myc* (transcribed from the retrovirus LTR plus neomycin resistant gene transcribed from an internal SV40 early promoter), PASK1.2, not only permitted the propagation of human neural crest stem cell clones by genetic means but also enabled confirmation of the monoclonal origin of all progeny. This vector was generated using the ecotropic retroviral vector encoding v-*myc* (Fig. 2), similar to that described for generating murine neural stem cell clone C-17-2 *(23)*, and used to infect the PA317 amphotropic packaging cell line. Successful infectants were selected and expanded. Supernatants from the new producer cell line contained replication-incompetent retroviral particles bearing an amphotropic envelope, which efficiently infected the human neural cells as indicated by G418 resistance.

Infection of human DRG cells in six-well plates was performed three times. Supernatant (2 mL, 4×10^5 CFUs) from the packaging cell line and 8 µg/mL polybrene were added to target cells in six-well plates and incubated for 4 h at 37°C; the medium was then replaced with fresh growth medium; infection was repeated 24 h and 48 h later. Seventy-two hours following the third infection, infected cells were selected with G418 for 14 d and large clusters of cells were individually isolated and grown in polylysine-coated six-well plates. Individual clones were generated by limited dilution and propagated further. Six clones were isolated and designated as HNC10 human neural crest stem cell lines. To provide an unambiguous identification of externally introduced HNC10 cells following transplantation into the mouse brain, HNC10 cells were infected with a replication incompetent retroviral vector encoding β-galactosidase (LacZ) and puromycin-resistant

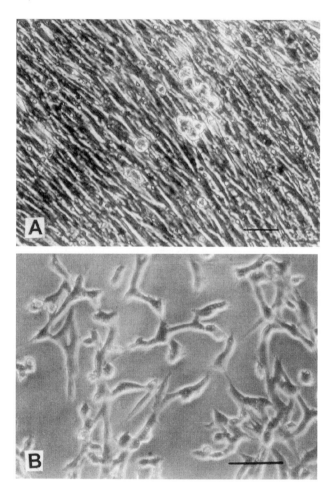

Fig. 1. (A) A live cell culture of human dorsal root ganglia isolated from a 15 wk gestation embryo and grown in vitro for 7 d. The culture contains a large number of neurons, Schwann cells, and a small number of neural crest stem cells and represents a starting material for the generation of human neural crest stem cell lines. Phase-contrast microscopy. Scale bar indicates 20 μm. **(B)** Immortalized human neural crest stem cells, HNC10, are grown in serum-free medium UBC1 containing basic FGF. When HNC10 cells are grown in serum-containing medium (5% fetal bovine serum and 5% horse serum), they differentiate into nerve cells, Schwann cells, adrenal chromaffin cells, and skeletal muscle cells.

genes. Six G418-resistant clones were isolated and expanded, and the cloned cells were tripolar or multipolar in morphology and were 10–12 μm in size (Fig. 1B). One of these clones was named HNC10.K10 and subjected to further study.

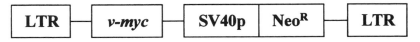

Fig. 2. PASK1.2 amphotropic replication-incompetent retroviral vector encoding v-*myc*.

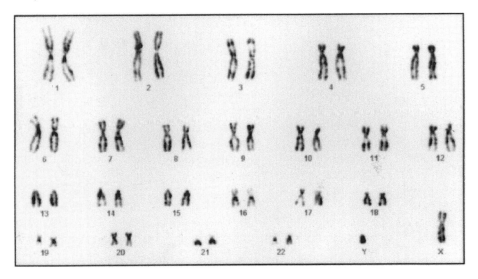

Fig. 3. Karyotype of HNC10 human neural crest stem cells at passage 5.

Cytogenetic analyses were performed on established cell lines at various passages. Cytogenetic analysis of HNC10 cells showed normal karyotype of human cells with a 46, XY karyotype without any chromosomal abnormality (Fig. 3). The doubling time for HNC human neural crest stem cells was determined by a population analysis by counting cell number at different time points for 3 d and was determined to be 23.67 h.

PHENOTYPIC CHARACTERIZATION OF IMMORTALIZED HUMAN NEURAL CREST STEM CELLS

HNC10.K10 cells grew in culture as single cells or large clusters that could be subcultured and passaged weekly for 6 mo. Immunochemical determination of cell-type-specific markers in human neural crest stem cell lines was performed using antibodies specific for neural crest stem cells, peripheral neurons, Schwann cells, and adrenal chromaffin cells (Table 1). HNC10 cells were grown on polylysine coated Aclar plastic coverslips (9 mm diameter) with serum-free medium for 3–7 d and processed for immunocytochemistry. Cultures grown in serum-free medium were fixed

Table 1
Markers of the Neural Crest Stem Cell Lineage Cells

Protein	Antibody	Cell type	Source
Nestin	Rabbit	Neural crest stem cells	Dr. K. Ikeda
Vimentin	mAb	NCSCs	Chemicon
FORSE-1	mAb	NCSCs	ATCC
LNGFR/p75	mAb	NCSCs	ATCC
NF-L	mAb	Neurons	Sigma
NF-M	mAb	Neurons	Dr. V. Lee
NF-H	mAb NE14	Neurons	Sigma
MAP-2	mAb AP14	Neurons	Sigma
Tubulin βIII	mAb	Neurons	Chemicon
S-100 protein	Rabbit	Schwann cells	Chemicon
P0 protein	Rabbit	Schwann cells	Dr. S. Uyemura
GFAP	Rabbit	Astrocytes	DAKO
Desmin	mAb	Muscle	Chemicon
Myosin	mAb	Muscle	Chemicon
Chromogranin	Rabbit	Adrenal chromaffin cells	Dr. R. Angeletti

in 4% paraformaldehyde for 3 min, washed twice with PBS, and incubated with an antibody specific for LNGFR/p75 (1:1, ATCC) for 30 min at room temperature (RT). For nestin immunostaining, cover slips were fixed in cold methanol for 15 min at −20°C, air-dried, incubated in rabbit polyclonal antibody specific for nestin (provided by Dr. K. Ikeda), followed by incubation in biotinylated secondary antibodies and avidin–biotin complex (ABC, Vector), and visualized with AEC (Sigma) chromagen development.

In order to induce differentiated cell types in neural crest stem cells, HNC10 cells were grown in serum-containing medium, which consisted of DMEM supplemented with 5% fetal bovine serum and 5% horse serum but without bFGF. For immunochemical characterization of neurons, the following antibodies were utilized: neurofilament low-molecular-weight protein (NF-L, 1:1000, mouse monoclonal, Sigma), NF-M (1:4, Dr. V. Lee), NF-H (1:1000, Sigma), microtubule-associated protein-2 (MAP-2, 1:1000, Sigma), tubulin βIII isoform (1:1000, Chemicon), and peripherin (1:1000, Chemicon). For Schwann cells, the following antibodies were utilized: S-100 protein (1:5000, DAKO) and glial fibrillary acidic protein (GFAP, 1:5000, DAKO). For myoblast/myotube, desmin (1:1000, Chemicon) and myosin (1:3000, Chemicon).

All of the HNC10 cells were immunopositive for nestin (Fig. 4A) *(24,25)*, vimentin *(26)*, and LNGFR/p75 (Fig. 4B) *(25,27,28)*, indicating that HNC10

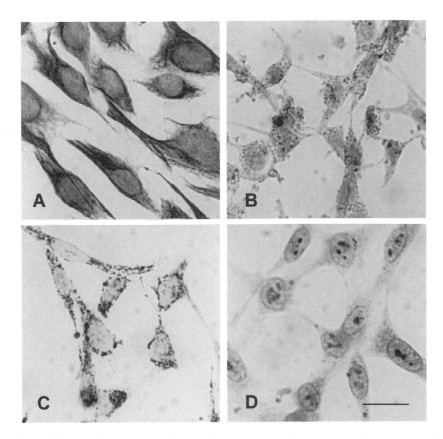

Fig. 4. Two cell-type-specific markers for neural crest stem cells are nestin, intermediate cytoskeletal protein only found in neural stem cells, and p75/ low-affinity neurotrophin receptor (LNGFR) protein. **(A)** HNC10 neural crest stem cells are shown to express a very strong nestin immunoreactivity; **(B)** HNC10 cells are reaction-positive for p75 protein; **(C)** HNC10 cells are reaction-positive for human mitochondria antigen; **(D)** Immunoreactivity for myc protein was found in nuclei of HNC10 cells, indicating that these cells are indeed transformed by myc oncogene.

cells are indeed neural crest stem cells *(19,25)*. In addition, HNC10 cells were positive for FORSE-1, a surface antigen marker for human neural progenitor cells *(29)*. HNC10 cells were also positive for antibody specific for human mitochondria (Fig. 4C) and for pan-myc antibody (Fig. 4D), indicating that HNC10 cells are uniquely human in origin and contain a copy or copies of *v-myc* encoding retrovirus. When HNC10 cells were grown in serum-containing medium, more than 10–30% of total cells expressed triplet neurofilament proteins (NF-L, NF-M, NF-H) (Fig. 5A), MAP2, tubulin βIII

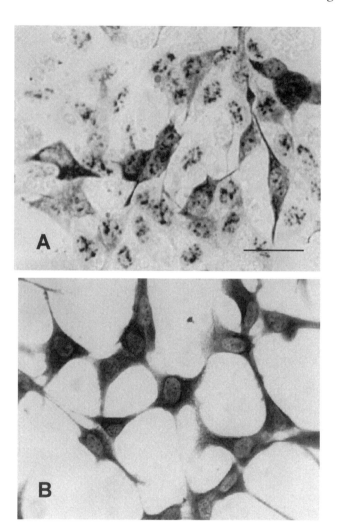

Fig. 5. (A) When HNC10 cells are grown in serum-containing medium, many of the cells differentiate into neurons as shown by a positive reaction for neurofilament (NF)-H. Scale bar indicates 10 μm. **(B)** HNC cells are positive for S-100 protein, a cell-type-specific antigen for Schwann cells.

isoform, and peripherin. These phenotypes are specifically and exclusively expressed by mammalian nerve cells, including human, indicating that HNC10 cells could differentiate into nerve cells under suitable culture conditions. HNC10 cells grown in serum-containing medium also expressed S-100 and P0 proteins, both cell-type-specific markers for Schwann cells, indicating differentiation of Schwann cells from HNC10 cells (Fig. 5B).

Table 2
Primers for RT-PCR Analyses

Nestin sense: 5'-CTCTGACCTG TCAGAAGAAT-3'
    antisense: 5'-GACGCTGACACTTACAGAAT-3'   (316 bp)
LNGFR/first sense: 5'-CTCACACCGGGGATGTG-3'
    antisense: 5'-GTGGGCCTTGTGGCCTAC-3'
LNGFR/second sense: 5'-TGTGGCCTACATAGCCTTC-3'
    antisense: ATGTGGCAGTGGACTCACT-3'   (476 bp)
NF-L sense: 5'-TCCTACTACACCAGCCATGT-3'
    antisense: 5'-TCCCCAGCACCTTCAACTTT-3'   (284 bp)
NF-M sense: 5'-TGGGAAATGGCTCGTCATTT-3'
    antisense: 5'-CTTCATGGAAGCGGCCAATT-3'   (333 bp)
NF-H sense: 5'-CTGGACGCTGAGCTGAGGAA-3'
    antisense: 5'-CAGTCACTTCTTCAGTCACT-3'   (316 bp)
P0 sense: 5'-TTCTGGTCCAGTGAGTGGGTCTCAG-3'
    antisense: 5- TCACTGTAGTCTAGGTTGTGTATGA-3'   (209 bp)
GFAP sense: 5'-GCAGAGATGATGGAGCTCAATGACC-3'
    antisense: 5',GTTTCATCCTGGAGCTTCTGCCTCA-3'   (266 bp)

There was also expression of chromogranin, a specific cell-type marker for adrenal chromaffin cells, in HNC10 cells grown in serum-containing medium. In addition, HNC10 cells grown in serum-containing medium for 2–3 wk started to show morphology of myotubes and these structures were reaction-positive for desmin and myosin, indicating that HNC10 cells were capable of developing into skeletal muscle phenotype (results not shown).

RT-PCR ANALYSIS OF HUMAN NEURAL CREST STEM CELLS

The primers used for the reverse transcriptase–polymerase chain reaction (RT-PCR) for human nestin, human LNGFR/p75, human NF-L, human NF-M, human NF-H, human P, and human GFAP are listed in Table 2. Results of RT-PCR analysis for mRNA isolated from HNC10 human neural crest stem cells are shown in Fig. 6. Transcripts for nestin and p75/LNGFR, cell-type-specific markers for neural crest stem cells, were clearly demonstrated in HNC10 cells grown in serum-free medium supplemented with bFGF. When HNC10 cells were grown in serum-containing medium, cells expressed transcripts for NF-L, NF-M, and NF-H, cell-type-specific markers for neurons. Transcripts for P0, structural protein of peripheral myelin and a cell-type-specific marker for Schwann cells, were also demonstrated. Transcripts for glial fibrillary acidic protein (GFAP), cytoskeletal protein specifically

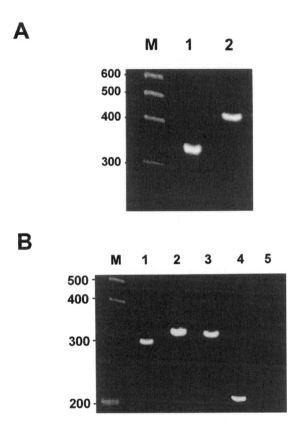

Fig. 6. RT-PCR analyses of genes expressed by HNC10 cells. **(A)** RNA transcripts for nestin and p75/low-affinity neurotrophin receptor protein (both cell-type-specific markers for neural crest stem cells) demonstrated in HNC10 cells. **(B)** RNA transcripts for differentiated cell-type markers for neurons (NF-L, NF-M, and NF-H) and Schwann cells (P0) were demonstrated in HNC10 cells grown in serum-containing medium. GFAP message in lane 5 was not detected in HNC10 cells.

expressed by central nervous system (CNS) astrocytes, were not detected by the RT-PCR analysis. Our RT-PCR analyses of HNC10 cells indicate that the cells express transcripts for nestin and p75/LNGFR, specific antigen markers of neural crest stem cells as grown in serum-free medium, whereas HNC10 cells grown in serum-containing medium expressed cell-type-specific markers for neurons (NF-L, NF-M, NF-H) and Schwann cells (P0).

TRANSPLANTATION INTO MOUSE BRAIN

HNC10 human neural crest stem cells were dissociated into single cells by brief trypsin treatment and suspended in PBS at 5×10^6 cells/0.1 mL plus 5 µg trypan blue and kept on ice until transplanted. Neonatal mice were cold-anesthetized by placing on ice and 2 µL volume of cell suspension was injected into lateral ventricle. All transplant recipients received daily cyclosporin A (10 mg/kg intraperitoneal injection, Novartis). Animals were sacrificed 2–4 wk postoperation and perfused with 4% paraformaldehyde solution. Brain sections were prepared in a cryostat and processed for β-galactosidase (X-gal) histochemical staining.

HNC10 cells integrated well within the subventricular layer, and from this region, an extensive migration of X-gal-positive HNC10 cells was found to neighboring anatomical sites such as the hippocampus and neostriatum (Fig. 7). Double immunostaining of grafted cells for β-galactosidase and neurofilament protein demonstrated that many, if not all, β-galactosidase-positive cells were also positive for neurofilament (results not shown). These results indicate that some of the grafted HNC10 cells differentiated into neurons by responding to developmental signals provided by the locus of implantation.

ESTABLISHED CELL LINES OF NEURAL CREST STEM CELLS

Established cell lines of human neurons/glial cells and neural stem cells are important and useful for three reasons: First, they provide human cells, which are exceedingly difficult to obtain; second, they are easy to maintain and expand, and they provide homogeneous population of cells by which identification and characterization of molecules ranging from neurotrophic factors to specific receptors, and the regulatory pathways, can be readily investigated; and third, human neuronal/neural stem cell lines should provide a ready and valuable source of human cells to be implanted into defined regions of nervous system to correct specific defects and eventual restoration of neurological function for various neurological disorders. The production of such stable, genetically homogeneous human neural crest stem cell lines should provide a valuable tool for studying genetic and epigenetic mechanisms regulating the development of the peripheral and autonomous nervous systems in vivo and in vitro. Previous investigators have attempted to produce established cell lines of neural crest progenitor/stem cells and the results were mixed. In cell lines obtained from retrovirally infected mouse neural crest explants, it was demonstrated that the cells differentiate spontaneously, are phenotypically unstable, and are probably a mixture of several different clones *(30)*. From Anderson's group, two cell lines of neural crest lineage cells have been generated; one is a sympathoadrenal progenitor cell line isolated from embryonic rat adrenal gland, immortalized by v-*myc* oncogene and giving rise to sympathetic neurons and adrenal

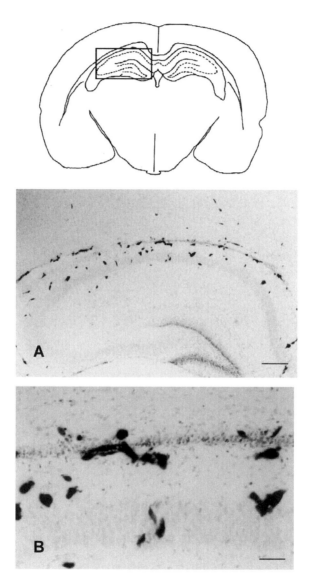

Fig. 7. (A) β-Galactosidase (X-gal)-expressing HNC10 cells were transplanted into cerebral lateral ventricle of newborn mouse brain. Four weeks postoperation, the X-gal-positive cells emerged from ventricle and migrated deeply into hippocampal region and integrated into neuronal layer. Scale bar = 100 μm. **(B)** Higher magnification of Fig. 7A.

chromaffin cells *(31)* and the other is of bipotential neural crest progenitor cells isolated from embryonic rat neural crest and also immortalized with a retrovirus containing a v-*myc (32)*. These murine cell lines were used to investigate late events in peripheral neurogenesis. To date there has been no report of the development of the human neural crest progenitor/stem cell line in the literature.

We have established, for the first time, immortalized cell lines of human neural crest stem cells via retrovirus-mediated v-*myc* transfer into progenitor cells initially derived from human embryonic dorsal root ganglia. HNC10 cells express phenotypes characteristic for neural crest stem cells (i.e., nestin and p75/LNGFR), indicating that HNC10 cells are of neural crest stem cell origin *(25,33)*. Neural crest stem cells can be distinguished from CNS stem cells by their morphology, by expression of low-affinity NGF receptor, by the progeny that they generate, and by their inability to generate CNS derivatives such as astrocytes *(28)*. HNC10 cells are self-renewing and multipotent in vitro, giving rise to four different lineage cells (i.e., neurons, Schwann cells, adrenal chromaffin cells, and skeletal muscle cells). When HNC10 cells were grown in serum-containing medium, a large number of cells expressed NF-L, NF-M, NF-H, MAP-2, tubulin βIII isoform, and peripherin, cell-type-specific markers for neurons. In addition, HNC10 cells grown in serum-containing medium were reaction-positive for S-100 (for Schwann cells), chromogranin (for adrenal chromaffin cells), and desmin/myosin (for skeletal muscle cells), indicating that HNC10 neural crest stem cells are capable of differentiation into non-neuronal cells such as Schwann cells, adrenal chromaffin cells, and muscle cells. Whether HNC10 cells could differentiate into other non-neuronal cell derivatives of neural crest stem cells such as melanocytes, chondrocytes, or osteocytes should await further studies.

The development of clonal cultures of avian and mammalian neural crest stem cells provides an assay for environmental signals that may influence neural crest development. Many such studies have been performed in avian systems, and growth factors such as bFGF, BDNF, TGF-β1, neuregulin, and bone morphogenic protein 2 (BMP2) have been found to influence neural crest development in vitro *(31–33)*. These observations indicate that the choice of each of several alternative fates available for neural crest stem cells can be instructively promoted by different environmental signals, including those growth factors previously reported by various investigators. In a preliminary study, we found that bFGF in the presence of serum promotes differentiation of HNC10 cells into neurons, whereas TGF-β1 induced HNC10 cells to differentiate into Schwann cells and blocked neuronal differentiation. Our results suggest that members of the FGF family, TGF superfamily, and neurotrophins could influence the fate of human neural crest stem cells at multiple levels for each and different lineage cells. Further studies on this line of inquiry are currently underway in our laboratory.

One to four weeks after implantation of HNC10 cells into the ventricle of newborn mice, an extensive migration of X-gal-positive HNC10 cells was demonstrated streaking out from ventricular lumen, the sites of implantation, to neighboring anatomical sites, including the hippocampus and striatum. These results suggest that when HNC10 cells are implanted into ventricular space, their migratory and differentiation potentials are strongly influenced by the environmental signals produced by neighboring brain sites. When neural stem cells and/or neural crest stem cells are implanted into different areas of the developing nervous system, they generate progeny that would normally be generated in that area at the time the cells are grafted. Thus, the same precursor cells can generate Purkinje cells when implanted into the cerebellum and hippocampal neurons when introduced into the hippocampus at the time that these cells are being generated from endogenous precursors *(23,34)*. In addition, the fate of multipotent neural stem cells/neural crest stem cells is influenced by molecules generated from the environment into which they were placed. Thus, multipotent progenitors introduced into the brains of myelin-deficient rats or mice generate oligodendrocytes *(12,35)*. It appears that HNC10 cells, human neural crest stem cells, could also be influenced by signals produced in the CNS anatomical sites where they were implanted and terminally differentiate into CNS neurons or glial cells. The spectacular property of the human neural crest stem cell lines to differentiate *in situ* in the recipient animal brain may eventually allow targeted introduction of neural crest stem cells into defined regions of the brain to correct specific defects and the eventual restoration of neurological function for many types of neurological diseases with widespread pathology.

EXPERIMENTAL AND CLINICAL APPLICATIONS

Cell therapy has become a most promising strategy for the treatment of many human diseases, including neurological disorders. The objective of cell therapy is to replace lost cells and restore the function of damaged cells/tissues following diseases or trauma. Transplantation of renewable, homogeneous, multipotent, and well-characterized neural crest stem cells into the damaged nervous system tissues should replace lost cells and restore

Table 3
Clinical Applications of Human Neural Crest Stem Cells

Disease	Cell	Function	References
Motor and sensory neuropathy	Sensory neurons, Schwann cells	Cell replacement	Aguayo et al. *(36,37)*
Multiple sclerosis	Schwann cells	Cell replacement	Honmou et al. 1996; Archer et al. *(38)*
Parkinson's disease	DA neurons, Adrenal chromaffin cells	Cell replacement	Lindvall et al. *(39)*; Olanow et al. *(40)*
Huntington's disease	GABA neurons	Cell replacement	Isacson et al. *(41)*
Spinal cord injury	Schwann cells	Factor delivery, axon guidance	Menei et al. *(42)*; Xu et al. *(43)*
Duchenne muscular dystrophy	Muscle cells	Cell replacement, protein delivery	Gussoni et al. *(44)*; Miller et al. *(45)*
Pain control	Adrenal chromaffin cells	Pain-killer delivery	Eaton et al. *(46)*

damaged function. HNC10 cells, immortalized human neural crest stem cells, should fulfill the above-described criteria as an ideal cell type to serve as donor cells for cell replacement therapy in various neurological diseases, including motor and sensory neuropathy *(36,37)*, multiple sclerosis *(35,38)*, Parkinson's disease *(39,40)*, Huntington's disease *(41)*, spinal cord injury *(42,43)*, Duchenne muscular dystrophy *(44,45)*, and pain control *(46)* (*see* Table 3). The stable immortalized human neural crest stem cell lines described here can be expanded readily and provide a renewable and homogeneous population of neuronal, glial, and muscle cells and will prove to be most valuable for future studies of fundamental questions in developmental neurobiology, cell and gene therapies, and research and development of new drugs and treatments.

ACKNOWLEDGMENTS

This study was supported by grants from Myelin Research Initiative of Canada, Multiple Sclerosis Society of Canada, and KOSEF/ BDRC Ajou University.

REFERENCES

1. Buc-Caron, M. H. (1995) Neuroepithelial progenitor cells explanted from human fetal brain proliferate and differentiate in vitro. *Neurobiol. Dis.* **2,** 37–47.
2. Sah, D. W., Ray, J., and Gage, F. H. (1997) Bipotent progenitor cell lines from the human CNS. *Nat. Biotechnol.* **15,** 574–580.
3. Flax, J. D., Aurora, S., Yang, C., Simonin, C., Willis, A., Billinghurst, L., et al. (1998) Engraftable human neural stem cells respond to developmental cues, replace neurons, and express foreign genes. *Nat. Biotechnol.* **16,** 1033–1039.
4. Brustle, O., Choudary, K., Karram, K., Huttner, A., Murray, K., Dalcq, M., et al. (1998) Chimeric brains generated by intraventricular transplantation of fetal human brain cells into embryonic rats. *Nat. Biotechnol.* **16,** 11,040–11,044.
5. Eriksson, P., Perfilieva, E., Bjork, T., Alborn. A., Nordborg, C., Peterson, D., et al. (1998) Neurogenesis in the adult human hippocampus. *Nat. Med.* **4,** 1313–1317.
6. Villa, A., Snyder, E. Y., Vescovi, A., and Martinez-Serrano, A. (2000) Establishment and properties of a growth factor-dependent, perpetual neural stem cell line from the human CNS. *Exp. Neurol.* **161,** 67–84.
7. Snyder, E. Y. (1994) Grafting immortalized neurons to the CNS. *Curr. Opin. Neurobiol.* **4,** 742–751.
8. Gage, F. H., Ray, J., and Fisher, L. (1995) Isolation, characterization and use of stem cells from the CNS. *Ann. Rev. Neurosci.* **18,** 159–192.
9. Brustle, O. and McKay, R. D. (1996) Neuronal progenitors as tools for cell replacement in the nervous system. *Curr. Opin. Neurobiol.* **6,** 688–695.
10. Martinez-Serrano, A. and Bjorklund, A. (1997) Immortalized neural progenitor cells for CNS gene transfer and repair. *Trends Neurosci.* **20,** 530–538.
11. Fricker, R., Carpenter, M., Winkler, C., Greco, C., Gates, M., and Bjorklund, A. (1999) Site-specific migration and neuronal differentiation of human progenitor cells after transplantation in the adult rat brain. *J. Neurosci.* **19,** 5990–6005.
12. Yandava, B. D., Billinghurst, L., and Snyder, E. Y. (1999) Global cell replacement is feasible via neural stem cell transplantation: evidence from the dysmyelinated shiverer mouse brain. *Proc. Natl. Acad. Sci. USA* **96,** 7029–7034.
13. Aboody, K., Brown, A., Rainov, N., Bower, K., Liu, S., Yang, W., et al. (2000) Neural stem cells display extensive tropism for pathology in adult brain: evidence from intracranial gliomas. *Proc. Natl. Acad. Sci. USA* **97,** 12,846–12,851.
14. Le Douarin, N. M. (1993) *The Neural Crest*, Cambridge University Press, Cambridge, 1982.

15. Anderson, D. J. (1993) Cell and molecular biology of neural crest cell lineage diversification. *Curr. Opin. Neurobiol.* **3**, 8–13.
16. Le Douarin, N. M., Ziller, C., and Couly, G. F. (1993) Patterning of neural crest derivatives in the avian embryo: in vivo and in vitro studies. *Dev. Biol.* **159**, 24–49.
17. Le Douarin, N. M. (1986) Cell line segregation during peripheral nervous system ontogeny. *Science* **231**, 1515–1522.
18. Patterson, P. (1990) Control of cell fate in a vertebrate neurogenic lineage. *Cell* **62**, 1035–1038.
19. Anderson, D. J. (1997) Cellular and molecular biology of neural crest cell lineage determination. *Trends Genet.* **13**, 276–280.
20. Kim, S. U., Kim, K. M., and Moretto, G. (1985) The growth of fetal human sensory ganglion neurons in culture: A scanning electron microscopic study. *Scan. Electron Microsc.* **2**, 837–848.
21. Kim, S. U., Yong, V. W., Watabe, K., and Shin, D. H. (1989) Human fetal Schwann cells in culture: phenotypic expressions and proliferative capability. *J. Neurosci. Res.* **22**, 50–59.
22. Kim, S. U., Stern, J., Kim, M. W., and Pleasure, D. E. (1983) Culture of purified rat astrocytes in serum-free medium supplemented with mitogen. *Brain Res.* **274**, 79–86.
23. Snyder, E. Y., Deitcher, D. L., Walsh, C., Arnold-Aldea, S., Hartwieg, E. A., and Cepko, C. L. (1992) Multipotent neural cell lines can engraft and participate in development of mouse cerebellum. *Cell* **68**, 33–51.
24. Lendahl, U., Zimmerman, L. B., and McKay, R. D. G. (1990) CNS stem cells express a new class of intermediate filament protein. *Cell* **60**, 585–595.
25. Stemple, D. L. and Anderson, D. J. (1992) Isolation of a stem cell for neurons and glia from the mammalian neural crest. *Cell* **71**, 973–985.
26. Houle, J. and Fedoroff, S. (1983) Temporal relationship between the appearance of vimentin and neural tube development. *Dev. Brain Res.* **9**, 189–195.
27. Chao, M. V. (1992) Neurotrophin receptors: a window into neuronal differentiation. *Neuron* **9**, 583–593.
28. Morrison, S. J., White, P., Zeck, C., and Anderson, D. J. (1999) Prospective identification, isolation by flow cytometry, and in vivo self-renewal of multipotent mammalian neural crest stem cells. *Cell* **96**, 737–749.
29. Tole, S. and Patterson, P. H. (1995) Regulation of the developing forebrain: a comparison of FORSE-1, Dlx-2, and BF-1. *J. Neurosci.* **15**, 970–980.
30. Stemple, D. L. and Anderson, D. J. (1993) Lineage diversification of the neural crest: in vitro investigations. *Dev. Biol.* **159**, 12–23.
31. Shah, N. M., Marchionni, M. A., Isaacs, I., Stroobant, P., and Anderson, D. J. (1994) Glial growth factor restricts mammalian neural crest stem cells to a glial fate. *Cell* **77**, 349–360.
32. Shah, N. M., Groves, A. K., and Anderson, D. J. (1996) Alternative neural crest cell fates are instructively promoted by TGF beta superfamily members. *Cell* **85**, 331–343.
33. Sieber-Blum, M. and Zhang, J. M. (1997) Growth factor action in neural crest diversification. *J. Anat.* **191**, 439–499.

34. Renfranz, P. J., Cunningham, M. G., and MacKay, R. D. (1991) Region-specific differentiation of the hippocampus stem cell line HiB5 upon implanatation into the developing mammalian brain. *Cell* **66,** 713–729.
35. Duncan, I. D. and Milward, E. A. (1995) Glial cell transplants: experimental therapies of myelin diseases. *Brain Pathol.* **5,** 301–310.
36. Aguayo, A. J., Epps, J., Charron, L., and Bray, G. M. (1976) Multipotentiality of Schwann cells in cross-anastomosed and grafted myelinated and unmyelinated nerves: quantitative microscopy and radioautography. *Brain Res.* **104,** 1–20.
37. Honmou, O., Felts, P., Waxman, S., and Kocsis, J. (1996) Restoration of normal conduction properties in demyelinated spinal cord axons in the adult rat by transplantation of exogenous Schwann cells. *J. Neurosci.* **16,** 3199–3208.
38. Archer, D. R., Cuddon, P. A., Lipsitz, D., and Duncan, I. D. (1997) Myelination of the canine central nervous system by glial cell transplantation: a model for repair of human myelin disease. *Nat. Med.* **3,** 54–59.
39. Lindvall, O., Sawle, G., and Widner, H. (1994) Evidence for long-term survival and function of dopaminergic grafts in progressive Parkinson's disease. *Ann. Neurol.* **35,** 172–180.
40. Olanow, C. W., Kordower, J. H., and Freeman, T. B. (1996) Fetal nigral transplantation as a therapy for Parkinson's disease. *Trends Neurosci.* **19,** 102–109.
41. Isacson, O., Dunnett, S. B., and Bjorklund, A. (1986) Graft-induced behavioral recovery in an animal model of Huntington disease. *Proc. Natl. Acad. Sci. USA* **83,** 2728–2732.
42. Menei, P., Montero-Menei, C., Whittemore, S. R., Bunge, R. P., and Bunge, M. B. (1998) Schwann cells genetically modified to secrete human BDNF promote enhanced axonal regrowth across transfected adult rat spinal cord. *Eur. J. Neurosci.* **10,** 607–621.
43. Xu, X. M., Zhang, S. X., Li, H., Aebischer, P., and Bunge, M. B. (1999) Regrowth of axons into the distal spinal cord through a Schwann-cell-seeded mini-channel implanted into hemisected adult rat spinal cord. *Eur. J. Neurosci.* **11,** 1723–1740.
44. Gussoni, E., Pavlath, G. K., Lanctot, A. M., Sharma, K. R., Miller, R. G., Steinman, L., et al. (1992) Normal dystrophin transcripts detected in Duchenne muscular dystrophy patients after myoblast transplantation. *Nature* **356,** 435–438.
45. Miller, R. G., Sharma, K. R., Pavlath, G. K., Gussoni, E., Mynhier, M., Yu, P., et al. (1997) Myoblast implantation in Duchenne muscular dystrophy: the San Francisco study. *Muscle Nerve* **20,** 469–478.
46. Eaton, M. J., Martinez, M., Karmally, S., Lopez, T., and Sagan, J. (2000) Initial characterization of the transplant of immortalized chromaffin cells for the attenuation of chronic neuropathic pain. *Cell Transplant.* **9,** 637–656.

5

The Search for Neural Progenitors in Bone Marrow and Umbilical Cord Blood

Shijie Song, Paul Sanberg, and Juan Sanchez-Ramos

INTRODUCTION

The study of hematopoiesis, the generation of blood cell lines throughout life, has provided conceptual, experimental, and therapeutic approaches useful to all stem cell biologists. From a clinical perspective, no other area of stem cell biology has been applied as successfully as has transplantation of bone marrow and cord blood for the treatment of blood diseases *(1,2)*. In the last few years, research in stem cell biology has rapidly expanded to include the study of stem cells from embryonic, fetal, and various adult tissues, engendering novel perspectives regarding the identity, origin, and full therapeutic potential of tissue-specific stem cells *(3–6)*. There are many similarities, as well as differences, between the stem cells that give rise to the nervous system (neuropoiesis) and those that generate the various blood cell lineages (hematopoiesis) *(see* Table 1). Just as bone marrow gives rise to blood cell lineages throughout life, the "brain marrow" is a proliferative central core of the central nervous system (CNS) comprised of the subependymal zone (SEZ), and other periventricular cell groups, that have the capacity to contribute to neurogenesis throughout life *(5,7)*. In this chapter, we shall examine the theoretical and experimental relationships between bone marrow and brain by reviewing published research that addresses the capacity of adult bone marrow and umbilical cord blood to generate neurons and glia.

DEFINITIONS AND ORIGINS

A stem cell is succinctly defined as an undifferentiated cell with the ability to proliferate, to self-renew, and to generate a large number of progeny *(2,4,5)*. The mother of all stem cells, the true totipotent cell

From: *Neural Stem Cells for Brain and Spinal Cord Repair*
Edited by: T. Zigova, E. Y. Snyder, and P. R. Sanberg © Humana Press Inc., Totowa, NJ

Table 1
Features of Hematopoiesis and Neuropoiesis

Hematopoiesis	Neuropoiesis
(a) Pluripotential stem/progenitor cells located in bone marrow	(a) Pluripotential neural stem/progenitor cells located in subependymal zone and hippocampus
(b) CD34+ cell as the hematopoeitic stem cell?	(b) Nestin+ cells as neural precursors?
(c) Stem cell replaces short-lived blood cell lineages throughout lifetime of organism	(c) Neural progenitor replaces only some adult neurons over the life of the organism
(d) Originates in embryonic endoderm?	(d) Originates in embryonic ectoderm
(e) Can give rise experimentally to non-endodermal tissues	(e) Can give rise experimentally to blood cell lineages
Extracellular matrix and supportive cells (marrow stromal cells) essential for regulating hematopoiesis	Extracellular matrix (e.g., tenascin) and radial glial cells essential in regulating migration of neural precursors
Migratory pathways: from marrow sinusoids to vasculature and specific organ targets (e.g., spleen)	Migratory pathways: from SEZ to olfactory bulb throughout adult life
Trophic factors, cytokines, and other molecules Instructive for proliferation and differentiation into blood cell lineages: IL-3, IL-4, IL-5, IL-6, colony-stimulating factors (CSF), including CSF-1, CSF-GM, CSF-G, c-kit, TGF-βs, BMPs, erythropoietin, etc.	Trophic factors, cytokines, and other molecules instructive for proliferation and differentiation into neuronal, glial, and oligodendroglia lineages: EGF, PDGF, BDNF, bFGF, GDNF, BMPs, steroids, retinoids, IL-1

capable of generating *all* cell types and constructing a complete organism, is the fertilized egg. During morphogenesis, cells proliferate, migrate, and differentiate, but throughout this process, a small residual of quiescent, uncommitted cells are retained. This is a result of asymmetric division of the uncommitted cells, ensuring self-renewal of the stem cell population. Stem cells can be isolated from the developing embryo and even from specific adult tissues, but their potential for differentiation into all cell types becomes gradually restricted to a more limited range of cells typical of the mature tissue in which the stem cell (or progenitor cell) resides. The term *progenitor cell* refers to the offspring of stem cells that have more restricted cell fates.

The stem cells that give rise to blood cell lineages reside primarily in adult bone marrow, but their origin can be traced to the embryonic yolk sac, with later migration to embryonic liver (both are derived from the endodermal layer of the gastrula) and, from there, to a final adult residence in the vascular sinuses of flat and long bones *(8)*. During early development, ectodermal cells are induced to become mesoderm by factors that emanate in the endodermal pole. Nucleated erythrocytes first appear in "blood islands" in the yolk sac, and yolk sac hematopoiesis is predominant during early development (through 6–8 wk of gestation in humans, 10–11 d of gestation in mice). During mid-gestation, soon after the development of organized hepatic structure, the liver becomes the major hematopoietic organ of the fetus. The spleen also participates in hematopoiesis during this period. In the latter phases of mammalian fetal development, the bone marrow becomes the main site of hematopoiesis. In the human, the bone marrow is the exclusive site of postnatal hematopoiesis under normal circumstances, whereas in the mouse, the spleen is also a hematopoietic organ throughout life. However, hematopoiesis requires the presence of instructive signals and the appropriate microenvironment provided by bone marrow stromal cells whose origin can be traced to the embryonic mesodermal germ layer. Some authors have wondered whether the hematopoietic stem cells of these various periods of development are the same cells that have migrated and seeded developing organs or whether they have originated within the pool of stem/progenitor cells that gave rise to and compose each organ system. Many studies on the origins of stem cells during development of amphibians, birds, and mammals suggest that there are probably separate origins of yolk sac, fetal, and adult hematopoietic and stromal stem cells *(9–13)*.

Hematopoietic stem cells (as well as neural-crest-derived melanoblasts and primordial germ cells) express c-kit on their cell membrane *(14)*. Stem cell factor (SCF), the ligand for c-kit, is expressed during embryogenesis in cells associated with both the migratory pathways and homing sites of hematopoietic stem cells, melanoblasts, and germ cells. Both SCF and c-kit, a member of the tyrosine kinase family, are also expressed in a variety of other tissues, including the brain and spinal cord, suggesting that this receptor–ligand system has additional roles in embryogenesis *(15)*. Moreover, the migration of bone marrow cells to the brain, skeletal muscle, or cardiac muscle under specific experimental conditions (described later in this chapter) may rely, in part, on the attraction exerted on c-kit expressing cells by the membrane bound SCF on the homing targets.

EXTRACELLULAR MATRIX AND SUPPORTIVE CELLS

The extracellular matrix and supportive cells play a critical role in both hematopoiesis and neuropoiesis. In the bone marrow, the stromal elements create a rich microenvironment in which to nourish the hematopoietic cells and determine their fate. The bone marrow extracellular matrix has been thought to affect marrow cells by regulating their adhesive and migratory properties and by binding and presenting cytokines and growth factors to the proliferating and differentiating marrow cells. The brain marrow also exhibits extracellular matrix consisting of a variety of glycoproteins and proteoglycans (16,17). Molecules such as tenascin have been found in progenitor cells themselves, including migrating primordial germ and hematopoietic progenitor cells (18). These extracellular matrix molecules have definite effects on cell migration and differentiation in both bone and brain marrow. Extracellular matrix proteins, for example, inhibit neurite outgrowth in embryonic dopaminergic neurons (19). Extracellular matrix proteins are highly expressed in the SEZ rostral migratory pathway, and this corresponds to the lack of neuritic growth by the immature neurons within the pathway (20). As pointed out by Steindler, many of the extracellular matrix molecules are related to migratory properties of the cells (16).

In bone marrow, the newly born blood cells need to migrate a short distance to reach the vascular spaces, whereas in the brain, cells need to travel relatively large distances within the neurogenic brain marrow and through established neural pathways such as the rostral migratory pathway to the olfactory bulb. Steindler and colleagues have suggested that both germinal areas (i.e., bone and brain marrow) relate to luminal surfaces; specialized astrocytes within the SEZ are similar to the stromal barrier cells of the bone marrow in that they form a tubular ensheathment around the mitotically active population of stem/progenitor cells (20).

GROWTH FACTORS AND CYTOKINES

Many of the cytokines and growth factors required for hematopoiesis are constitutively expressed by bone marrow stromal cells (21). Two cytokine superfamilies produced by bone marrow stromal cells [transforming growth factor-β (TGF-β) and the hematopoietins] mediate a range of developmental events in the nervous system that rivals that of the classic neurotrophins. The bone morphogenetic proteins (BMPs), a subclass of the TGF-β superfamily, is present in bone matrix and is well known for their significant effects on the development of osteocytes and chondrocytes in bone. However, BMP ligands and receptor subunits are also present throughout neural development within discrete regions of the embryonic brain and within neural-crest-derived premigratory and postmigratory zones *(22,23)*. BMPs exhibit a broad range of cellular and context-specific effects during multiple stages of neural development (see review, ref. *23*). For example, BMPs initially inhibit the formation of neuroectoderm during gastrulation, whereas within the neural tube, they act as gradient morphogens to promote the differentiation of dorsal cell types and intermediate cell types through cooperative signaling. In the peripheral nervous system, BMPs serve as instructive signals for neural lineage commitment and promote graded stages of differentiation. However, within the CNS, these same factors promote astroglial lineage elaboration from SVZ progenitor cells, with concomitant suppression of neuronal or oligodendroglial lineages. In addition, synergistically with basic fibroblast growth factor (bFGF), BMP-2 acts on more lineage-restricted embryonic CNS progenitor cells to induce the expression of the dopamine neuronal marker, tyrosine hydroxylase *(24)*. Various hematopoietic cytokines were recently shown to enhance the number of dopaminergic neurons in mesencephalic cultures, but only interleukin-1 (IL-1) induced the expression of the DA neuronal marker, tyrosine hydroxylase, in the progenitor cells *(25)*.

Other regulatory proteins that influence blood cell differentiation have effects on neuronal development. Erythropoietin, in addition to its function in promoting the production of red blood cells, also has trophic effects on central cholinergic and cortical neurons *(26)*. Erythropoietin receptor mRNA is expressed in the mouse brain and in the CNS of the developing human fetus. Neuronal cell lines including PC12 and SN6, also express a functional erythropoietin receptor. Campana et al. have confirmed the neurotrophic activity of erythropoietin in neural cells *(27)*. They also identified a 17-mer peptide sequence (EPO peptide) in erythropoietin with activity similar to that of the holoprotein. This peptide induced differentiation and prevented the proliferation of erythropoeitic cell lines or mouse primary spleen cells. When the EPO peptide or erythropoietin was locally injected into mice, the frequency of motor end plate sprouting in adjacent muscles increased in a manner similar to that induced by ciliary neurotrophic factor (CNTF). Because neural cells and not hematological cells respond to a specific peptide sequence within erythropoietin, it is likely that the holoprotein has separate domains for neurotrophic and hematotrophic function.

NEURAL PROGENITORS CAN REPOPULATE
THE BONE MARROW

Bjornson et al. were the first to investigate and demonstrate the potential of neural stem cells to develop into blood cell lineages. Their approach was to test the capacity of neural stem cells (cultured from either the embryonic or adult forebrain) to repopulate bone marrow of sublethally irradiated Balb/c recipient mice *(28)*. As a comparison, they transplanted unfractionated adult bone marrow cells into the irradiated mice. To mark and trace the donor cells, the transplanted neural stem cells or bone marrow cells were obtained from ROSA26 mice. The ROSA26 animals were chosen because they were of a different immunological background than the recipient Balb/c mice and were transgenic for *lacZ*, which encodes for the *Escherichia coli* enzyme β-galactosidase. To eliminate possible contamination of neural stem cells with cells of mesodermal origin, clonally derived adult ROSA26 neural stem cells were used in some of the experiments. To detect the presence of *lacZ* gene in splenic DNA in recipient mice, a polymerase chain reaction assay was used. Five to 12 months after transplantation, a strong *lacZ* mRNA signal was observed in splenic DNA taken from both untreated ROSA26 animals and irradiated Balb/c mice that had received ROSA26 bone marrow cells, demonstrating that the donor bone marrow cells could reconstitute hematopoiesis in the recipient mice. Especially striking was the detection of a strong *lacZ* mRNA in splenic DNA taken from animals injected with embryonic, adult, or clonally derived adult neural stem cells. To determine whether the engrafted neural stem cells had given rise to distinct blood cell lineages, three methods were utilized: in vitro clonogenic assays, immunocytochemistry and flow cytometric analysis. Two distinct markers were used to verify the origin of the blood cell lineages from the ROSA26 neural stem cells. The first was the histochemical reaction product generated by X-gal in β-galactosidase-expressing cells and the second marker was based on antibody recognition of the distinct cell surface antigens expressed by ROSA26 mice. Bone marrow isolated from the grafted animals when treated with the appropriate cytokines formed colonies that reacted to X-gal. Moreover, a range of blood cell lineages was expressed from these colonies, including granulocyte, granulocyte–macrophage, and pure macrophage lineages, as well as mixed colonies. Megakaryocytic and B-cell colonies were also present. The authors were careful to point out that none of the neural stem cell cultures either proliferated or formed colonies when used in the same cytokine-stimulated clonogenic assays done in parallel to the grafting experiments. Confirmation of these results was provided by flow cytometry studies using antibodies specific to H-2K^b (expressed in ROSA26 mice). These elegant experiments demonstrated that neural stem cells (isolated from either the embryonic or adult murine forebrain) could engraft into the hematopoietic system of irradiated hosts to produce a spectrum of blood cell lineages. Clearly, the potential cell fate of adult neural stem cells is much broader than previously surmised. As mentioned earlier, the genomic totipotentiality of differentiated cells has been demonstrated in the cloning of the ewe, Dolly *(8)*. The work of Bjornson et al. suggests that the reactivation of "dormant" genetic programs may not require nuclear transfer or experimental modification of the genome *(28)*. It appears that the neural stem cell is endowed with the capacity to express an otherwise silent genomic potential in response to instructive signals in the appropriate microenvironment.

BONE MARROW CELLS GENERATE NONHEMATOPOIETIC CELLS

Nonhematopoietic precursors from bone marrow stroma are also known as colony-forming-unit (CFU) fibroblasts, mesenchymal stem cells, or bone marrow stromal cells (BMSC). The presence of nonhematopoietic progenitor cells in bone marrow that contribute fibroblasts for the wound-healing process was first suggested in the late 19th century. The isolation and differentiation of marrow stromal cells into osteoblasts, chondroblasts, adipocytes, and myoblasts has only recently been demonstrated *(29)*. Although BMSC can naturally be expected to be a source of surrounding tissue of bone, cartilage, and fat, several recent reports demonstrate that these cells, under specific experimental conditions, can differentiate into skeletal and cardiac muscle, hepatocytes, glia, and neurons *(30–34)*.

BONE MARROW TO MUSCLE AND LIVER CELLS

Rat bone-marrow-derived mesenchymal stem cells in cell culture were induced to differentiate into a muscle cell phenotype *(35)*. Multinucleated myotubes were observed in some culture dishes 7–11 d after 24-h incubation with 5-azacytidine. Adipocytes, evidenced by Sudan black-positive cytoplasmic droplets, were also observed. This experiment was the first to demonstrate that culture-propagated rat bone marrow mesenchymal stem cells had the capacity to differentiate in vitro into a cellular phenotype not normally found in bone *(35)*. Based on these observations, Ferrari et al. tested the capacity of bone marrow stromal cells to regenerate muscle in vivo *(30)*. A resident population of mononuclear myogenic precursors known as satellite cells normally mediates the repair of skeletal muscle fibers. In response to injury or disease, the satellite cells divide and fuse to repair or replace the damaged fibers. To induce muscle regeneration, the researchers injected cardiotoxin (25 µL of 1 mM) into the anterior tibialis muscle of a strain of immunodeficient mice (scid/bg). Unfractionated bone marrow cells obtained from C57/MlacZ transgenic mice were injected into the damaged muscle. These genetically labeled bone marrow cells underwent myogenic differentiation and participated in the regeneration of the damaged muscle fibers. In addition, the investigators fractionated the genetically labeled bone marrow cells into adherent and nonadherent components. These components were then separately injected into the regenerating muscles of immunodeficient mice. Marrow-derived cells were found in regenerating muscle in all six mice injected with the adherent cell population (bone marrow stromal cells) and in three of the six mice injected with the nonadherent cells. This suggests that the marrow stromal cells were the source of the myogenic progenitors. To determine whether the myogenic progenitors could be physiologically recruited from bone marrow and target a site of muscle regeneration, genetically labeled bone marrow cells were transplanted into 12 irradiated scid/bg mice. Five weeks after bone marrow transplantation, muscle regeneration was chemically induced in skeletal muscles of nine surviving mice. Histochemical review of the muscles showed regenerating fibers containing bone-marrow-derived cells in five of the six marrow reconstituted animals. This work by Ferrari et al. elegantly demonstrated that bone-marrow-derived myogenic progenitors could migrate into a degenerating muscle, participate in the regeneration process, and give rise to fully differentiated muscle fibers *(30)*. The bone marrow cells appeared to be recruited by long-range, possibly inflammatory signals originating from the degenerating tissue, and they appeared to access the damaged muscle from the circulation. Interestingly, the time-course of differentiation of bone-marrow-derived progenitors differed from the differentiation of committed adult myogenic precursors. Injected satellite cells fused into muscle fibers within 5 d, whereas bone-marrow-derived cells were not detected in regenerating fibers before 2 wk after induction of muscle damage. This suggested that bone-marrow-derived progenitors undergo a longer, multistep differentiation process, which may comprise migration, cell division, commitment to the myogenic lineage, and eventual terminal maturation and fusion.

In similar experiments, bone marrow cells were shown to give rise to hepatic oval cells *(32)*. Hepatic oval cells proliferate under conditions when hepatocytes are prevented from proliferating in response to liver damage. Oval hepatocytes may be stem cells for hepatocytes and bile duct cells or the intermediate progeny of a hepatic stem cell. Oval cells originate from either the cells present in the canals of Herring or from blastlike cells located next to bile ducts. Oval cells were shown to arise from a cell population originating in bone marrow by using three approaches for tracking the transplanted bone marrow cells *(32)*. Bone marrow was transplanted from male rats into lethally irradiated syngeneic females and the donor cells were detected in the recipients by means of DNA probes to the Y chromosomes. The second approach was based on transplanting bone marrow from dipeptidyl peptidase IV-positive (DPPIV+) male rats into DPPIV– syngeneic females and detecting the DPPIV+ cells of the donor in the recipient animals. The third approach involved whole-liver transplantation with Lewis rats that express the L.21.6 antigen as recipients and Brown–Norway rats that do not express this antigen as allogeneic donor. This approach was used to demonstrate that an extrahepatic source (L21.6+ cells) could repopulate the transplanted (L21.6– cells) liver. Oval cells that originated from an extrahepatic source would be L21.6+, whereas those originating *in situ* would be negative. The results from this approach indicated that some oval cells were derived from an extrahepatic source and others (L21.6–), were derived *in situ* from the donor liver. These data support the possibility that a cell associated with bone marrow may act, under certain pathophysiological conditions, as the progenitor of several types of liver cells. In a broader view, these data added to growing evidence that bone marrow cells in the adult organism have a greater potential than previously estimated.

BONE MARROW TO CARDIAC MUSCLE

Bone marrow also contains a progenitor cell capable of generating cardiac myocytes. A select population of bone marrow cells was injected into the contracting heart wall bordering the infarct in a mouse model of myocardial infarction *(34)*. After several weeks, newly formed myocardium was seen in the infracted portion of the ventricular wall of the heart. Although repair of myocardium was only obtained in 40% of the mice transplanted with bone marrow cells, these results demonstrate yet another potential fate for these cells. It is important to note that the researchers utilized a defined population of marrow cells, depleted of committed blood cell lineages but positive for c-kit receptor. They were selected on the basis of absence of surface markers for blood cell lineages CD4 and CD8 (T-lymphocytes), B-220 (B-lymphocytes), Mac-1 (macrophages), GR-1 (granulocytes), and TER119 (erythrocytes). Monoclonal antibodies to these markers were incubated with the bone marrow cells, followed by incubation with magnetic beads coated with goat anti-rat immunoglobulin. The lineage negative (lin⁻) cells were removed by a biomagnet and then further sorted on the basis of the presence of c-kit receptor (c-kit^{pos}). The newly formed myocardium occupied up to 68% of the infracted portion of the ventricle 9 d after transplanting the marrow cells. The tissue comprised proliferating myocytes and vascular structures. These elegant studies convincingly demonstrate that injected bone marrow cells can generate *de novo* myocardium and improve heart muscle function *(34)*.

It is known that fetal and neonatal hearts express SCF, the ligand for c-kit receptor *(15)*, and the authors suggested that it is the signaling between c-kit^{pos} cells and SCF that may be responsible for migration of the injected marrow cells to the site of infarct. However, it is not clear whether adult myocytes express SCF. During development, membrane-bound SCF, the ligand for c-kit receptor, appears to mediate migration of hematopoeitic stem cells to their target organs in yolk sac and later in the fetal liver. Indeed, other cells in addition to hematopoietic cells rely on the interaction of c-kit receptor with SCF for homing function, and these include germ cells, melanoblasts from the neural crest, and the interstitial cells of Cajal *(15)*.

BONE MARROW AS A SOURCE OF GLIAL CELLS

Astrocytes and oligodendroglia are derived from embryonic neuroecto-derm and are developmentally distinct from microglia, which are believed by many to derive from a hematopoietic line (monocytes) *(36,37)*. Some microglia, however, may have a neuroectodermal origin *(38)*. Eglitis et al. sought to determine the extent to which cells outside the CNS contribute to the maintenance of microglia in adult mice *(39)*. Bone marrow cells were labeled with a retroviral vector carrying the gene for neomycin resistance (*neo*^{R}). A second approach for tracking marrow-derived cells in the recipient relied on *in situ* hybridization with a probe specific to the Y chromosome of male donor marrow cells. They then infused (by tail vein) the *neo*^{R}-labeled male bone marrow cells into sublethally irradiated female mice (WBB6F1/ J-Kit^W/Kit^{W-v}). Over the next few days to weeks later, there was an influx of labeled cells into the brain of the recipients *(39)*. Marrow-derived cells were found throughout all regions of the brain, from cortex to brainstem. They appeared to reside with the parenchyma because perfusion with phosphate-buffered saline did not remove them. Occasional marrow-derived cells were found in association with vascular structures. The densities of donor cells in the recipient brain parenchyma paralleled the vascularity of a given region. Cortex, with few capillaries, had a lower cell density of marrow-derived cells than the more vascularized choroid plexus. Area postrema had the highest density of marrow-derived cells within the parenchyma. Some bone marrow-derived cells were positive for the microglial antigenic marker F4/80. Other marrow-derived cells expressed the astroglial marker glial fibrillary acidic protein (GFAP). Approximately 10% of the marrow-derived cells in the brain expressed either the microglial F4/80 antigen or GFAP. The identity of the remaining 90% of the marrow-derived cells in the recipient brain is unknown. These results indicated that some microglia and astroglia arose from a precursor that is a normal constituent of adult bone marrow. The authors considered the appearance of marrow-derived astroglia a normal process because the numbers of marrow-derived cells detected in the brain increased over time, and their appearance did not appear to be a consequence of the transplantation procedure. In addition, the radiation did not appear necessary for the marrow cells to migrate to the brain. Male donor cells engrafted and persisted for greater than 2 mo in recipients that had received no irradiation. Furthermore, there were as many Y chromosome/GFAP double-stained cells seen in the animals without radiation as seen in animals with irradiation. An interesting observation made by the authors was the appearance of cells with marrow markers in the ependymal layer of the ventricles. The finding of bone-marrow-derived cells suggests that these cells home in and differentiate in response to signals from the subependymal zone.

In another set of experiments using the male bone marrow to female recipient paradigm, Eglitis et al. demonstrated a preferential homing of marrow-derived progenitors to the site of a hypoxic/ischemic injury in rat brain *(40)*. In an acute unilateral middle cerebral artery occlusion model, 2.8% of the total DAPI-stained nuclei counted in the ischemic lesioned-side of the brain were derived from bone marrow (i.e., Y chromosome+), whereas 1.8% of the total nuclei were bone marrow derived in the intact unlesioned side. Thus, the ischemic side of the brain had attracted 55% more marrow-derived cells than the nonischemic side. No such difference was found between the two hemispheres in brain sections obtained from two intact animals grafted with bone marrow cells. The percentage of total GFAP+ cells that were double labeled (Y chromosome+/GFAP+) was 161% greater in the lesioned hemisphere compared to the unlesioned side. Of the total GFAP+ cells on the lesioned side, 4.7% were bone marrow derived, and on the unlesioned side, 1.8% were bone marrow derived. There was clearly a preferential targeting of the marrow-derived astrocytes to a region of cerebral injury. Astrocytic proliferation has been shown to occur in both the ischemic regions of the brain as well as in regions that are undamaged by ischemia. The findings of Eglitis et al. suggest that an additional source of astrocytes is related to increased migration and differentiation of cells derived from bone marrow.

Other researchers have reported that infusion of human BMSC into rodent brain resulted in engrafting, migration, and survival of cells *(31,41)*. A subset of human marrow cells (separated on the basis of adherence to plastic) labeled with bisbenzamide was injected directly into the corpus striatum of rat. From 5 to 72 d later, brain sections were examined for the presence of donor cells. Approximately 20% of the infused cells had engrafted in the host brain. The cells had migrated from the injection site to corpus callosum, contralateral cerebral cortex and ipsilateral temporal lobe. After engraftment, these cells lost markers typical of marrow stromal cells in culture, such as immunoreactivity to antibodies against collagen and fibronectin. BMSC developed many of the characteristics of astrocytes, and their engraftment and migration markedly contrasted with fibroblasts that continue to produce collagen and undergo gliosis after implantation.

Grafting of a subset of BMSC into the lateral ventricle of neonatal mice also resulted in their migration throughout the forebrain and cerebellum without disruption of host brain architecture *(41)*. In these experiments, the bone marrow stromal cells were depleted of cells that express the cell surface receptor CD11b, a marker of myelopoietic cells. The grafted BMSC were labeled with bisbenzamide or BrdU to track the fate of the cells. In the forebrain, a large number of donor cells were found ipsilateral to the injection site throughout the striatum, from the anterior commissure to the cingulate cortex. BMSC were also reported to line white-matter tracts, including the corpus callosum and the external capsule, suggesting that their distribution throughout the forebrain was an ordered process of migration. Many BMSC were detected lining the ependyma throughout the ventricles. The presence of cells double labeled for BrdU and GFAP suggested that some of the BMSC within the corpus striatum, the molecular layer of the hippocampus and the cerebellum had differentiated into astrocytes. Interestingly, BMSC were also located in areas undergoing active postnatal neurogenesis, including the islands of Calleja in the ventral forebrain and the subependyma of the olfactory bulb. A large number of BrdU-labeled cells were found integrated within the folia of the cerebellum. The majority of the cells were localized to the external granular layer, the internal granular layer, and, to a lesser extent, the molecular layer. The Purkinje cells in the cerebellum were not labeled with BrdU, consistent with their earlier maturation during embryogenesis. Most of the BrdU-labeled cells in the cerebellum were found ipsilateral to the side of injection, but some were found on the contralateral side. Many BrdU-labeled BMSC uniformly lined the fourth ventricle, and small foci were seen in the white-matter tracts adjacent to the dorsal horns of the fourth ventricle. The authors suggested that the BMSC gained access to the external granular layer and the reticular formation of the brainstem by following a pathway similar to that used by neural progenitors at the time when they emigrate out of the rhombic lip into the primordial external granular layer during embryogenesis.

In rare sections, occasional neurofilament-positive BMSC were found in the brainstem, suggesting that some BMSC differentiated into a neuronal phenotype *(41)*. An alternative explanation for these findings was not put forth by the authors. In grafting experiments, it is typical that many of the grafted cells do not survive. It is possible that the BrdU from the dead cells was released into a general metabolic pool and taken up by cells undergoing proliferation in the neonatal host brain such as the cerebellum and subependymal zone of the ventricles, accounting for the observed distribution of BrdU-labeled cells. Unless the proper control experiments are done with lysed BrdU-labeled BMSC and lysed bisbenzamide-labeled cells, these results remain unclear.

BONE MARROW AS A SOURCE OF NEURONS IN VITRO

Recent studies by several researchers have demonstrated that a subset of both human and murine bone marrow has the capacity to differentiate into neural cells and, in particular, to express neuronal markers in vitro *(33,42)*. BMSC were separated from whole bone marrow by adherence to polyethylene culture flasks. Proliferation of cells was maintained by use of epidermal growth factor (EGF). Prior to induction of differentiation, cultures were enriched in fibronectin-ir cells and depleted of mouse hematopoietic stem cell (Sca1) or human hematopoietic stem cells (CD34+). Treatment of the cultures with retinoic acid and brain derived neurotrophic factor (BDNF) resulted in a decreased number of fibronectin-ir cells. This was due to gradual loss of the large, flat fibronectin-ir cells and the appearance of smaller ovoid or spindle-shaped cells. Analysis of BMSC lysates, prepared from cultures treated with either proliferation medium or differentiation medium, demonstrated the presence of nestin, NeuN, and GFAP protein. Treatment with RA or RA + BDNF decreased the expression of nestin protein. Microscopic examination of the cultures, following immunocyto-chemical processing, revealed a small proportion of β-tubulin, NeuN-ir and GFAP-ir cells (0.5% and 1%, respectively, of the BMSC) (*see* Fig. 1). Coculture of the BMSCs with rat fetal brain cells increased the proportion of NeuN cells.

In an independent study by another group of researchers, adult rat stromal cells were expanded as undifferentiated cells in culture for more than 20 passages, indicating their proliferative capacity. A simple treatment protocol induced the stromal cells to exhibit a neuronal phenotype, expressing neuron-specific enolase, NeuN, neurofilament-M (NF-M), and tau *(42)*. With an optimal differentiation protocol (β-mercaptoethanol followed by dimethylsulfoxide and butylated hydroxyanisole), almost 80% of the cells

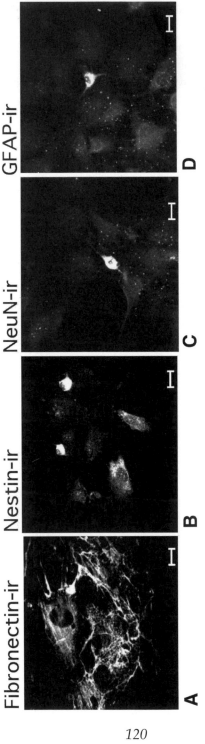

Fig. 1. Human BMSCs were cultured for 7 d in all-*trans*-retinoic acid (0.5 μ*M*) and BDNF (10 ng/mL). Fluorescence immunocytochemistry was used to detect the presence of (**A**) fibronectin, (**B**) nestin, (**C**) neuron-specific nuclear protein (NeuN) and (**D**) glial fibrillary acidic protein (GFAP). [Reproduced with permission from *Experimental Neurology*, **164**, 247–256 (2000).]

expressed NSE and NF-M. The refractile cell bodies extended long processes terminating in typical growth cones and filopodia. The differentiating cells expressed nestin, characteristic of neuronal precursor stem cells, at 5 h, but the trait was undetectable at 6 d. In contrast, expression of trkA, the nerve growth factor receptor, persisted from 5 h through 6 d. Clonal cell lines, established from single cells, proliferated, yielding both undifferentiated and neuronal cells. Human marrow stromal cells subjected to this protocol also differentiated into neurons *(42)*.

These results, along with those discussed above, suggest that adult BMSC have a potentially larger developmental repertoire than previously appreciated. It should not be surprising that BMSC can give rise to neural cells because the marker for neural precursors, nestin, was expressed in BMSC even in the absence of the differentiation factors retinoic acid and BDNF. Moreover, BMSC, as described in this review, have the capacity to develop into cell lineages beyond their normal fates.

BONE MARROW AS A SOURCE OF NEURONS IN VIVO

Adult bone marrow cells from male mice were transplanted into females of a mouse strain (PU.1 knockouts) that lacks macrophages, neutrophils, mast cells, osteoclasts, B-cells, and T-cells at birth *(43)*. These animals require a bone marrow transplant within 48 h of birth if they are to survive. Within 24 h after birth, PU.1 homozygous recipients were given intraperitoneal injections of unprocessed bone marrow cell suspensions (containing 10^7 cells) from wild-type male mice. Between 1 and 4 mo after transplantation, marrow-derived cells were present in the brains of all the transplanted mice examined. Between 2.3% and 4.6% of all cells (all identifiable nuclei) were Y chromosome positive (derived from donor bone marrow). The Y-chromosome-bearing cells were evenly distributed throughout different brain regions. The Y chromosome was present in 0.3–2.3% of the NeuN-immunoreactive nuclei. In the brains of transplanted female mice, all of the Y chromosome positive, NeuN-immunopositive nuclei also expressed neuron-specific enolase (NSE). These studies demonstrate that bone marrow cells administered systemically can migrate into the brain and differentiate into cells that express neuron-specific antigens.

In a similar study published back to back in the same journal as the above-described study *(43)*, another group of researchers found that bone marrow cells infused into irradiated mice migrated into the brain and differentiated into cells that expressed neuronal antigens. Adult marrow was harvested from transgenic mice that ubiquitously express enhanced green fluorescent protein (GFP). GFP-expressing (GFP[pos]) bone marrow was administered by the tail vein (6 million cells per recipient) into lethally irradiated, isogenic mice. Brains harvested several months after the transplant and examined by light microscopy revealed the presence of GFP[pos] cells throughout the brain, including the olfactory bulb, hippocampus, cortical areas, and cerebellum. Examination of dissociated brain and bone marrow cells from the recipients revealed that essentially all of the GFP[pos] cells that engrafted in the host bone marrow also expressed CD45 (surface marker of all nucleated mature blood lineages). However, a significant subset (up to 20%) of the GFP[pos] cells that engrafted in the brain lacked both CD45 and CD11b (surface marker expressed by all myelo-monocytic cells). These findings suggested that exposure to a brain microenvironment led a subpopulation of bone-marrow-derived cells to acquire novel phenotypes. Using confocal microscopy, the researchers determined that individual cells coexpressed GFP and neuron-specific antigens. The olfactory bulb was selected for in-depth quantification and revealed that 0.2–0.3% of the total number of neurons were derived from bone marrow by 8–12 wk after transplantation. A substantial proportion of marrow-derived cells coexpressed multiple neuron-specific gene products, including two neuronal proteins, 200-kDa neurofilament (NF-H) and class III β-tubulin, but did not express the glial cell marker GFAP.

Taken together, these in vitro and in vivo studies shed new light on the pluripotentiality of bone marrow stem cells and open doors to novel therapeutic applications for bone marrow transplants.

EXPERIMENTAL THERAPY OF STROKE AND SPINAL CORD INJURY WITH BONE MARROW CELLS

Recent work has demonstrated that bone marrow cells can hasten recovery of neurologic deficits in rodent models of stroke *(44,45)*. Adult mouse bone marrow, previously labeled in cell culture with BrdU, was transplanted into the striatum after embolic middle cerebral artery occlusion (MCAO). Twenty-eight days after stroke, many BrdU reactive cells survived and migrated a distance of approx 2.2 mm from the grafting areas toward the ischemic areas. BrdU reactive cells expressed the neuron-specific protein NeuN in 1% of BrdU-stained cells and the astrocytic-specific protein GFAP in 8% of the BrdU-stained cells. Functional recovery from a rotorod test and modified neurologic severity score tests (including motor, sensory, and reflex) were significantly improved in the mice receiving bone marrow nonhematopoietic cells compared with MCAO alone. The improved functional recovery of adult mice occurred even though infarct volumes did not change significantly *(44)*. The mechanism by which marrow cells improved neurologic function is not yet clear, but the authors suggest that the release of growth factors and cytokines by the transplanted marrow cells may play a role in the enhanced recovery. In a variation of this study, rat bone marrow labeled with BrdU was grafted into the ischemic border zone along with the growth factor BDNF. The combination of bone marrow cells and BDNF was shown to enhance differentiation of marrow-derived cells into neural cells and to improve recovery *(45)*.

Transplantation of BMSC into the spinal cord after a contusion injury was also reported to enhance recovery based on standardized assessments *(46)*. Rats were subjected to a weight-driven implant injury. BMSC or phosphate-buffered saline was injected into the spinal cord 1 wk after injury. Sections of tissue were analyzed by double-labeled immunohistochemistry for BMSC identification. Functional outcome measurements were performed weekly up to 5 wk postinjury. The data indicate significant improvement in functional outcome in animals treated with BMSC transplantation compared to control animals. Scattered cells derived from BMSCs expressed neural protein markers, but like the stroke studies, it is unlikely that reconstruction of injured spinal cord by differentiating marrow cells is responsible for the improvement. It is, the authors suggest, most likely mediated by humoral factors released by marrow cells that enhance endogenous processes of recovery *(46)*.

UMBILICAL CORD BLOOD AS A SOURCE OF NEURAL CELLS

Recently, human umbilical cord blood cells treated with retinoic acid (RA) and nerve growth factor (NGF) were shown, in a cell culture system, to exhibit a change in phenotype, to extend neuritic fibrillar processes, and to express molecular markers usually associated with neurons and glia *(47)*. Musashi 1 and β-tubulin III, proteins found in early neuronal development, were expressed in the induced cord blood cell cultures (*see* Figs. 2 and 3). Other molecules specifically linked to neurons such as glypican-4, pleiotrophin, and neuronal-growth-associated protein 43 (GAP-43) were detected using DNA microarray analysis and confirmed independently with reverse transcriptase–polymerase chain reaction (RT-PCR). GFAP and its mRNA were also detected in the induced cells (Fig. 4). Musashi 1 is an RNA-binding protein that has been found in the developing and/or adult CNS tissues of frogs, birds, rodents, and humans *(48)*. The anti-Musashi-1 monoclonal antibody has been shown to react with undifferentiated, proliferative cells of the subventricular zone in the CNS of all vertebrates tested. β-Tubulin III is one of the most specialized tubulins specific for neurons *(49)*. Both the upregulation and the posttranslational

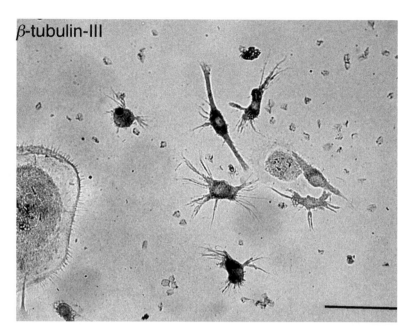

Fig. 2. Human umbilical cord blood (HUCB) mononuclear cells were cultured for 2 d in DMEM plus fetal calf serum and then 5 d in N5 medium supplemented with RA (0.5 µM) and NGF (50 ng/mL). Several cells in this panel exhibit β-tubulin-III immunoreactivity, a neurofilament found in early neuronal development.

processing of class-III β-tubulin are believed to be essential throughout neuronal differentiation (50,51). The cord blood cultures treated with RA+NGF also increased expression of other genes specific for neurons, including glypican-4, pentraxin II, and GAP43. Glypican-4 has been reported to be expressed in cells immunoreactive for nestin and the D1.1 antigen, other known markers of neural precursor cells, but it has not been detected in early postmitotic or fully differentiated neurons (52). Its presence was confirmed by independent RT-PCR analysis of mRNA from the cultures. Neuronal pentraxin II is a member of a new family of proteins identified through interaction with a presynaptic snake venom toxin taipoxin. Neuronal pentraxin-II may function during synapse formation and remodeling (53). Neuronal-growth-associated protein 43 is considered a specific neuronal marker but may also be expressed in developing myocytes (54). Other genes indicative of neurogenesis that were expressed following treatment included the neurofilament subunits NF-L and NF-M, MAP-2, the vesicular acetyl choline transporter, and neuronal DHP-sensitive, voltage-dependent, calcium-channel α-1D-subunit. Cord blood cells expressed mRNA for

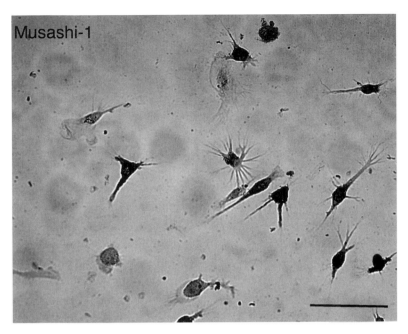

Fig. 3. Human umbilical cord blood mononuclear cells were cultured for 2 d in DMEM plus fetal calf serum and then 5 d in N5 medium supplemented with RA (0.5 μ*M*) and NGF (50 ng/mL). Several cells in this panel show Mushashi 1 immunoreactivity, an early developmental marker of neural cells.

neuronal specific enolase, but this protein is also expressed by many cells in bone marrow, especially megakaryocytes.

Cord blood transplantation for bone marrow replacement has been shown to have many advantages over bone marrow as the donor tissue. It is readily available, can be stored in cord blood banks for decades, and is less immunogenic than bone marrow cell. The size of the potential donor pool is much larger for cord blood than for bone marrow. There are 4 million births in the United States each year, each of which is a potential opportunity to collect cord blood. Cord blood is advantageous in terms of speed of matching the donor to the recipient. Identifying a suitable unrelated bone marrow donor is a time-consuming process that takes an average of 4 mo. In contrast, cord blood is readily available from a bank's freezer and has already undergone viral testing and tissue typing. An umbilical cord blood match can be made in as few as 3 or 4 d. Cord blood is virtually free of cytomegalovirus (CMV) that in the past has been responsible for 10% of deaths following bone marrow transplants. Given the above-presented findings regarding the capacity of bone marrow and umbilical cord to

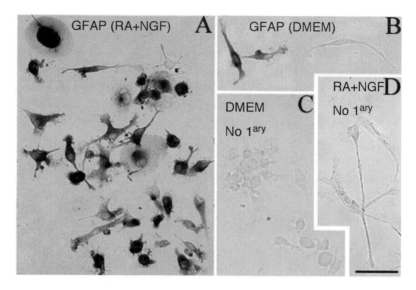

Fig. 4. Human umbilical cord blood mononuclear cell cultures: **(A)** NGF+RA-treated culture immunostained for GFAP; **(B)** DMEM-treated culture stained for GFAP; **(C)** DMEM-treated culture, primary antibody was deleted as a control; **(D)** NGF+RA-treated culture, primary antibody was deleted as a control.

differentiate into neural cells, umbilical cord blood has great therapeutic potential for neuronal replacement or gene delivery in neurodegenerative diseases, trauma, and genetic disorders.

FUTURE DIRECTIONS

A number of critical issues should guide ongoing and future research on the development of bone marrow and umbilical cord blood cells as a source of neurons. The true multipotent stem cell in the bone marrow has not yet been identified, primarily because stem cells are elusive and bear no known specific marker. It is unlikely that hematopoietic stem cells are the source of the neural precursors. The BMSC utilized by Sanchez-Ramos et al. were depleted of CD34+ cells, yet gave rise to cells that expressed several neural markers *(33)*. Lineage-negative marrow cells that express c-kit receptor have been shown to give rise to cardiac myocytes *(34)*, but this population of marrow cells has not yet been tested for neurogenic potential. Given the very small proportion of BMSC that were induced to express neuronal or glial markers in most of the studies reviewed, improved procedures need to be developed to enrich the neural precursor population.

Unassailable evidence that the neuronlike cells derived from bone marrow are true neurons is not yet available. The expression by BMSC of a more mature neuronal protein (neuron-specific nuclear protein or NeuN) is insufficient proof that BMSC become neurons. The expression of one or even several neuronal proteins does not prove that the cell bearing these "neuronal markers" is capable of all of the complex functions of a neuron. Re-establishment of synaptic contacts with host neurons and electrophysiological evidence of neuronal function needs to be gathered. Proof of the reversal of neurologic deficits following transplantation has been presented, but cynics will argue that measures of recovery following stroke are inadequate given the normal propensity of ischemic injuries in rodents to recover spontaneously.

Bone marrow stromal cells are amenable to genetic manipulations in vitro and can be a way to correct genetic defects involving neuronal tissues. For example, bone marrow transplantation has been demonstrated to correct the enzyme defect in neurons of the CNS in a lysosomal storage disease α-minidisks *(55)*. Recently, rat marrow cells were transduced to synthesize L-DOPA, demonstrating a potential application of these cells for gene transfer into the brain *(56)*.

In the context of the rapidly expanding field of stem cell biology, bone marrow and umbilical cord blood are invaluable resources. Understanding the molecular mechanisms responsible for neuronal differentiation of these cells will ultimately yield a readily available source of neural cells for cellular therapies ranging from gene therapeutics to neural reconstruction in neurodegenerative diseases, stroke, and trauma.

ACKNOWLEDGMENTS

This work was supported by Layton Biosciences, Helen E. Ellis Endowment, and the Parkinson's Disease Research Fund, and a USF and VA Merit Review Grant to JSR.

REFERENCES

1. Amos, T. A. S. and Gordon, M. Y. (1995) Sources for human hematopoietic stem cells for transplantation—a review. *Cell Transplant.* **4,** 547–569.
2. Gordon, M. Y. and Blackett, N. M. (1998) Reconstruction of the hematopoietic system after stem cell transplantation. *Cell Transplant.* **7,** 339–344.
3. Reynolds, B. A. and Weiss, S. (1992) Generation of neurons and astrocytes from isolated cells of the adult mammalian central nervous system. *Science* **255,** 1707–1709.
4. McKay, R. (1997) Stem cells in the nervous system. *Science* **276,** 66–71.

5. Scheffler, B., Horn, M., Blumcke, I., Laywell, E. E., Coomes, D., Kukekov, V. G., et al. (1999) Marrow-mindedness: a perspective on neuropoiesis. *Trends Neurosci.* **22,** 348–356.

6. Quesenberry, P. J., Hulspas, R., Joly, M., Benoit, B., Engstrom, C., Rielly, J., et al. (1999) Correlates between hematopoiesis and neuropoiesis: Neural stem cells. *J. Neurotrauma* **16,** 661–666.

7. Sanchez-Ramos, J., Song, S., Daadi, M., and Sanberg, P. R. (2000) The potential of bone marrow as a source of neural precursors. *Neurosci. News* **3,** 32–43.

8. Bondurant, M. C. and Koury, M. J. (1999) Origin and development of blood cells, in *Wintrobe's Clinical Hematology,* 10th ed. (Lee, G. R., Foerster, J., Luken, J., Paraskervas, F., Greer, J. P., and Rodgers, G. M., eds.), Lippincott, Williams and Wilkins, Philadelphia, pp. 147–149.

9. Charbord, P., Tavian, M., Coulombel, L., et al. (1995) Early ontogeny of the human hematopoietic system. *C.R. Seances Soc. Biol. Fil.* **189,** 617–627.

10. Tavian, M., Coulombel, L., Luton, D., et al. (1996) Aorta-associated CD34+ hematopoietic cells in the early human embryo. *Blood* **87,** 67–72.

11. Dzierzak, E. and Medvinsky, A. (1995) Mouse embryonic hematopoiesis. *Trends Genet.* **11,** 359–366.

12. Medvinsky, A. L., Gan, O. T., Semenova, M. L., and Samoylina, N.L. (1996) Development of day 8 colony-forming unit-spleen hematopoietic progenitors during early murine embryogenesis: spatial and temporal mapping. *Blood* **87,** 557–566.

13. Zon, L. I. (1995) Developmental biology of hematopoiesis. *Blood* **86,** 2876–2891.

14. Matsui, Y., Zsebo, K. M., and Hogan, B. L. (1990) Embryonic expression of a haematopoietic growth factor encoded by the SI locus and the ligand for c-kit. *Nature* **347,** 667–669.

15. Kunisada, T., Yoshida, H., Yamazaki, H., Miyamoto, A., Hemmi, H., Nishimura, E., et al. (1998) Transgene expression of steel factor in the basal layer of epidermis promotes survival, proliferation, differentiation and migration of melanocyte precursors. *Development* **125,** 2915–2923.

16. Steindler, D. A., Kukekov, V. G., Thomas, L. B., Fillmore, H., Suslov, O., Scheffler, B., et al. (1998) Boundary molecules during brain development, injury, and persistent neurogenesis—in vivo and in vitro studies. *Prog. Brain Res.* **117,** 179–196.

17. Yoder, M. C. and Williams, D. A. (1995) Matrix molecule interactions with hematopoietic stem cells. *Exp. Hematol.* **23,** 961–967.

18. Anstrom, K. K. and Tucker, R. P. (1996) Tenascin-C lines the migratory pathways of avian primordial germ cells and hematopoietic progenitor cells. *Dev. Dynam.* **206,** 437–446.

19. Gates, M. A., Fillmore, H., and Steindler, D. A. (1996) Chondroitin sulfate proteoglycan and tenascin in the wounded adult mouse neostriatum in vitro: dopamine neuron attachment and process outgrowth. *J. Neurosci.* **16,** 8005–8018.

20. Thomas, L. B., Gates, M. A., and Steindler, D. A. (1996) Young neurons from the adult subependymal zone proliferate and migrate along an astrocyte, extracellular matrix-rich pathway. *Glia* **17,** 1–14.
21. Kittler, E. L., McGrath, H., Temeles, D., Crittenden, R. B., Kister, V. K., and Quesenberry, P. J. (1992) Biologic significance of constitutive and subliminal growth factor production by bone marrow stroma. *Blood* **79,** 3168–3178.
22. Flanders, K. C., Ludecke, G., Engels, S., Cissel, D. S., Roberts, A. B., Kondaiah, P., et al. (1991) Localization and actions of transforming growth factor-βs in the embryonic nervous system. *Development* **113,** 183–191.
23. Mehler, M. F., Mabie, P. C., Zhang, D., and Kessler, J. A. (1997) Bone morphogenetic proteins in the nervous system. *Trends Neurosci.* **20,** 309–317.
24. Daadi, M., Arcellana-Panlilio, M. Y., and Weiss, S. (1998) Activin co-operates with fibroblast growth factor 2 to regulate tyrosine hydroxylase expression in the basal forebrain ventricular zone progenitors. *Neuroscience* **86,** 867–880.
25. Ling, Z. D., Potter, E. D., Lipton, J. W., and Carvey, P. M. (1998) Differentiation of mesencephalic progenitor cells into dopaminergic neurons by cytokines. *Exp. Neurol.* **149,** 411–423.
26. Konishi, Y., Chui, D.-H., Hirose, H., Kunishita, T., and Tabira, T. (1993) Trophic effects of erythropoietin and other hematopoietic factors on central cholinergic neurons in vitro and in vivo. *Brain Res.* **609,** 29–35.
27. Campana, W. M., Misasi, R., and O'Brien, J. S. (1998) Identification of a neurotrophic sequence in erythropoietin. *Int. J. Mol. Med.* **1,** 235–241.
28. Bjornson, C. R. R., Rietze, R. L., Reynolds, B. A., Magli, M. C., and Vescovi, A. L. (1999) Turning brain into blood: a hematopoietic fate adopted by adult neural stem cells in vivo. *Science* **283,** 534–537.
29. Prockop, D. J. (1997) Marrow stromal cells as stem cells for non-hematopoietic tissues. *Science* **276,** 71–74.
30. Ferrari, G., Cusella-DeAngelis, G., Coletta, M., Paolucci, E., Stornaiuolo, A., Cossu, G., et al. (1998) Muscle regeneration by bone marrow-derived myogenic precursors. *Science* **279,** 1528–1530.
31. Azizi, S. A., Stokes, D., Augelli, B. J., DiGirolamo, C., and Prockop, D. J. (1998) Engraftment and migration of human bone marrow stromal cells implanted in the brains of albino rats-similarities to astrocyte grafts. *Proc. Nat. Acad. Sci. USA* **95,** 3908–3913.
32. Petersen, B. E., Bowen, W. C., Patrene, K. D., Mars, W. M., Sullivan, A. K., Murase, N., et al. (1999) Bone marrow as a source of hepatic oval cells. *Science* **284,** 1168–1170.
33. Sanchez-Ramos, J., Song, S., Cardozo-Pelaez, F., Hazzi, C., Stedeford, T., Willing, A., et al. (2000) Adult bone marrow stromal cells differentiate into neural cells in vitro. *Exp. Neurol.* **164,** 247–256.
34. Orlic, D., Kajstura, J., Chimenti, S., Jakoniuk, I., Anderson, S. M., Li, B., et al. (2001) Bone marrow cells regenerate infarcted myocardium. *Nature* **410,** 701–705.

35. Wakitani, S., Saito, T., and Caplan, A. I. (1995) Myogenic cells derived from rat bone marrow mesenchymal stem cells exposed to 5-azacytidine. *Muscle Nerve* **18,** 1417–1426.

36. Ling, E. A. (1994) Monocytic origin of ramified microglia in the corpus callosum in postnatal rat. *Neuropathol. Appl. Neurobiol.* **20,** 182–183.

37. Ling, E. A. and Wong, W. C. (1993) The origin and nature of ramified and amoeboid microglia: a historical review and current concepts. *Glia* **7,** 9–18.

38. Neuhaus, J. and Fedoroff, S. (1994) Development of microglia in mouse neopallial cell cultures. *Glia* **11,** 11–17.

39. Eglitis, M. A. and Mezey, E. (1997) Hematopoietic cells differentiate into both microglia and macroglia in the brains of adult mice. *Proc. Natl. Acad. Sci. USA* **94,** 4080–4085.

40. Eglitis, M. A., Dawson, D., Park, K. W., and Mouradian, M. M. (1999) Targeting of marrow-derived astrocytes to the ischemic brain. *Neuroreport* **10,** 1289–1292.

41. Kopen, G. C., Prockop, D. J., and Phinney, D. G. (1999) Marrow stromal cells migrate throughout forebrain and cerebellum, and they differentiate into astrocytes after injection into neonatal mouse brains. *Proc. Natl. Acad. Sci. USA* **96,** 10,711–10,716.

42. Woodbury, D., Schwarz, E. J., Prockop, D. J., and Black, I. B. (2000) Adult rat and human bone marrow stromal cells differentiate into neurons. *J. Neurosci. Res.* **61,** 364–370.

43. Mezey, E., Chandross, K. J., Harta, G., Maki, R. A., and McKercher, S. R. (2000) Turning blood into brain: cells bearing neuronal antigens generated in vivo from bone marrow. *Science* **290,** 1779–1782.

44. Li, Y., Chopp, M., Chen, J., Wang, L., Gautam, S. C., Xu, Y. X., et al. (2000) Intrastriatal transplantation of bone marrow nonhematopoietic cells improves functional recovery after stroke in adult mice. *J. Cerebr. Blood Flow Metab.* **20,** 1311–1319.

45. Chen, J. L., Li, Y., and Chopp, M. (2000) Intracerebral transplantation of bone marrow with BDNF after MCAO in rat. *Neuropharmacology* **39,** 711–716.

46. Chopp, M., Zhang, X. H., Li, Y., Wang, L., Chen, J., Lu, D., et al. (2000) Spinal cord injury in rat: treatment with bone marrow stromal cell transplantation. *Neuroreport* **11,** 3001–3005.

47. Sanchez-Ramos, J., Song, S., Kamath, S. G., Zigova, T., Willing, A., Cardozo-Pelaez, F., et al. (2001) Expression of neural markers in human umbilical cord blood. *Exp. Neurol.* **171,** 109–115.

48. Kaneko, Y., Sakakibara, S., Imai, T., Suzuki, A., Nakamura, Y., Sawamoto, K., et al. (2000) Musashi1: an evolutionarily conserved marker for CNS progenitor cells including neural stem cells. *Dev. Neurosci.* **22,** 139–153.

49. Fanarraga, M. L., Avila, J., and Zabala, J. C. (1999) Expression of unphosphorylated class III beta-tubulin isotype in neuroepithelial cells demonstrates neuroblast commitment and differentiation. *Eur. J. Neurosci.* **11,** 517–527.

50. Laferriere, N. B. and Brown, D. L. (1996) Expression and posttranslational modification of class III beta-tubulin during neuronal differentiation of P19 embryonal carcinoma cells. *Cell Motil. Cytoskeleton* **35,** 188–199.

51. Laferriere, N. B., MacRae, T. H., and Brown, D. L. (1997) Tubulin synthesis and assembly in differentiating neurons. *Biochem. Cell Biol.* **75,** 103–117.
52. Hagihara, K., Watanabe, K., Chun, J., and Yamaguchi, Y. (2000) Glypican-4 is an FGF2-binding heparan sulfate proteoglycan expressed in neural precursor cells. *Dev. Dynam.* **219,** 353–367.
53. Kirkpatrick, L. L., Matzuk, M. M., Dodds, D. C., and Perin, M.S. (2000) Biochemical interactions of the neuronal pentraxins. Neuronal pentraxin (NP) receptor binds to taipoxin and taipoxin-associated calcium-binding protein 49 via NP1 and NP2. *J. Biol. Chem.* **275,** 17,786–17,792.
54. Moos, T. and Christensen, L. R. (1993) GAP43 identifies developing muscle cells in human embryos. *Neuroreport* **4,** 1299–1302.
55. Walkley, S. U., Thrall, M. A., Dobrenis, K., Huang, M., March, P. A., Siegel, D. A., et al. (1994) Bone marrow transplantation corrects the enzyme defect in neurons of the central nervous system in a lysosomal storage disease. *PNAS* **91,** 2970–2974.
56. Schwarz, E. J., Alexander, G. M., Prockop, D. J., and Azizi, S. A. (1999) Rat marrow stromal cells can be transduced to synthesize L-DOPA, *American Society for Neural Transplantation and Repair*, 1999, p. 48, No.B-02.

II

In Vitro and/or In Vivo Manipulations of Stem/Progenitor Cells for the CNS

6

Signal Transduction Pathways That Regulate Neural Stem Cell Division and Differentiation

Luciano Conti and Elena Cattaneo

INTRODUCTION

Following the identification and isolation of putative neural stem cells, a plethora of studies have addressed the potential of soluble growth factors to control their proliferation, maturation, and survival (1–4). Mitogenic growth factors, neurotrophins, and cytokines have been shown to play key roles in these events (5–8). Instead, much less investigated are the gene transcription events and signaling proteins activated during central nervous system (CNS) stem cell division and differentiation.

In recent years, it has become clear that intracellular signaling processes are critical mediators of the responses of CNS cells to growth factors. Analysis of the mechanisms of signal transduction has begun to reveal the way by which signaling molecules act to specify the responsiveness of the cell to its environment. This analysis has lead to the striking finding that a handful of conserved signaling pathways appear to be used in different combinations to specify a wide variety of tissues or cells. However, the pleiotropic nature of these pathways has prompted the question of how they can preserve the specificity of a signal in order to ultimately elicit distinct cell-specific responses.

This chapter will focus on the mechanisms by which signaling molecules control and influence the transition from proliferation to differentiation in the brain.

SIGNALING NETWORKS INVOLVING TYROSINE KINASES: GROWTH FACTORS ACTIVATING TYROSINE KINASE SIGNALING IN THE BRAIN

New data continuously accumulate indicating the pivotal role played by growth factors and cytokines in controlling the proliferation of CNS

From: *Neural Stem Cells for Brain and Spinal Cord Repair*
Edited by: T. Zigova, E. Y. Snyder, and P. R. Sanberg © Humana Press Inc., Totowa, NJ

stem cells and in the modulation of the survival and differentiation of their postmitotic descendants *(1,4)*.

As a consequence of this large body of evidence, it is now clear that a single factor is not endowed with a predetermined biological activity but, rather, that the final effect is dependent on the cellular context and on the degree of maturation of that given cell type, thus unveiling the complexity of the signal transduction events.

Platelet-derived growth factor (PDGF), for example, stimulates proliferation and prevents premature differentiation of glial progenitors of oligodendrocytes/type-2 astrocytes in rat optic nerve cultures *(9)*. However, the same factor also stimulates neuronal differentiation by uncommitted neuroepithelial cells in vivo *(10)* and in vitro *(11)*.

Fibroblast growth factor (FGF) and epidermal growth factor (EGF) molecules are potent mitogens for the immature brain cells and are also involved in differentiation and migration of neuronal cells *(12–18)* or gliogenesis *(19)*.

All of these diffusible factors exert their action by interaction with specific surface receptors and subsequent ignition of intracellular signaling pathways in which tyrosine kinase (TK) activity represents a key step. TKs catalyze the transfer of the phosphate of ATP to a tyrosine residue in a protein substrate. TK catalytic activity is a highly regulated process, as indicated by the numerous TKs that have been identified as oncogenes *(20,21)*.

Cell surface receptors are divided into two major classes: (1) receptors presenting intrinsic tyrosine kinase activity, also known as receptor tyrosine kinases (RTKs) *(4,22,23)* and exemplified by the receptors for EGF (EGFR), acidic and basic FGF (FGFR), platelet-derived growth factor (PDGFR), and Trk neurotrophin receptors; (2) receptors like the cytokine receptors that lack intrinsic catalytic kinase domains and recruit nonreceptor tyrosine kinases (NRTKs). Among these, there are the receptors for ciliary neurotrophic factor (CNTFR) and leukemia inhibitory factor (LIFR) *(24–28)*.

The Eph (ephrine) receptor tyrosine kinases are another class of RTKs (the largest) that are emerging as the molecules that guide migration of cells and growth cones during embryonic development *(29,30)*. A notable feature of Eph receptor signaling is that, upon receptor binding, responses may also be elicited in the ligand-expressing cells. In these cells, the ligands are anchored to the membrane via either a phospholipid (PI) anchor or a transmembrane region (TM) with a highly conserved cytoplasmic region and multiple potential sites for tyrosine phosphorylation *(31,32)*. Analyses of Nuk function, a murine Eph receptor that binds TM ligands, revealed that both receptor and ligands are associated with a tyrosine kinase and that their interaction mediates bidirectional cell signaling *(33)*. As a consequence, lack of *Nuk* in mice causes abnormal axonal connections *(34)*.

Notably, to complicate the receptors scene further, it has been demonstrated that the exclusive binding of a single growth factor to an individual receptor is more likely the exception than the rule. Indeed, the oligomeric nature of activated receptors allows the formation of receptor complexes composed of distinct, albeit closely related subunits, which can have different signaling potentials *(35)*.

A crucial point in the understanding of receptor tyrosine kinase signaling is whether different receptors activate different signal transduction pathways and whether there are distinct pathways for proliferation and differentiation in the brain.

Receptor Tyrosine Kinases

Receptor tyrosine kinases (RTKs) are transmembrane glycoproteins that are activated by the binding of their cognate ligands (for a review, see ref. *36*). The ligand–receptor interaction leads to the transduction of the extracellular signal to the cytoplasm by phosphorylating tyrosine residues on the receptors themselves (autophosphorylation) and on downstream signaling proteins. RTKs activate several signaling pathways within cells, leading to cell proliferation, differentiation, migration, or metabolic changes *(37)*. The RTK family includes the receptors for EGF and for many growth factors, such as, FGF, NGF, PDGF, and insulin.

Structure of Receptor Tyrosine Kinases

Receptor tyrosine kinases consist of an extracellular domain involved in the specific binding with polypeptide ligands, a hydrophobic transmembrane helix, and a cytoplasmic portion endowed with tyrosine kinase catalytic activity *(36)*. The typical RTK structure exhibits a single polypeptide chain. Exceptions include the insulin receptor and its family members, which display a heterotetrameric structure. Most polypeptide ligands for RTKs are soluble. Ephrins, the ligands for the Eph receptor family, represent an exception because either they span the cell membrane or they are anchored to the membrane through a glycosyl phosphatidylinositol (GPI) linkage *(38,39)*. The extracellular portion of RTKs typically contains a diverse array of discrete globular domains such as immunoglobulin (Ig)-like domains, EGF-like domains, and cysteine-rich domains. In contrast, the domain organization in the cytoplasmic portion of RTKs is simpler, consisting of a juxtamembrane region, followed by the tyrosine kinase catalytic domain and a carboxy terminal region. The juxtamembrane and carboxy terminal

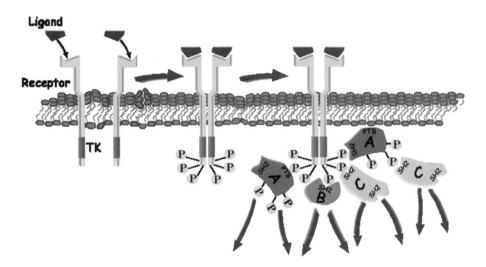

Fig. 1. Activation of RTK. Interaction of the RTK with the specific ligand induces enhancement of intrinsic tyrosine kinase (TK) activity with subsequent autophosphorylation on tyrosine residues and generation of binding sites for recruitment of downstream signaling proteins (A, B, C), which contain the phosphotyrosine-binding domain (SH2, PTB) (see the text for details).

regions vary in length among RTKs and contain regulatory tyrosine residues that are autophosphorylated upon ligand binding.

Activation of Receptor Tyrosine Kinases

Activation of RTKs is typically a three-step process that involves enhanced intrinsic catalytic activity, autophosphorylation on tyrosine residues (a consequence of ligand-mediated oligomerization) and creation of binding sites specifically recognized by downstream signaling proteins (Fig. 1). In general, autophosphorylation of tyrosines in the activation loop within the kinase domain results in stimulation of kinase activity. Autophosphorylation of tyrosines in the juxtamembrane, kinase insert, and carboxy-terminal regions generates docking sites for cytoplasmic molecules carrying phosphotyrosine-binding domains (SH2 and the PTB domains; see below) that recognize phosphotyrosine in specific sequence contexts. For example, the juxtamembrane region of TrkA contains a tyrosine in an NPXY motif that, upon autophosphorylation, behaves as an interaction site for the PTB domain of Shc *(40)*. Shc recruitment via this autophosphorylation site leads to its phosphorylation, binding of Grb2 and Sos to phosphorylated Shc, and Ras activation.

All RTKs thus far identified contain between one and three tyrosines in the kinase activation loop *(41)*. Phosphorylation of these tyrosines has been shown to be important for the regulation of catalytic activity and biological function from various RTKs, including FGF receptor *(42)*, PDGF receptor *(43)*, and TrkA *(44)*.

Ligand-induced oligomerization of RTKs is the mechanism by which tyrosine autophosphorylation is triggered *(45,46)*. Binding of RTKs to their specific extracellular ligands mediates the noncovalent oligomerization of monomeric receptors or may induce a structural rearrangement in heteromeric receptors, facilitating tyrosine autophosphorylation in the cytoplasmic domains. Ligand binding is thought to stabilize a dimeric configuration of the extracellular domains of RTKs with the cytoplasmic domains that transiently associate, acting as enzyme and substrate for each other.

Receptor autophosphorylation can occur in cis (within a receptor) or in trans (between receptors) *(36)*. In the first case, ligand-induced dimerization causes a conformational change in the receptor that facilitates cis autophosphorylation. In the second case, no conformational change needs to occur upon dimerization; a simple proximity effect would provide sufficient opportunity for trans autophosphorylation.

Downregulation of RTKs may occur via several processes, including receptor-mediated endocytosis *(47)*, ubiquitin-directed proteolysis *(48)*, and the action of protein tyrosine phosphatases (PTPs) *(49)*. Although it has proven somewhat difficult to determine whether specific PTPs are responsible for dephosphorylating specific RTKs in vivo, examples are beginning to emerge. Disruption of the gene-encoding PTP1B in mice results in a phenotype that indicates that at least one of the functions of PTP1B is to dephosphorylate the insulin receptor *(50)*.

Nonreceptor Tyrosine Kinases

In addition to the RTKs, there exists a large family of nonreceptor tyrosine kinases (NRTKs), which includes Src, the Janus kinases (JAKs), and Abl, among others. The NRTKs are integral components of the signaling cascades triggered by RTKs and by other cell surface receptors such as interleukines receptors and G-protein-coupled receptors.

The largest subfamily of NRTKs, with nine members, is the Src family *(51)*. Src family members participate in a variety of signaling processes, including mitogenesis and cytoskeleton restructuring. Multiple in vivo substrates have been described for Src and include, among others, the PDGF and EGF receptors and the NRTK focal adhesion kinase (Fak). Src has also been implicated in several human carcinomas, including breast, lung, and colon cancer *(52)*. Among the NRTKs, in the last few years, the JAK family has assumed a leading role for signal transduction events in the brain *(28)*, and for this reason, we will better describe their structure and activation modalities.

The JAK family of NRTKs are noncovalently associated with the cytoplasmic domain of cytokine receptors, such as receptors for interleukins, LIF, and CNTF *(53)*. Ligand–receptor interaction-induced receptor oligomerization leads one or more of the four known JAK kinases (Jak1, Jak2, Jak3 and Tyk2) to be recruited to the membrane *(54)*. Activated JAKs then phosphorylate the cytokine receptors with which they are associated, providing binding sites for the STAT family of transcription factors. Phosphorylation of STATs by JAKs leads to STATs dimerization, translocation to the nucleus, and transcription of specific genes.

Structure of JAK Kinases

Four members of the JAK kinases family have been discovered to date (for a review, see ref. *55*). Their discovery, along with several other members of the tyrosine kinase family, resulted in the usage of the acronym JAK (just another kinase). The JAK family of protein tyrosine kinases (PTKs) differs markedly from other classes of PTKs by the presence of an additional kinase domain. To denote this unique structural feature, these kinases were renamed "Janus kinases" in reference to an ancient two-faced Roman god of gates and doorways *(53–65)*. The most intriguing feature of these proteins is the presence of two domains (JH1 and JH2), with extensive homology to the tyrosine kinase domains *(55)*. A second interesting feature is the lack of any SH2 or SH3 domains. Instead, these proteins encode a group of well-conserved domains termed as JAK homology (JH1–JH7) domains that follow a nonconserved amino-terminus of about 30–50 amino acids. Of the dual kinase domains identified, only the JH1 domain appears to be functional. The JH2 domain, which harbors considerable homology to tyrosine kinase domains, lacks certain critical amino acids required for a functional kinase and does not appear to be associated with a kinase activity. Both the tyrosine kinase domain (JH1) and the pseudokinase domain (JH2) are housed at the carboxy-terminus of the protein. The other conserved regions (JH3–JH7) that are characteristic to members of the JAK family comprise approx 600 N-terminal amino acid residues. The precise functions of the JH3–JH7 domains as well as the pseudokinase domain (JH2) are currently under investigation. Considering the variety of interactions and functions performed by the JAK kinase family members, it seems plausible

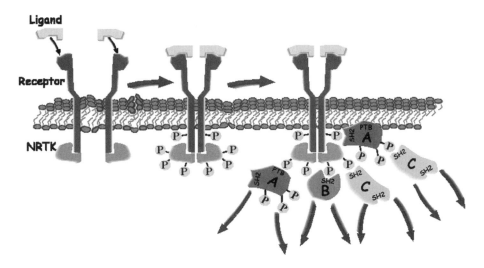

Fig. 2. Activation of NRTK. Ligand–receptor interaction induces receptor oligomerization leading to NRTKs recruitment to the membrane and enhancement of tyrosine kinase activity. Activated NRTKs then phosphorylate the receptors they are associated with, providing binding sites for the cytoplasmic effector/adaptor molecules as described in Fig. 1.

that these domains facilitate some key functions like protein–protein interactions, recruitment of substrates, and so forth.

Activation of JAK Kinases

Janus kinases are generally believed to be present in unstimulated cells in an inactive form. Ligand-induced receptor oligomerization, such as cytokine interaction with its specific receptor, serves as a trigger to signal JAKs activation (Fig. 2). Although the precise mechanism by which ligand binding results in the activation of JAKs is not known, various studies suggest a model to explain the mechanism of activation of JAK kinases. JAK kinases form complexes with native ligand-free receptors, in a catalytically inactive latent state. Receptor dimerization/oligomerization resulting from ligand binding results in the juxtapositioning of the JAKs, which are in the vicinity either through homodimeric or heterodimeric interactions. This recruitment of the JAK kinases appears to result in their phosphorylation either via autophosphorylation and/or cross-phosphorylation by other JAK kinases or other tyrosine kinase family members. This activation is presumed to result in an increased JAK kinase activity. The activated JAKs then phosphorylate receptors on target tyrosine sites. The phosphotyrosine sites on the receptors can then serve as docking sites that allow the binding of other SH2-domain-containing signaling molecules such as STATs, Src kinases, protein phosphatases, and other adaptor signaling proteins such as Shc, Grb2, and Cbl *(55)*.

Protein–Protein Interactions: Phospho-tyrosine Binding Domains

Over the last two decades, considerable understanding has been achieved of the mechanisms by which signals are conveyed from receptors at the plasma membrane to their targets in the cytoplasm and nucleus. However, only in the last few years molecular recognition processes have been demonstrated to represent a central issue in this scenario. In fact, specificity in signaling means specificity of interaction of receptors with particular targets and recognition of proteins of one pathway from related signaling components. Furthermore, crosstalk between distinct signaling pathways clearly represents the crux of the matter, because the entire cell must ultimately function as a single unit, so that different elements respond in an coordinated fashion to external cues.

As discussed above, mechanisms by which activated growth factor receptors stimulate cytoplasmic and nuclear events are largely based on phosphorylation cascades, which first occur at the cytoplasmic face of transmembrane receptors, and rapidly propagate onto specific signaling molecules inside the cell *(4,66)*. Tyrosine phosphorylation, in particular, comprises, by far, a large proportion of all the phosphorylation events occurring in a given cell, the remainder being serine or threonine phosphorylation. This simple substitution of a phosphate group for a hydroxyl group on tyrosine residues results in modification of the activity, life-span, or cellular localization of proteins. Protein dephosphorylation is of equal importance for the regulation of cellular behavior, and the functions of protein phosphatases are strictly controlled by protein–protein interactions like those for protein kinases.

Phosphotyrosil (pTyr) residues on proteins are, however, only one component of the scenario. These pTyr residues function as binding sites for intracellular signaling proteins containing domains of 100–180 amino acids (aa) known as the SH2 (from Src homology 2) or phosphotyrosil-binding (PTB) domains *(66)*.

In addition to domains involved in the recognition of pTyr, there are a series of modules that selectively recognize and interact with other targets. Among them, there are the SH3 domains that recognize specific proline-rich peptide motifs *(67,68)*, the EH domains that recognize Asn–Pro–Phe sequences, commonly found in polypeptides involved in protein trafficking

(69), and the PDZ domains, which bind short peptide motifs at the extreme carboxyl termini of proteins, typically transmembrane receptors *(70)*.

SH2 Domains: Structure and Specificity

The SH2 domains are protein modules of about 100 aa that show the ability to recognize phosphotyrosine-residue-containing peptides. The involvement of SH2-containing molecules in protein–protein interactions in signaling events from cell surface receptors was first characterized in the context of RTKs and it is currently well known that these domains are not restricted to particular types of signal transduction proteins *(71–75)*. In fact, they are found in protein kinases, phosphatases, transcription factors, as well as adaptor molecules with no enzymatic function, such as Shc or Grb2. Several pieces of evidence show that different SH2s bind to distinct phosphotyrosine-containing regions of the RTK. In fact, specificity of recognition is dictated by the fact that SH2 domains recognize the pTyr along with the three to six carboxy-terminal amino acids *(76–78)*. Therefore, although all SH2 domains require phosphorylation of the peptide ligand for high-affinity binding, they display the ability to bind preferentially to a specific phosphorylated motif *(75)*. Thus, in the case of activated RTKs, their ability to stimulate cytoplasmic signaling pathways is, to some extent, determined by the sequence contexts of their autophosphorylation sites, which, in turn, dictates which SH2-containing proteins will engage the autophosphorylated receptor.

The relevance of specific SH2-domain-mediated interactions to biological signaling pathways has also been tested by introducing mutations into SH2 docking sites on receptors. Examples are provided by the EGF receptor homolog in *Caenorhabditis elegans*, LET23, which is required for vulval differentiation, viability, and ovulation *(79)*, or by the Trk receptors *(80)*.

Currently, three classes of SH2 domains can be distinguished:

1. PLCγ-like SH2 domains, preferentially recognizing the pTyr-hydrophobic–X-hydrophobic motif. They bind phosphopeptides as an extended strand, with carboxy-terminal residues fitting into a hydrophobic cleft.
2. Src-like SH2 domains are similar, but have a flat binding surface that selects for charged residues at the +1 and +2 positions, whereas the side chain of the +3 residue fits into a hydrophobic pocket.
3. Grb2–SH2 domain, which, different from the Src-like class, has a bulky Trp side chain that blocks the progress of the phosphopeptide ligand, which is forced into a β-turn, best accommodated by a +2 Asn.

Interestingly, single amino acid substitution is sufficient to convert the specificity of one SH2 domain class to another one. For example, substitution of a Thr with a Trp can convert a Src-like SH2 domain to a Grb2-like specificity *(81)*. This apparent flexibility may have an evolutionary advantage, in the sense that SH2 domain-binding specificity might change rather rapidly, allowing the formation of new signaling connections with more complex regulation. In fact, despite the existence of such strong specificity, it should be kept in mind that the recognition is not absolute and that there may be more than one SH2-containing protein within a cell that has a high affinity for a particular ligand. Therefore, in vivo, the ability of an SH2 domain to engage a particular phosphoprotein may be critically dependent on the local concentration of proteins and on the modulating effect of other domains found on interacting proteins.

PTB Domains: Structure and Specificity

The PTB domain was first described in the protein Shc *(82,83)*, as a domain that could bind tyrosine-phosphorylated growth factor receptors. Since the discovery of the Shc PTB domain, many proteins have been identified that contain PTB domains. These proteins fall into two major groups. The first group contains PTB domains that have primary sequence similarity to the Shc PTB domain. Well-studied proteins belong to this class, such as Numb, Fe65, and mDab (mouse Disabled).

Other proteins belonging to this class are Jip-1 (Jnk-interacting protein 1, which scaffolds members of the Jun kinase signaling pathway), Capon (a protein that binds neuronal nitric oxide synthase), Ced-6 (a protein involved in the phagocytosis of apoptotic cells in *C. elegans*), and RGS-12 (a GTPase-activating protein for heterotrimeric G-proteins). The second group contains proteins such as insulin receptor substrate (IRS) proteins, which contain PTB domains with limited sequence similarity to the Shc PTB domain but similar binding characteristics.

Unlike the SH2 domain, which is crucial for many aspects of tyrosine kinase signaling, the PTB domain has a much more limited role in tyrosine kinase signaling. In fact, PTB domains specifically recognize pTyr-containing motifs, although in an entirely different way than SH2 domains *(84)*. In Shc, the PTB domain binds a CXNPXpY motif on receptor tyrosine kinases *(85)*, where C is a hydrophobic residue, X is any amino acid residue, N is asparagine, P is proline, and pY is phosphotyrosine. Interestingly, several studies indicate that the PTB domains are able to bind NPXY or related motifs but do not require phosphorylation for high-affinity binding *(86)*. This has led to the notion that PTB domains originally evolved to recognize nonphosphorylated peptide motifs and subsequently developed a capacity for pTyr binding in specific cases.

One of the interesting features of the structure of the Shc PTB domain is its remarkable similarity to the structures of the pleckstrin homology (PH) domains, despite the absence of amino-acid-sequence homology. PH domains have been shown to bind acidic phospholipids in vitro and have been implicated in playing a role in membrane localization of proteins (87). To this respect, several studies have suggested that the Shc PTB domain can interact with phosphotyrosine-containing residues and phospholipids and that both of these binding activities might be important for Shc function (88). However, the Shc PTB domain interacts with phospholipids and peptides at distinct sites on its surface. Studies have indicated that phospholipids can bind to other PTB domains, but there is also additional evidence that peptide binding is a crucial function of PTB domains.

Strikingly, PTB domains display a similar structural fold to the EVH1 domain involved in proline-rich motifs binding (89), as well as to a protein that binds the Ran GTPase (90). These observations suggest that the PTB may represent a versatile scaffold that has been exploited for several different protein and phospholipid recognition events.

Driving Intracellular Signaling Through Adaptor Molecules

Cellular responses largely lie on both molecular recognition and activation. A genetic test of this concept came from transgenic mice where the pTyr residues on the Met receptor, which normally bind a large number of SH2 molecules, were converted to Phe residues. This resulted in the phenotype that was a null mutation, despite the fact that the activity of the kinase domain is unaltered (91).

Many protein kinases have relatively broad substrate specificities and may be used in several combinations to achieve distinct biological responses. Thus, mechanisms must exist to organize the correct repertoires of enzymes into individual signaling pathways. The specific assembly and spatial localization of signaling proteins into biochemical pathways or networks represents one of these mechanisms. The classical example is from auto-phosphorylated receptor tyrosine kinases that engage specific interaction with cytoplasmic proteins containing specialized protein modules that mediate formation of signaling complexes. Cytoplasmic signaling adaptor proteins possess multiple protein–protein and protein–phospholipid interaction domains, covalently linked in various combinations. Furthermore, they can exhibit multiple sites for tyrosine phosphorylation and SH2 binding (92,93). The presence of these modular domains and phosphorylation sites serves multiple functions as indicated. They are important for the following:

1. *The assembly of complexes of signaling proteins around an activated cell surface receptor leading to activation of multiple pathways.* Separate domains, indeed, may interact with distinct partners, as observed for adaptors with SH2 and SH3 domains, such as Grb2 *(94).* In this case, SH2 domains bind specific phosphotyrosyl residues on Shc or on activated receptors, and SH3 domains bind to polyproline motifs on a separate set of target proteins. This permits simultaneous association of a single protein containing both SH2 and SH3 domains with two or more binding partners, thus generating complexes of signaling proteins around an activated cell surface receptor. On the other hand, two domains may interact with different sites on the same target, as commonly occurs with polypeptides that possess tandem SH2 domains, thereby increasing both the affinity and specificity of the interaction *(95).*

2. *The specific amplification of one pathway.* A simple way by which receptors may amplify their signaling is to use adaptor proteins that provide additional docking sites for modular signaling proteins. Once it is associated with the appropriate activated receptor, the adapter protein becomes phosphorylated at multiple sites that interact with specific SH2 domains of signaling proteins. Notably, the signaling properties of these adapter proteins likely depend on the sequences of their SH2-binding motifs. Phosphorylation of IRS1 by the insulin receptor creates multiple binding sites for PI3K, whereas Shc primarily engage Grb2 and thus, are principally involved in activating the mitogen-activated protein kinase pathway.

3. *The generation of signaling cores specifically containing the right players in the correct spatial sequence, allowing a coordinated signaling cascade.* The adapter proteins may lead to the juxtaposition of cytoplasmic proteins that act at successive stages of a pathway. An example is provided by the SH2-containing protein SLP76 that functions downstream of the T-cell antigen receptor, acting as a scaffold that juxtaposes members of a pathway involved in cytoskeletal reorganization *(96).*

DOWNSTREAM SIGNALING PATHWAYS CONTROLLING PROLIFERATION, SURVIVAL, AND DIFFERENTIATION OF NEURAL CELLS

The MAPK Family

Mitogen-activated protein (MAP) kinases, found in all eukaryotes, are common participants in signal transduction pathways from the membrane to the nucleus *(97,98).* In *Xenopus* eggs, MAPKs are involved in the activation of cyclin-dependent kinase cdc2 and the formation of mitotic microtubules during normal progression through mitosis *(99).* MAP kinases were first characterized as enzymatic activities phosphorylating MAP2 and the model substrate myelin basic protein in the late 1980s *(100,101).* They signal in a wide range of processes from the pheromone control of cell cycle arrest and mating in

Saccharomyces cerevisiae, in which these enzymes were first discovered, to proliferation and differentiation in metazoans *(102)*. These kinases are activated through multistep protein kinase cascades by dual phosphorylation on tyrosine and threonine residues (reviewed in ref. *103*). The mammalian MAP kinase family includes ERK (extracellular regulated kinases) 1 and 2, often referred to as p44 and p42 MAP kinases, 4 isoforms of p38 MAP kinase, 10 or more splice variants of the cJun N-terminal kinase/stress-activated protein kinases (JNK/SAPKs), and at least 3 forms of ERK3.

The first three mammalian MAP kinases, ERK1, 2, and 3, were cloned in the early 1990s and since that time, it has become clear that ERK1 and ERK2 are among the protein kinases most commonly activated in signal transduction pathways *(104,105)*. They have been particularly linked to cell growth *(106)*, but have important roles in many other events including long-term potentiation in neurons *(107–109)*. The three ERK3 isoforms known are most similar in sequence to ERK2. No ERK3 substrates are known. Unlike ERK2, which translocates to the nucleus following activation, at least one form of ERK3 is constitutively localized to the nucleus, despite the lack of an obvious localization sequence *(110)*.

A second subgroup of MAP kinases, the JNK/SAPKs, was discovered and characterized for its activity to bind to and phosphorylate the N-terminal sites of c-Jun following exposure of cells to ultraviolet (UV) light *(111)*. cDNA cloning revealed three genes that encode the 46- and 54-kDa isoforms of JNK/SAPK. These enzymes are activated by antibiotics, cytokines, and other environmental stresses and, to a lesser degree, by growth factors *(112)*. They have been found associated with both cell apoptosis and survival *(113)*. JNK knockout mice indicate that JNK1 and JNK2 are required for apoptosis in distinct regions of the brain *(114)*. To this regard, it has been shown that nerve growth factor (NGF) withdrawal from cultured sympathetic neurons triggered cell death via activation of the JNK and phosphorylation of the c-Jun transcription factor. On the other hand, in the adult nervous system, c-Jun is also involved in neuroprotection and regeneration. It has thus been proposed that expression of the c-Jun transcription factor is an early and consistent component in the effective response of neurons to various forms of injury, and it functions as a prerequisite for other transcriptional components to exert their specific decision about death, survival, or regeneration *(115)*.

The p38 subgroup was discovered as a lipopolysaccharide (LPS)-induced tyrosine phosphoprotein *(116)*. Four p38 like MAP kinases are now known *(117)*. Like the JNK/SAPKs, they are often activated by cellular stresses and, as a result, are also referred to as stress-activated protein kinases *(118)*.

The MAPK Signaling Cascade

Mitogen-activating protein kinase signaling cascades contain at least three protein kinases that work in series *(103)*. In the case of ERK1/2, the first of the three kinases is a Raf isoform, most commonly Raf-1. Activation of Raf-1 usually requires Ras, although other mechanisms have been proposed *(119)*. Raf is phosphorylated on several residues and the accumulated phosphorylations lead to multiple activity states *(120,121)*. Once activated, Raf-1 phosphorylates two residues, either serine or threonine, to activate the protein kinases that directly activate ERKs. These kinases known as MAPK/ERK kinases (MEKs), or MAP kinase kinases, are a family of dual-specificity protein kinases that phosphorylate two residues, a tyrosine and a threonine, to activate their MAP kinase targets *(103)*. Although they are dual-specificity kinases, MEKs are highly selective for their MAP kinase targets. MEK1 and MEK2 phosphorylate and activate only ERK1 and ERK2 and no other known MAP kinase family members *(103,122)*. Activation requires both phosphorylations; dual phosphorylation, in the case of most MAP kinases, is an on–off switch, increasing the activity of ERK2, for example, by over 1000-fold *(123)*. Notably, Raf isoforms are not abundant enzymes, whereas both ERK1/2 and MEK1/2 are much more abundant and are roughly equimolar in many cell types. Thus, only the first step of this cascade amplifies the signal to ERKs.

This three-kinase cascade has been successfully reconstituted in vitro, with complete activation of ERK1/2. Nevertheless, additional components are necessary for the efficient activation of this cascade induced by growth factors stimulation.

The classical, and first elucidated, mechanism for activation of ERK1/2 cascade was the one by ligands that utilized RTK signaling, such as the EGF receptor *(4)* (Fig. 3). Induction of RTK activity by the appropriate ligand causes enhanced tyrosine kinase autophosphorylation, creating docking sites for signal transducers and adapters. The Shc (from Src homolog and collagen homolog) adapter molecule binds to the activated RPTK and becomes phosphorylated in at least three sites *(4)*. Phosphotyrosines on Shc then function as docking sites for the Grb2 adaptor, which is bound constitutively to a proline-rich region of the protein named Son of Sevenless (Sos) *(124)*. Sos catalyzes GDP release and GTP binding to Ras leading to its activation. The GTP-bound form of Ras binds to its effectors, notably Raf-1 *(119)*, bringing it to the plasma membrane. At this location its protein kinase activity is increased by one or several mechanisms including phosphorylation, thereby activating the first step of the MAPK cascade.

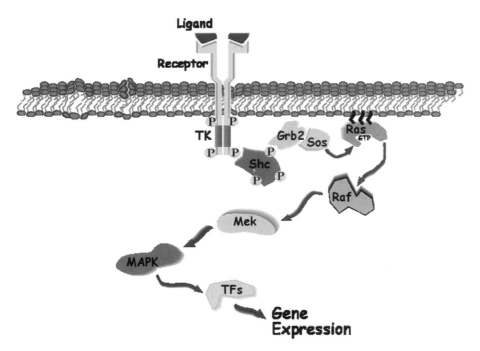

Fig. 3. The MAPK pathway. Activated RTKs transduce via adaptors like Shc and Grb2. The latter is constitutively associated with Sos, an exchanger of GDP/GTP, which activates Ras. Subsequent steps involve three kinases levels (Raf1, MEK, and, finally, the MAPKs). MAPKs ultimately translocate to the nucleus where they phosphorylate transcription factors (TFs) to generate specific gene transcription responses (see the text for details).

Mechanisms of MAPK pathway activation different from the classical RTK have been described. Ligands that bind to G-protein-coupled receptors have been shown to stimulate MAP kinases by multiple mechanisms. This mechanistic diversity is the result, in part, of receptor coupling and functional differences among the classes of G-proteins. Like receptor tyrosine kinases, all G-proteins activate ERK1/2 through the kinase module; however, different mechanisms may be involved in communicating to Raf by each class. Agents that activate G_s-coupled receptor, as the A2A-adenosine receptor, may activate or inhibit ERK1/2 by increasing the activity of cAMP-dependent protein kinase (PKA) *(125,126)*. To this regard, Rap1, a PKA substrate, has been implicated in positive regulation of ERK1/2 by cAMP. Stork and colleagues proposed that PKA activates Rap1, which, although it negatively regulates Raf-1, positively regulates another Raf isoform, B-Raf *(127)*.

Notably, B-Raf has a more restricted distribution than the ubiquitously expressed Raf-1 and is a major Raf isoform in brain *(128)*. Stimulation of ERK1/2 by βγ subunits may also occur through the Src family tyrosine kinase *(129)*. Src phosphorylates a membrane-associated scaffolding protein, either another nonreceptor tyrosine kinase such as PYK2 or a receptor tyrosine kinase such as the EGF receptor, to recruit Shc–Grb2–Sos complexes to the membrane *(130)*. Sos then stimulates the exchange of GDP for GTP in Ras, which, in turn, activates Raf-1. This mechanism has several steps in common with that utilized by receptor tyrosine kinases. Notably, ligands for G_i-coupled receptors can stimulate the formation of complexes of Shc–Grb2–Sos with the EGF receptor *(131)*.

The second-messenger calcium has also been proved to induce the activation of ERK1/2. For example, increased intracellular free Ca^{2+} following depolarization in PC12 cells has been shown to induce activation of ERK1/2 in a PKC-independent way *(132)*. Two calcium-activated Ras guanine nucleotide exchange factors directly coupling the calcium signal to Ras without utilizing tyrosine phosphorylation pathways have been described *(133,134)*. The expression of these exchangers is more restricted than Sos, suggesting that this direct mechanism may be restricted to neuronal tissue. Furthermore, in several cell types, an essential role for calcium-calmodulin-dependent protein kinase (CaMK) II has been shown for ERK1/calcium-mediated activation *(135)*. CaMKII overexpression leads to tyrosine phosphorylation and has been proposed to activate ERK1/2 by a tyrosine kinase-dependent mechanism, perhaps using the EGF receptor. Notably, the tyrosine kinases Src and PYK2 have been implicated in calcium-dependent activation of ERKs *(136)*.

Proliferation or Differentiation: Ras-MAPK-Driven Events

Several studies have indicated the MAPK to control a plethora of functions in the brain *(108)*. MAPK activation has been described in the proliferative response of CNS stem cells to factors such as EGF and bFGF. In a recent study by Learish et al., it has been demonstrated that inhibition of EGF/bFGF-induced MAPK activation in CNS stem cells cultures coincided with a downregulation in the proliferation of these cells and in a decrease of the number of nestin-positive cells *(137)*.

Mitogen-activated protein kinase activation has been observed also in response to neurotrophins, factors involved in several responses, including promotion of neuronal differentiation, stimulation of nerve outgrowth and spouting, modulation of synaptic activity, and gene regulation *(138)*. Analyses of growth-factors-activated MAPK signaling in PC12 chromaffin neuronal cell line indicated that EGF promotes tyrosine phosphorylation of the receptor and of Shc adaptors, leading to activation of the Ras/Raf/ MEK1/MAPK cascade and to proliferative activity by the cells. In an apparently contrasting manner, the same signaling pathway was found to be activated by NGF, a factor that is known to induce neurite outgrowth in PC12 cells *(139)*. In these cells, blocking the activity of MAPK kinase (MEK1) by using the synthetic inhibitor PD98059 disrupts the mitogenic response to growth factors by preventing the induction of cell cycle regulatory genes such as c-*myc* and cyclin D1 *(140)*. Conversely, inhibition of MEK1 has been shown to block MAPK activation and neurite outgrowth in NGF treated PC12 cultures *(141)*.

This controversy about the different biological effects elicited by activation of the same pathway began to be solved with the discovery of quantitative differences in the way the Ras–MAPK pathway is activated by EGF or NGF. It is now well established that EGF activates MAPK transiently and promotes proliferation while NGF produces sustained MAPK activation, leading to differentiation and neurite outgrowth in PC12 cells *(106)*. To this regard, overexpression of a constitutively active isoform of MAPK kinase (MEK1) resulted in increased phosphorylation of MAPK and morphologic differentiation *(142)*.

On the basis of these data, a model has been proposed that accounts for the different consequences of transient versus sustained activation of the ERKs *(106)*. This model considers the subcellular localization of the MAPK. Indeed, location of an enzyme in the cell has been demostrated to be determinant for its activities. In many cells, although the underlying processes are still elusive, it has been proved that MAPK stimulation causes their translocation to the nucleus. This nuclear localization is essential for certain functions such as morphological transformation of fibroblasts and differentiation of PC12 cells *(143)*. The duration of nuclear retention caused by different ligands may thus indicate distinctions in functional consequences. According to this view, sustained MAPK activation leads to their persistent nuclear accumulation, resulting in phosphorylation of transcription factors and changes in gene expression, which causes differentiation *(98)*.

Several mechanisms have been hypothesized in order to explain the differences in the duration of MAPK activation. One of these considers the number of receptors expressed on the cell membrane. Indeed, overexpression of the EGFR in PC12 cells overrides the limited ERK activation by EGF and produced differentiative effects. Similarly, NGF elicites cell proliferation in PC12 subclones that only present a limited number of NGFR. Thus, although a high EGFR number leads to proliferation of PC12 cells, activation of fewer EGFRs is associated with cell differentiation. Burrows and collaborators *(19)* have extended to the developing brain the finding that different receptor levels may change cellular responsiveness. The authors introduced extra EGFRs into early cortical progenitor cells of the ventricular zone and found that they could modulate responsiveness of these cells to EGF *(19)*. These data indicate, therefore, that the ability of a cell to proliferate or differentiate in response to a given factor may be dictated by the number of receptors expressed on its membrane, this being translated into different degrees of intracellular tyrosine phosphorylation and, ultimately, of ERK activation.

More recent work on NGF-triggered neurite outgrowth in PC12 cells has demonstrated that the differentiative effects of NGF rely on two phases of ERK activation, one that is Ras-dependent and a second that is dependent on Rap1 and recruits B-Raf *(144)*.

Nevertheless, the ability of different receptors to recruit multiple pathways leading to MAPK activation must be considered. For example, TrkA exhibits different tyrosine residues that are phosphorylated following NGF binding and may activate the ERKs. It has been found that mutations within TrkA that abrogate its interaction with Shc have very little effect on the ability of NGF to reduce MAPK activation and neurite outgrowth in PC12 cells. In contrast, mutant TrkAs that do not bind Shc and phospholipase Cγ (PLCγ) specifically failed to elicit MAPK activation and neurite outgrowth *(40,145)*.

Importantly, it has also been shown that the activity and availability of signaling proteins can change during brain maturation. For example, in cultured cortical neurons, the degree of activation of MAPK in response to a single factor varies as a function of maturation of the donor tissue *(146)*. This may be explained by qualitative or quantitative changes at the receptor level *(19,147)*, but may also imply a dynamic regulation of the expression and activity of upstream signaling molecules that occurs as cells mature and that, in turn, influences the activity of downstream components of the cascade. Among these, dramatic changes in Shc(s) adaptor proteins have been described at the transition from proliferation to differentiation in the brain *(4,148,149)*.

PI3K-AKT Signaling: Survival Players for Neural Stem Cells and Neurons

The long-established ability of growth factors to promote the survival of peripheral and central nervous system neurons both in vitro and in vivo (e.g., *see* refs. *150–152*) has held interest in understanding the intracellular path-

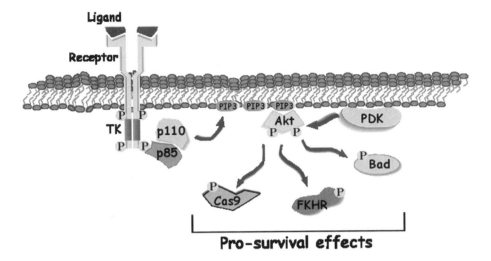

Fig. 4. The PI3K signaling cascade. Following receptor activation, the dimeric enzyme PI3K (consisting of the p85 and the p110 subunits) catalyzes the production of 3′-phosphorylated phosphoinositides (PIP$_3$) resulting in Akt translocation to the plasma membrane and activation by PDK. Activated Akt promotes cell survival mainly via three mechanisms: (1) inhibition of caspase 9 activation; (2) phosphorylation of pro-apoptotic Bad; (3) phosphorylation of FKH transcription factor (see the text for details).

ways that mediate these responses. Studies using small phosphatidylinositol 3-kinase (PI3K) inhibitor molecules (such as wortmannin or LY294098) revealed that the ability of trophic factors to promote survival is driven through the PI3K pathway activation. PI3K was first implicated in the suppression of apoptosis in a study by Yao and Cooper *(153)* that demonstrated that inhibition of PI3K activity abrogated the ability of NGF to promote cell survival in PC12 cells. In addition, transfection experiments using PDGF receptor mutants indicated that the PDGFR tyrosine residues that bind to and activate PI3K were both necessary and sufficient for the PDGF mediated survival of these cells.

PI3K Activation and Substrates

Phosphatidylinositol 3-kinases are SH2-containing enzymes associated with a variety of receptor and nonreceptor TKs *(154)*. The enzymes are heterodimers (constituted by a p110 catalytic and a p85 regulatory subunits) that phosphorylate the 3′ position on a variety of inositol lipids and serines, on protein substrates (Fig. 4). PI3Ks can be activated following the binding of the p85 subunit (via SH2 domains) to tyrosine phosphorylated motifs present in a variety of activated growth factor receptors, such as the PDGF receptor, the colony-stimulating factor receptor, the insulin receptor, and the NGF receptor *(73,155,156)*.

Activation of these receptors results in the recruitment of PI3K isoforms to the inner surface of the plasma membrane as a result of ligand-regulated protein–protein interactions (for review, *see* refs. *157* and *158*). PI3K can also be activated by direct interaction with the Ras protooncogene *(20,159–163)*. Once localized to the plasma membrane, PI3Ks generate 3′ phosphorylated phosphoinositides that act as signaling intermediates regulating downstream signal transduction cascades. Additional evidence of a role for the products of PI3K in survival signaling comes from studies of the lipid phosphatase PTEN (phosphatase and tensin homolog deleted from chromosome 10) suppressor. Overexpression of PTEN is sufficient to lower basal 3′-phosphorylated phosphoinositide levels in cells, thus inhibiting PI3K action. PTEN knockout (KO) mice die during embryogenesis as a result of a failure in developmental apoptosis *(164,165)* as a result of an elevation in the basal level of 3′-phosphorylated phosphatidylinositides.

PI3K Downstream Targets in Cell Survival

The serine/threonine kinase Akt (PKB) has been shown to be the immediate downstream effector of PI3K in the survival cascade, and in this context, it is has been shown that PI3K activity is necessary and sufficient for growth-factor-dependent activation of Akt *(166,167)*.

Protein kinase B (PKB)/Akt was discovered by three groups in 1991 *(168–170)*, and in the intervening years, this serine/threonine kinase has been increasingly linked to enhanced cell survival. At least three isoforms of PKB are encoded at distinct genetic loci in mammals *(171)*. PKB is activated downstream of PI3K in signaling cascades effected by a wide range of receptors, including those for growth and survival factors *(172)*. Experiments using constitutively active forms of PI3K and Akt support the view that the activation of this cascade can fully account for growth factor survival responses *(173,174)*. The activation of PI3K results in the production of 3′-phosphorylated phosphatidylinositides that bind to the pleckstrin homology domain of PKB and induce translocation to the plasma membrane, where it is phosphorylated on Thr and Ser by PDK (3-phosphoinositide-dependent protein kinase) I and II *(175,176)* (Fig. 4). Akt then influences cell survival by phosphorylation of a variety of substrates that affect apoptosis directly.

Downstream actions of Akt appear to include phosphorylation of proteins involved in the apoptotic cascade and regulation of the expression of apoptotic proteins. Akt can, indeed, phosphorylate the proapoptotic bcl-2 family member Bad at ser136 causing its binding to 14-3-3 family proteins and blocking its proapoptotic function *(177,178)*. Akt can also inhibit apoptosis by directly regulating caspase activation. Indeed, Akt can phosphorylate caspase-9 at Ser196, leading to a reduction in the protease activity of this caspase in vitro *(179)*. Thus, Akt activation may influence the mitochondrial phase of the apoptotic cascade at two distinct stages: (1) the phosphorylation of Bad will lead to reduced release of cytochrome-*c* and indirect inhibition of caspase 9 by prevention of the formation of the apoptosome complex and (2) activation of caspase 9 will also be influenced directly by Akt-mediated phosphorylation *(172)* (Fig. 4). More recently, studies have shown that Akt was able to prevent the loss of mitochondrial membrane integrity and release of cytochrome-*c* independently of Bad phosphorylation, in a Raf-1-dependent way *(180)*.

Furtheremore, in the presence of survival factors, Akt has been shown to phosphorylate (at Thr32 and Ser253) FKHRL1, a member of the forkhead family of transcription factors (Fig. 4). As a result, phosphorylated FKHRL1 is sequestered in the cytoplasm in association with 14-3-3 proteins, thus remaining apart from its nuclear transcriptional targets. In the absence of survival factors, the PI3K/Akt pathway is inactivated and FKHRL1 is unphosphorylated at its Akt sites and accumulates in the nucleus, where it can activate death genes like the Fas ligand *(172,181)*.

Finally, Akt has also been shown to phosphorylate and activate IKK-α *(182)*, the kinase that regulates I-kB and leads to nuclear translocation and activation of the transcription factor NF-κB. NF-κB regulates the transcription of a variety of survival factors, including proinflammatory cytokines *(183)*.

Clearly, Akt is able to act at several points along the apoptotic cascade, and this is unlikely to represent only functional redundancy. Clearly, Akt is a key mediator of cell survival and it is therefore not surprising that it is cleaved and inactivated by caspase 3 early in the apoptotic program *(184)*.

JAK/STAT Signaling: The Glial Pathway

Cytokine receptors distinguish themselves for the absence of the tyrosine kinase domain in their cytoplasmic tail *(24)*. As previously described, this class of receptors uses the nonreceptor tyrosine kinases JAKs to transduce

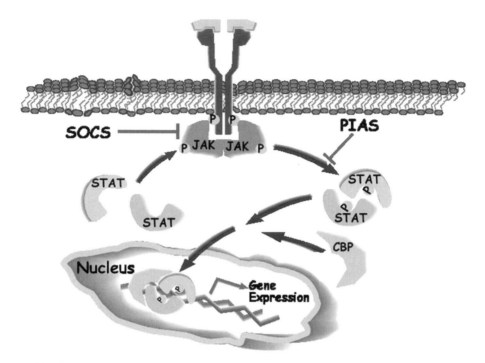

Fig. 5. The JAK/STAT signaling molecules and their inhibitors. The JAK/STAT transduce from activated cytokine receptors, which do not possess an intrinsic tyrosine kinase domain and therefore recruit the cytoplasmic JAKs. STATs are found in the cytoplasm in monomeric form and, following tyrosine phosphorylation by the activated JAKs, they associate with the receptor, homodimerize or heterodimerize (or associate with STAT-interacting proteins, such as CBP; see the text for details), translocate to the nucleus, and bind to specific DNA elements (STAT-recognition sites) situated upstream of genes transcribed by the specific ligand. Points of action of inhibitors of the JAK/STAT pathway (Socs and PIAs) are indicated (see the text for details).

its signal. The activated JAKs phosphorylate a family of transcription factors named STATs (from signal transducer and activator of transcription), that then dimerize and translocate into the nucleus where they activate the transcription of genes carrying STATs binding sites in their promoters (Fig. 5). Seven mammalian members of the STAT family are known to date, namely Stat1, Stat2, Stat3, Stat4, Stat5a, and Stat5b, and Stat6 (reviewed in ref. *28*).

The STATs share several highly homologous domains. To form dimers, the SH2 domain of one STAT monomer binds to the opposing phosphotyrosine; this tyrosine, near the C-terminus, is highly conserved and is phosphorylated in response to ligands. The linker domain, N-terminal to the SH2 domain, has a less well-defined function. However, mutation of two residues within this domain of STAT1 abolishes transcriptional responses to interferon (IFN)-γ but not to IFN-α. The DNA-binding domain is closer to the N-terminus, and the N-terminus itself is involved in many protein–protein interactions and also in the nuclear translocation of the STATs.

Regulation of this pathway is allowed by specific molecular inhibitors of JAKs recently discovered (Fig. 5). These are molecules that block the interaction between activated intracellular tyrosine kinases and the STATs. One, identified with various names, is a JAK-binding protein that interacts with the JH1 domain of all of the four JAKs, thereby reducing their tyrosine kinase activity *(185–187)*. Transcription of these inhibitory molecules is positively modulated by cytokines, suggesting that they may function in a negative-feedback loop to regulate JAK/STAT signaling. Specific inhibitors of the STATs have also been identified *(188–190)*.

Despite their initial identification as cytokine signaling proteins, the JAKs and STATs also participate in signaling from activated RTKs *(191–193)*. Among them are receptors for molecules with well known effects on developing and mature glia and neurons, like bFGF, PDGF, and EGF. Notably, it has been also found that the functions of the STATs may also be influenced by serine phosphorylation, which is necessary to achieve maximal STAT activation and DNA-binding activity *(194–196)*. Interestingly, the sites of serine phosphorylation present in the Stat1 and Stat3 molecules matches with the known MAP kinase recognition sequence *(197)*, indicating that there may be crosstalk between the STATs and the MAPK pathway at this phosphorylation site *(195)*. Furthermore, cytokines may also signal through the Ras pathway *(198–200)*. Although the role of Ras in cytokine signaling is not fully understood, it may be assumed that the pleiotropic effects of these ligands require recruitment of more than one pathway.

Despite their original characterization as "hematopoietic molecules," it is now well established that this signal transduction pathway may serve multiple functions in different tissues. Interestingly, it has been demonstrated that Stat3 activation is necessary for the maintenance of pluripotency and the proliferation of embryonic stem cells *(201)*.

The findings that the cytokine receptors and JAK/STAT are widely distributed in the developing and mature brain *(202)* indicate that there may be a wide range of activities in the brain in which these signaling proteins are involved *(28)*. Analyses of JAK/STAT presence and function in CNS cells have revealed that they are present at various stages of brain development. Some of these transcription factors and their upstream kinases are expressed with a certain degree of spatiotemporal specificity *(202)*. For example, whereas Stat3 levels were remarkably constant at different embryonic and postnatal stages in various brain regions in vivo, levels of Stat6 were progressively decreased.

Among the different STATs, Stat1 and Stat3 have been extensively investigated in CNS cells, because they are involved in transducing survival and differentiative signals from activated CNTFR (reviewed in ref. *28; see* ref. *200*), a cytokine that elicits important roles in the brain. The biological effects of this cytokine are mediated by CNTFRα on responsive cells, which, once bound to CNTF, triggers sequential heterodimerization of the two signal transducing subunits of the CNTFR–LIF receptor β (LIFRβ) and gp130 (130-kDa glicoprotein). More recently, a role for Stat3 and Stat1 has been identified during gliogenesis, whereby differentiation of cerebral cortical precursor cells into GFAP-expressing cells following CNTF stimulation was found to require Stat3 and Stat1 signaling. Transfection of these cells with dominant-negative variants of these transcription factors abolished the expression of GFAP *(199)*. The promoter for GFAP is known to contain seven consensus sequences for activated STATs *(203)*. Deletion of these sequences prevented GFAP expression by the activated STATs *(199)*. These data indicate, therefore, that GFAP expression is under the control of phosphorylated STATs and CNTF and that the JAK/STAT signaling pathway critically controls cell fate during mammalian brain development.

In addition to the JAK/STAT, CNTF was also shown to influence the MAPK, although the role of MAPK activation by CNTF is less clear. One study showed that transfection of a dominant interfering form of MAPK kinase (MEK) augmented CNTF induction of the GFAP promoter *(199)*. In another study, however, activation of both MAPKs and JAK/STAT pathways appeared to be positively coupled to astrocytic differentiation in vitro *(200)*. It is possible that differences in the culture conditions and in CNTF doses used in the two studies may have accounted for the opposite results observed. Recent data do indeed indicate that different CNTF doses may elicit or inhibit GFAP expression, depending on quantitative differences in recruitment of members of these two pathways *(204)*.

More recently, cytokine-induced STAT activation was found to act in synergy with BMP2 (bone morphogenetic protein-2) signals (which involves the Smads transcription factors) to promote astrocytes differentiation from fetal neuronal progenitor cells. This synergy occurs at the level of the p300 transcriptional coactivator to which both Stat3 and Smad1 physically bind *(205)*.

The role of the JAK/STAT pathway has also been investigated in interleukin (IL)-6 exposed PC12 cells, where IL-6-induced phosphorylation of Stat3 was a negative regulator of MAPK-dependent neurite extension *(206,207)*.

Thus, it appears that the final biological response to a given factor is the result of a complex interplay between signaling pathways and interacting growth factors. Given the large number of JAKs and STATs members known, it is likely that new functions for these signal transducers in CNS cells will soon emerge.

MECHANISMS INSTRUCTING SIGNAL TRANSDUCTION SPECIFICITY IN THE BRAIN

A fundamental question about signal transduction is how these various signaling pathways interact to communicate intracellularly the message from an activated RTK. Particularly, this is a key matter for the proliferation and the differentiation of neural stem cells, as an overlapping of external cues may act to regulate both of these events.

In this scenario, proper regulation of the activity and interactions between different signaling pathways is highly important. In the previous sections, we have described the players of signal transduction events; in this section, we will show two recently described mechanisms that may explain how the activity of multiple signaling pathways may be modulated at the transition from proliferation to differentiation in the brain.

Proliferation or Differentiation: A Shc(s) Decision?

A "simple" mechanism that the cell has developed to generate diversity in biological responses to growth factors lies on the regulation of the expression and activity of Shc(s) adapter molecules, which couple the signal from the activated receptor to the downstream effectors. Shc(s) proteins, indeed appear to play a role in the control of the maturation of mitotically active neural stem/progenitor cells into postmitotic neurons *(4,148,149)*.

As with many other intracellular signaling proteins, Shc (recently renamed ShcA) proved to be only one member of a family of genes, which is now known to include two new homologs, ShcB/Sli and ShcC/Rai/N-Shc (reviewed in ref. *4*). These three Shc(s) are characterized by the presence of phosphotyrosine regulatory residues and the PTB, CH1 (a proline-rich domain), and SH2 domains in the presented order. Three isoforms are known for ShcA (of 66, 52, and 46 kDa), two isoforms for ShcB (of 52 and 47 kDa), and two for ShcC (of 54 and 69 kDa). Some isoforms, such as the p66[ShcA], display a further N-terminal CH domain (CH2) that contains important regulatory serine residues. They share elevated homology in both the C-terminus SH2 domain and the N-terminus PTB domain, the most divergent sequence being in the proline- and glycine-rich CH1 (collagene homology 1) region.

ShcA proteins have been extensively characterized and shown to be widely expressed outside the CNS. Their importance is indicated by (1) the early embryonic lethal phenotype of p52^{ShcA} null mutation *(208)* and (2) by the increase in life-span and resistance to stress stimuli of p66^{ShcA} knockout animals *(209)*.

Despite the apparently constitutive presence of ShcA in extraneural tissues, ShcA expression and activity within the brain appears to be tightly regulated during development and maximal at the early developmental stages (Fig. 6). Particularly, it has been previously demonstrated that ShcA proteins and mRNAs are sharply downregulated in coincidence with neurogenesis in the brain *(148)*. In the embryonic rodent brain, ShcA is localized to the germinal epithelium where mitotically active immature CNS stem cells are located (Fig. 6). In the areas of the embryonic or postnatal brain where postmitotic neurons are present, ShcA is very low or not detectable. The adult brain is mostly devoid of ShcA, the only exception being the olfactory epithelium, which is the only tissue that still undergoes active cell renewal in the adult. These changes in the expression and activity of ShcA as a function of neuronal maturation were confirmed in vitro in differentiating neuronal cultures *(148)*.

It was also found that, in vivo, immunoprecipitated ShcA from the telencephalic vesicles of embryonic brains injected intraventricularly with mitogens (EGF) exhibited a higher phosphorylation of the p52^{ShcA} isoform with respect to vehicle-injected control animals *(148)*. Particularly, in treated samples, Grb2 coimmunoprecipitation was observed, indicating that ShcA is not only present in the germinal epithelium but it is also able to elicit a functional response to mitogens by recruitment of Grb2 and activation of the downstream Ras–MAPK pathway.

The demonstration that ShcA availability is tighly regulated and that, particularly, ShcA presence becomes limited during neural stem cell maturation in vivo and in vitro leads to the proposition that these changes may influence as well the activity of the Ras-MAPK pathway during development and/or that other Shc-like proteins may substitute for ShcA function in mature neurons.

Given the existence of two new Shc members, ShcB and ShcC, the latter being selectively expressed in the brain, it has been recently suggested that one or both of them could replace ShcA in mature neurons *(4)*. Analyses of ShcC expression indeed showed that ShcA is replaced by ShcC, which is not expressed in neural progenitors, but it appears in postmitotic neurons and reaches maximal levels in the adult brain where it is localized only in

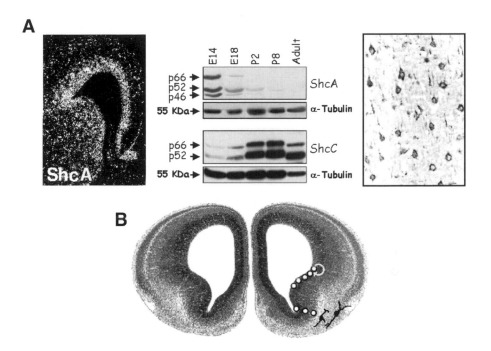

Fig. 6. (A) Expression of ShcA/C during forebrain maturation. Left panel: Expression of ShcA mRNAs coincides with the proliferative epithelium. *In situ* analysis performed on embryonic day 14.5 brain section (reprinted with permission from ref. *148*). Middle panel: Western-blot analyses showing maximal expression of the three ShcA isoforms during early stages of neurogenesis, whereas ShcC expression increases at late maturation stages (reprinted with permission from ref. *149*). Right panel: ShcC is expressed in mature neurons in the adult rodent brain (reprinted with permission from ref. *149*). Immunohistochemistry performed on adult cortex section. (B) Schematic drawing. The scheme indicates the switch from ShcA to ShcC during the transition from proliferation (open circles indicate stem cells that are ShcA positive) to differentiation (black drawings indicate postmitotic neurons that are ShcC positive).

neurons (Fig. 6). Similar changes in ShcA and ShcC levels during neuronal maturation have been observed in several mammalian species (rat, mouse, and human) *(149)*. Notably, ShcC is found in neurons from various regions of the adult brain, thus predicting a key general role played by ShcC in these cells (Fig. 6). Particularly, given the above-described central roles of ShcA in signal transduction, ShcC appearance in differentiating neural stem cells has been hypothesized to serve different "connector functions" compared with ShcA *(149)*. To this regard, a recent study from Lai and Pawson *(208)* demonstrated the existence of a strict link between Shc levels and cell responsiveness. The authors showed that ShcA expression and activity are required in cells of the cardiovascular system to make them responsive to low concentrations of growth factors. Indeed, whereas a low concentration of growth factors is necessary to activate the MAPK pathway in mouse embryo fibroblasts (MEF), cells from ShcA knockout mice require an higher concentration of growth factors to activate the same signaling cascade. Transfection experiments in primary neural stem cells and in postmitotic neurons revealed that ShcC acts to promote neuronal differentiation and improve survival of these cells *(149)*. It was also found that ShcC elicits these effects through a different kinetic of activation of downstream effector molecules with respect to ShcA. Indeed, ShcC was found to elicit neuronal differentiation via prolonged stimulation of the MAPK. This behavior is reminiscent of that described in PC12 cells exposed to NGF, where persistent activation of MAPK is required for neuronal differentiation. On the contrary, ShcC-driven prosurvival effect occurs via recruitment of the PI3K–Akt pathway, as demonstrated by the fact that its pharmacological or molecular inhibition markedly abolishes this effect. To this respect, ShcC-induced Akt activation was found to cause phosphorylation (with inhibition) of Bad, a proapoptotic member of the Bcl2 family *(149)*.

Single and double ShcB/C null mice have been recently described *(210)*. ShcB-deficient mice exhibit a loss of peptidergic and nonpeptidergic nociceptive sensory neurons. ShcC null mice do not appear to show gross anatomical abnormalities. Noteworthy, mice lacking both ShcB and ShcC exhibit a significant additional loss of neurons within the superior cervical ganglia. This aspect may emphasize that the lack of phenotype in ShcC null mice could be the result of a partial compensation by the other ShcB or other Shc members during development, thus masking ShcC real function in neural tissues. Further analyses will be required to elucidate ShcC role in vivo in the mice.

Taken together, these results unveil a new scenario within which physiological changes in the availability of ShcA and ShcC adaptors during brain development act to modify neural stem/progenitor cell responsiveness as a function of the new and developing environment (Fig. 6).

Neuronal or Glial Differentiation? A Balance Between Activating and Inhibiting Signals

A single progenitor can give rise to the three major cell types of the brain: neurons, astrocytes, and oligodendrocytes *(211–213)*. In vivo, the events that generate these cell types are temporally regulated, with neurogenesis that precedes gliogenesis and oligodendrogenesis. The same kind of temporal regulation is observed in vitro. Cultures of cortical progenitors isolated at different embryonic stages behave in a manner that recapitulates these developmental processes *(19,214)*. Indeed, it has been described that cortical progenitor cells from rat embryonic day 14 give rise to neurons and to dividing precursor cells *(146)*, with astrocytes that are generated only after several days in vitro. Conversely, progenitor cells derived from embryonic day 17 are able to give rise to astrocytes immediately *(199)*.

In the last few years, the signaling mechanisms by which neuronal, glial, or oligodendrocytic cell fates are specified have become clearer. Several studies have demonstrated that CNTF and LIF exert their effects by the activation of Jak1 that in turn phosphorylates STAT1 and STAT3, promoting their dimerization and translocation to the nucleus where they activate gene expression *(215)* (Fig. 7). BMP has been shown to promote both neuronal and astrocyte differentiation, depending on the age of the stimulated cortical progenitors *(216,217)*. It acts by binding a multimeric receptor, which, in turn, results in the direct phosphorylation of Smad1 *(218)*. This induces Smad1–Smad4 dimerization and nuclear translocation *(219)*, allowing them to cooperate with STATs to activate glial-specific programs of gene expression *(205)* (Fig. 7). Several studies have demonstrated the ability of the ubiquitously expressed transcription factors CBP (CREB-binding protein)/p300 to act as STATs and Smads coactivators in this process. To this respect, the STAT/p300/Smad complex has been shown to facilitate the transcriptional activation of glial-specific genes, such as GFAP *(220)* (Fig. 7). Recently, Nakashima et al. demonstrated that the STAT/p300/Smad complex is particularly effective at inducing astrocyte differentiation in neural stem cells, by acting at the STAT-binding element in the astrocyte-specific GFAP promoter *(205)*.

Recent knockout studies *(221–223)* have also demonstrated that the actions of several basic helix–loop–helix (bHLH) transcription factors play an important role in the specification of neuronal or oligodendrocytic cell fates during brain maturation. These bHLH factors include neurogenin1 and 2 (Ngn1 and Ngn2) for neuronal differentiation *(224,225)* and oligo1 and oligo2 for oligodendrocyte specification *(226,227)*. Notably, during CNS development, Ngn1 and Ngn2 were shown to be exclusively expressed in neuroepithelial precursor cells *(228,229)*. Ngn1 and Ngn2 dimerize with ubiquitous bHLH proteins, such as E12 or E47. These heterodimers then bind via their positively charged basic domains to DNA sequences that contain the E box consensus motif *(228)*. E box binding has been found to

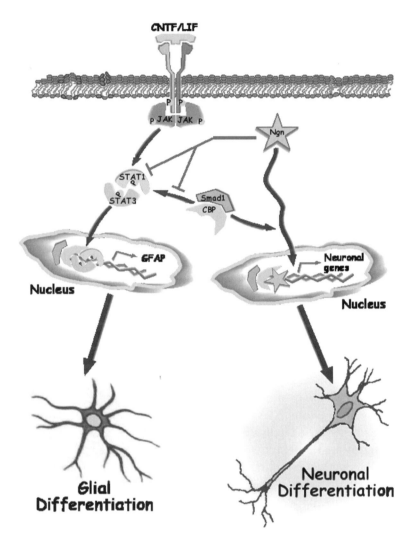

Fig. 7. Double action of neurogenin in promoting neuronal maturation and inhibiting glial differentiation. CNTF/LIF treatment of cortical progenitors induces activation of Jak1, resulting in Stat1/3 phosphorylation, dimerization, and interaction with CBP to activate glial-specific promoters (GFAP). Neurogenin acts to activate neuronal-specific gene transcription and suppresses glial fate by inhibiting both STAT phosphorylation and interaction with CBP (see the text for details).

be critical for bHLH proteins to activate tissue-specific gene expression that promotes neuronal differentiation *(230)*.

A key question in developmental neurobiology is that cell fate specification involves the reciprocal activation of genes related to a particular cell fate and the suppression of genes of alternative fates, so that inducers of neuronal fate must somehow suppress glial differentiation and vice versa. Sun et al. have demonstrated recently the existence of such a mechanism during glial differentiation of CNS stem cell *(231)*. They found that Ngn1 exerts the double function of activating neuronal differentiation genes and suppressing glial-specific genes (Fig. 7). This double action is supported by the fact that knockout mice exhibit both inhibition of neurogenesis and increased glial differentiation *(232)*. Nevertheless, as neurogenin levels are high during cortical neurogenesis and low during gliogenesis, neurogenin's ability to suppress only glial differentiation may explain why astrocytes fail to develop during the period of neurogenesis, even in the presence of glial-inducing cues. Furthermore, it was demonstrated that Ngn1 effectively inhibits astrocyte differentiation in a manner that is independent of its ability to induce neuronal differentiation and that was occurring through a double action: (1) by binding and sequestering the transcriptional cofactors CBP and Smad1 and (2) by blocking the activation of Stat1 and Stat3 signaling molecules that are critical inducers of astrocyte differentiation *(231)* (Fig. 7). The mechanism by which Ngn1 reduces the level of phospho-Stat1 and Stat3 has not yet understood.

Neurogenin's ability to both promote neurogenesis and inhibit gliogenesis may allow neurogenins to function as neuronal traffic lights, allowing the differentiating progenitor cells to correctly respond to extracellular stimuli.

CONCLUDING REMARKS

It is now well established that the final biological response to a given factor is the result of a complex and highly dynamic interplay between different actors at different levels. In addition to the complexity of the overlapping ligand–receptor interactions, it appears that each receptor is capable of alternatively recruiting different combinations of signaling systems, depending on the developmental stage, the cell cycle point, and its spatial position, thereby generating multiple responses.

Taken together, a new idea has now to be considered; that is, it is not only the message that the factor carries, but also how a cell is set to respond to that extracellular cue that will finally instruct proliferative or differentiative events. During development, the dynamics of cell–cell interaction, morphogenesis, and growth are too rapid to permit the exchange of an entire signaling cascade at each proliferative/differentiative state. As a result, the modulation of each transduction system at different stages of development becomes fundamentally important. Changing adaptor molecules or switching from one signaling pathway to another (already in use in the cell) permits the rapid responses that are required to successfully drive the complex machinery of development.

ACKNOWLEDGMENTS

The work of the authors is supported by Associazione Italiana Ricerca Cancro, Telethon (Italy, #E840) to E.C.; Telethon (Italy, #E1024), and UK Wellcome Trust to L.C.

REFERENCES

1. Gage, F. H., Ray, J., and Fisher, L. J. (1995) Isolation, characterization and use of stem cells from the CNS. *Annu. Rev. Neurosci.* **18,** 158–192.
2. Gage, F. H. (1998) Stem cells of the central nervous system. *Curr. Opin. Neurobiol.* **8,** 671–676.
3. McKay, R. (1997) Stem cells in the central nervous system. *Science* **276,** 6671–6674.
4. Cattaneo, E. and Pelicci, P. G. (1998) Emerging roles for SH2/PTB-containing Shc adaptor proteins in the developing mammalian brain. *Trends Neurosci.* **21,** 476–481.
5. Mehler, M. F. and Kessler, J. A. (1994) Growth factor regulation of neuronal development. *Dev. Neurosci.* **16,** 180–195.
6. Mehler, M. F. and Kessler, J. A. (1997) Hematolymphopoietic and inflammatory cytokines in neural development. *Trends Neurosci.* **20,** 357–365.
7. Reichardt, L. F. and Farinas, I. (1997) Neurotrophic factors and their receptors: roles in neuronal development and function, in *Molecular and Cellular Approaches to Neural Development* (Cowan, W. M., Jessel, T. M., and Zipursky, S. L., eds.), Oxford University Press, New York, pp. 220–263.
8. Eide, F. F., Lowenstein, D. H. and Reichardt, L. F. (1993) Neurotrophins and their receptors: current concepts and implications for neurologic disease. *Exp. Neurol.* **121,** 200–214.
9. Barres, B. A, Schmid, R., Sendtner, M., and Raff, M. C. (1993) Multiple extracellular signals are required for long-term oligodendrocyte survival. *Development* **118,** 283–295.
10. Williams, B. P., Park, J. K., Alberta, J. A., Muhlebach, S. G., Hwang, G. Y., Roberts, T. M., et al. (1997) A PDGF-regulated immediate early gene response initiates neuronal differentiation in ventricular zone progenitor cells. *Neuron* **18,** 553–562.
11. Johe, K. K., Hazel, T. G., Muller, T., Dugich-Djordjevic, M. M., and McKay, R. D. G. (1996) Single factors direct the differentiation of stem cells from the fetal and adult central nervous system. *Genes Dev.* **10,** 3129–3140.
12. Morrison, R. S., Sharma, A., de Vellis, J., and Bradshaw, R. A. (1986) Basic fibroblast growth factor supports the survival of cerebral cortical neurons in primary culture. *Proc. Natl. Acad. Sci. USA* **83,** 7537–7541.

13. Gensburger, C., Labourdette, G., and Sensenbrenner, M. (1987) Brain basic fibroblast growth factor stimulates the proliferation of rat neuronal precursor cells in vitro. *FEBS Lett.* **217,** 1–5.
14. Cattaneo, E. and McKay, R. D. G. (1990) Proliferation and differentiation of neuronal stem cells regulated by nerve growth factor. *Nature* **347,** 762–765.
15. Reynolds, B. A. and Weiss, S. (1991) Generation of neurons and astrocytes from isolated cells of the adult mammalian central nervous system. *Science* **255,** 1646–1649.
16. Reynolds, B. A., Tetzlaff, W., and Weiss, S. (1992) A multipotent EGF-responsive striatal embryonic progenitor cell produces neurons and astrocytes. *J. Neurosci.* **12,** 4565–4574.
17. DeHamer, M. K., Guevara, J. L., Hannon, K., Olwin, B. B., and Calof, A. L. (1994) Genesis of olfactory receptor neurons in vitro: regulation of progenitor cell divisions by fibroblast growth factors. *Neuron* **13,** 1083–1097.
18. Gritti, A., Parati, E. A., Cova, L., Frolichsthal, P., Galli, R., Wanke, E., et al. (1996) Multipotential stem cells from the adult mouse brain proliferate and self-renew in response to basic fibroblast growth factor. *J. Neurosci.* **16,** 1091–1100.
19. Burrows, R. C., Wancio, D., Levitt, P., and Lillien, L. (1997) Response diversity and the timing of progenitor cell maturation are regulated by developmental changes in EGFR expression in the cortex. *Neuron* **19,** 251–267.
20. Rodriguez-Viciana, P., Warne, P. H., Dhand, R., Vanhaesebroeck, B., Gout, I., Fry, M. J., et al. (1994) Phosphatidylinositol-3-OH kinase as a direct target of Ras. *Nature* **370,** 527–532.
21. Hunter, T. (1997) Oncoprotein networks. *Cell* **88,** 333–346.
22. Van der Geer, P., Hunter, T., and Lindberg, R. A. (1994) Receptor protein-tyrosine kinases and their signal transduction pathways. *Annu. Rev. Cell Biol.* **10,** 251–337.
23. Segal, R. A. and Greenberg, M. E. (1996) Intracellular signaling pathways activated by neurotrophic factors. *Annu. Rev. Neurosci.* **19,** 463–489.
24. Miyajima, A., Kitamura, T., Harada, N., Yokota, T., and Arai, K. (1992) Cytokine receptors and signal transduction. *Annu. Rev. Immunol.* **10,** 295–331.
25. Ihle, J. N. (1995) Cytokine receptor signalling. *Nature* **377,** 591–594.
26. Ip, N. Y. and Yancopoulos, G. D. (1996) The neurotrophins and CNTF: two families of collaborative neurotrophic factors. *Annu. Rev. Neurosci.* **19,** 491–515.
27. Taga, T. and Kishimoto, Y. (1997) Gp130 and the interleukin-6 family of cytokines. *Annu. Rev. Immunol.* **15,** 797–815.
28. Cattaneo, E., Conti, L. and De-Fraja, C. (1999) Signalling through the JAK/STAT pathway in the developing brain. *Trends Neurosci.* **22,** 365–369.
29. Lavigne, M. and Goodman, C. S. (1996) The molecular biology of axon guidance. *Science* **274,** 1123–1133.
30. Goodman, C. S. and Tessier-Lavigne, M. (1997) Molecular mechanism of axon guidance and target recognition, in *Molecular and Cellular Approaches to*

Neural Development (Cowan, W. M., Jessel, T. M., and Zipursky, S. L., eds.), Oxford University Press, New York, pp. 108–178.

31. Kalo, M. S. and Pasquale, E. B. (1999) Multiple in vivo tyrosine phosphorylation sites in EphB receptors. *Biochemistry* **38,** 14,396–14,408.

32. Binns, K., Taylor, P. P., Sicheri, F., Pawson, T., and Holland, S. J. (2000) Phosphorylation of tyrosine residues in the kinase domain and juxtamembrane region regulates the biological and catalytic activities of Eph receptors. *Mol. Cell. Biol.* **20,** 4791–4805.

33. Holland, S. J., Gale, N. W., Mbamalu, G., Yancopoulos, J. D., Henkemeyer, M., and Pawson, T. (1996) Bi-directional signalling through the Eph family Nuk and its transmembrane ligands. *Nature* **383** 722–725.

34. Orioli, D., Henkemeyer, M., Lemke, G., Klein, R., and Pawson, T. (1996) Sek4 and Nuk receptors cooperate in guidance of commissural axons and in palate formation. *EMBO J.* **15,** 6035–6049.

35. Pinkas-Kramarski, R., Shelly, M., Guarino, B. C., Wang, L. M., Lyass, L., Alroy, I., et al. (1998) ErbB tyrosine kinases and the two neuregulin families constitute a ligand-receptor network. *Mol. Cell. Biol.* **18,** 6090–6101.

36. Hubbard, S. R. (1999) Structural analysis of receptor tyrosine kinases. *Prog. Biophys. Mol. Biol.* **71,** 343–358.

37. Schlessinger, J. and Ullrich, A. (1992) Growth factor signaling by receptor tyrosine kinases. *Neuron* **9,** 383–391.

38. Flanagan, J. G. and Vanderhaeghen, P. (1998) The ephrins and Eph receptors in neural development. *Annu. Rev. Neurosci.* **21,** 309–345.

39. Holland, S. J., Peles, E., Pawson, T., and Schlessinger, J. (1998) Cell-contact-dependent signalling in axon growth and guidance: Eph receptor tyrosine kinases and receptor protein tyrosine phosphatase beta. *Curr. Opin. Neurobiol.* **8,** 117–127.

40. Stephens, R. M., Loeb, D. M., Copeland, T. D., Pawson, T., Greene, L. A., and Kaplan, D. R. (1994) Trk receptors use redundant signal transduction pathways involving Shc and PLC-gamma 1 to mediate NGF responses. *Neuron* **12,** 691–705.

41. Hanks, S. K., Quinn, A. M., and Hunter, T. (1988) The protein kinase family: conserved features and deduced phylogeny of the catalytic domains. *Science* **241,** 42–52.

42. Mohammadi, M., Schlessinger, J., and Hubbard, S. R. (1996) Structure of the FGF receptor tyrosine kinase domain reveals a novel autoinhibitory mechanism. *Cell* **86,** 577–587.

43. Fantl, W. J., Escobedo, J. A., and Williams, L. T. (1989) Mutations of the platelet-derived growth factor receptor that cause a loss of ligand-induced conformational change, subtle changes in kinase activity, and impaired ability to stimulate DNA synthesis. *Mol. Cell. Biol.* **9,** 4473–4478.

44. Mitra, G. (1991) Mutational analysis of conserved residues in the tyrosine kinase domain of the human trk oncogene. *Oncogene* **6,** 2237–2241.

45. Ullrich, A. and Schlessinger, J. (1990) Signal transduction by receptors with tyrosine kinase activity. *Cell* **61,** 203–212.

46. Heldin, C. H. (1995) Dimerization of cell surface receptors in signal transduction. *Cell* **80,** 213–223.
47. Sorkin, A. and Waters, C. M. (1993) Endocytosis of growth factor receptors. *BioEssays* **15,** 375–382.
48. Mori, S., Claesson-Welsh, L., Okuyama, Y., and Saito, Y. (1995) Ligand-induced polyubiquitination of receptor tyrosine kinases. *Biochem. Biophys. Res. Commun.* **213,** 32–39.
49. Tonks, N. K. and Neel, B. G. (1996) From form to function: signaling by protein tyrosine phosphatases. *Cell* **87,** 365–368.
50. Elchebly, M., Payette, P., Michaliszyn, E., Cromlish, W., Collins, S., Loy, A. L., et al. (1999) Increased insulin sensitivity and obesity resistance in mice lacking the protein tyrosine phosphatase-1B gene. *Science* **283,** 1544–1548.
51. Tatosyan, A. G. and Mizenina, O. A. (2000) Kinases of the Src family: structure and functions. *Biochemistry* **65,** 49–58.
52. Biscardi, J. S., Tice, D. A., and Parsons, S. J. (1999) c-Src, receptor tyrosine kinases, and human cancer. *Adv. Cancer Res.* **76,** 61–119.
53. Schindler, C. and Darnell, J. E. (1995) Transcriptional responses to polypeptide ligands: the JAK–STAT pathway. *Annu. Rev. Biochem.* **64,** 621–651.
54. Wilks, A. F. and Harpur, A. G. (1996) in *Intracellular Signal Transduction: the JAK–STAT Pathway* (Wilks, A. F. and Harpur, A. G., eds.), Springer-Verlag, New York.
55. Rane, S. G. and Reddy, E. P. (2000) Janus kinases: components of multiple signaling pathways. *Oncogene* **19,** 5662–5679.
56. Darnell, J. E., Jr. (1998) Studies of IFN-induced transcriptional activation uncover the Jak–Stat pathway. *J. Interferon Cytokine Res.* **18,** 549–554.
57. Darnell J. E., Jr., Kerr, I. M., and Stark, G. R. (1994) Jak–STAT pathways and transcriptional activation in response to IFNs and other extracellular signaling proteins. *Science* **264,** 1415–1421.
58. Ihle, J. N., Witthuhn, B. A., Quelle, F. W., Yamamoto, K., and Silvennoinen, O. (1995) Signaling through the hematopoietic cytokine receptors. *Annu. Rev. Immunol.* **13,** 369–398.
59. Ihle, J. N., Nosaka, T., Thierfelder, W., Quelle, F. W., and Shimoda, K. (1997) Jaks and Stats in cytokine signaling. *Stem Cells* **15,** 105–112.
60. Pellegrini, S. and Dusanter-Fourt, I. (1997) The structure, regulation and function of the Janus kinases (JAKs) and the signal transducers and activators of transcription (STATs). *Eur. J. Biochem.* **248,** 615–633.
61. Heim, M. H. (1999) The Jak–STAT pathway: cytokine signalling from the receptor to the nucleus. *J. Recept. Signal Transduct. Res.* **19,** 75–120.
62. Leonard, W. J. and O'Shea, J. J. (1998) Jaks and STATs: biological implications. *Annu. Rev. Immunol.* **16,** 293–322.
63. Schindler, C. (1999) Cytokines and JAK–STAT signaling. *Exp. Cell. Res.* **253,** 7–14.
64. Leonard, W. J. and Lin, J. X. (2000) Cytokine receptor signaling pathways. *J. Allergy Clin. Immunol.* **105,** 877–888.

65. Ward, A. C., Touw, I., and Yoshimura, A. (2000) The Jak–Stat pathway in normal and perturbed hematopoiesis. *Blood* **95,** 19–29.

66. Pawson, T. (1995) Protein modules and signalling networks. *Nature* **373,** 573–580.

67. Nguyen, J. T., Turck, C. W., Cohen, F. E., Zuckermann, R. N., and Lim, W. (1998) Exploiting the basis of proline recognition by SH3 and WW domains: fesign of N-substituted inhibitors. *Science* **282,** 2088–2092.

68. Buday, L. (1999) Membrane-targeting of signalling molecules by SH2/SH3 domain-containing adaptor proteins. *Biochem. Biophys. Acta* **1422,** 187–204.

69. Salcini, A. E., Confalonieri, S., Doria, M., Santolini, E., Tassi, E., Minenkova, O., et al. (1997) Binding specificity and in vivo targets of the EH domain, a novel protein–protein interaction module. *Genes Dev.* **11,** 2239–2249.

70. Songyang, Z., Fanning, A. S., Fu, C., Xu, J., Marfatia, S. M., Chishti, A. H., et al. (1997) Recognition of unique carboxyl-terminal motifs by distinct PDZ domains. *Science* **275,** 73–77.

71. Sadowski, I., Stone, J. C., and Pawson, T. (1986) A noncatalytic domain conserved among cytoplasmic protein-tyrosine kinases modifies the kinase function and transforming activity of Fujinami sarcoma virus P130gag-fps. *Mol. Cell. Biol.* **6,** 4396–4408.

72. Mayer, B. J., Hamaguchi, M., and Hanafusa, H. (1988) A novel viral oncogene with structural similarity to phospholipase C. *Nature* **332,** 272–275.

73. Koch, C. A., Anderson, D., Moran, M.F., Ellis, C., and Pawson, T. (1991) SH2 and SH3 domains: elements that control interactions of cytoplasmic signaling proteins. *Science* **252,** 668–674.

74. Olivier, J. P., Raabe, T., Henkemeyer, M., Dickson, B., Mbamalu, G., Margolis, B., et al. (1993) A Drosophila SH2-SH3 adaptor protein implicated in coupling the sevenless tyrosine kinase to an activator of Ras guanine nucleotide exchange, Sos. *Cell* **73,** 179–191.

75. Songyang, Z., Shoelson, S. E., Chaudhuri, M., Gish, G., Pawson, T., Haser, W. G., et al. (1993) SH2 domains recognize specific phosphopeptide sequences. *Cell* **72,** 767–778.

76. Eck, M. J., Atwell, S. K., Shoelson, S. E., and Harrison, S. C. (1993) Recognition of a high-affinity phosphotyrosyl peptide by the Src homology-2 domain of p56lck. *Nature* **362,** 87–91.

77. Waksman, G., Shoelson, S. E., Pant, N., Cowburn, D., and Kuriyan, J. (1993) Binding of a high affinity phosphotyrosyl peptide to the Src SH2 domain: Crystal structures of the complexed and peptide-free forms. *Cell* **72,** 779–790.

78. Pascal, S. M., Singer, A. U., Gish, G., Yamazaki, T., Shoelson, S. E., Pawson, T., et al. (1994) Nuclear magnetic resonance structure of an SH2 domain of phospholipase C-gamma1 complexed with a high affinity binding peptide. *Cell* **77,** 461–472.

79. Lesa, G. M. and Sternberg, P. W. (1997) Positive and negative tissue-specific signaling by a nematode epidermal growth factor receptor. *Mol. Cell. Biol.* **8,** 779–793.

80. Minichiello, L., Casagranda, F., Tatche, R. S., Stucky, C. L., Postigo, A., Lewin, G. R., et al. (1998) Point mutation in trkB causes loss of NT4-dependent neurons without major effects on diverse BDNF responses. *Neuron* **21,** 335–345.

81. Marengere, L. E. M., Songyang, Z., Gish, G. D., Schaller, M. D., Parsons, T., Stern, M. J., et al. (1994) SH2 domain specificity and activity modified by a single residue. *Nature* **369,** 502–505.

82. Blaikie, P., Immanuel, D., Wu, J., Li, N., Yajnik, V., and Margolis, B. (1994) A region in Shc distinct from the SH2 domain can bind tyrosine-phosphorylated growth factor receptors. *J. Biol. Chem.* **269,** 32,031–32,034.

83. Kavanaugh, W. M. and Williams, L. T. (1994) An alternative to SH2 domains for binding tyrosine-phosphorylated proteins. *Science* **266,** 1862–1865.

84. Zhou, M. M., Ravichandran, K. S., Olejniczak, E. T., Petros, A. M., Meadows, R. P., Sattler, M., et al. (1995) Structure and ligand recognition of the phosphotyrosine binding domain of Shc. *Nature* **378,** 584–592.

85. Van der Geer, P., Wiley, S., Lai, V. K. M., Olivier, J. P., Gish, G. D., Stephens, T., et al. (1995) A conserved amino-terminal SHC domain binds to activated growth factor receptors and phosphotyrosine-containing peptides. *Curr. Biol.* **5,** 404–412.

86. Ong, S. H., Guy, G. R., Hadari, Y. R., Laks, S., Gotoh, N., Schlessinger, J., et al. (2000) FRS2 proteins recruit intracellular signaling pathways by binding to diverse targets on fibroblast growth factor and nerve growth factor receptors. *Mol. Cell. Biol.* **20,** 979–989.

87. Lemmon, M. A. and Ferguson, K. M. (1998) Pleckstrin homology domains. *Curr. Top. Microbiol. Immunol.* **228,** 39–74.

88. Collins, L. R., Ricketts, W. A., Yeh, L., and Cheresh, D. (1999) Bifurcation of cell migratory and proliferative signaling by the adaptor protein Shc. *J. Cell. Biol.* **47,** 1561–1568.

89. Prehoda, K. E., Lee, D. J., and Lim, W. A. (1999) Structure of the enabled/VASP homology 1 domain-peptide complex: a key component in the spatial control of actin assembly. *Cell* **97,** 471–480.

90. Vetter, I. R., Nowak, C., Nishimoto, T., Kuhlmann, J., and Wittinghofer, A. (1999) Structure of a Ran-binding domain complexed with Ran bound to a GTP analogue: implications for nuclear transport. *Nature* **398,** 39–46.

91. Maina, F., Casagranda F., Andero E., Simeone A., Comoglio P. M., Klein, G., et al. (1996) Uncoupling of Grb2 from the Met receptor in vivo reveals complex roles in muscle development. *Cell* **87,** 531–542.

92. Van der Geer, P., Wiley, S., Gish, G. D., Lai, V. K., Stephens, R., White, M. F., et al. (1996) Identification of residues that control specific binding of the Shc phosphotyrosine-binding domain to phosphotyrosine sites. *Proc. Natl. Acad. Sci. USA* **93,** 963–968.

93. Kouhara, H., Hadari, Y. R., Spivak-Kroizman, T., Schilling, J., Bar-Sagi, D., Lax, I., et al. (1997) A lipid-anchored Grb2-binding protein that links FGF-receptor activation to the Ras/MAPK signaling pathway. *Cell* **89,** 693–702.

94. Rozakis-Adcock, M., Fernley, R., Wade, J., Pawson, T., and Bowtell, D. (1993) The SH2 and SH3 domains of mammalian Grb2 couple the EGF-receptor to mSos1, an activator of Ras. *Nature* **363**, 83–85.
95. Ottinger, E. A., Botfield, M. C., and Shoelson, S. E. (1998) Tandem SH2 domains confer high specificity in tyrosine kinase signaling. *J. Biol. Chem.* **273**, 729–735.
96. Bubeck-Wardenburg, J., Pappu, R., Bu, J. Y., Mayer, B., Chernoff, J., Straus, D., et al. (1998) Regulation of PAK activation and the T cell cytoskeleton by the linker protein SLP-76. *Immunity* **9**, 607–616.
97. Lewis, T. S., Shapiro, P. S., and Ahn, N. G. (1998) Signal transduction through MAP kinase cascades. *Adv. Cancer Res.* **74**, 49–139.
98. Cobb, M. H. (1999) MAP kinase pathways. *Prog. Biophys. Mol. Biol.* **71**, 479–500.
99. Guadagno, T. M. and Ferrell, J. E. (1998) Requirement for MAPK activation for normal mitotic progression in Xenopus egg extracts. *Science* **282**, 1312–1315.
100. Ray, L. B. and Sturgill, T. W. (1987) Rapid stimulation by insulin of a serine/threonine kinase in 3T3-L1 adipocytes that phosphorylates microtubule-associated protein 2 in vitro. *Proc. Natl. Acad. Sci. USA* **84**, 1502–1506.
101. Hoshi, M., Nishida, E., and Sakai, H. (1988) Activation of a Ca^{2+} inhibitable protein kinase that phosphorylates microtubule-associated protein 2 in vitro by growth factors, phorbo esters, and serum in quiescent cultured human fibroblasts. *J. Biol. Chem.* **263**, 5396–5401.
102. Herskowitz, I. (1995) MAP kinase pathways in yeast: For mating and more. *Cell* **80**, 187–197.
103. Dhanasekaran, N. and Reddy, E. P. (1998) Signaling by dual specificity kinases. *Oncogene* **17**, 1447–1455.
104. Boulton, T. G., Yancopoulos, G. D., Gregory, J. S., Slaughter, C., Moomaw, C., Hsu, J., et al. (1990) An insulin-stimulated protein kinase similar to yeast kinases involved in cell cycle control. *Science* **249**, 64–67.
105. Boulton, T. G., Nye, S. H., Robbins, D. J., Ip, N. Y., Radziejewska, E., Morgenbesser, S. D., et al. (1991) ERKs: a family of protein-serine/threonine kinases that are activated and tyrosine phosphorylated in response to insulin and NGF. *Cell* **65**, 663–675.
106. Marshall, C. J. (1995) Specificity of receptor tyrosine kinase signaling: transient versus sustained extracellular signal-regulated kinase activation. *Cell* **80**, 179–185.
107. Martin, K. C., Michael, D., Rose, J. C., Barad, M., Casadio, A., Zhu, H., et al. (1997) MAP kinase translocates into the nucleus of the presynaptic cell and is required for long-term facilitation in Aplysia. *Neuron* **18**, 899–912.
108. Fukunaga, K. and Miyamoto, E. (1998) Role of MAP kinase in neurons. *Mol. Neurobiol.* **16**, 79–95.
109. Blum, S., Moore, A. N., Adams, F., and Dash, P. K. (1999) A mitogen-activated protein kinase cascade in the CA1/CA2 subfield of the dorsal hippocampus is essential for long-term spatial memory. *J. Neurosci.* **19**, 3535–3544.

110. Cheng, M., Boulton, T. G., and Cobb, M. H. (1996) ERK3 is a constitutively expressed nuclear protein kinase. *J. Biol. Chem.* **271,** 8951–8958.

111. Kyriakis, J. and Woodgett, J. R. (1994) The stress-activated protein kinase subfamily of c-Jun kinases. *Nature* **369,** 156–160.

112. Kyriakis, J. M. and Avruch, J. (1996) Sounding the alarm: protein kinase cascades activated by stress and inflammation. *J. Biol. Chem.* **271,** 24,313–24,316.

113. Xia, Z., Dickens, M., Raingeaud, J., Davis, R. J., and Greenberg, M. E. (1995) Opposing effects of ERK and JNK-p38 MAP kinases on apoptosis. *Science* **270,** 1326–1331.

114. Kuan, C. Y., Yang, D. D., Samanta, R. D., Davis, R. J., Rakic, P., and Flavell, R. A. (1999) The Jnk1 and Jnk2 protein kinases are required for regional specific apoptosis during early brain development. *Neuron* **22,** 667–676.

115. Herdegen, T., Skene, P., and Bahr, M. (1997) The c-Jun transcription factor-bipotential mediator of neuronal death, survival and regeneration. *Trends Neurosci.* **20,** 227–231.

116. Han, J., Bibbs, L., and Ulevitch, R. J. (1994) A MAP kinase targeted by endotoxin and hyperosmolarity in mammalian cells. *Science* **265,** 808–811.

117. Rouse, J., Cohen, P., Trigon, S., Morange, M., Alonso-Llama-zares, A., Zamanillo, D., et al. (1994) A novel kinase cascade triggered by stress and heat shock that stimulates MAPKAP kinase-2 and phosphorylation of the small heat shock proteins. *Cell* **78,** 1027–1037.

118. Ichijo, H. (1999) From receptor to stress-activated MAP kinases. *Oncogene* **18,** 6087–6093.

119. Morrison, D. K. and Cutler, R. E. (1997) The complexity of Raf-1 regulation. *Curr. Opin. Cell Biol.* **9,** 174–179.

120. Morrison, D. K., Heidecker, G., Rapp, U. R., and Copeland, T. D. (1993) Identification of the major phosphorylation sites of the Raf-1 kinase. *J. Biol. Chem.* **268,** 17,309–17,316.

121. Mason, C. S., Springer, C. J., Cooper, R. G., Superti-Furga, G., Marshall, C. J., and Marais, R. (1999) Serine and tyrosine phosphorylations cooperate in Raf-1, but not B-Raf activation. *EMBO J.* **18,** 2137–2148.

122. Wu, J., Harrison, J. K., Vincent, L. A., Haystead, C., Haystead, T. A. J., Michel, H., et al. (1993) Molecular structure of a protein-tyrosine/threonine kinase activating p42 mitogen-activated protein (MAP) kinase: MAP kinase kinase. *Proc. Natl. Acad. Sci. USA* **90,** 173–177.

123. Robbins, D. J., Zhen, E., Owaki, H., Vanderbilt, C., Ebert, D., Geppert, T. D., et al. (1993) Regulation and properties of extracellular signal-regulated protein kinases 1 and 2 in vitro. *J. Biol. Chem.* **268,** 5097–5106.

124. Li, N. A., Batzer, R., Daly, E., Skolnik, P., Chardin, D., Bar-Sagi, B., et al. (1993) Guanine nucleotide releasing factor hSos1 binds to Grb2 and links receptor tyrosine kinases to Ras signaling. *Nature* **363,** 85–88.

125. Daaka, Y., Luttrell, L. M., and Lefkowitz, R. J. (1997) Switching of the coupling of the beta2-adrenergic receptor to different G proteins by protein kinase A. *Nature* **390,** 88–91.

126. Seidel, M. G, Klinger, M., Freissmuth, M., and Iler, C. (1999) Activation of mitogen-activated protein kinase by the A(2A)-adenosine receptor via a rap1-dependent and via a p21(ras)-dependent pathway. *J. Biol. Chem.* **274,** 25,833–25,841.

127. Vossler, M. R., Yao, H., York, R. D., Pan, M. G., Rim, C. S., and Stork, P. J. (1997) cAMP activates MAP kinase and Elk-1through a B-Raf- and Rap1-dependent pathway. *Cell* **89,** 73–82.

128. Hunter, T. (1995) Protein kinases and phosphatases: the yin and yang of protein phosphorylation and signaling. *Cell* **80,** 225–236.

129. Dikic, I., Tokiwa, G., Lev, S., Courtneidge, S. A., and Schlessinger, J. (1996) A role for Pyk2 and Src in linking G-protein-coupled receptors with MAP kinase activation. *Nature* **383,** 547–550.

130. Daub, H., Weiss, F. U., Wallasch, C., and Ullrich, A. (1996) Role of transactivation of the EGF receptor in signalling by G-protein-coupled receptors. *Nature* **379,** 557–560.

131. Daub, H., Wallasch, C., Lankenau, A., Herrlich, A., and Ullrich, A. (1997) Signal characteristics of G protein-transactivatedEGF receptor. *EMBO J.* **16,** 7032–7044.

132. Rosen, L. B., Ginty, D. D., Weber, M. J., and Greenberg, M. E. (1994) Membrane depolarization and calcium influx stimulate MEK and MAP kinase via activation of Ras. *Neuron* **12,** 1207–1221.

133. Farnsworth, C. L., Freshney, N. W., Rosen, L. B., Ghosh, A., Greenberg, M. E., and Feig, L. A. (1995) Calcium activation of Ras mediated by neuronal exchange factor Ras-GRF. *Nature* **376,** 524–527.

134. Chen, H. J., Rojassoto, M., Oguni, A., and Kennedy, M. B. (1998) A synaptic Ras-GTPase activating protein (p135 SYN-GAP) inhibited by CaM kinase II. *Neuron* **20,** 895–904.

135. Krueger, K. A., Bhatt, H., Landt, M., and Easom, R. A. (1997) Calcium-stimulated phosphorylation of MAP-2 in pancreatic βTC3 cells is mediated by Ca^{2+}/calmodulin-dependent kinase II. *J. Biol. Chem.* **272,** 27,464–27,469.

136. Dellarocca, G. J., Vanbiesen, T., Daaka, Y., Luttrell, D. K., Luttrell, L. M., and Lefkowitz, R. J. (1997) Ras-dependent mitogen-activated protein kinase activation by G protein-coupled receptors–convergence of G(i)- and G(q)-mediated path-wayson calcium/calmodulin, PYK2, and Src kinase. *J. Biol. Chem.* **272,** 19,125–19,132.

137. Learish, R. D., Bruss, M. D., and Haak-Frendscho, M. (2000) Inhibition of mitogen-activated protein kinase kinase blocks proliferation of neural progenitor cells. *Dev. Brain Res.* **122,** 97–109.

138. Kaplan, D. and Miller, F. (1997) Signal transduction by neurotrophin receptors. *Curr. Opin. Cell Biol.* **9,** 213–221.

139. Greene, L. A. and Kaplan, D. R. (1995) Early events in neurotrophin signalling via Trk and p75 receptors. *Curr. Opin. Neurobiol.* **5,** 579–587.

140. Aziz, N., Cherwinski, H., and McMahon, M. (1999) Complementation of defective colony-stimulating factor 1 receptor signaling and mitogenesis by Raf and v-Src. *Mol. Cell. Biol.* **19,** 1101–1115.

141. Pang, L. Sawada, T. Decker, S. J., and Saltiel, A. R. (1995) Inhibition of MAP kinase kinase blocks the differentiation of PC-12 cells induced by nerve growth factor. *J. Biol. Chem.* **270,** 13,585–13,588.

142. Schramek, H., Feifel, E., Healy, E., and Pollack, V. (1997) Constitutively mutant of the mitogen-activated protein kinase kinase MEK1 induces epithelial dedifferentiation and growth inhibition in Darby canine kidney C7 cells. *J. Biol. Chem.* **272,** 11,426–11,433.

143. Robinson, M. J., Stippec, S. A., Goldsmith, E., White, M. A., and Cobb, M. H. (1998) Constitutively active ERK2 MAP kinase is sufficient for neurite outgrowth and cell transformation when targeted to the nucleus. *Curr. Biol.* **8,** 1141–1150.

144. York, R. D., Yao, H., Dillon, T., Ellig, C. L., Eckert, S. P., McCleskey, E. W., et al. (1998) Rap1 mediates sustained MAP kinase activation induced by nerve growth factor. *Nature* **392,** 622–626.

145. Obermeier, A., Bradshaw, R. A., Seedorf, K., Choidas, A., Schlessinger, J., and Ullrich, A. (1994) Neuronal differentiation signals are controlled by nerve growth factor receptor/Trk binding sites for Shc and PLC gamma. *EMBO J.* **13,** 1585–1590.

146. Ghosh, A. and Greenberg, M. E. (1995) Distinct roles for bFGF and NT-3 in the regulation of cortical neurogenesis. *Neuron* **15,** 89–103.

147. Knusel, B., Rabin, S. J., Hefti, F., and Kaplan, D. R. (1994) Regulated neurotrophin receptor responsiveness during neuronal migration and early differentiation. *J. Neurosci.* **14,** 1542–1554.

148. Conti, L., De-Fraja, C., Gulisano, C., Migliaccio, M., Govoni, S., and Cattaneo, E. (1997) Expression and activation of SH2/PTB containing Shc A adaptor protein reflects the pattern of neurogenesis in the mammalian brain. *Proc. Natl. Acad. Sci. USA* **94,** 8185–8190.

149. Conti, L., Sipione, S., Magrassi, M., Bonfanti, L., Rigamonti, D., Pettirossi, V., et al. (2001) Shc(s) signaling in differentiating neural progenitor cells. *Nat. Neurosci.* **6,** 579–586.

150. Thoenen, H. and Barde, Y. A. (1980) Physiology of nerve growth factor. *Physiol. Rev.* **60,** 1284–1335.

151. Barde, Y. A., Edgar, D., and Thoenen, H. (1982) Purification of a new neurotrophic factor from mammalian brain. *EMBO J.* **1,** 549–553.

152. Oppenheim, R. W. (1991) Cell death during development of the nervous system. *Annu. Rev. Neurosci.* **14,** 453–501.

153. Yao, R. and Cooper, G. M. (1995) Requirement for phosphatidylinosotol-3 kinase in the prevention of apoptosis by nerve growth factor. *Science* **267,** 2003–2006.

154. Fruman, D. A., Meyers, R. A., and Cantley, L. C. (1998) Phosphoinositide Kinases. *Annu. Rev. Biochem.* **67,** 481–507.

155. Cantley, L. C., Auger, K. R., Carpenter, C., Duckworth, B., Graziani, A., Kapeller, R., et al. (1991). Oncogenes and signal transduction. *Cell* **64,** 281–302.

156. Piccione, E., Case, R. D., Domchek, S. M., Hu, P., Chaudhuri, M., Backer, J. M., et al. (1993) Phosphatidylinositol 3-kinase p85 SH2 domain specificity defined by direct phosphopeptide/SH2 domain binding. *Biochemistry* **32,** 3197–3202.

157. Toker, A and Cantley, L. C. (1997) Signalling through the lipid products of phosphoinositide-3-OH kinase. *Nature* **387,** 673–676.
158. Rameh, L. E. and Cantley, L. C. (1999) The role of phosphoinositide-3-kinase lipid products in cell function. *J. Biol. Chem.* **274,** 8347–8350.
159. Marte, B. M., Rodriguez-Viciana, P., Wennstrom, S., Warne, P. H., and Downward, J. (1996) R-Ras can activate the phosphoinositide 3-kinase but not the MAP kinase arm of ras effector pathways. *Curr. Biol.* **7,** 63–70; May, M. J. and Ghosh, S. 1997. Rel/ NF-B and IB proteins: an overview. *Semin. Cancer Biol.* **8,** 63–73.
160. Kauffmann-Zeh, A., Rodriguez-Vicana, P., Ulrich, E., Gilbert, C., Coffer, P., Downward, J., et al. (1997) Suppression of c-Myc-induced apoptosis by Ras signalling through PI(3)K and PKB. *Nature* **385,** 544–548.
161. Khwaja, A., Rodriguez-Vicana, P., Wennstrom, S., Warne, P. H., and Downward, J. (1997) Matrix adhesion and Ras transformation both activate a phosphoinositide 3-OH kinase and protein kinase B/Akt cellular survival pathway. *EMBO J.* **16,** 2783–2793.
162. Liu, X. X., Testa, J. R., Hamilton, T. C., Jove, R., Nicosia, S. V., and Cheng, J. Q. (1998) Akt2, a member of the protein kinase B family, is activated by growth factors, v-Ha-ras, and v-src through a phosphatidylinositol 3-kinase in human ovarian epithelial cancer cells. *Cancer Res.* **58,** 2973–2977.
163. Murga, C., Laguinge, L., Wetzker, R., Cuadrado, A., and Gutkind, J. S. (1998) Activation of Akt/protein kinase b by G protein-coupled receptors. *J. Biol. Chem.* **273,** 19,080–19,085.
164. Stambolic, V., Suzuki, A., de la Pompa, J. L., Brothers, G. M., Mirtsos, C., Sasaki, T., et al. (1998) Negative regulation of PKB/Akt-dependent cell survival by the tumor suppressor PTEN. *Cell* **95,** 29–39.
165. Suzuki, A., de la Pompa, J. L., Stambolic, V., Elia, A. J., Sasaki, T., del Barco Barrantes, I., et al. (1998) High cancer susceptibility and embryonic lethality associated with mutation of the PTEN tumor suppressor gene in mice. *Curr. Biol.* **8,** 1169–1178.
166. Burgering, B. M. and Coffer, P. J. (1995) Protein kinase B (c-Akt) in phosphatidylinosotol-3-OH kinase signal transduction. *Nature* **376,** 599–602.
167. Franke, T. F., Yang, S. I., Chan, T. O., Datta, K., Kazlauskas, A., Morrison, D. K., et al. (1995) The protein kinase encoded by the Akt proto-oncogene is a target of the PDGF-activated phosphatidylinositol 3-kinase. *Cell* **81,** 727–736.
168. Bellacosa, A., Testa, J. R., Staal, S. P., and Tsichlis, P. N. (1991) A retroviral oncogene, akt, encoding a serine threonine kinase containing an SH2-like region. *Science* **254,** 274–277.
169. Coffer, P. J. and Woodgett, J. R. (1991) Molecular cloning and characterization of a novel putative protein-serine kinase related to the cAMP-dependent and protein kinase C families. *Eur. J. Biochem.* **201,** 475–481.
170. Jones, P. F., Jakubowicz, T., Pitossi, F. J., Maurer, F., and Hemmings, B. A. (1991) Molecular cloning and identification of a serine/threonine protein kinase of the second messenger subfamily. *Proc. Natl. Acad. Sci. USA* **88,** 4171–4175.

171. Kandel, E. S. and Hay, N. (1999) The regulation and activities of the multi-functional serine/threonine kinase Akt/PKB. *Exp. Cell Res.* **253,** 210–229.
172. Datta, S. R., Brunet, A., and Greenberg, M. E. (1999) Cellular survival: a play in three Akts. *Genes Dev.* **13,** 2905–2927.
173. Kennedy, S. G., Wagner, A. J., Conzen, S. D., Jordan, J., Bellacosa, A., Tsichlis, P. N., et al. (1997) The PI3-Kinase/Akt signaling pathway delivers an anti-apoptotic signal. *Genes Dev.* **11,** 701–713.
174. Franke, T. F., Kaplan, D. R., Cantley, L. C., and Toker, A. (1997) Direct regulation of the Akt proto-oncogene product by phosphatidylinositol-3,4-bisphosphate. *Science* **275,** 665–668.
175. Alessi, D. R., Deak, M., Casamayor, A., Caudwell, F. B., Morrice, N., Norman, D. G., et al. (1997) 3-phosphoinositide-dependent protein kinase-1 (PDK-1): structural and functional homology with the Drosophila DSTPK61 kinase. *Curr. Biol.* **7,** 776–789.
176. Balendran, A., Casamayor, A., Deak, M., Paterson, A., Gaffney, P., Currie, R., et al. (1999) PDK1 acquires PDK2 activity in the presence of a synthetic peptide derived from the carboxyl terminus of PRK2. *Curr. Biol.* **9,** 393–404.
177. Datta, S. R., Dudek, H., Tao, X. Masters, S., Fu, H., Gotoh, Y., et al. (1997) Akt phosphorylation of BAD couples survival signals to the cell-intrinsic death machinery. *Cell* **91,** 231–241.
178. Del Peso, L., Gonzalez-Garcia, M., Page, C., Herrera, R., and Nuñez, G. (1997) Interleukin-3-induced phosphorylation of BAD through the protein kinase Akt. *Science* **278,** 687–689.
179. Cardone, M. H., Roy, N., Stennicke, H. R., Salvesen, G. S., Franke, T. F., Stanbridge, E., et al. (1998) Regulation of cell death protease caspase-9 by phosphorylation. *Science* **282,** 1318–1321.
180. Kennedy, S. G., Kandel, E. S., Cross, T. K., and Hay, N. (1999) Akt/PKB inhibits cell death by preventing the release of cytochrome c from mitochondria. *Mol. Cell. Biol.* **19,** 5800–5810.
181. Brunet, A., Bonni, A., Zigmond, M. J., Lin, M. Z., Juo, P., Hu, L. S., et al. (1999) Akt promotes cell survival by phosphorylating and inhibiting a forkhead transcription factor. *Cell* **96,** 857–868.
182. Romanshkova, J. A. and Makarov, S. S. (1999) NF-B is a target of Akt in anti-apoptotic PDGF signaling. *Nature* **401,** 86–89.
183. Zuckerman, S. H., Evans, G. F., and Guthrie, L. (1991) Transcriptional and post-transcriptional mechanisms involved in the differential expression of LPS-induced IL-1 and TNF messenger RNA. *Immunology* **73,** 460–465.
184. Widmann, C., Gibson, S., and Johnson, G. L. (1998) Caspase-dependent cleavage of signaling proteins during apoptosis. *J. Biol. Chem.* **273,** 7141–7147.
185. Starr, R., Willson, T. A., Viney, E. M., Murray, L. J., Rayner, J. R., Jenkins, B. J., et al. (1997) A family of cytokine-inducible inhibitors of signalling. *Nature* **387,** 917–921.
186. Endo, T. A., Masuhara, M., Yokouchi, M., Suzuki, R., Sakamoto, H., Mitsui, K., et al. (1997) A new protein containing an SH2 domain that inhibits JAK kinases. *Nature* **387,** 921–924.

187. Naka, T., Narazaki, M., Hirata, M., Matsumoto, T., Minamoto, S., Aono, A., et al. (1997) Structure and function of a new STAT-induced STAT inhibitor. *Nature* **387,** 924–929.

188. Yoshimura, A., Ohkubo, T., Kiguchi, T., Jenkins, N. A., Gilbert, D. J., Copeland, N. G., et al. (1995) A novel cytokine-inducible gene CIS encodes an SH2-containing protein that binds to tyrosine-phosphorylated interleukin 3 and erythropoietin receptors. *EMBO J.* **14,** 2816–2826.

189. Chung, C. D., Liao, J., Liu, B., Rao, X., Jay, P., Berta, P., et al. (1997) Specific inhibition of Stat3 signal transduction by PIAS3. *Science* **278,** 1803–1805.

190. Patel, B. K. R., Pierce, J. H., and LaRochelle, W. J. (1998) Regulation of interleukin 4-mediated signaling by naturally occurring dominant negative and attenuated forms of human Stat6. *Proc. Natl. Acad. Sci. USA* **95,** 172–177.

191. Patel, B. K., Wang, L. M., Lee, C. C., Taylor, W. G., Pierce, J. H., and LaRochelle, W. J. (1996) Stat6 and Jak1 are common elements in platelet-derived growth factor and interleukin-4 signal transduction pathways in NIH 3T3 fibroblasts. *J. Biol. Chem.* **271,** 22,175–22,182.

192. Chin, Y. E., Kitagawa, M., Su, W. C., You, Z. H., Iwamoto, Y., and Fu, X. Y. (1996) Cell growth arrest and induction of cyclin-dependent kinase inhibitor p21 WAF1/CIP1 mediated by STAT1. *Science* **272,** 719–722.

193. Faris, M., Ensoli, B., Stahl, N., Yancopoulos, G., Nguyen, A., Wang, S., et al. (1996) Differential activation of the extracellular signal-regulated kinase, Jun kinase and Janus kinase-Stat pathways by oncostatin M and basic fibroblast growth factor in AIDS-derived Kaposi's sarcoma cells. *AIDS* **10,** 369–378.

194. Eilers, A., Georgellis, D., Klose, B., Schindler, C., Ziemiecki, A., Harpur, A. G., et al. (1995) Differentiation-regulated serine phosphorylation of STAT1 promotes GAF activation in macrophages. *Mol. Cell. Biol.* **15,** 3579–3586.

195. Wen, Z., Zhong, Z., and Darnell, J. E. (1995) Maximal activation of transcription by Stat1 and Stat3 requires both tyrosine and serine phosphorylation. *Cell* **82,** 241–250.

196. Zhang, X., Blenis, J., Li, H. C., Schindler, C., and Chen-Kiang, S. (1995) Requirement of serine phosphorylation for formation of STAT-promoter complexes. *Science* **267,** 1990–1994.

197. Clark-Lewis, I., Sanghera, J. S., and Pelech, S. L. (1991) Definition of a consensus sequence for peptide substrate recognition by p44mpk, the meiosis-activated myelin basic protein kinase. *J. Biolog. Chem.* **266,** 15,180–15,184.

198. Sato, N., Sakamaki, K., Terada, N., Arai, K., and Miyajima, A. (1993) Signal transduction by the high-affinity GM–CSF receptor: two distinct cytoplasmic regions of the common beta subunit responsible for different signaling. *EMBO J.* **12,** 4181–4189.

199. Bonni, A., Sun, Y., Nadal-Vicens, M., Bhatt, A., Frank, D. A., Rozovsky, I., et al. (1997) Regulation of gliogenesis in the central nervous system by the Jak-Stat signaling pathway. *Science* **278,** 477–483.

200. Rajan, P., Symes, A. J., and Fink, J. S. (1996) STAT proteins are activated by ciliary neurotrophic factor in cells of central nervous system origin. *J. Neurosci. Res.* **43,** 403–411.

201. Niwa, H., Burdon, T., Chambers, I., and Smith, A. (1998) Self-renewal of pluripotent embryonic stem cells is mediated via activation of STAT 3. *Genes Dev.* **12,** 2048–2060.
202. De-Fraja, C., Conti, L., Magrassi, L., Govoni, S., and Cattaneo, E. (1998) Members of the Jak/Stat proteins are expressed and regulated during development in the mammalian forebrain. *J. Neurosci. Res.* **54,** 320–330.
203. Kahn, M. A., Huang, C. J., Caruso, A., Barresi, V., Nazarian, R., Condorelli, D. F., et al. (1997) Ciliary neurotrophic factor activates Jak/Stat signal transduction cascade and induces transcriptional expression of glial fibrillary acidic protein in glial cells. *J. Neurochem.* **68,** 1413–1423.
204. Monville, C., Coulpier, M., Conti, L., De-Fraja, C., Riche, D., Fages, C., et al. (2001) Ciliary neurotrophic factor activates mature astrocytes via binding with the LIF receptor. *Mol. Cell. Neurosci.* **17,** 373–384.
205. Nakashima, K., Yanagisawa, M., Arakawa, H., Kimura, N., Hisatsune, T., Kawabata, M., et al. (1999) Synergistic signaling in fetal brain by STAT3-Smad1 complex bridged by p300. *Science* **248,** 479–482.
206. Wu, Y. Y. and Bradshaw, R. A. (1996) Induction of neurite outgrowth by interleukin-6 is accompanied by activation of Stat3 signaling pathway in a variant PC12 cell (E2) line. *J. Biol. Chem.* **271,** 13,023–13,032.
207. Ihara, S., Nakajima, K., Fukada, T., Hibi, M., Nagata, S., Hirano, T., et al. (1997) Dual control of neurite outgrowth by STAT3 and MAPkinase in PC12 cells stimulated with interleukin-6. *EMBO J.* **16,** 5345–5352.
208. Lai, K. M. V. and Pawson, T. (2000) The ShcA phosphotyrosine docking protein sensitizes cardiovascular signaling in the mouse embryo. *Genes Dev.* **14,** 1132–1145.
209. Migliaccio, E., Giorgio, M., Mele, S., Pelicci, G., Reboldi, P., Pandolfi, P. P., et al. (1999) The p66(shc) adapter protein controls oxidative stress response and life span in mammals. *Nature* **402,** 309–313.
210. Sakai, R., Henderson, J. T., O'Bryan, J. P., Elia, A. J., Saxton, T. M., and Pawson, T. (2000) The mammalian ShcB and ShcC phosphotyrosine docking proteins function in the maturation of sensory and sympathetic neurons. *Neuron* **28,** 819–833.
211. Turner, D. L. and Cepko, C. L. (1987) A common progenitor for neurons and glia persists in rat retina late in development. *Nature* **328,** 131–136.
212. Luskin, M. B., Pearlman, A. L., and Sanes, J. R. (1988) Cell lineage in the cerebral cortex of the mouse studied in vivo and in vitro with a recombinant retrovirus. *Neuron* **1,** 635–647.
213. Price, J. and Thurlow, L. (1988) Cell lineage in the rat cerebral cortex: a study using retroviral-mediated gene transfer. *Development* **104,** 473–482.
214. Qian, X., Shen, Q., Goderie, S. K., He, W., Capela, A., Davis, A. A., et al. (2000) Timing of CNS cell generation: a programmed sequence of neuron and glial cell production from isolated murine cortical stem cells. *Neuron* **28,** 69–80.
215. Bonni, A., Frank, D. A., Schindler, C., and Greenberg, M. E. (1993) Characterization of a pathway for ciliary neurotrophic factor signaling to the nucleus. *Science* **262,** 1575–1579.

216. Gross, R. E., Mehler, M. F., Mabie, P. C., Zang, Z., Santschi, L., and Kessler, J. A. (1996) Bone morphogenetic proteins promote astroglial lineage commitment by mammalian subventricular zone progenitor cells. *Neuron* **17,** 595–606.

217. Li, W., Cogswell, C. A., and LoTurco, J. J. (1998) Neuronal differentiation of precursors in the neocortical ventricular zone is triggered by BMP. *J. Neurosci.* **18,** 8853–8862.

218. Hoodless, P. A., Haerry, T., Abdollah, S., Stapleton, M., O'Connor, M. B., Attisano, L., et al. (1996) MADR1, a MAD-related protein that functions in BMP2 signaling pathways. *Cell* **85,** 489–500.

219. Massague, J. (1998) TGF-beta signal transduction. *Annu. Rev. Biochem.* **67,** 753–791.

220. Goodman, R. H. and Smolik, S. (2000) CBP/p300 in cell growth, transformation, and development. *Genes Dev.* **14,** 1553–1577.

221. Ben-Arie, N., Bellen, H. J., Armstrong, D. L., McCall, A. E., Gordadze, P. R., Guo, Q., et al. (1997) Math1 is essential for genesis of cerebellar granule neurons. *Nature* **390,** 169–172.

222. Fode, C., Ma, Q., Casarosa, S., Ang, S. L., Anderson, D. J., and Guillemot, F. (2000) A role for neural determination genes in specifying the dorsoventral identity of telencephalic neurons. *Genes Dev.* **14,** 67–80.

223. Tomita, K., Moriyoshi, K., Nakanishi, S., Guillemot, F., and Kageyama, R. (2000) Mammalian achaete-scute and atonal homologs regulate neuronal versus glial fate determination in the central nervous system. *EMBO J.* **19,** 5460–5472.

224. Guillemot, F., Lo, L. C., Johnson, J.E., Auerbach, A., Anderson, D. J., and Joyner, A. L. (1993) Mammalian achaete-scute homolog 1 is required for the early development of olfactory and autonomic neurons. *Cell* **75,** 463–476.

225. Ma, Q., Kintner, C., and Anderson, D. J. (1996) Identification of neurogenin, a vertebrate neuronal determination gene. *Cell* **87,** 43–52.

226. Lu, Q. R., Yuk, D., Alberta, J. A., Zhu, Z., Pawlitzky, I., Chan, J., et al. (2000) Sonic hedgehog-regulated oligodendrocyte lineage genes encoding bHLH proteins in the mammalian central nervous system. *Neuron* **25,** 317–329.

227. Zhou, Q., Wang, S., and Anderson, D. J. (2000) Identification of a novel family of oligodendrocyte lineage-specific basic helix–loop–helix transcription factors. *Neuron* **25,** 331–343.

228. Gradwohl, G., Fode, C., and Guillemot, F. (1996) Restricted expression of a novel murine atonal-related bHLH protein in undifferentiated neural precursors. *Dev. Biol.* **180,** 227–241.

229. Sommer, L., Ma, Q., and Anderson, D. J. (1996) Neurogenins, a novel family of atonal-related bHLH transcription factors, are putative mammalian neuronal determination genes that reveal progenitor cell heterogeneity in the developing CNS and PNS. *Mol. Cell. Neurosci.* **8,** 221–241.

230. Cau, E., Gradwohl, G., Fode, C., and Guillemot, F. (1997) Mash1 activates a cascade of bHLH regulators in olfactory neuron progenitors. *Development* **124,** 1611–1621.

231. Sun, Y., Nadal-Vicens, M., Misono, M., Lin, M. Z., Zubiaga, A., Hua, X., et al. (2001) Neurogenin promotes neurogenesis and inhibits glial differentiation by independent mechanisms. *Cell* **104,** 365–376.
232. Nieto, M., Schuurmans, C., Britz, O., and Guillemot, F. (2001) Neural bHLH genes control the neuronal versus glial fate decision in cortical progenitors. *Neuron* **29,** 401–413.

7

Neural Stem/Progenitor Cell Clones or "Neurospheres"

A Model for Understanding Neuromorphogenesis

Dennis A. Steindler, Bjorn Scheffler , Eric D. Laywell, Oleg N. Suslov, Tong Zheng, Thomas Reiniger, and Valery G. Kukekov

INTRODUCTION: A NEW STEM CELL BIOLOGY

Stem cell biology has contributed an impressive list of new and important findings in the last decade that are anticipated to lead to potentially powerful therapeutics for debilitating human diseases. The generation of human embryonic stem cell (ES cell) lines *(1,2)* is among this list of crucial technological breakthroughs, not to mention the applications of genetic, cell, and molecular biology to the first-time cloning of an entire organism *(3)* that has since been achieved in numerous species, including the mouse *(4)*. Utilizing insights and approaches from these fields, as well as from developmental biology, the field of developmental neurobiology has been astonished in recent years by numerous "reversals of dogma" related to neurogenesis in the mature mammalian brain [defined as the generation of neurons, or the shortened form of "neuronogenesis," versus "gliogenesis," which is the production of astroglia and oligodendroglia; "neuromorphogenesis" is the combined events of neurogenesis and gliogenesis that ultimately generate a nervous system *(5–8)*]. In particular, despite the work of Allen *(9)*, Altman and Das *(10)*, and others supporting the existence of persistent neurogenesis in the adult rat olfactory and hippocampal systems, these were considered highly specialized cases that by no means supported a notion of neuropoiesis (persistent neurogenesis) in the adult central nervous system. The in vitro propagation of a putative stem cell population from the adult rat brain by the Weiss and Bartlett groups *(11,12)* suggested that there may be neuropoiesis; studies that followed have established the source of these stem/progenitor cells [a term used because it not only sidesteps the

From: *Neural Stem Cells for Brain and Spinal Cord Repair*
Edited by: T. Zigova, E. Y. Snyder, and P. R. Sanberg © Humana Press Inc., Totowa, NJ

contentious issue of stemness *(5,8,13)* of these cells, but also because it encompasses the entire spectrum of proliferative neurogenic cells that can generate all cells in the nervous system] as the subependymal zone, ependyma, and hippocampus, even within the aged human brain *(6,14–16)*. It is possible to clone stem/progenitor cells from these adult brain regions *(7)* and even from cadaver specimens *(17,18)* with surprisingly long postmortem intervals [up to 5 d from cadaver specimens *(17)*]. For the purpose of discussion, these persistently neurogenic regions have been amalgamated under one term—"brain marrow" *(8,14,19)*. The analogy of a brain neuropoietic core to the hematopoietic bone marrow has been substantiated by the recent surprising finding that adult brain-derived stem/progenitor cells are considerably more pluripotent than ever expected [e.g., giving rise to blood cells after homing to bone marrow following systemic grafting *(20)*, as well as to muscle *(21,22)* and even multiple organ systems *(23)*]; it also has recently been shown that non-neural stem cells can also be coaxed into nerve cell phenotypes following different neuralizing conditions *(24–27)*. However, to date, most transplant or other in vivo studies of stem/progenitor cells derived from brain marrow suggest that these cells have a rather limited fate potential in the central nervous system (CNS) [e.g., glia *(28)*]—but also giving rise to granule cells and other interneuron populations although fetal neural stem cells and immortalized neural stem cell lines do appear to be more plastic and potent, incorporating into a variety of brain circuits *(29–31)*.

BRAIN STEM/PROGENITOR CELLS
AND THEIR PROGENY FORM NEUROSPHERES

"Neurospheres" are artificial in vitro-generated structures of clonal origin (i.e., each neurosphere represents the progeny of a single proliferative stem/progenitor cell) first produced by Reynolds and Weiss *(11)* following dissociations of adult mouse brain *(see* Figs. 1 and 2). However, we are still uncertain about the exact nature of such pluripotent cells, including their in vivo identification and localization (except see refs. *15* and *16*). The population of cells that constitute a neurosphere is pluripotent, capable of generating the three types of neural cell found in the mammalian brain (neurons, astrocytes, and oligodendrocytes) as well as cells of other tissues; that is, the neurosphere begins as a single proliferative stem/progenitor cell, and through proliferation and differentiation within the neurosphere, we end up with a heterogeneous population of cells, including immature neurons and glia. Individual clones of neural stem/progenitor cells have been studied

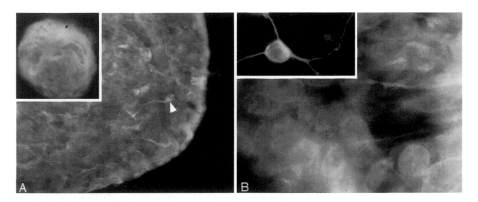

Fig. 1. Neurospheres can be generated from different neural stem/progenitor, "neurosphere-forming" cells. In this figure, multipotent astrocytic stem cells, as described by Laywell et al. *(17)* were used to generate neurosphere clones. In **(A)**, such a neurosphere has been immunostained for microtubule-associated protein-2 (MAP-2, green), a cytoskeletal protein present in neurons. Many neurons are present in this clone (e.g., arrowhead) derived from an astrocytic stem cell, as well as other cells (propidium iodide [PI], counterstained). The inset in this figure is another similar clone immunostained for neuronal βIII tubulin (green), again showing large numbers of neurons. This immature neurosphere clone is 100 μm in diameter. **(B)** Another astrocyte-derived neurosphere also immunostained for neuronal βIII tubulin, with an inset showing an isolated neuron from this clone also stained for βIII tubulin and its nucleus counterstained with PI. This neuron is 10 μm in diameter.

from dissociated embryonic and adult mouse and human brain *(7,11,32–35)*. It can be hypothesized that during their growth under specific tissue culture conditions, cells in each neurosphere undergo differentiation through a variety of stages, which mirror, to an extent, cell growth and differentiation seen in the developing brain in vivo. Thus, the study of a neurosphere could uncover stages and signals involved in the basic processes, leading to neuronal or glial generation from a single stem/progenitor cell. Each neurosphere can be considered an isolated developing microsystem, the study of which could provide insights into cell–cell, cell–substrate, and basic genetic and molecular interactions underlying neuromorphogenesis. Furthermore, the neurosphere as an isolated developing neural microsystem can be subjected to different microenvironments (e.g., the addition of particular factors, including growth factors, as well as drug screenings) to evaluate the consequences of varied molecular conditions in this model neurogenic system.

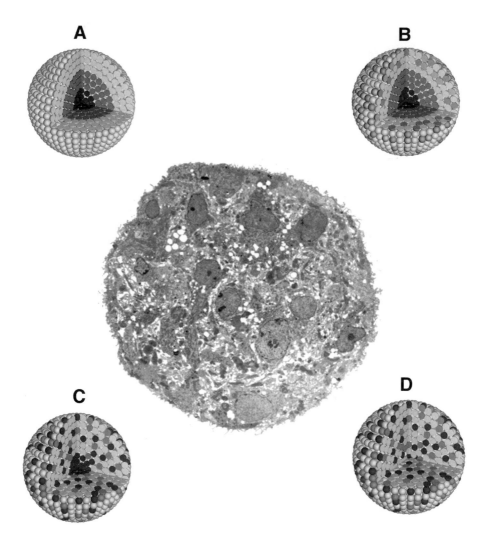

Fig. 2. What the architecture of a neurosphere could tell us. A future goal of neuropoiesis research is the purification of neural stem cells, in order to uncover their nature and direct their fate. Current methods allow the identification of putative stem/progenitor cells, relying on, for example, their ability to form neurospheres in culture. Neurospheres are made up of hundreds to thousands of cells, and when dissociated, there is only a subfraction (although with long-term culturing or serial dissociations, this subfraction increases) of cells that maintain the key feature of a neurosphere-forming cell (NFC)—the ability to give rise to secondary multipotent spheres (e.g., see refs. *32–34*). This observation suggests that neurospheres are composed of a mixture of cells in varying stages of differentiation *(7,35,38)*. Furthermore, once a neurosphere appears, it seems to be impossible to identify the original clonally expanding cell (NFC), even when using electron microscopy (as

NEUROSPHERES AS HETEROGENEOUS STRUCTURES

There is evidence that neurospheres are heterogeneous structures that reflect distinct spatiotemporal origins *(8)* or experimental conditions (e.g., see refs. *36* and *37*) that can affect their growth and differentiation. Little attention has been paid to date on the potential heterogeneity of cells within a single as well as across a population of neurospheres (except see refs. *33,35,37, and 38*). Because neurospheres are made up of cells in many different states of differentiation, there is a need for experimental approaches that afford the morphological and phenotypic analyses of each clone. It

---

Fig. 2. *(continued)* seen in the center of this figure—a neurosphere from adult mouse brain, revealing a mixture of cells exhibiting different nuclear morphologies, but it is not possible to recognize if or where there is an NFC present). The illustrations **(A–D)** offer four possible plans for neurosphere composition, and in light of a study (using a novel neurosphere-sectioning procedure) that revealed one type of neurosphere organization with the most differentiated cells forming an outer rim *(38)*, it is therefore reasonable to propose scenarios for neurosphere composition and organization. **(A)** Perhaps the most reasonable model would be that sheets of progressively maturing cells surround a central core of quiescent stem cells that exhibit the highest level of immaturity (black cells, NFCs). This core population would give rise to descendants that are highly proliferative cells (progenitors, gray), which, in turn, produce more committed cells (white cells in the outer layer). However, the structure of neurospheres could also be more dynamic, with the outer sheets of cells being composed of a mixture of progenitors (gray) and more mature cells (white cells), as indicated in **(B)**. They could also be randomly distributed within the neurosphere, possessing varied degrees of proliferation and maturation [as suggested in **(C)**]. It is also possible that there is no core of the most primitive stemlike cells **(D)**, but, instead, the full range of stem/progenitor and differentiated cells is distributed rather uniformly throughout the neurosphere. It also seems possible that there would be no stem cells in the neurosphere, however, previous neurosphere subcloning studies *(33,45)* indicate they are able to form secondary neurospheres, suggesting the presence of clonogenic NFCs. In each of these scenarios, there can be a dynamic or purposeful migration [e.g., as seen in vivo in brain marrow *(14,19)*] or a more random population dispersion following stochastic patterns of cell division of stem/progenitor cells that results in the displacement of more mature neighbors.

Thus, the in vitro activities of neurospheres might mirror the behaviors of immature cells in brain marrow, their natural habitat. Neurospheres, even as culture-generated artifacts, in essence represent a microversion of the brain germinal matrix. They are a starting point for understanding neurogenesis in general, and knowledge of their composition could offer insights into mechanisms of stem/progenitor cell propagation and differentiation.

would seem useful to establish particular growth conditions that favor controlled proliferation and differentiation of the different cellular morphotypes based on distinct temporal expressions of developmentally regulated genes and their products. Thus, there are key questions related to (1) stem/progenitor cell propagation and differentiation and (2) factors that affect the behaviors, including fate choice and commitment of these cells. Can particular stem/progenitor cell populations be selectively expanded? Can they be stimulated to differentiate into just one "desired" type of cell? Can the isolation, expansion, and manipulation of the most primitive cells of neurospheres be used as a source of uncommitted cells for further in vitro analyses and perhaps even large-scale expansion for cell replacement/gene therapy approaches for debilitating neurological disease? It is, therefore, potentially useful to propose models for neurosphere growth and cellular architecture (*see* Figs. 2 and 3) based on previous culture studies of neurospheres (e.g., see refs. *32,33,* and *38*). Such models can focus on dynamics of cellular distribution or positioning that accompany particular culture conditions, including the use of permissive substrates [or other conditions, e.g., feeder layers *(40)* or conditioned media (*see* refs. *36* and *41*)] that lead to migration and differentiation of neurosphere cells (e.g., *see* Fig. 3). A type of neurosphere heterogeneity also has been described with regard to the predominance toward glial verus neuronal or mixed phenotypy; using conditioned medium from B104 cell lines, oligodendrocyte spheres or "oligospheres" have been generated by Duncan and collaborators *(41)* that have the potential to produce large numbers of myelinating oligodendrocytes. These cells have tremendous clinical potential for use in demyelinating diseases, and recent studies from this group indicate the ability to isolate and cultivate these cells from the adult rodent brain *(42)*. This is in accordance with a study *(7)* that has shown the ability to clone significant numbers of neurospheres from the subependymal zone and hippocampus of the mature human brain. The heterogeneity of neuroshere-forming cells includes multipotent astrocytic stem cells *(16,43)* (*see* Fig. 1), ependymal cells *(15)*, and possibly a dedifferentiation potential of mature glial cells *(43,44)* following priming with particular growth factors.

LOOKING FOR THE NATURE
OF BRAIN STEM/PROGENITOR CELLS

It has proved difficult to identify stem cells as single cells *(13)*, even when investigating enzymatically dissociated cells of brain marrow regions *(19)*, which are known to contain a highly enriched fraction of cells with proliferative or sphere-generating potential *(33,39,45)*. It seems that by

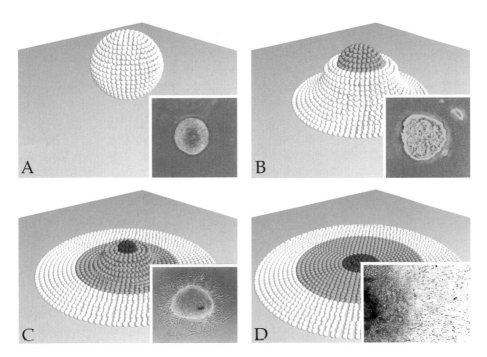

Fig. 3. Model of neurosphere architecture based on long-term observations of individual clones of brain cell clones in suspension, plated on plastic or laminin, or on different feeder layers. (**A**) and inset: A floating neurosphere prior to plating looks like a rather uniform mass under the phase microscope. (**B**) and inset: Minutes after attachment, a neurosphere begins to flatten, and the model reflects a sheet of cells (white cells) that have begun to disperse from their parental cells (darker core). (**C**) and inset: The first generation of cells begin to differentiate (dispersed cells with processes at edge of sphere in inset), whereas the younger cells (darker cells in cartoon) remain in an amorphic core. (**D**) and inset: Weeks after plating, the most primitive cells of the neurosphere occupy a central core region (they are nestin positive, as seen in the inset of an immunochemical preparation), and progressively more mature generations are found surrounding this core in a bull's-eye, targetlike manner.

proliferating into a neurosphere, stem/progenitor cells might abandon their propensity to differentiate in vitro and be maintained or even perpetuated within a neurosphere under particular culture conditions. Hence, neurospheres seem to be a major tool available for studying neural stem cells in vitro. However, as recently emphasized, "There are also basic technical issues relating to the growth and propagation of [such] cells in culture that need to be overcome . . . Specifically, the reported difficulty in dissociating . . . human ES/EG [embryonic stem and embryonic germ cell lines] cell clusters into viable single cells is problematic, particularly for gene-targeting experiments" *(46)*. ES-cell-generated cell structures, like neurospheres, thus pose a similar obstacle for gene-discovery studies. Feeder layers are a reasonable approach for capturing stem/progenitor cells of neurospheres in a particular state that affords perpetuation of the cells for further study (as has been exploited in hematopoietic bone marrow studies). Preliminary studies from our lab indicate that there are specific requirements for observing neurospheres on feeder layers, including knowing that your neurospheres are clonal in origin and determining that the original or parental neurosphere-forming cell is multipotent and perhaps even pluripotent. We have observed different effects of different tissue-derived feeder layer cells on the growth and differentiation of stem/progenitor cells and neurospheres, including a propensity for certain feeder layer cells to coax neuronal versus glial phenotypy. Specifically, we have seen that endothelial cells support the differentiation of neurons, and other cell feeder layers may support the growth of primarily glia (Scheffler and Kukekov, unpublished observations). This notion is reinforced by work from the Goldman lab, where a preliminary report *(47)* has described endothelial cell feeder layers supporting the growth of newly generated neurons.

NEUROSPHERE ARCHITECTURE

In preliminary studies of neurosphere architecture, we have found that spheres which are plated on plastic seem to reveal a similar blueprint or basic organizational plan. Shifting from a three-dimensional neurosphere (Fig. 2) into a flatter, two-dimensional structure following plating, there is a recurrence of a typical "bull's-eye" pattern (*see* Fig. 3) of a flattening neurosphere. Phase-microscopic and preliminary immunocytochemical observations have been used to generate graphic diagrams of neurosphere organization or cytoarchitecture as shown in Fig. 3. Basic immunophenotypic analyses of neurospheres has been accomplished now in many studies *(7,11,32–35,37,39,48)* and confirm the premise that these structures are derived from multipotent stem/progenitor cells that can give rise to both neurons and macroglia. Numerous genes and proteins have now been identified in neurospheres, including a variety of developmental and cell phenotype markers [e.g., nestin, vimentin, GFAP, substance P, GABA, choline acetyl transferase, glutamate, tyrosine hydroxylase, neuron-specific enolase, tau, MAP-2, neurofilament, c-*fos*, βIII tubulin, 04, galactocerebroside, the L1 adhesion molecule, bcl-2, tenascin, and growth factors receptors, including FGFR1 *(7,32–36,39,45,48)*]. It is suggested that, in flattened spheres, for the most part sheets of cells in advanced stages of maturity surround a core of the most immature cells of a neurosphere [Figs. 2 and 3, and refer to Reynolds et al. *(32)*, who described a similar organization in long-term neurosphere cultures, refering to this core as a "proliferative core" that can also appear phase dark in some of their preparations as well]. However, neurosphere architecture studies by the van der Kooy group *(37)* suggest, based on nestin staining, that the most immature cells may be distributed rather randomly throughout the neurosphere; however, we have also observed nestin-positive cells throughout adult human brain neurospheres *(7)*, and it should be noted that nestin also labels the progeny of stem/progenitor cells, including young astrocytes, adding some degree of complication to this interpretation. We do not know how many proliferative stem cells might reside in such a primitive core, nor the number of doublings or asymmetric divisions they may sustain, nor what the exact nature of the "mother" stem cell is and what factors or genes are crucial for its retention of stemness and potential quiescence. The fact that we are able to generate neurosphere clones from postmortem specimens with impressively long postmortem survival times (e.g., 5 d; *see* ref. *17*) suggests that these clonogenic cells might possess extremely low metabolic requirements in vivo and, therefore, they are able to survive potentially harsh conditions (e.g., limited oxygen and nutrients resulting from constraints of cell-packing density) within the center of a neurosphere. When a neurosphere attaches, the subsequent morphological changes indicate an outgrowth (or "melting down"; see Fig. 3) of cells that appear to make up the outer layers of the neurosphere. These are the cells that might be first observed to migrate radially away from the neurosphere's center and, eventually, these cells are the quickest to mature. Although the attached structure flattens, a proliferative core becomes visible and, ultimately, gives rise to an intermediate population of cells that may remain quiescent until the first fully differentiated cells on the edge of the flattening sphere disappear. Eventually, the core will disappear and all that remains are differentiated cells. Immunolabeling with antibodies to the putative neural stem cell marker nestin *(13)* reveals large numbers of densely packed labeled cells within the often phase-dark core, in addition to scattered labeled cells within the surrounding rings (Fig. 3D). The immature phase-dark core eventually appears to become quiescent (e.g., nestin expression disappears). These preliminary observations indicate that neurosphere models shown in Figs. 2A and 2B are possible candidates for neurosphere architecture, and the idea of concentric sheets of maturing cells, with the innermost representing the least committed and the outermost representing the most differentiated has been reinforced by studies from the Svendsen group *(38)*, which found, using a living neurosphere sectioning and subcloning method, that the outermost rim of neurosphere cells is the most differentiated. However, as already pointed out earlier, it is possible that there is a great deal of neurosphere heterogeneity *(7,8,35)*, possibly related to stem/progenitor cell heterogeneity based on the time of generation of the ancestoral "mother" stem cell, regional variation (related to different CNS brain marrow regions), and/or variations in response to different culture conditions. Because neurospheres are not a pure population of stem/progenitor cells, but rather a mixture of cells in various stages of differentiation, future experiments might exploit comparative population assays, including comparing fetal, postnatal, and adult mouse brain clones, because previous studies indicate that the age of the brain affects the extent of differentiation of stem/progenitor cells, with the youngest specimens (e.g., embryonic brain) containing the largest numbers of multipotent cells *(7,49)*. Additionally, long-term culturing approaches (including the aforementioned use of feeder layers), as developed for hematopoietic stem cells and their various colony-forming units *(50)*, as well as designer media that facilitates the growth of particular neural stem/progenitor cells and their different progeny, also will likely be exploited. Distinct feeder layer cells and other novel culture conditions might acerbate the isolation and growth of different stem/progenitor cells and encourage the selective propagation and differentiation of these cells and lead to increased numbers of certain populations of their more commited descendants. There are, however, other approaches for achieving this, including immunopanning with antibodies to developmentally regulated cell surface proteins [e.g., PSA-NCAM *(51)*; AC133 *(52)*] that allow the isolatation of particular populations of stem/progenitor cells for the production of enriched primary cultures as well as cell lines.

ISOLATING THE NEUROSPHERE-FORMING CELL: NEW FACTORS AND MARKERS

Once distinct stem/progenitor cells can be propagated over longer periods of time and subpopulations of these cells can be induced to proliferate and differentiate via the application of particular factors, including growth factors like EGF, bFGF, ciliary neurotrophic factor (CNTF), and leukemia-inhibitory factor (LIF), steroid hormones such as T3 and retinoic acid, or other morphogenetic molecules, including transcription factors, developmentally regulated genes and their products *(13,32–34,36,37,39,53,54)*, then we will have established a methodology with which to study the maturational process of a stem/progenitor cell all the way to fully differentiated cells. Future studies should define which factors are responsible for discrete effects on stem/progenitor cell proliferation, growth, and differentiation. This includes the characterization of factors present in media, derived from particular feeder layers, or naturally present in the microenvironment of the developing nervous system (that feeder layers are meant, at least in part, to mimic) that give rise to selective effects on expansion, potency, fate choice, and terminal differentiation. To date, molecular studies of neurospheres have focused on genetic analyses of populations rather than individual neurospheres, perhaps because of difficulties in obtaining mRNA from mechanical or chemical disruption of single neurospheres. Reverse transcription–polymerase chain reaction (RT-PCR) has been applied to single and populations of neurospheres (e.g., *see* Fig. 4) for the confirmation of cell phenotype- and growth factor-related molecules associated with these unique structures *(54–57)*. There is an ongoing search for "neuropoietins"—factors that have actions like erythropoietin and GMCSF in the expansion of primitive hematopoietic stem/progenitor cells—but these and other known hemapoietic factors have thus far failed to affect neural stem/progenitor cells in culture *(54)*. Factors that stimulate asymmetrical divisions of stem/progenitor cells also have yet to be fully characterized, although there are candidates genes and their products, such as NOTCH and other developmentally regulated molecules [e.g., NUMB *(58)*], that have been implicated in cell proliferation and cell fate choices. Molecular analyses of stem/progenitor cells and neurospheres have been carried out, mostly with the goal of confirming the presence of particular transcripts for developmental and neurotransmitter-related genes in populations of neurospheres (*see* refs. *13,32,34,48,54,59,60,* and *61*). Studies also have characterized the expression of genes involved in regional specification and neuronal differentiation and survival in adult human brain-derived neurospheres [e.g., *Pax, bcl*-2 *(7)*. However, to further probe stem/progenitor cell and neurosphere heterogeneity as described earlier, there is a need to begin characterizing gene expression in single neurospheres (see Fig. 4, which shows a comparison of single versus multiple neurosphere RT-PCR for a variety of developmental and cell phenotype gene transcripts) in order to begin to determine differential patterns of gene expression that may underly heterogeneity. At the same time, because neurospheres are a model of neuromorphogenesis as it may occur in vivo, the characterization of new genes or even discrete temporal patterns of expression of particular gene families may lead to the discovery of factors responsible for the spatiotemporal variance in CNS neuromorphogenesis. So far, however, there is only a short window of opportunity available to study the effects of potential agents in the differentiating/maturing process of NFCs (that is, the time

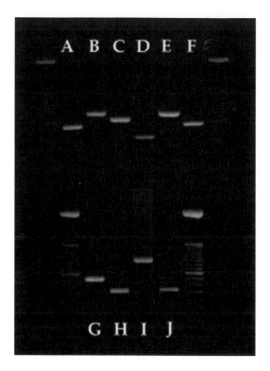

Fig. 4. RT-PCR products from an individual **(A–F)** compared to a population **(G–J)** of neurospheres looking at a variety of precursor (e.g., nestin), extracellular matrix (e.g., tenascin), neuronal (e.g., MAP-2, neurofilament), and glial (e.g., glial fibrillary acidic protein, GFAP) gene markers. On either side of **(A)** and **(J)** are DNA ladders. **(A)** β-actin; **(B)** GFAP; **(C)** tenascin (nested primer); **(D)** nestin (nested primer); **(E)** neurofilament-m; **(F)** HuD (a neuron and neuron progenitor marker); **(G)** β-actin; **(H)** tenascin; **(I)** GFAP (nested primer); and **(J)** nestin (nested primer).

between floating [i.e., in suspension] and attaching [which results in their immediate switching from a proliferation to a differentiation mode]) and additional methods and approaches are needed that protract the growth and maturation of stem/progenitor cells and neurospheres in culture (as referred to earlier regarding long-term culturing), in order, for example, to bioassay the efficacy of new neuromorphogenetic factors. Although the process of neurosphere formation remains a matter of speculation and need further investigation, one can assume that the NFC is always present in a neurospheres, unless it has been transformed during the process of in vitro propagation. In order to specifically modify stem/progenitor cells, to drive them toward a desired phenotype as well as perhaps perform "gene therapy" on these cells, it would be fortuitous to first localize the most primitive stem/progenitor cells within the neurosphere. As an example, before narrowing their commitment, the most immature subset of proliferating hippocampal progenitor cells seems to express a characteristically ambiguous set of markers [as seen with hematopoietic stem cells (e.g., CD34bright-, Thy1-, CD38-, etc.); see ref. *8* for review), including nestin, NCAM, Map2c, NSE, and O4 *(61)*. In this study, lineage ambiguity was resolved only after complete differentiation in culture when concensus lineage specific markers were expressed (i.e., neuronal markers such as Map2ab, Map5, NeuN, or NF200, astrocyte markers, including GFAP, and oligodendrocyte immunomarkers, including GalC and MBP). As of now, there seems to be no marker available that uniquely labels the quiescent/slow-cycling neural stem/progenitor cell, and there is hope that, through the application of new molecular and genetic dissections, such markers will become available not only for identifying the most primitive stem/progenitor cell both in neurospheres as well as in vivo but also to disclose biogenic factors associated with these markers that might allow us to study their origins and nature and, ultimately, manipulate their fate. Of course, there are a variety of markers and conditions that already have been exploited to enrich, for example, neuronal progenitors [e.g., PSA-NCAM *(51)*; GFP-tubulin *(62)*] and even distinct classes of neurons [e.g., dopamine neurons, using reduced oxygen *(63)*, and astrocyte feeder layers in combination with dopaminizing factors *(64)*].

CONCLUSION: THE BIOTECHNOLOGY OF NEUROSPHERES

How might neurospheres further challenge our understanding of neuromorphogenesis? There is a need for new techniques for defining cell–cell interactions and factors that support discrete aspects of neural cell growth and differentiation. Because we at least have insights gained from the 50-yr-old hematopoiesis field, where, again, stem/progenitor cells form distinct spherelike colony units in culture that are amenable to cell and molecular biological analyses (see ref. *8* for a review of bone marrow hematopoiesis versus brain marrow neuropoiesis), then there is hope that new technologies can be developed for the unique issues surrounding the expansion and controlled differentiation of neural stem/progenitor cells. An important issue is how we should describe the heterogeneity of cell morphotypes that are present along the continuous pathway from the most primitive ancestoral stem/progenitor cell to fully differentiated cells. From a biotechnological standpoint, this question can be rephrased to be "What are the most reliable markers that can be used to characterize the cells along this pathway, alone or in combinations?" Another challenge is to establish technology for the separation of particular subtypes of cells in order to obtain pure or enriched cultures of cells at different stages of their development or that represent distinct lineages. A third challenge is to determine what combination of factors derived from different sources (e.g., feeder layers or other culture conditions) can encourage the propagation of distinct neuropoietic morphotypes; subsequent studies could focus on differentiation factors that push the stem/progenitor cell population toward a particular fate. Along this line, neurospheres might also be used in coculture paradigms to neuralize stem/progenitor cells from other tissues (e.g., hematopoietic stem cells), producing different populations of neurons for possible cell therapeutics for neurological disease.

Neurospheres can be used to solve some of these challenges; furthermore, they can be used as a source or model system for gene discovery that could ultimately address all of these challenges by providing data on the temporal patterns of gene expression seen during the maturation of individual neurospheres. A new research direction in this field will be the establishment of new approaches for the discovery and profiling of genes expressed during neuropoiesis. This could be a powerful approach, as many previous studies defining gene expression during stages of neuromorphogenesis have relied on the isolation of nucleic acids using in vivo models. Neurospheres originate from neural stem/progenitor cells that carry transcripts of genes involved in neural cell proliferation, survival/death, and differentiation, and they may represent different stages of brain development because stem/progenitor cells may have been suspended in time from different stages of brain development (e.g., in response to the cessation of growth factor expression at the end of postnatal neurogenesis, leaving stem/progenitor cells in different states including quiescence), and the in vitro growth of a single neurosphere also may mirror many aspects of neuromorphogenesis (perhaps a virtual history of the germinal matrix of the developing brain, compressed within diverse populations of stem/progenitor cells and neurospheres; see ref. 8). It is anticipated that the development of new technology for the study of these brain cell clones, including the discovery of factors that support their long-term growth and survival, may eventually lead to therapeutics that stimulate intrinsic regenerative programs, even in the compromised, aged brain.

ACKNOWLEDGMENTS

This work was supported by NIH/NINDS grant NS37556, the McKnight Brain Institute and Shands Cancer Center of the University of Florida, and a gift from the Harter Family.

REFERENCES

1. Thomson, J. A., Itskovitz-Eldor, J., Shapiro, S. S., Waknitz, J. A., Swiergiel, J. J., Marshall, V. S., et al. (1998) Embryonic stem cell lines derived from human blastocysts. *Science* **282,** 1145–1147.
2. Shamblott, M. M., Axelman, J., Wang, S., Bugg, E. M., Littlefield, J. W., Donovan, P. J., et al. (1998) Derivation of pluripotent stem cells from cultured human primordial germ cells. *Proc. Natl. Acad. Sci. USA* **95,** 13,726–13,731.
3. Wilmut, I., Schnieke, A. E., McWhir, J., Kind, J., and Campell, K. H. (1997) Viable offspring derived from fetal and adult mammalian cells. *Nature* **385,** 810–813.
4. Wakayama, T., Rodriguez, I., Perry, A. C., Yanagimachi, R., and Mombaerts, P. (1999) Mice cloned from embryonic stem cells. *Proc. Natl. Acad. Sci. USA* **96,** 14,984–14,989.
5. Gage, F., Coates, P. W., Palmer, T. D., Kuhn, H. G., Fisher, L. J., et al. (1995) Survival and differentiation of adult neuronal progenitor cells transplanted to the adult brain. *Proc. Natl. Acad. Sci. USA* **92,** 11,879–11,883.
6. Eriksson, P. S., Perfilieva, E., Bjork-Eriksson, T., Alborn, A.-M., Nordborg, C., Peterson, D. A., et al. (1998) Neurogenesis in the adult human hippocampus. *Nat. Med.* **4,** 1313–1317.
7. Kukekov, V. G., Laywell, E. D., Thomas, L. B., Scheffler, B., Davies, K., O'Brien, T. F., et al. (1999) Multipotent stem/progenitor cells with similar properties arise from two neurogenic regions of adult human brain. *Exp. Neurol.* **156,** 333–344.
8. Scheffler, B., Horn, M., Blumcke, I., Laywell, E. D., Coomes, D., Kukekov, V. G., et al. (1999) Marrow-mindedness: a perspective on neuropoiesis. *Trends Neurosci.* **22,** 348–357.
9. Allen, E. (1912) Cessation of mitosis in central nervous system of the albino rat. *J. Comp. Neurol.* **22,** 547–568.
10. Altman, J. and Das, G. D. (1965) Autoradiographic and histological evidence of postnatal hippocampal neurogenesis in rats. *J. Comp. Neurol.* **124,** 319–335.
11. Reynolds, B. A. and Weiss, S. (1992) Generation of neurons and astrocytes from isolated cells of the adult mammalian central nervous system. *Science* **255,** 1707–1710.
12. Richards, L. J., Kilpatrick, T. J., and Bartlett, P. F. (1992) De novo generation of neuronal cells from the adult mouse brain. *Proc. Natl. Acad. Sci. USA* **89,** 8591–8595.
13. McKay, R. D. G. (1997) Stem cells in the central nervous system. *Science* **276,** 66–71.
14. Thomas, L. B., Gates, M., and Steindler, D. A. (1996) Young neurons from the adult mouse subependymal zone migrate and proliferate along an astrocyte, extracellular matrix-rich pathway. *Glia* **17,** 1–14.
15. Johansson, C. B., Momma, S., Clarke, D. L., Risling, M., Lendahl, U., and Frisen, J. (1999) Identification of a neural stem cell in the adult mammalian central nervous system. *Cell* **96,** 25–34.

16. Doetsch, F., Caille, I., Lim, D. A., Garcia-Verdugo, J. M., and Alvarez-Buylla, A. (1999) Subventricular zone astrocytes are neural stem cells in the adult mammalian brain. *Cell* **97,** 703–716.

17. Laywell, E. D., Kukekov, V. G., and Steindler, D. A. (1999) Multipotent neurospheres can be derived from forebrain subependymal zone and spinal cord of adult mice after protracted postmortem intervals. *Exp. Neurol.* **156,** 430–433.

18. Palmer, T. D., Schwartz, P. H., Taupin, P., Kaspar, B., Stein, S. A., and Gage, F. H. (2001) Cell culture. Progenitor cells from human brain after death. *Nature* **411,** 42–43.

19. Steindler, D. A., Kadrie, T., Fillmore, H., and Thomas, L. B. (1996) The subependymal zone: "brain marrow." *Prog. Brain Res.* **108,** 349–363.

20. Bjornson, C. R. R., Rietze, R. L., Reynolds, B. A., Magli, M. C., and Vescovi, A. L. (1999) Turning brain into blood: a hematopoietic fate adopted by adult neural stem cells in vivo. *Science* **283,** 534–537.

21. Galli, R., Borelo, U., Gritti, A., Minasi, G., Bjornson, C., Coletta, M., et al. (2000) Skeletal myogenic potential of human and mouse neural stem cells. *Nat. Neurosci.* **3,** 986–991.

22. Tsai, R. Y. L. and McKay, R. D. G. (2000) Cell contact regulates fate choice by cortical stem cells. *J. Neurosci.* **20,** 3725–3735.

23. Clarke, D. L., Johansson, C. B., Wilbertz, J., Veress, B., Nilsson, E., Karlstrom, H., et al. (2000) Generalized potential of adult neural stem cells. *Science* **288,** 1660–1663.

24. Sanchez-Ramos, J., Song, S., Cardozo-Pelaez, F., Hazzi, C., Stedeford, T., Willing, A., et al. (2000) Adult bone marrow stromal cells differentiate into neural cells in vitro. *Exp. Neurol.* **164,** 247–256.

25. Woodbury, D., Schwarz, E. J., Prockop, D. J., and Black, I. (2000) Adult rat and human bone marrow stromal cells differentiate into neurons. *J. Neurosci. Res.* **61,** 364–370.

26. Mezey, E., Chandross, K. J., Harta, G., Maki, R. A., and McKercher, S. R. (2000) Turning blood into brain: cells bearing neuronal antigens generated in vivo from bone marrow. *Science* **290,** 1779–1782.

27. Brazelton, T. R., Rossi, F. M., Keshet, G. I., and Blau, H. M. (2000) From marrow to brain: expression of neuronal phenotypes in adult mice. *Science* **290,** 1775–1779.

28. Cao, Q. L., Zhang, Y. P., Howard, R. M., Walters, W. M., Tsoulfas, P., and Whittemore, S. R. (2001) Pluripotent stem cells engrafted into the normal or lesioned adult rat spinal cord are restricted to a glial lineage. *Exp. Neurol.* **167,** 48–58.

29. Brustle, O., Maskos, U., and McKay, R. D. (1995) Host-guided migration allows targeted introduction of neurons into the embryonic brain. *Neuron* **15,** 1275–1285.

30. Snyder, E. Y., Yoon, C., Flax, J. D., and Macklis, J. D. (1997) Multipotent neural precursors can differentiate toward replacement of neurons undergoing targeted apoptotic degeneration in adult mouse neocortex. *Proc. Natl. Acad. Sci. USA* **94,** 11,663–11,668.

31. Auerbach, J. M., Eiden, M. V., and McKay, R. D. (2000) Transplanted cells form functional synapses in vivo. *Eur. J. Neurosci.* **12**, 1696–1704.
32. Reynolds, B., Tetzlaff, W., and Weiss, S. (1992) A multipotent EGF-responsive striatal embryonic progenitor cell produces neurons and astrocytes. *J. Neurosci.* **12**, 4565–4574.
33. Gritti, A., Parati, E. A., Cova, L., Frolichsthal, P., Galli, R., Wanke, E., et al. (1996) Multipotential stem cells from the adult mouse brain proliferate and self-renew in response to basic fibroblast growth factor. *J. Neurosci.* **16**, 1091–1100.
34. Weiss, S., Dunne, C., Hewson, C. J., et al. (1996) Multipotent CNS stem cells are present in the adult mammalian spinal cord and ventricular neuroaxis. *J. Neurosci.* **16**, 7599–7609.
35. Kukekov, V. G., Laywell, E. D., Thomas, L. B., and Steindler, D. A. (1997) A nestin-negative precursor cell from the adult mouse brain gives rise to neurons and glia. *Glia* **21**, 399–407.
36. Daadi, M. and Weiss, S. (1999) Generation of tyrosine hydroxylase-producing neurons from precursors of the embryonic and adult forebrain. *J. Neurosci.* **19**, 4484–4497.
37. Tropepe, V., Sibilia, M., Ciruna, B. G., Rossant, J., Wagner, E. F., and van der Kooy, D. (1999) Distinct neural stem cells proliferate in response to EGF and FGF in the developing mouse telencephalon. *Dev. Biol.* **208**, 166–188.
38. Svendsen, C. N., ter Borg, M. G., Armstrong, R. J., Rosser, A. E., Chandran, S., Ostenfeld, T., et al. (1998) A new method for the rapid and long term growth of human neural precursor cells. *J. Neurosci. Meth.* **85**, 141–152.
39. Reynolds, B. A. and Weiss, S. (1996) Clonal and population analyses demonstrate that an EGF-responsive mammalian embryonic CNS precursor is a stem cell. *Dev. Biol.* **175**, 1–13.
40. Temple, S. and Davis, A. A. (1994) Isolated rat cortical progenitor cells are maintained in division in vitro by membrane-associated factors. *Development* **120**, 999–1008.
41. Zhang, S. C., Lipsitz, D., and Duncan, I. D. (1998) Self-renewing canine oligodendroglial progenitor expanded as oligospheres. *J. Neurosci. Res.* **54**, 181–190.
42. Zhang, S. C., Ge, B., and Duncan, I. D. (1999) Adult brain retains the potential to generate oligodendroglial progenitors with extensive myelination capacity. *Proc. Natl. Acad. Sci. USA* **96**, 4089–4094.
43. Laywell, E. D., Rakic, P., Kukekov, V. G., Holland, E. C., and Steindler, D. A. (2000) Identification of a multipotent astrocytic stem cell in the immature and adult mouse brain. *Proc. Natl. Acad. Sci. USA* **97**, 13,883–13,888.
44. Kondo, T. and Raff, M. (2000) Oligodendrocyte precursor cells reprogrammed to become multipotential CNS stem cells. *Science* **289**, 1754–1756.
45. Davis, A. and Temple, S. (1994) A self-renewing multipotential stem cell in embryonic rat cerebral cortex. *Nature* **372**, 263–266.
46. Keller, G. and Snodgrass, H. R. (1999) Human embryonic stem cells: the future is now. *Nat. Med.* **5**, 151–152.

47. Leventhal, C., Rafii, S., Shahar, A., Rafii, D., Grafstein, B., and Goldman, S. A. (1998) Endothelial neurotrophic support of adult neurogenesis. *Neurosci. Abstr.* **24,** 1276.
48. Vescovi, A. L., Reynolds, B. A., Fraser, D. D., and Weiss, S. (1993) bFGF regulates the proliferative fate of unipotent (neuronal) and bipotent (neuronal/ astroglial) EGF-generated CNS progenitor cells. *Neuron* **11,** 952–966.
49. Kuhn, H., Dickinson-Anson, H., and Gage, F. (1996) Neurogenesis in the dentate gyrus of the adult rat: age-related decrease of neuronal progenitor proliferation. *J. Neurosci.* **16,** 2027–2033.
50. Gartner, S. and Kaplan, H. S. (1980) Long-term culture of human bone marrow cells. *Proc. Natl. Acad. Sci. USA* **77,** 4756–4759.
51. Mayer-Proschel, M., Lalyani, A. J., Mujtaba, T., and Rao, M. S. (1997) Isolation of lineage restricted neuronal precursors from multipotent neuroepithelial stem cells. *Neuron* **19,** 773–785.
52. Uchida, N., Buck, D. W., He, D., Reitsma M. J., Masek, M., Phan, T. V., et al. (2000) Direct isolation of human central nervous system stem cells. *Proc. Natl. Acad. Sci. USA* **97,** 14,720–14,725.
53. Koblar, S. A., Turnley, A. M., Classon, B. J., Reid, K. L., Ware, C. B., Cheema, S. S., et al. (1998) Neural precursor differentiation into astrocytes requires signaling through the leukemia inhibitory factor receptor. *Proc. Natl. Acad. Sci. USA* **95,** 3178–3181.
54. Johe, K. K., Hazel, T. G., Muller, T., Dugich-Djordjevic, M. M., and McKay, R. D. G. (1996) Single factors direct the differentiation of stem cells from the fetal and adult central nervous system. *Genes Dev.* **10,** 3129–3140.
55. Suslov, O. N., Kukekov, V. G., Laywell, E. D., Scheffler, B., and Steindler, D. A. (2000) RT-PCR Amplification of mRNA from individual brain clones or "neurospheres" using sonication. *J. Neurosci. Methods* **96,** 57–61.
56. Suslov, O. N., Steindler, D. A., Juravleva, M., and Kukekov, V. G. (2000) Gene discovery and temporal profiling using uncloned cDNA libraries from neural stem/progenitor cell clones. *Neurosci. Abstr.* **26,** 832.
57. Geschwind, D. H., Ou, J., Easterday, M. C., Dougherty, J. D., Jackson, R. L., Chen, Z., et al. (2001) A genetic analysis of neural progenitor differentiation. *Neuron* **29,** 325–339.
58. Wakamatsu, Y., Maynard, T. M., Jones, S. U., and Weston, J. A. (1999) NUMB localizes in the basal cortex of mitotic avian neuroepithelial cells and modulates neuronal differentiation by binding to notch-1. *Neuron* **23,** 71–81.
59. Arsenijevic, Y. and Weiss, S. (1998) Insulin-like growth factor-I is a differentiation factor for postmitotic CNS stem cell-derived neuronal precursors: distinct actions from those of brain-derived neurotrophic factor. *J. Neurosci.* **18,** 2118–2128.
60. Ahmed, S., Reynolds, B. A., and Weiss, S. (1995) BDNF enhances the differentiation but not the survival of CNS stem cell-derived neuronal precursors. *J. Neurosci.* **15,** 5765–5778.
61. Palmer T. D., Takahashi, J., and Gage, F. H. (1997) The adult rat hippocampus contains primordial neural stem cells. *Mol. Cell. Neurosci.* **8,** 389–404.

62. Wang, S., Wu, H., Jiang, J., Delohery, T. M., Isdell, F., and Goldman, S. A. (1998) Isolation of neuronal precursors by sorting embryonic forebrain transfected with GFP regulated by the T alpha1 tubulin promoter. *Nat. Biotechnol.* **16,** 196–201.
63. Studer, L., Csete, M., Lee, S. H., Kabbani, N., Walikonis, J., Wold, B., et al. (2000) Enhanced proliferation, survival, and dopaminergic differentiation of CNS precursors in lowered oxygen. *J. Neurosci.* **20,** 7377–7383.
64. Wagner, J., Akerud, P., Castro, D. S., Holm, P. C., Canals, J. M., Snyder, E. Y., et al. (1999) Induction of a midbrain dopaminergic phenotype in Nurr1-overexpressing neural stem cells by type 1 astrocytes. *Nat. Biotechnol.* **17,** 653–659.

8

Neural Stem Cells and Specification of Cell Fates

Biological and Therapeutical Perspectives

Marcel M. Daadi

INTRODUCTION

Stem cells, whether blastocyst or organ/tissue derived, are increasingly becoming of interest to the general public, politicians, and scientists. This is reflected by the popularity of stem cell research in the media and the dramatic increase of scientific productivity in the field. Stem cells provide an exciting opportunity to understand the basic mechanisms involved in histogenesis, mechanisms that may translate into prospective applications in tissue engineering and biomedicine in general.

In this chapter, we will be discussing some of the recent advances made toward answering the fascinating questions of how a neural stem cell (NSC) maintains itself and how its progeny choose their fate. We will discuss current principles of cellular differentiation with a particular interest in the generation of dopaminergic neurons from dopaminergic and nondopaminergic sources.

IDENTITY OF NEURAL STEM CELLS

There is extensive literature on the definition of NSCs. Classically, NSCs are defined by the ability to self-renew and generate a large number of progeny able to differentiate into the principal central nervous system (CNS) cell types: neurons, astrocytes, and oligodendrocytes. NSCs have the ability to maintain themselves in culture under genetic or epigenetic stimulation and to generate a large number of progeny. This definition will be in use until a specific cellular marker, necessary and sufficient to identify a NSC, is determined. This identification will also shed light on the current debate regarding the origin and identity of NSCs in the CNS.

From: *Neural Stem Cells for Brain and Spinal Cord Repair*
Edited by: T. Zigova, E. Y. Snyder, and P. R. Sanberg © Humana Press Inc., Totowa, NJ

It has long been suggested that ependymal cells lining the ventricular system may undergo neurogenesis, especially after CNS injury in the lower vertebrate, and replace lost neurons *(1–3)*. In the mammalian CNS, however, conflicting data have been published. Initially, a study demonstrated that ependymal cells are able to proliferate in vitro and generate clusters of undifferentiated cells that have the ability to differentiate into all CNS cell types *(4)*. Thus, the claim is that ependymal cells are NSCs. A second study demonstrated that ependymal cells have limited potential in vitro and give rise to an astrocytic lineage only *(5,6)*. A third study demonstrated that ependymal cells are incapable of proliferating in vitro *(7)* and that NSCs are, in fact, a separate well-defined subpopulation of astrocytes residing within the subependymal zone *(7)*. Using a clonogenic stem cell assay, a recent study *(6)* has simultaneously examined the multipotentiality of both ependymal and astrocyte-generated clones. To identify the founder astrocyte of all the in vitro-generated progeny, Laywell and colleagues *(6)* took advantage of the *Gtv-a* mice that carry a transgene-encoding receptor for the avian leukosis virus under control of the glial fibrillary acidic protein (GFAP) promoter. Astrocytes from these mice were then infected with replication-competent avian leukosis-virus-vector-encoding human alcaline phosphatase. The selected astrocytic clone expressing the reporter gene was found to generate both neurons and glia, thus confirming that the subependymal zone (SEZ)-derived astrocytes are multipotent, whereas the ependymal-derived clones generated astrocytes only. The common astrocytic future of the SEZ-derived NSC, ependymal cells, and radial glia and their correlative function in the developing brain led the authors to suggest a possible phylogenetic relation to the invertebrates ependymoglial cells *(6)*. In the cerebral cortex of lizards, ependymocytes show a high degree of plasticity evident during lesion-induced regeneration *(8)*. In this model, ependymal cells appear to generate neurons *in situ*, remove postlesion debris, and form GFAP immunoreactive radial glia-ependymocytes that insulate blood capillaries, thus monitoring or enforcing the blood-brain barrier. More intriguing is the finding that neural, endothelial and hematopoetic cells express the surface antigen CD34 that is specific for hematopoetic stem cells *(9,10)*. As a common marker between neural and hematopoetic tissue, this suggests that endothelial and/or ependymal cells may act as a potential interface between blood and neural tissues, which may offer an explanation of how bone-marrow-derived precursors enter the nervous system and differentiate into neurons and glia *(11,12)*. In the subgranular zone (SGZ) of the adult hippocampus, continuous regeneration of neurons is known to take place throughout adult life, presumably from a vestigial NSC *(13–15)*.

A recent study *(16)* brought a twist into the origin of this continuous neurogenesis by demonstrating that 37% of cells proliferating in the SGZ are endothelial precursors. These angioblasts may be associated with neural precursors and give rise to clusters of cells formed in a mixed population of endothelial cells, neurons and glia. Alternatively, these angioblasts may transdifferentiate, especially in brain injury or a diseased situation, into neural precursors that give rise to neural progeny. Together, these observations favor the hypothesis of a common stem cell for multiple tissues/organs trafficking throughout the circulatory system *(17)*.

From these studies, an interesting point was addressed: In vitro culture conditions may induce dedifferentiation of astrocytes, leading them to an "archaic" immature and multipotent stage. In this regard, there is growing attention on the potential transformation/dedifferentiation NSCs may undergo during extensive in vitro expansion paradigm *(18)*. Under in vivo physiologic conditions, one may expect that NSCs will rarely undergo cell division while their probability of self-maintenance remains high. By dividing asymmetrically, these stem cells will give rise to transitory generations that migrate out of the stem cell niche to the subependymal layer, where they actively proliferate. In addition, as documented earlier, converging evidence suggests that the default fate of endogenous or in vitro isolated NSCs is astroglial cells *(6,7,19)*. If it holds, this phenomenon will recapitulate the ontogenesis of neurons and glia as it happens early during development. The astroglial phenotype and dynamic nature of a stem cell within the brain make it challenging to identify and track its behavioral patterns.

The cardinal property of NSCs that is decisive of their potential therapeutic application, is the ability to repopulate an injured or diseased area of the brain after implantation. This main attribute can only and effectively be defined by transplanting the cells into an environment that provides them with the opportunity to express their potential. For instance, transplantation studies have demonstrated that NSCs survive, engraft, and differentiate according to local environmental cues *(20–23)*. In addition, in vivo work has shown that NSCs express foreign genes *(21)*, do not form tumors *(24)*, and repopulate diseased or injured regions of the CNS *(25–28)*. In the mutant shiverer mouse model of global myelin basic protein (MBP) dysfunction, intracerebroventricular implantation of clonal neural stem cells resulted in widespread engraftment, differentiation into oligodendrocytes, and production of MBP *(28)*. These transplanted animals showed improvement in their behavioral deficits *(28)*. In the adult mouse, neocortex undergoing targeted photolitic cell death, transplanted NSCs differentiated into neurons with morphological characteristics similar to the degenerated pyramidal neurons *(27)*. In a recent study, Snyder and colleagues have demonstrated that transplanted NSCs have the ability to target tumor cells *(29)*. Furthermore, when genetically engineered to produce cytosine deaminase, an enzyme that converts 5-fluorocytosine to the oncolytic drug 5-fluorouracil, NSCs were able to kill tumor cells and reduce tumor mass in vivo *(29)*. This migratory ability of NSCs has been recently confirmed in primates *(30)*. In this report, a clone of undifferentiated NSCs were implanted into the lateral cerebral ventricle of a fetal monkey. After a 1-mo posttransplantation survival time, NSCs engrafted, migrated into different regions of the primate brain, and assumed the fate of the local host cells. In the cortex, for instance, NSCs migrated into different cortical layers and differentiated appropriately into neurons and glial cells in layers II/III and IV–VI, respectively *(30)*.

Together, these in vitro and in vivo data suggest that NSCs are a suitable alternative source of cells for neural transplantation therapy. For focal neurodegenerative diseases, NSC progeny may need to be directed *a priori* into a specific neuronal differentiation pathway aimed to replace missing neurons in the injured or diseased CNS area. This concept leads us to a bifurcating situation that is to either manipulate stem cells ex vivo and direct them to a specific fate before implantation or to implant them as a quasihomogeneous nestin-positive population into the diseased brain and let the later influence their destiny. Both strategies may be suitable depending on the indicated destination. In a comparison to differentiated progeny, undifferentiated NSCs are endowed with a great migratory potential both in vitro and after implantation and will be suitable for global diseases. Alternatively, especially for focal diseases, the differentiation of stem cell progeny may be driven to a committed transit generation toward the desired phenotype. This may be achieved by permanently inducing early and critical intracellular determinant(s) that basically sets the direction to the specific neuronal lineage, then, after implantation, letting the host tissue achieve the last steps of maturation that bring the progeny to their functional competence. However, variations in the preclinical efficacy of a given neural stem cell line may exist as a result of conditions of derivation, clonality, regionality, or even the animal model used. Thus, each cell line or clone of NSCs may or may not require ex vivo manipulation prior to implantation. Given the tremendous advances during the last decade in tissue engineering technology, we are starting to understand and master these fleeting mechanisms regulating neuronal development. One concept agreed upon is that understanding the inherent mechanisms of the developing brain in generating different specialized cell types will help us design successful cellular engineering strategies. However, these mechanisms in generating specialized neurons may not be the only way, as we are starting to experiment with both neural and embryonic stem cells.

DIFFERENTIATION POTENTIALS OF NEURAL STEM CELLS

Cellular differentiation may be defined as a multistep process driving a given cell from a precursor stage to functional competence. These differentiation steps often are manifested by changes in cellular morphology and by the appearance of new gene products. Each differentiation step is timely orchestrated and often depends on an interplay between the cell's intrinsic and extrinsic programs. Knowledge in both extrinsic differentiation signals and the molecular machinery underlying the intrinsic events are rapidly progressing.

Extrinsic cues may regulate neuronal diversity by selectively rescuing a specific subpopulation of neuronal precursors committed to express a specific neurotransmitter phenotype or by instructing the neuronal precursors during a narrow developmental window to adopt a specific fate. However, it is not as simple as it sounds; other factors including physiologic inhibitors and potentiators, biological sensors, cell–cell interaction, and gradient concentration of signals may come into play and modify the outcome of the inductive action. It is important at this stage to at least attempt to define what may be a cellular biological sensor. The biological sensor may be defined as the process by which a stem cell measures and stores information about time, space, and mitotic history, which, as discussed below, may regulate the stem cell's functional properties over time. This intrinsic information may be coupled to extrinsic cues and dictate the cell fate.

Early work with the glial progenitor cells derived from the rat optic nerve that have the ability to differentiate into either an oligodendrocyte or type-2 astrocyte (O-2A) *(31)* have highlighted the convergence of cell intrinsic programs and extracellular signals in the differentiation process. Platelet-derived growth factor (PDGF) was identified as a critical factor in the differentiation of oligodendrocytes from the O-2A progenitors *(32)*. In these early studies, this single growth factor exerted what we now may define as an instructive action on the O-2A-derived progeny to differentiate into oligodendrocytes. Interestingly, the generation of oligodendrocytes seems to occur according to an internal biological clock of the O-2A *(33,34)* that may be regulated by stochastic and deterministic mechanisms *(35)*. The inductive action of PDGF was overridden by the simultaneous addition of basic fibroblast growth factor (FGF2), which alone induced a premature differentiation into oligodendrocytes *(36)*. In the presence of both factors, the O-2A progenitors acquired an unlimited potential of self-renewal with no apparent differentiation. Surprisingly, the oligodendrocyte-restricted progenitors purified from the postnatal day 6 rat optic nerve by sequential immunopanning *(36)*, thus termed oligodendrocyte precursor cells (OPC), were shown to be reprogrammed into multipotential CNS stem cells with broader developmental potential *(37)*. This reprogrammation was induced first by cultivating the cells in fetal calf serum (FCS) for 3 d and, subsequently, by removing FCS and PDGF and then supplementing the medium with FGF2. In the presence of adherent substrate, the OPCs reverted to a bipolar morphology characteristic of the prenatal O-2A, whereas in the absence of an adherent substrate, the OPCs generated a free-floating cluster of cells with the ability to differentiate into neurons, astrocytes, and oligodendrocytes *(37)*.

Neurogenesis Versus Gliogenesis

Recent studies by Temple and colleagues illustrate the effects of gradient concentration, developmental stage, and biological sensor on the functional properties of the cortex-derived stem cells *(38–40)*. Using standardized clonal growth conditions in Terasaki wells and time-lapse video microscopy, the authors were able to examine, at the cellular and chronological levels, the neurogenic and gliogenic capacities of the embryonic cortical stem cells *(40)*. After 6 d in vitro, more than 95% of the stem-cell-generated clones contained neurons and undifferentiated cells, but no astrocytes or oligodendrocytes. By 10 d, >90% of clones contained GFAP-positive astrocytes and >70% contained O4-positive oligodendrocytes. Furthermore, the neurogenic versus gliogenic potential of cortical stem cells seems to depend on both the developmental stage and the concentration of FGF2. When grown in 0.1 ng/mL of FGF2, E10 stem cells generate mostly neuronal clones with no glia-only clones, whereas at 10 ng of FGF2, 15% of the generated clones were glia-only clones. The percentage of pure glial clones is further potentiated when stem cells were derived from E12 cortex and reached 20% and 75% in 0.1 and 10 ng/mL of FGF2, respectively *(40)*.

Few studies address the mechanisms underlying developmental changes of NSCs over time and/or in response to the same stimuli. Among conditions that may be responsible for these behavioral changes of NSCs is the ligand–receptor relationship. In this scenario, changes may be the result of variations in the thresholds of receptor stimulation, to the dose of ligand available or to the changes in receptor number or density, which may influence the level and duration of receptor activation and result in different cellular responses. For instance, increased virally transduced epidermal growth factor (EGF) receptor (EGFR) expression in E15 rat cortical progenitors resulted in enhanced migration typical of E20 progenitors *(41)*. Moreover, EGFR-infected progenitors express glial markers prematurely and their probability to generate multipotential spheres increases with decreasing concentration of EGF in the culture medium. Interestingly, the in vivo expression of EGFR normally increases during the period of gliogenesis onset at E18 and the probability of generating spheres of cells increases with age *(42)*. Bone morphogenetic protein-4 (BMP-4), on the other hand, reversibly inhibits changes in EGFR expression and responsiveness, resulting in a delay of cortical progenitor development *(43)*. These studies suggest that temporal changes in stem cell behavioral may be modulated by changes in the number/density of EGFR during development.

Neuronal Neurotransmitter Phenotype Choices

If we move a step further in the arborization tree of fate specification and examine how neurotransmitter phenotype is specified within the neuronal population, we encounter the same inductive mechanisms, whereby fate decisions of neuronal progenitors are a function of the history, time, space, and the surrounding microenvironment.

The relative distribution of the in vivo environmental cues is thought to play a critical role in directing fate choices of stem cell progeny. In the peripheral nervous system (PNS), neural crest stem cells (NCSCs) derived from the E10.5 neural tube behave differently from the E14.5 sciatic-nerve-derived NCSC. The later became less sensitive to the autonomic instructive action of BMP-2 and, consequently, their potential of fate choices became limited to a cholinergic one. This time-dependent decrease in the BMP-2 sensitivity may have resulted from a combination of an in vivo selection and developmental conversion of the NCSC.

We have been interested in understanding how stem cell progeny of the mammalian forebrain are specified to become dopaminergic. A well-defined subpopulation of neurons born in the ventricular zone of the ganglionic eminencies of the developing forebrain have the potential to become dopaminergic when exposed to the synergistic action of activin or BMP-2 and FGF2 *(44)*. The expression of mRNA encoding both BMP and activin receptors has been extensively studied during rodent embryogenesis. The striking finding in these studies is that BMP and activin receptors are expressed particularly in the ependymal layer of the neural tube and the brain, where stem cells reside. In the corpus striatum, activin/BMP receptor transcripts are present in the ependymal layer as well as the mantle layer *(45–48)*. When treated with activin, neuroblasts of the E14 mouse ventricular

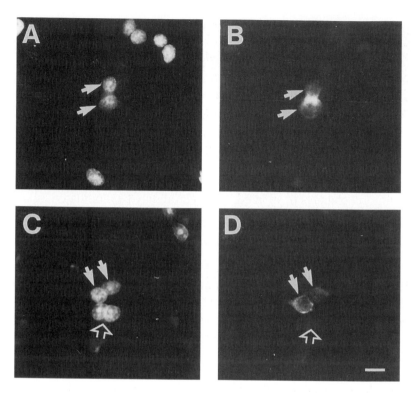

Fig. 1. Symmetrical and asymmetrical divisions of VZ-derived progenitors. Dissociated cells (5×10^5) of the E14 dorsal-most aspect of the medial and lateral ganglionic eminencies (MGE and LGE) were cultured on poly-L-ornithine-coated glass cover slips in serum-free medium with activin (50 ng/mL) and FGF2 (20 ng/mL) for 24 h. All culture wells received 1 μ*M* of BrdU 2 h after plating. Fixed cells were processed for BrdU-immunocytochemistry (**A,C**) and for the enzyme TH immunocytochemistry (**B,D**). The twin arrows show newly generated doublet of cells that had incorporated BrdU (**A**) and both expressed TH (**B**), suggesting that both sister cells are competent to respond to the TH inductive signals and thus a symmetrical division of the parent precursor, whereas the twin arrows in (**C**) and (**D**) show only one sister cell of the newly generated doublet expressing TH and thus suggesting an asymmetrical division of the parent precursor cell. The open arrow in (**C**) and (**D**) shows a BrdU+ doublet incompetent to respond to the TH-inducing conditions. Scale bar = 10 μm.

zone (VZ) (which, in essence, may be termed "stem cell progeny") did not show significant differences in tyrosine hydroxylase (TH) induction, when compared to control culture condition, whereas FGF2 induced a very small number (210 ± 29 cells/cm^2) of striatal neurons to express TH. When combined, the two growth factors were clearly synergistic in inducing TH in

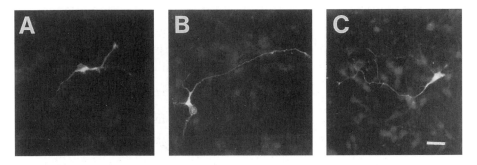

Fig. 2. Induction of TH expression in neurons cultured from noncatecholamin-ergic brain regions. Neurons from the spinal cord (**A**), cerebellum (**B**), and thalamus (**C**) were cultured for 3 d in the presence of activin and FGF2 and processed for TH immunoreactivity. Scale bar = 20 μm.

striatal neurons, resulting in a 25-fold increase (to 4930 ± 806 cells/cm^2) over that seen with FGF2 alone. Activin actions were dose dependent *(44)*. At a fixed concentration of FGF2 (20 ng/mL) activin actions were first detected at 1 ng/mL and maximal activation, which induced 4000–5000 cells/cm^2 to express TH, was achieved with 50–100 ng/mL. Double labeling with anti-body to β-tubulin showed that maximal induction with activin and FGF2 resulted in the expression of TH in 4–5% of the total numbers of striatal neurons detected after 24 h in culture. Frequently, doublets of bromodeoxy-uridine (BrdU, a DNA replication marker)-positive cells expressed TH, suggesting that they may be generated by the same ancestor precursor cell (Fig. 1). This observation suggested that newly generated progenitor cells may require epigenetic signals during the first moment of birth to make their commitment to a given neuronal cell type. Furthermore, delayed addition of the TH-inducing factors demonstrated that the intrinsic plasticity of these neuronal progenitor cells to express TH is progressively lost in aged cultures *(44)*.

We also asked whether TH induction was restricted to the striatal neurons of the E14 mouse CNS. First, we tested whether dissociated cultures of the ventral mesencephalon would show enhanced expression of TH in the presence of activin or BMP-2 and FGF2 and found no differences between control and treated cultures. This was not surprising, as dopaminergic neurons are born much earlier than E14 *(49,50)*. In contrast to cultures of the ventral mesencephalon, neurons cultured from the embryonic day 14 cortex, thalamus, spinal cord, and cerebellum showed TH induction in neurons, albeit at numbers significantly lower that those obtained with striatal neurons. Figure 2 shows examples of TH-immunoreactive neurons in dissociated cultures of embryonic day 14 spinal cord, cerebellum, and thalamus. In another set of experiments, we asked whether other growth factors might mimic the cooperative action of activin, BMP2, and FGF2 (Fig. 3). Brain-derived neurotrophic factor (BDNF), ciliary neurotrophic factor (CNTF), PDGF or transforming growth factor-α (TGF-α), glial-derived neurotrophic factor (GDNF), TGF-β_2 and TGF-β_3 could not mimic the synergistic actions of activin or BMP-2 with FGF2. Glial-cell-derived conditioned media, however, mimicked BMP-2 and activin's ability to cooperate with FGF2 in inducing TH. FGF4 mimicked FGF2's ability to cooperate with either activin or BMP-2 in inducing TH, whereas FGF1 and FGF7 were less effective and ineffective, respectively (Fig. 3).

How does BMP-2 mimic activin's action in turning on the TH gene? Why do we have this redundancy? The answer to the second question is beyond the scope of this chapter. For the first question, it is worthy to first document both ligands and their receptor superfamilies. Activins are dimeric proteins belonging to the TGF-β superfamily. They are present in vivo in at least three different forms: the homodimeric activin A (βAβA) and activin B (βBβB) and the heterodimeric activin AB (βAβB); for review, *see* ref. *52*. BMP-2 is a member of a second major subgroup, BMP/decapentaplegic (BMP/dpp), of the TGF-β superfamily *(53)*. These cytokines signal their actions through two types of transmembrane serine/threonine kinase known as receptor type I and type II *(54)*. TGF-βs and activin-related factors initiate their actions by binding type-II receptors (TGF-βR-II, ActR-II, and ActR-IIB), then type-I receptors (TGF-βR-I, ActR-I, and ActR-IB) are recruited into the complex and phosphorylated by type-II receptors at serine and threonine residues *(55)*. However, the signaling scenario for BMP/dpp family members is different. They are able to bind either type-I (BMPR-IA and BMPR-IB) or type-II receptors (Daf4 and BMPR-II) *(23,56–60)* independently, but the heteromeric kinase receptor complex formation is required for signal transduction *(61–63)*.

Both activin and BMP repress the proliferation induced by FGFs and stimulate TH gene expression in the striatal precursors *(44)*. These observations are consistent with the possibility of crossreactivity between BMP and activin receptor interactions. Indeed, drosophila *punt* gene product, an activin type-II receptor *(64,65)* can bind either BMP-2 or activin and form a heteromeric complex with either dpp type-I receptor (tkv or sax) or activin type-I receptor (atr-I) *(66–68)*. Overexpression studies of dominant negative ActR-II receptors in *Xenopus* animal caps suggest that ActR-II and ActR-IIB may act as a type-II receptor for Vg1, a BMP/dpp-related factor *(69)*.

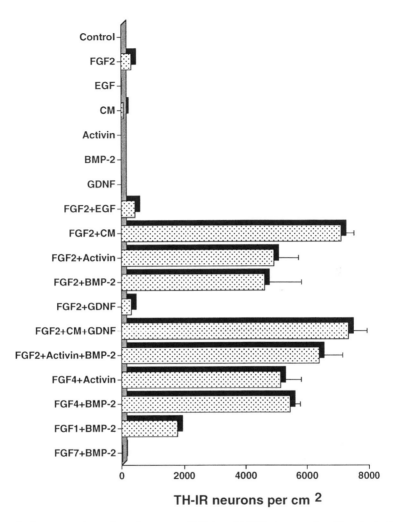

Fig. 3. Synergestic actions between FGF and TGF-β superfamilies in inducing TH expression in cultured VZ progenitors. Dissociated cells of the dorsal-most aspect of the medial and lateral ganglionic eminencies of E14 mouse embryo were cultured on poly-L-ornithine-coated glass cover slips in serum-free medium. The indicated growth factors were added 2 h after plating: FGF2 at 20 ng/mL, CM at 75% of the defined media, and all other growth factors at 50 ng/mL. Cultures were fixed after 24 h and processed for TH immunocytochemistry. (Reproduced with permission from refs. *44* and *51*; Copyright © 1999 by the Society for Neuroscience.)

Receptor cDNAs transfection studies have shown that BMP-2 can bind a newly cloned human BMP type II receptor, heterodimerize with ActR-I, and elicit the same transcriptional response induced by activin *(61)*. This unique mechanism of crosstalk between different TGF-β superfamily receptors may underlie the functional redundancy of their ligands *(61,66,70)*.

Interestingly, the rapidly proliferating stem cell progeny of the ganglionic eminencies ventricular and periventricular zones express activin, BMP-2 ligands and activin, BMP-2 receptors from E12 until birth *(46,47)*. In studies of neural tissue, the action of activin has been related to neurotransmitter phenotype regulation, whereby activin induced somatostatinlike immunore-activity in ciliary ganglion neurons *(71,72)*. In the sympathetic neurons, activin induced acetylcholine synthetic enzyme and neuropeptide gene expression *(73,74)*. In studies of the P19 embryonic stem cell line, activin was found to promote survival and neuronal differentiation *(75)*. Other studies have suggested that the activin induction of mesoderm require cooperative actions of FGF2 *(76,77)*. Furthermore, expression studies have demonstrated the presence of both FGFs and their receptors in the brain, particularly in the periventricular zone *(38,78)*, suggesting potential interactions between FGF2 and activin/BMP-2 in the forebrain histogenesis.

Examples illustrating the complex interaction that may take place between different superfamily growth factors may be taken from morphogenesis happening during the chicken limb bud development or from mesoderm induction in *Xenopus*. In the limb bud development, FGF4 RNA is detected in the posterior half of the apical ectodermal ridge (AER: a specialized epithelium at the limb tip) and has been proposed to be the AER endogenous signal that stimulates the proliferation of cells located in the progress zone. In combination with retinoic acid, FGF4 induces sonic hedgehog (Shh) expression in the zone of polarizing activity (ZPA) *(79)*. Once induced, Shh expression can be maintained by FGF4 alone. Shh is considered as the ZPA endogenous polarizing signal that organizes patterns by directing the expression of secondary signals (i.e., BMP-2 in the mesoderm and FGF4 in the ectoderm), therefore establishing a positive feedback loop with FGF4 *(79)*. In addition, cooperative actions between Shh and FGF4 signals are required for controlling the expression pattern of the BMP-2 gene in the mesoderm.

In the developing brain, the floor plate and notochord influence the development of multiple neuronal classes along the rostrocaudal body through soluble secreted factors and contact-dependent mechanisms *(80)*. In the midbrain, Shh has been shown to be involved in the specification of dopaminergic (DA) neurons *(81)*. A previous study has demonstrated that

FGF8 is expressed in the neural plate at embryonic stage E8–8.5 and in the rhombic isthmus, the region that lies at the midbrain–hindbrain junction at E9–9.5 and plays a role as the signaling center for the midbrain patterning during this period *(82)*. Interestingly, dopaminergic neurons are born in the rhombic isthmus at E8.5–9 in the mouse *(83)* and at E12–13 in the rat, and then migrate in a radial pattern toward the ventral mesencephalon and rostrally to their final site *(49,50,84)*. These findings suggested that FGFs might be involved in the early phase of dopaminergic specification. Indeed, when E9 ventral mid/hindbrain explants were cultured for 5 d in the presence of the soluble form of FGF receptor 3c, a high-affinity blocking receptor for FGF8, the DA neurons failed to develop *(85)*; whereas, in explant cultures incubated with soluble T-cell receptor CD4 or FGF receptor 1c, DA neurons developed normally *(85)*. These data suggest that members from the FGF superfamily are required for midbrain DA phenotype induction. Gene knockout studies of the nuclear receptor transcription factor Nurr1 lead to the agenesis of the midbrain DA neurons *(86,87)*. These studies demonstrated that Nurr1 is essential for the induction of the DA phenotype and for the survival and the maturation of DA neurons *(86,88)*, but it is not required for the induction of two homeobox transcription factors engrailed and Ptx-3 that are specifically expressed by newly generated DA neuroblasts *(86,88–90)*. The Nurr1 null mutant showed a normal expression of Ptx3 in DA neurons at E12.5, but this expression was lost in neonates. Concomitant with this loss is the increase of TUNEL+ nuclei and, thus, dying cells in the ventral midbrain region *(91)*. Both Ptx3 and Nurr1 are expressed by the midbrain DA neurons throughout adulthood suggesting their implication in the normal function or maintenance of other critical transcription factors required for the existence of DA neurons. Engrailed genes are another set of homeodomain transcription factors controlling the fate of DA neurons. Two engrailed homologs, *En-1* and *En-2*, are highly expressed by the DA neurons from their generation to adulthood *(89)*. In *En-1/En-2* double mutant TH+, DA neurons develop normally but are lost by E14 *(89)*. Interestingly, a recent report demonstrated that a specific subpopulation of the midbrain DA neurons require a Nurr1-independant pathway for their specification *(92)*. These authors reported that LIM homeodomain transcription factor Lmx1b is necessary for the induction of Ptx3 in this subpopulation of DA neurons, a subpopulation that is lost during embryonic maturation in Lmx1b null mutant *(92)*. Future studies should address potential interaction networks between these transcription factors in the induction, differentiation, maturation, and possibly normal function of the midbrain DA neurons.

STEM-CELL-DERIVED DOPAMINERGIC NEURONS AND THEIR THERAPEUTIC APPLICATION IN PARKINSON'S DISEASE

Parkinson's disease (PD) is a neurodegenerative disorder characterized by the loss of dopamine neurons in the substantia nigra pars compacta, resulting in decreased dopaminergic input to the striatum. Symptoms include tremors, rigidity, bradykinesia, and instability. Existing therapies for PD are only palliative and treat the symptoms but do not address the underlying cause or prevent the progression of the disease. Levodopa (L-dopa), the gold-standard pharmacological treatment to restore dopamine, is compromised over time by decreased efficacy and by increased side effects. Neurosurgical treatments, such as pallidotomy, thalamotomy, and deep electrical stimulation, are only considered after the failure of pharmacological treatment. A reliable long-term treatment to halt the progression of the disease and restore function remains elusive.

Neural transplantation is a promising strategy for improving dopaminergic dysfunction in PD. Over 20 yr of research using fetal mesencephalic tissue as a source of DA neurons has demonstrated the therapeutic potential of cell transplantation therapy in rodents and nonhuman primate animal models and in human patients. Grafts survive, form synaptic connections, and improve motor function in many patients (93,94). However, there are limitations associated with human fetal tissue transplantation, including high tissue variability, limited availability, ethical concerns, and inability to obtain tissue in an epidemiologically meaningful quantity. Thus, the control of the identity, purity, and potency of these cells become impossible and jeopardizing for both the safety of the patient and the efficacy of the therapy. Relying on fetal tissue as a source of neurons, cell replacement therapy cannot develop into a widely available treatment option for patients with neurodegenerative diseases. These critical issues render the search and development of alternative sources of cells a very worthwhile goal with societal importance and commercial application. Other explored alternative sources of natural dopamine-expressing cells are the adrenal medulla cells (95), PC12 cells (96), the glomus cells of the carotid bodies (97), and the porcine fetal tissue (98). Most of these sources were abandoned because of their poor survival, inefficiency, or their health risks for the patient (99,100). A second strategy in generating an unlimited supply of cells for neural transplantation is the epigenetic generation of dopamine-expressing neurons from NSCs. The third approach consists of gene therapy whereby critical dopamine biosynthesis enzymes and cofactor gene sequences are inserted in vectors and transferred in vivo into striatal cells. Alternatively,

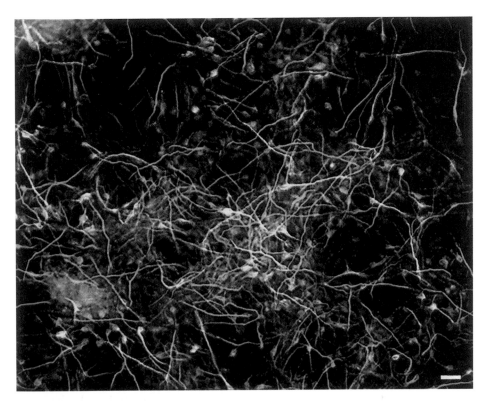

Fig. 4. Forebrain-derived human NSC progeny were induced to differentiate into DA lineage in vitro. NSCs derived from fetal human tissue were dissociated and plated on poly-L-ornithine-coated glass cover slips in the presence of 20 ng/mL of bFGF2 and the active fraction of glial-derived conditioned medium for 3 d. Fixed cultures were then stained with rhodamine-labeled antibody to TH (red). Green flurescence results from staining with the neuronal cell type-specific marker anti-β-tubulin class III. Scale bar = 70 μm.

stable cell lines may be genetically modified in vitro and then implanted into the host CNS to alleviate PD deficiency.

Using a combination of glial-derived conditioned medium and FGF2, we have been able to induce TH expression in multiple stable human NSCs (hNSCs) that we have established. The percentage of TH was about 10% of the total viable cells and represented about 30% of the total neuronal population, as identified using β-tubulin class III marker (Fig. 4). These hNSC-derived TH+ neurons also expressed dopamine transporter characteristic of functional DA neurons. The induction of these phenotypical characteristics was also confirmed at the mRNA level using RT-PCR and Northern blot.

More importantly, these established hNSCs responded consistently after each passage to the TH-inductive action, suggesting that under our culture growth conditions, there is no drift in the properties of our hNSCs. Thus, this assay constitutes a reliable tool to monitor the functionality of hNSCs progeny over time.

A second potential stem cell source for DA neurons arc the embryonic stem (ES) cells. Pluripotent ES cells are derived from the inner cell mass of the developing blastocyst. These cells appear to have the potential to generate DA neurons *(101,102)*. In the presence of serum, ES cells form clusters of floating cells or embryoid bodies (EBs) containing ectodermal, mesodermal, and endodermal derivatives. When these EBs are treated with FGF2, FGF8, and Shh, 71% of the cells differentiated into neurons, as identified with the neuronal marker class III β-tubulin and 33% of these neurons displayed characteristics of the midbrain DA neurons *(102)*. A second group of investigators proceeded first to generate a homogenous neural lineage from the ES cells *(101)* before inducing DA phenotype. This was achieved by coculturing ES cells with the stromal cells PA6, which induced the pan-neural marker NCAM in 92% of the ES cells colonies. PA6-derived conditioned media were inefficient in inducing neural differentiation, but they were not blocked by a 0.4-μm membrane barrier. Paradoxically, paraformaldehyde-fixed PA6 cells retained the inductive activity. Under these culture conditions, 52% of differentiated cells expressed neuronal markers and 30% of these neurons assumed midbrain DA phenotype. These DA-induced neurons appear to engraft after implantation and to improve behavioral deficits of 6-OHDA lesioned mice.

There is also a great effort in trying to generate NSC lines from the midbrain in the aim that the progeny will be destined or at least inducible to become typical midbrain projecting DA neurons. There has been a concerted effort to isolate a stable, expandable stem cell from the midbrain that its neuronal progeny will differentiate exclusively into nigrostriatal-like DA neurons. Early studies have demonstrated that EGF-responsive precursor cells do exist within the midbrain; however, these progeny did not, consistently or robustly, differentiate into DA neurons either in vitro *(103–105)* or after implantation into the rat striatum *(106,107)*. Interestingly, interleukin (IL)-1 induced TH expression in the midbrain-derived progenitors *(104)*. In addition, membrane-bound factors potentiated the TH induction and stimulated morphological maturation in these progenitors *(108,109)*. The continuous generation of DA neurons from a long-term expandable midbrain-derived stem cell will require the development of processes for proliferation and maintenance of DA phenotype. Ascorbic acid and lowered oxygen concentration, for instance, appear to support survival and proliferation of DA neurons, respectively *(110,111)*. The effect of the low O_2 level ($3 \pm 2\%$) was partially mimicked by erythropoietin (Epo). In an innovative set of experiments *(112)*, mesencephalic precursors were FACS sorted according to their expression of green fluorescent protein (GFP) that is driven by the nestin enhancer. Nestin is a neurofilament, expressed by neuroepithelial stem cells *(113)*. The nestin–GFP+ precursors were analyzed clonally and shown to have the ability to self-renew and give rise to clusters of progeny able to differentiate into neurons, astrocytes, and oligodendrocytes. Among the neuronal population TH+ neurons were identified with no particular treatment. Importantly, 5 wk after implantation of the sorted GFP+ into the striatum of 6-OHDA hemiparkinsonian rats, the animals showed reduction in amphetamine induced rotation.

CONCLUSIONS

Stem cells offer us a great tool for understanding the basic biology of cell fate choices and allow us to explore novel inducing factors and new developmental networks of gene cascades that may not necessarily occur under in vivo physiological conditions. Developing technologies that allow a consistent and efficient generation of specific neuronal subpopulations is the goal of many neuroscientists. The next and ultimate step is to put the cellular product or the therapeutic factor to the in vivo test and to determine its preclinical efficacy. Concerning generation of DA neurons, we are still far from using a stem-cell-derived product for the treatment of Parkinson's disease. Among basic issues that need to be addressed before a consistent source of cells may be considered for product development is the cell stability, viability, cryo-preservation, recovery, identity, purity, potency, and the in vivo fate of each implanted cell. All these issues are currently targeted aggressively and give us hope that cell replacement therapy will make a difference in the life of patients suffering from neurological disorders.

REFERENCES

1. Anderson, M. J., Choy, C. Y., and Waxman, S. G. (1986) Self-organization of ependyma in regenerating teleost spinal cord: evidence from serial section reconstructions. *J. Embryol. Exp. Morphol.* **96,** 1–18.
2. Anderson, M. J., Waxman, S. G., and Laufer, M. (1983) Fine structure of regenerated ependyma and spinal cord in Sternarchus albifrons. *Anat. Rec.* **205,** 73–83.
3. Molowny, A., Nacher, J., and Lopez-Garcia, C. (1995) Reactive neurogenesis during regeneration of the lesioned medial cerebral cortex of lizards. *Neuroscience* **68,** 823–836.

4. Johansson, C. B., Momma, S., Clarke, D. L., Risling, M., Lendahl, U., and Frisen, J. (1999) Identification of a neural stem cell in the adult mammalian central nervous system. *Cell* **96,** 25–34.
5. Chiasson, B. J., Tropepe, V., Morshead, C. M., and van der Kooy, D. (1999) Adult mammalian forebrain ependymal and subependymal cells demonstrate proliferative potential, but only subependymal cells have neural stem cell characteristics. *J. Neurosci.* **19,** 4462–4471.
6. Laywell, E. D., Rakic, P., Kukekov, V. G., Holland, E. C., and Steindler, D. A. (2000) Identification of a multipotent astrocytic stem cell in the immature and adult mouse brain. *Proc. Natl. Acad. Sci. USA* **97,** 13,883–13,888.
7. Doetsch, F., Caille, I., Lim, D. A., Garcia-Verdugo, J. M., and Alvarez-Buylla, A. (1999) Subventricular zone astrocytes are neural stem cells in the adult mammalian brain. *Cell* **97,** 703–716.
8. Nacher, J., Ramirez, C., Palop, J. J., Molowny, A., Luis de la Iglesia, J. A., and Lopez-Garcia, C. (1999) Radial glia and cell debris removal during lesion-regeneration of the lizard medial cortex. *Histol. Histopathol.* **14,** 89–101.
9. Brown, J., Greaves, M. F., and Molgaard, H. V. (1991) The gene encoding the stem cell antigen, CD34, is conserved in mouse and expressed in haemopoietic progenitor cell lines, brain, and embryonic fibroblasts. *Int. Immunol.* **3,** 175–184.
10. Lin, G., Finger, E., and Gutierrez-Ramos, J. C. (1995) Expression of CD34 in endothelial cells, hematopoietic progenitors and nervous cells in fetal and adult mouse tissues. *Eur. J. Immunol.* **25,** 1508–1516.
11. Brazelton, T. R., Rossi, F. M., Keshet, G. I., and Blau, H. M. (2000) From marrow to brain: expression of neuronal phenotypes in adult mice. *Science* **290,** 1775–1779.
12. Mezey, E., Chandross, K. J., Harta, G., Maki, R. A., and McKercher, S. R. (2000) Turning blood into brain: cells bearing neuronal antigens generated in vivo from bone marrow. *Science* **290,** 1779–1782.
13. Gould, E., Beylin, A., Tanapat, P., Reeves, A., and Shors, T. J. (1999) Learning enhances adult neurogenesis in the hippocampal formation. *Nat. Neurosci.* **2,** 260–265.
14. Kempermann, G., Kuhn, H. G., and Gage, F. H. (1997) More hippocampal neurons in adult mice living in an enriched environment. *Nature* **386,** 493–495.
15. Stanfield, B. B. and Trice, J. E. (1988) Evidence that granule cells generated in the dentate gyrus of adult rats extend axonal projections. *Exp. Brain Res.* **72,** 399–406.
16. Palmer, T. D., Willhoite, A. R., and Gage, F. H. (2000) Vascular niche for adult hippocampal neurogenesis. *J. Comp. Neurol.* **425,** 479–494.
17. Blau, H. M., Brazelton, T. R., and Weimann, J. M. (2001) The evolving concept of a stem cell: entity or function? *Cell* **105,** 829–841.
18. Bjorklund, A. and Svendsen, C. N. (2001) Chimeric stem cells. *Trends Mol. Med.* **7,** 144–146.
19. Alvarez-Buylla, A., Garcia-Verdugo, J. M., and Tramontin, A. D. (2001) A unified hypothesis on the lineage of neural stem cells. *Nat. Rev. Neurosci.* **2,** 287-293.

20. Brustle, O., Choudhary, K., Karram, K., Huttner, A., Murray, K., Dubois-Dalcq, M., et al. (1998) Chimeric brains generated by intraventricular transplantation of fetal human brain cells into embryonic rats. *Nat. Biotechnol.* **16,** 1040–1044.
21. Flax, J. D., Aurora, S., Yang, C., Simonin, C., Wills, A. M., Billinghurst, L. L., et al. (1998) Engraftable human neural stem cells respond to developmental cues, replace neurons, and express foreign genes. *Nat. Biotechnol.* **16,** 1033–1039.
22. Fricker, R. A., Carpenter, M. K., Winkler, C., Greco, C., Gates, M. A., and Bjorklund, A. (1999) Site-specific migration and neuronal differentiation of human neural progenitor cells after transplantation in the adult rat brain. *J. Neurosci.* **19,** 5990–6005.
23. Penton, A., Chen, Y., Staehling-Hampton, K., Wrana, J. L., Attisano, L., Szidonya, J., et al. (1994) Identification of two bone morphogenetic protein type I receptors in Drosophila and evidence that Brk25D is a decapentaplegic receptor. *Cell* **78,** 239–250.
24. Ostenfeld, T., Caldwell, M. A., Prowse, K. R., Linskens, M. H., Jauniaux, E., and Svendsen, C. N. (2000) Human neural precursor cells express low levels of telomerase in vitro and show diminishing cell proliferation with extensive axonal outgrowth following transplantation. *Exp. Neurol.* **164,** 215–226.
25. Bjorklund, A. and Lindvall, O. (2000) Cell replacement therapies for central nervous system disorders. *Nat. Neurosci.* **3,** 537–544.
26. Snyder, E. Y., Taylor, R. M., and Wolfe, J. H. (1995) Neural progenitor cell engraftment corrects lysosomal storage throughout the MPS VII mouse brain. *Nature* **374,** 367–370.
27. Snyder, E. Y., Yoon, C., Flax, J. D., and Macklis, J. D. (1997) Multipotent neural precursors can differentiate toward replacement of neurons undergoing targeted apoptotic degeneration in adult mouse neocortex. *Proc. Natl. Acad. Sci. USA* **94,** 11,663–11,668.
28. Yandava, B. D., Billinghurst, L. L., and Snyder, E. Y. (1999) "Global" cell replacement is feasible via neural stem cell transplantation: evidence from the dysmyelinated shiverer mouse brain. *Proc. Natl. Acad. Sci. USA* **96,** 7029–7034.
29. Aboody, K. S., Brown, A., Rainov, N. G., Bower, K. A., Liu, S., Yang, W., et al. (2000) From the cover: neural stem cells display extensive tropism for pathology in adult brain: evidence from intracranial gliomas [in process citation]. *Proc. Natl. Acad. Sci. USA* **97,** 12,846–12,851.
30. Ourednik, V., Ourednik, J., Flax, J. D., Zawada, M., Hutt, C., Yang, C., et al. (2001) Segregation of human neural stem cells in the developing primate forebrain. *Science* **26,** 26.
31. Raff, M. C., Miller, R. H., and Noble, M. (1983) A glial progenitor cell that develops in vitro into an astrocyte or an oligodendrocyte depending on culture medium. *Nature* **303,** 390–396.
32. Raff, M. C., Lillien, L. E., Richardson, W. D., Burne, J. F., and Noble, M. D. (1988) Platelet-derived growth factor from astrocytes drives the clock that times oligodendrocyte development in culture. *Nature* **333,** 562–565.
33. Abney, E. R., Bartlett, P. P., and Raff, M. C. (1981) Astrocytes, ependymal cells, and oligodendrocytes develop on schedule in dissociated cell cultures of embryonic rat brain. *Dev. Biol.* **83,** 301–310.

34. Temple, S. and Raff, M. C. (1986) Clonal analysis of oligodendrocyte development in culture: evidence for a developmental clock that counts cell divisions. *Cell* **44,** 773–779.
35. Ibarrola, N., Mayer-Proschel, M., Rodriguez-Pena, A., and Noble, M. (1996) Evidence for the existence of at least two timing mechanisms that contribute to oligodendrocyte generation in vitro. *Dev. Biol.* **180,** 1–21.
36. Barres, B. A., Hart, I. K., Coles, H. S., Burne, J. F., Voyvodic, J. T., Richardson, W. D., et al. (1992) Cell death and control of cell survival in the oligodendrocyte lineage. *Cell* **70,** 31–46.
36a. Bogler, O., Wren, D., Barnett, S. C., Land, H., and Noble, M. (1990) Cooperation between two growth factors promotes extended self-renewal and inhibits differentiation of oligodendrocyte-type-2 astrocyte (O-2A) progenitor cells. *Proc. Natl. Acad. Sci. USA* **87,** 6368–6372.
37. Kondo, T. and Raff, M. (2000) Oligodendrocyte precursor cells reprogrammed to become multipotential CNS stem cells. *Science* **289,** 1754–1757.
38. Qian, X., Davis, A. A., Goderie, S. K., and Temple, S. (1997) FGF2 concentration regulates the generation of neurons and glia from multipotent cortical stem cells. *Neuron* **18,** 81–93.
39. Qian, X., Goderie, S. K., Shen, Q., Stern, J. H., and Temple, S. (1998) Intrinsic programs of patterned cell lineages in isolated vertebrate CNS ventricular zone cells. *Development* **125,** 3143–3152.
40. Qian, X., Shen, Q., Goderie, S. K., He, W., Capela, A., Davis, A. A., et al. (2000) Timing of CNS cell generation: a programmed sequence of neuron and glial cell production from isolated murine cortical stem cells. *Neuron* **28,** 69–80.
41. Burrows, R. C., Lillien, L., and Levitt, P. (2000) Mechanisms of progenitor maturation are conserved in the striatum and cortex. *Dev. Neurosci.* **22,** 7–15.
42. Burrows, R. C., Wancio, D., Levitt, P., and Lillien, L. (1997) Response diversity and the timing of progenitor cell maturation are regulated by developmental changes in EGFR expression in the cortex. *Neuron* **19,** 251–267.
43. Lillien, L. and Raphael, H. (2000) BMP and FGF regulate the development of EGF-responsive neural progenitor cells. *Development* **127,** 4993–5005.
44. Daadi, M., Arcellana-Panlilio, M. Y., and Weiss, S. (1998) Activin cooperates with fibroblast growth factor 2 to regulate tyrosine hydroxylase expression in the basal forebrain ventricular zone progenitors. *Neuroscience* **86,** 867–880.
45. Dewulf, N., Verschueren, K., Lonnoy, O., Moren, A., Grimsby, S., Vande Spiegle, K., et al. (1995) Distinct spatial and temporal expression patterns of two type I receptors for bone morphogenetic proteins during mouse embryogenesis. *Endocrinology* **136,** 2652–2663.
46. Feijen, A., Goumans, M. J., and van den Eijnden-van Raaij, A. J. (1994) Expression of activin subunits, activin receptors and follistatin in postimplantation mouse embryos suggests specific developmental functions for different activins. *Development* **120,** 3621–3637.
47. Roberts, V. J. and Barth, S. L. (1994) Expression of messenger ribonucleic acids encoding the inhibin/activin system during mid- and late-gestation rat embryogenesis. *Endocrinology* **134,** 914–923.

48. Verschueren, K., Dewulf, N., Goumans, M. J., Lonnoy, O., Feijen, A., Grimsby, S., et al. (1995) Expression of type I and type IB receptors for activin in midgestation mouse embryos suggests distinct functions in organogenesis. *Mech. Dev.* **52,** 109–123.

49. Lauder, J. M. and Bloom, F. E. (1974) Ontogeny of monoamine neurons in the locus coeruleus, Raphe nuclei and substantia nigra of the rat. I. Cell differentiation. *J. Comp. Neurol.* **155,** 469–481.

50. Specht, L. A., Pickel, V. M., Joh, T. H., and Reis, D. J. (1981) Light-microscopic immunocytochemical localization of tyrosine hydroxylase in prenatal rat brain. II. Late ontogeny. *J. Comp. Neurol.* **199,** 255–276.

51. Daadi, M. M. and Weiss, S. (1999) Generation of tyrosine hydroxylase-producing neurons from precursors of the embryonic and adult forebrain. *J. Neurosci.* **19,** 4484–4497.

52. Vale, W., Bilezikjian, L. M., and Rivier, C. (1994) Reproductive and other roles of inhibins and activins, in *The Physiology of Reproduction* (Knobil, E. and Neill, J. D., eds.), Raven, New York, pp. 1861–1878.

53. Kingsley, D. M. (1994) The TGF-beta superfamily: new members, new receptors, and new genetic tests of function in different organisms. *Genes Dev.* **8,** 133–146.

54. Attisano, L., Wrana, J. L., Lopez-Casillas, F., and Massague, J. (1994) TGF-beta receptors and actions. *Biochim. Biophys. Acta* **1222,** 71–80.

55. Wrana, J. L., Attisano, L., Wieser, R., Ventura, F., and Massague, J. (1994) Mechanism of activation of the TGF-beta receptor. *Nature* **370,** 341–347.

56. Brummel, T. J., Twombly, V., Marques, G., Wrana, J. L., Newfeld, S. J., Attisano, L., et al. (1994) Characterization and relationship of Dpp receptors encoded by the saxophone and thick veins genes in Drosophila. *Cell* **78,** 251–261.

57. Graff, J. M., Thies, R. S., Song, J. J., Celeste, A. J., and Melton, D. A. (1994) Studies with a Xenopus BMP receptor suggest that ventral mesoderm-inducing signals override dorsal signals in vivo. *Cell* **79,** 169–179.

58. Koenig, B. B., Cook, J. S., Wolsing, D. H., Ting, J., Tiesman, J. P., Correa, P. E., et al. (1994) Characterization and cloning of a receptor for BMP-2 and BMP-4 from NIH 3T3 cells. *Mol. Cell. Biol.* **14,** 5961–5974.

59. Suzuki, A., Thies, R. S., Yamaji, N., Song, J. J., Wozney, J. M., Murakami, K., et al. (1994) A truncated bone morphogenetic protein receptor affects dorsal-ventral patterning in the early Xenopus embryo. *Proc. Natl. Acad. Sci. USA* **91,** 10,255–10,259.

60. ten Dijke, P., Yamashita, H., Ichijo, H., Franzen, P., Laiho, M., Miyazono, K., et al. (1994) Characterization of type I receptors for transforming growth factor-beta and activin. *Science* **264,** 101–104.

61. Liu, F., Ventura, F., Doody, J., and Massague, J. (1995) Human type II receptor for bone morphogenic proteins (BMPs): extension of the two-kinase receptor model to the BMPs. *Mol. Cell. Biol.* **15,** 3479–3486.

62. Nohno, T., Ishikawa, T., Saito, T., Hosokawa, K., Noji, S., Wolsing, D. H., et al. (1995) Identification of a human type II receptor for bone morphogenetic protein-4 that forms differential heteromeric complexes with bone morphogenetic protein type I receptors. *J. Biol. Chem.* **270,** 22,522–22,526.

63. Rosenzweig, B. L., Imamura, T., Okadome, T., Cox, G. N., Yamashita, H., ten Dijke, P., et al. (1995) Cloning and characterization of a human type II receptor for bone morphogenetic proteins. *Proc. Natl. Acad. Sci. USA* **92,** 7632–7636.

64. Childs, S. R., Wrana, J. L., Arora, K., Attisano, L., O'Connor, M. B., and Massague, J. (1993) Identification of a Drosophila activin receptor. *Proc. Natl. Acad. Sci. USA* **90,** 9475–9479.

65. Mathews, L. S. and Vale, W. W. (1991) Expression cloning of an activin receptor, a predicted transmembrane serine kinase. *Cell* **65,** 973–982.

66. Letsou, A., Arora, K., Wrana, J. L., Simin, K., Twombly, V., Jamal, J., et al. (1995) Drosophila Dpp signaling is mediated by the punt gene product: a dual ligand-binding type II receptor of the TGF beta receptor family. *Cell* **80,** 899–908.

67. Ruberte, E., Marty, T., Nellen, D., Affolter, M., and Basler, K. (1995) An absolute requirement for both the type II and type I receptors, punt and thick veins, for dpp signaling in vivo. *Cell* **80,** 889–897.

68. Wrana, J. L., Tran, H., Attisano, L., Arora, K., Childs, S. R., Massague, J., et al. (1994) Two distinct transmembrane serine/threonine kinases from *Drosophila melanogaster* form an activin receptor complex. *Mol. Cell. Biol.* **14,** 944–950.

69. Schulte-Merker, S., Smith, J. C., and Dale, L. (1994) Effects of truncated activin and FGF receptors and of follistatin on the inducing activities of BVg1 and activin: does activin play a role in mesoderm induction? *EMBO J.* **13,** 3533–3541.

70. Yamashita, H., ten Dijke, P., Huylebroeck, D., Sampath, T. K., Andries, M., Smith, J. C., et al. (1995) Osteogenic protein-1 binds to activin type II receptors and induces certain activin-like effects. *J. Cell Biol.* **130,** 217–226.

71. Coulombe, J. N., Schwall, R., Parent, A. S., Eckenstein, F. P., and Nishi, R. (1993) Induction of somatostatin immunoreactivity in cultured ciliary ganglion neurons by activin in choroid cell-conditioned medium. *Neuron* **10,** 899–906.

72. Darland, D. C. and Nishi, R. (1998) Activin A and follistatin influence expression of somatostatin in the ciliary ganglion in vivo. *Dev. Biol.* **202,** 293–303.

73. Fann, M. J. and Patterson, P. H. (1994) Depolarization differentially regulates the effects of bone morphogenetic protein (BMP)-2, BMP-6, and activin A on sympathetic neuronal phenotype. *J. Neurochem.* **63,** 2074–2079.

74. Fann, M. J. and Patterson, P. H. (1994) Neuropoietic cytokines and activin A differentially regulate the phenotype of cultured sympathetic neurons. *Proc. Natl. Acad. Sci. USA* **91,** 43–47.

75. Schubert, D., Kimura, H., LaCorbiere, M., Vaughan, J., Karr, D., and Fischer, W. H. (1990) Activin is a nerve cell survival molecule. *Nature* **344,** 868–870.

76. Cornell, R. A. and Kimelman, D. (1994) Activin-mediated mesoderm induction requires FGF. *Development* **120,** 453–462.

77. LaBonne, C. and Whitman, M. (1994) Mesoderm induction by activin requires FGF-mediated intracellular signals. *Development* **120,** 463–472.

78. Wanaka, A., Milbrandt, J., and Johnson, E. M., Jr. (1991) Expression of FGF receptor gene in rat development. *Development* **111,** 455–468.

79. Niswander, L., Jeffrey, S., Martin, G. R., and Tickle, C. (1994) A positive feedback loop coordinates growth and patterning in the vertebrate limb. *Nature* **371**, 609–612.
80. Edlund, T. and Jessell, T. M. (1999) Progression from extrinsic to intrinsic signaling in cell fate specification: a view from the nervous system. *Cell* **96**, 211–224.
81. Hynes, M., Porter, J. A., Chiang, C., Chang, D., Tessier-Lavigne, M., Beachy, P. A., et al. (1995) Induction of midbrain dopaminergic neurons by Sonic hedgehog. *Neuron* **15**, 35–44.
82. Crossley, P. H. and Martin, G. R. (1995) The mouse Fgf8 gene encodes a family of polypeptides and is expressed in regions that direct outgrowth and patterning in the developing embryo. *Development* **121**, 439–451.
83. Di Porzio, U., Zuddas, A., Cosenza-Murphy, D. B., and Barker, J. L. (1990) Early appearance of tyrosine hydroxylase immunoreactive cells in the mesencephalon of mouse embryos. *Int. J. Dev. Neurosci.* **8**, 523–532.
84. Marchand, R. and Poirier, L. J. (1983) Isthmic origin of neurons of the rat substantia nigra. *Neuroscience* **9**, 373–381.
85. Ye, W., Shimamura, K., Rubenstein, J. L., Hynes, M. A., and Rosenthal, A. (1998) FGF and Shh signals control dopaminergic and serotonergic cell fate in the anterior neural plate. *Cell* **93**, 755–766.
86. Saucedo-Cardenas, O., Quintana-Hau, J. D., Le, W. D., Smidt, M. P., Cox, J. J., De Mayo, F., et al. (1998) Nurr1 is essential for the induction of the dopaminergic phenotype and the survival of ventral mesencephalic late dopaminergic precursor neurons. *Proc. Natl. Acad. Sci. USA* **95**, 4013–4018.
87. Zetterstrom, R. H., Solomin, L., Jansson, L., Hoffer, B. J., Olson, L., and Perlmann, T. (1997) Dopamine neuron agenesis in Nurr1-deficient mice. *Science* **276**, 248–250.
88. Wallen, A., Zetterstrom, R. H., Solomin, L., Arvidsson, M., Olson, L., and Perlmann, T. (1999) Fate of mesencephalic AHD2-expressing dopamine progenitor cells in NURR1 mutant mice. *Exp. Cell Res.* **253**, 737–746.
89. Simon, H. H., Saueressig, H., Wurst, W., Goulding, M. D., and O'Leary, D. D. (2001) Fate of midbrain dopaminergic neurons controlled by the engrailed genes. *J. Neurosci.* **21**, 3126–3134.
90. Smidt, M. P., van Schaick, H. S., Lanctot, C., Tremblay, J. J., Cox, J. J., van der Kleij, A. A., et al. (1997) A homeodomain gene Ptx3 has highly restricted brain expression in mesencephalic dopaminergic neurons. *Proc. Natl. Acad. Sci. USA* **94**, 13,305–13,310.
91. Saucedo-Cardenas, O., Kardon, R., Ediger, T. R., Lydon, J. P., and Conneely, O. M. (1997) Cloning and structural organization of the gene encoding the murine nuclear receptor transcription factor, NURR1. *Gene* **187**, 135–139.
92. Smidt, M. P., Asbreuk, C. H., Cox, J. J., Chen, H., Johnson, R. L., and Burbach, J. P. (2000) A second independent pathway for development of mesencephalic dopaminergic neurons requires Lmx1b. *Nat. Neurosci.* **3**, 337–341.
93. Barker, R. A. and Dunnett, S. B. (1999) Functional integration of neural grafts in Parkinson's disease. *Nat. Neurosci.* **2**, 1047–1048.

94. Olanow, C. W., Freeman, T. B., and Kordower, J. H. (1997) Neural transplantation as a therapy for Parkinson's disease. *Adv. Neurol.* **74,** 249–269.
95. Schueler, S. B., Ortega, J. D., Sagen, J., and Kordower, J. H. (1993) Robust survival of isolated bovine adrenal chromaffin cells following intrastriatal transplantation: a novel hypothesis of adrenal graft viability. *J. Neurosci.* **13,** 4496–4510.
96. Ono, T., Date, I., Imaoka, T., Shingo, T., Furuta, T., Asari, S., et al. (1997) Evaluation of intracerebral grafting of dopamine-secreting PC12 cells into allogeneic and xenogeneic brain. *Cell Transplant.* **6,** 511–513.
97. Espejo, E. F., Montoro, R. J., Armengol, J. A., and Lopez-Barneo, J. (1998) Cellular and functional recovery of Parkinsonian rats after intrastriatal transplantation of carotid body cell aggregates. *Neuron* **20,** 197–206.
98. Deacon, T., Schumacher, J., Dinsmore, J., Thomas, C., Palmer, P., Kott, S., et al. (1997) Histological evidence of fetal pig neural cell survival after transplantation into a patient with Parkinson's disease. *Nat. Med.* **3,** 350–353.
99. Isacson, O. and Breakefield, X. O. (1997) Benefits and risks of hosting animal cells in the human brain [see comments]. *Nat. Med.* **3,** 964–969.
100. Yurek, D. M. and Sladek, J. R., Jr. (1990) Dopamine cell replacement: Parkinson's disease. *Annu. Rev. Neurosci.* **13,** 415–440.
101. Kawasaki, H., Mizuseki, K., Nishikawa, S., Kaneko, S., Kuwana, Y., Nakanishi, S., et al. (2000) Induction of midbrain dopaminergic neurons from ES cells by stromal cell-derived inducing activity. *Neuron* **28,** 31–40.
102. Lee, S. H., Lumelsky, N., Studer, L., Auerbach, J. M., and McKay, R. D. (2000) Efficient generation of midbrain and hindbrain neurons from mouse embryonic stem cells. *Nat. Biotechnol.* **18,** 675–679.
103. Mytilineou, C., Park, T. H., and Shen, J. (1992) Epidermal growth factor-induced survival and proliferation of neuronal precursor cells from embryonic rat mesencephalon. *Neurosci. Lett.* **135,** 62–66.
104. Potter, E. D., Ling, Z. D., and Carvey, P. M. (1999) Cytokine-induced conversion of mesencephalic-derived progenitor cells into dopamine neurons. *Cell Tissue Res.* **296,** 235–246.
105. Svendsen, C. N., Fawcett, J. W., Bentlage, C., and Dunnett, S. B. (1995) Increased survival of rat EGF-generated CNS precursor cells using B27 supplemented medium. *Exp. Brain Res.* **102,** 407–414.
106. Svendsen, C. N., Caldwell, M. A., Shen, J., ter Borg, M. G., Rosser, A. E., Tyers, P., et al. (1997) Long-term survival of human central nervous system progenitor cells transplanted into a rat model of Parkinson's disease. *Exp. Neurol.* **148,** 135–146.
107. Svendsen, C. N., Clarke, D. J., Rosser, A. E., and Dunnett, S. B. (1996) Survival and differentiation of rat and human epidermal growth factor-responsive precursor cells following grafting into the lesioned adult central nervous system. *Exp. Neurol.* **137,** 376–388.
108. Ling, Z. D., Potter, E. D., Lipton, J. W., and Carvey, P. M. (1998) Differentiation of mesencephalic progenitor cells into dopaminergic neurons by cytokines. *Exp. Neurol.* **149,** 411–423.

109. Ptak, L. R., Hart, K. R., Lin, D., and Carvey, P. M. (1995) Isolation and manipulation of rostral mesencephalic tegmental progenitor cells from rat. *Cell Transplant.* **4,** 335–342.
110. Studer, L., Csete, M., Lee, S. H., Kabbani, N., Walikonis, J., Wold, B., et al. (2000) Enhanced proliferation, survival, and dopaminergic differentiation of CNS precursors in lowered oxygen. *J. Neurosci.* **20,** 7377–7383.
111. Yan, J., Studer, L., and McKay, R. D. (2001) Ascorbic acid increases the yield of dopaminergic neurons derived from basic fibroblast growth factor expanded mesencephalic precursors. *J. Neurochem.* **76,** 307–311.
112. Sawamoto, K., Nakao, N., Kakishita, K., Ogawa, Y., Toyama, Y., Yamamoto, A., et al. (2001) Generation of dopaminergic neurons in the adult brain from mesencephalic precursor cells labeled with a nestin-GFP transgene. *J. Neurosci.* **21,** 3895–3903.
113. Lendahl, U., Zimmerman, L. B., and McKay, R. D. (1990) CNS stem cells express a new class of intermediate filament protein. *Cell* **60,** 585–595.

9

Manipulation of Neural Precursors *In Situ*

Potential for Brain Self-Repair

Sanjay S. Magavi and Jeffrey D. Macklis

INTRODUCTION

Over the past three decades, research exploring potential neuronal replacement therapies has focused on replacing lost neurons by transplanting cells or grafting tissue into diseased regions of the brain *(1)*. Over most of the past century of modern neuroscience, it was thought that the adult brain was completely incapable of generating new neurons. However, in the last decade, the development of new techniques has resulted in an explosion of new research showing that neurogenesis, the birth of new neurons, normally occurs in two limited and specific regions of the adult mammalian brain and that there are significant numbers of multipotent neural precursors in many parts of the adult mammalian brain *(2)*. Recent findings demonstrate that it is possible to induce neurogenesis *de novo* in the adult mammalian brain and that it may become possible to manipulate endogenous multipotent precursors *in situ* to replace lost or damaged neurons *(3,4)*.

Neural cell replacement therapies are based on the idea that neurological function lost to injury or neurodegenerative disease can be improved by introducing new cells that can replace the function of lost neurons. Theoretically, the new cells can do this in one of two general ways *(1)*. New neurons can anatomically integrate into the host brain, becoming localized to the diseased portion of the brain, receiving afferents, expressing neurotransmitters, and forming axonal projections to relevant portions of the brain. Such neurons would function by integrating into the microcircuitry of the nervous system and replacing lost neuronal circuitry. Alternatively, newly introduced cells could more simply constitutively secrete neurotransmitters into local central nervous system (CNS) tissue or they could be engineered to produce growth factors to support the survival or regeneration of existing neurons. Growing knowledge about the normal role of endogenous neural precursors,

From: *Neural Stem Cells for Brain and Spinal Cord Repair*
Edited by: T. Zigova, E. Y. Snyder, and P. R. Sanberg © Humana Press Inc., Totowa, NJ

their potential differentiation fates, and their responsiveness to a variety of cellular and molecular controls suggest that neuronal replacement therapies based on manipulation of endogenous precursors either *in situ* or ex vivo may be possible.

Neuronal replacement therapies based on the manipulation of endogenous precursors *in situ* may have advantages over transplantation-based approaches, but they may have several limitations as well. The most obvious advantage of manipulating endogenous precursors *in situ* is that there is no need for external sources of cells. Cells for transplantation are generally derived from embryonic tissue, nonhuman species (xenotransplantation), or cells grown in culture. Use of embryonically derived tissue aimed toward human diseases is complicated by limitations in availability and by both political and ethical issues; for example, current transplantation therapies for Parkinson's disease require tissue from several embryos. Xenotransplantation of animal cells carries potential risks of introducing novel diseases into humans, and there are questions about how well xenogenic cells will integrate into the human brain. In many cases, cultured cells need to be immortalized by oncogenesis, increasing the risk that the cells may become tumorigenic. In addition, transplantation of cells from many of these sources risk immune rejection and may require immunosuppression if they are not derived from the recipient.

However, there are also possible limitations to the potential of manipulating endogenous precursor cells as a neuronal replacement therapy. First, such an approach may be practically limited to particular regions of the brain, because multipotent neural precursors are densely distributed only in particular subregions of the adult brain (e.g., the subventricular zone [SVZ] and hippocampal subgranular zone). In some cases, it is possible that there simply may not be a sufficient number of precursor cells to effect functional recovery. In addition, the potential differentiation fates of endogenous precursors may be too limited to allow their integration into varied portions of the brain. Another potential difficulty is that it may be difficult to provide the precise combination and sequence of molecular signals necessary to induce endogenous precursors to efficiently and precisely proliferate and differentiate into appropriate types of neurons deep in the brain.

The substantial amount of prior research regarding constitutively occurring neurogenesis provides insight into the potential and limitations of endogenous precursor-based neuronal replacement therapies. Recent work has partially elucidated the normal behavior of endogenous adult precursors, including their ability to migrate to select brain regions, differentiate into neurons, integrate into normal neural circuitry, and, finally, functionally integrate into the adult brain. Research is also beginning to elucidate bio-chemical and behavioral controls over constitutively occurring neurogenesis. The location, identity, and differentiation potential of endogenous adult precursors are beginning to be understood. In this chapter, we will first review research from a variety of labs regarding normally occurring neurogenesis and, then, review some of our experiments demonstrating that endogenous neural precursors can be induced to differentiate into neurons in regions of the adult brain that do not normally undergo neurogenesis.

CONSTITUTIVELY OCCURRING ADULT MAMMALIAN NEUROGENESIS

Ramon y Cajal wrote of the brain, "In the adult centers the nerve paths are something fixed, ended and immutable. Everything may die, nothing may be regenerated." The relative lack of recovery from CNS injury and neurodegenerative disease and the relatively subtle and extremely limited distribution of neurogenesis in the adult mammalian brain resulted in the entire field reaching the conclusion that neurogenesis does not occur in the adult mammalian brain. Altman was the first to use techniques sensitive enough to detect the ongoing cell division that occurs in adult brain. Using tritiated thymidine as a mitotic label, he published evidence that neurogenesis constitutively occurs in the hippocampus *(5)* and olfactory bulb *(6)* of the adult mammalian brain. These results were later replicated using tritiated thymidine labeling followed by electron microscopy *(7)*. However, the absence of neuron-specific immunocytochemical markers at the time resulted in identification of putatively newborn neurons being made on purely morphological criteria. These limitations led to a widespread lack of acceptance of these results and made research in the field difficult.

The field of adult neurogenesis was rekindled in 1992, when Reynolds and Weiss showed that precursor cells isolated from the forebrain *(8)* can differentiate into neurons in vitro. These results and technical advances, including the development of immunocytochemical reagents that could more easily and accurately identify the phenotype of various neural cells, led to an explosion of research in the field. Normally occurring neurogenesis in the olfactory bulb, olfactory epithelium, and hippocampus have now been well characterized in the adult mammalian brain.

OLFACTORY BULB NEUROGENESIS

The cells contributing to olfactory bulb neurogenesis originate in the anterior periventricular zone and, thus, undergo a fascinating and intricate path of migration to reach their final position in the olfactory bulb. Adult olfactory bulb neurogenesis has been most extensively studied in the rodent, although there is in vitro *(9,10)* and in vivo *(11)* evidence suggesting that such neuronal precursors exist in humans. Several experiments show that the precursors that contribute to olfactory bulb neurogenesis reside in the anterior portion of the subventricular zone. When retroviruses *(12)*, tritiated thymidine *(12)*, vital dyes *(12,13)*, or virally labeled SVZ cells *(13,14)* are microinjected into the anterior portion of the SVZ of postnatal animals, labeled cells are eventually found in the olfactory bulb. Upon reaching the olfactory bulb, these labeled neurons differentiate into interneurons specific to the olfactory bulb, olfactory granule cells and periglomerular cells. To reach the olfactory bulb, the neuroblasts undergo tangential chain migration through the rostral migratory stream (RMS) into the olfactory bulb *(15)*. Once in the olfactory bulb, the neurons migrate along radial glia away from the RMS and differentiate into interneurons.

Of considerable interest have been the factors that contribute to the direction of migration of the neuroblasts, as well as factors involved in initiating and controlling migration itself. In vitro experiments show that caudal septum explants secrete a diffusible factor, possibly the molecule Slit *(16)* that repels olfactory bulb neural precursors *(17)*. Consistent with the idea that SVZ precursor migration is directed by repulsion is the finding that SVZ precursors migrate anteriorly along the RMS even in the absence of the olfactory bulb *(18)*. The tangential migration of the cells seems to be at least partially mediated by PSA-NCAM, which is expressed by the neuroblasts themselves *(19)*. This migration may be modified by tenascin and chondroitin sulfate proteoglycans that are located near the SVZ *(20)*. Neuroblasts undergoing chain migration along the RMS do not travel along radial glia, although glia may play a role in their migration. Garcia-Verdugo et al. presented anatomical evidence that SVZ neuroblasts migrate within sheaths of slowly dividing astrocytes *(21)*. However, the astrocyte sheaths may not be necessary for tangential migration, because a great deal of tangential migration occurs in the first postnatal week, before astrocytes can be found in the RMS *(22)*. Understanding the factors that contribute to normal SVZ precursor migration could be important in developing approaches to induce such precursors to migrate to injured or degenerating regions of the brain.

The effort to identify the neural precursors that contribute to olfactory bulb neurogenesis has generated a great deal of controversy. Two potential sources of olfactory bulb neuroblasts have been suggested: astrocytes in the subventricular zone and ependymal cells lining the ventricles. It has been reported that single ependymal cells are capable of producing neurospheres *(23)*, free-floating spheres of cultured multipotent neural precursors, neurons, and glia. In contrast, it has also been reported that ciliary ependymal cells form spheres that consist of only astrocytes *(24)*. Other investigators have been unable to generate neurospheres from single ciliary ependymal cells and, instead, suggest that the multipotent neural precursors found in the adult brain are a type of astrocyte, expressing astrocytic morphology and glial fibrillary acidic protein (GFAP) *(25)*. Another, independent report provides support for the concept that multipotent neural precursors with similarities to astrocytes contribute to adult neurogenesis *(26)*. Although the majority of currently available evidence suggests that GFAP-expressing cells in the SVZ are the source of olfactory bulb neurogenesis, it is important to distinguish between true astroglia and a distinct class of precursor cells that simply express GFAP. GFAP, although generally a reliable marker for activated astrocytes, has been used as a sole phenotypic marker in reports suggesting that astrocytes are multipotent neural precursors, or stem cells. It is quite possible that at least some the multipotent neural precursors simply express GFAP while remaining distinct from true astroglia. Further clarifying the identity of potentially multiple classes of multipotent neural precursors that contribute to adult neurogenesis will increase our ability to manipulate such cells.

Although the identity of the adult multipotent neural precursors in the SVZ is still controversial, a number of experiments have been performed to manipulate their fate and examine their potential, both in vitro and in vivo. These results will guide attempts to manipulate endogenous precursors for brain repair. In vitro, subventricular zone precursors have been exposed to a number of factors to determine their responses. Generally, precursor cells have been removed from the brain, dissociated, and cultured in epidermal growth factor (EGF) and/or basic fibroblast growth factor (bFGF). The EGF and/or bFGF is then removed and the cells are exposed to growth factors of interest. The details of this process, including the particular regions the cells are derived from, the media they are plated in, and the substrate they are plated on, can have significant effects on the fate of the precursors *(27)*. EGF and bFGF *(28–30)* both induce the proliferation of subventricular zone precursors and can influence their differentiation. EGF tends to direct cells to a glial fate, and bFGF more toward a neuronal fate *(27)*. Bone morphogenetic proteins (BMPs) promote differentiation of SVZ precursors into astroglial fates *(31)*, whereas platelet-derived growth factor (PDGF)

(27,32) and insulin-like growth factor-1 (IGF-1) *(33)* promote SVZ precursors to differentiate into neurons. There are conflicting results regarding whether brain-derived neurotrophic factor (BDNF) promotes the survival *(34,35)* or differentiation *(33,36)* of SVZ precursors in vitro *(36)*. In vitro results show that it may be possible to influence the proliferation and differentiation of adult SVZ precursors.

The effects of several growth factors have also been tested in vivo to investigate their effects under physiological conditions. Intracerebroventricularly (ICV) infused EGF or transforming growth factor (TGF)-α induces a dramatic increase in SVZ precursor proliferation, and bFGF induces a smaller increase in proliferation *(37,38)*. Even subcutaneously delivered bFGF can induce the proliferation of SVZ precursors in adult animals *(39)*. However, despite the fact that newborn, mitogen-induced cells disperse into regions of the brain surrounding the ventricles, it is generally accepted that none of the newborn cells differentiate into neurons *(38)*. Intraventricularly infused BDNF increased the number of newly born neurons found in the olfactory bulbs of adult animals *(40)*. These results extend the in vitro results and suggest that it may be possible to use growth factors to manipulate adult endogenous precursors in vivo for brain repair.

Several reports have attempted to establish the differentiation potential of SVZ multipotent precursors, but these have yielded conflicting results. Postnatal mouse SVZ precursors can differentiate into neurons in a number of regions in the developing neuraxis *(41)*, whereas their fate is more limited to astroglia when they are transplanted into adult brain *(42)*. Adult mouse SVZ precursors injected intravenously into sublethally irradiated mice have been reported to differentiate into hematopoetic cells, interpreted as demonstrating the broad potential for differentiation and interlineage "transdifferentiation" these neural precursors seem to possess. However, it is possible that a chance transformation of cultured SVZ cells led to a single transformant precursor, accounting for this finding. In this event, it could be concluded that this result is not generally the case. However, labeled multipotent neural precursors derived from adult mouse and transplanted into stage 4 chick embryos or developing mouse morulae or blastocysts can integrate into the heart, liver, and intestine and express proteins specific for each of these sites *(43)*. Adult multipotent neural precursors may not be totipotent, but they appear to be capable of differentiating into a wide variety of cell types under appropriate conditions. These results indicate that the local cellular and molecular environment in which SVZ neural precursors are located can play a significant role in their differentiation. Providing the cellular and molecular signals for appropriate differentiation and integration of new neurons will be critical for neuronal replacement therapies in which endogenous neural precursors are either transplanted or manipulated *in situ*.

OLFACTORY EPITHELIUM NEUROGENESIS

Sensory neurons in the olfactory epithelium are continually generated in adult rodents. The globose basal cells of the olfactory epithelium divide, differentiate into neurons, and send their axons through the olfactory nerve to the olfactory bulb *(44,45)*. Of all the neurons in the mammalian body, olfactory epithelium sensory neurons are most directly exposed to potentially damaging influences, quite likely necessitating their continual replacement. The constant flow of air over the epithelium brings a continuously varying combination of new odorants and potential insults to the olfactory sensory neurons. Despite the precarious position of olfactory receptor neurons, the population of olfactory receptor neurons is maintained, and mammals maintain a fairly consistent sense of smell throughout life.

Neurogenesis in the olfactory epithelium is strongly modulated by neuronal death in the epithelium. Olfactory nerve lesion or olfactory bulb lesion, which lead to the degeneration of axotomized neurons *(46,47)*, result in an increase in proliferation of precursors *(45–49)*. The new neurons that form in the adult olfactory epithelium send axons through the olfactory nerve and into the mature olfactory bulb *(50,51)*. The ability of the newborn neurons to re-form long-distance projections is probably the result of both their immature state and the environment through which they extend their axons.

In vitro experiments suggest that mature olfactory sensory neurons produce a signal that inhibits neurogenesis. Olfactory epithelial precursor cells undergo dramatically reduced neurogenesis in the presence of mature olfactory receptor neurons in vitro *(52)*, suggesting that when olfactory sensory neurons are lost in vivo, the factors inhibiting neurogenesis are reduced, allowing the formation of new sensory neurons. Understanding the signals that inhibit neurogenesis and neuronal integration may be as important as understanding the signals that foster neurogenesis.

However, the factors that inhibit neurogenesis in the adult olfactory epithelium have not yet been discovered, and research in the field has focused on many of the growth factors that influence neurogenesis in the SVZ and olfactory bulb. Many of the same factors influence both olfactory bulb and olfactory epithelial neurogenesis. IGF-1 increases the rate of neurogenesis of olfactory epithelium precursors, both in vivo and in vitro *(53)*. In vitro, FGF-2 stimulates the proliferation of globose basal cells, TGF-α induces their differentiation, and PDGF promotes their survival *(54)*.

Research in the olfactory epithelium highlights the role of inhibitory factors in controlling neurogenesis; understanding these signals in degenerating portions of the brain may be instrumental in developing neuronal replacement therapies.

HIPPOCAMPAL NEUROGENESIS

Neurogenesis in the adult hippocampus has been extensively studied, resulting at least partially from the tantalizing connection between the hippocampus and the formation of memory. Does hippocampal neurogenesis play a part in memory formation? This question has only begun to be answered, but our understanding of hippocampal neurogenesis is already quite significant. Of particular interest is the fact that hippocampal neurogenesis can be modulated by physiological and behavioral events such as aging, stress, seizures, learning, and exercise. These properties of hippocampal neurogenesis may provide novel methods of studying neurogenesis and may elucidate broader influences that may be relevant to neuronal replacement therapies.

Hippocampal neurogenesis has been described in vivo in adult rodents *(5)*, monkeys *(55–57)*, and adult humans *(58)*. Newborn cells destined to become neurons are generated along the innermost aspect of the granule cell layer, the subgranular zone, of the dentate gyrus of the adult hippocampus. The cells migrate a short distance into the granule cell layer, send dendrites into the molecular layer of the hippocampus, and send their axons into the CA3 region of the hippocampus *(59–61)*. Adult-born hippocampal granule neurons are morphologically indistinguishable from surrounding granule neurons *(62)*. The precursor cells appear to mature rapidly and they extend their processes into the CA3 region as early as 4 d after division *(61)*. The properties of both the precursor cells and the hippocampal environment likely contribute to the rapid maturation observed.

The majority of research on hippocampal precursors has been performed in vivo, but many in vitro results are also useful for understanding the effects of growth factors on the differentiation of hippocampal precursors. Hippocampal precursor cells are studied in vitro much like SVZ precursors: They are removed from the brain, dissociated, and cultured in either EGF or bFGF; the mitogen is then removed and the cells are exposed to the growth factors of interest. Hippocampal precursors proliferate in response to FGF-2 and can differentiate into astrocytes, oligodendrocytes, and neurons in vitro *(62)*. BDNF increases both neuronal survival and differentiation, whereas NT-3, NT-4/5, and ciliary neurotrophic factor (CNTF) have more limited effects *(63)*. Further demonstrating the existence of precursors in the adult human, multipotent precursors derived from the adult human brain can be cultured in vitro *(64,65)*.

Hippocampal neurogenesis occurs throughout adulthood, but declines with age *(66)*. This age-related decline could be the result of a depletion of multipotent precursors with time, a change in precursor cell properties, or a change in the levels of molecular factors that influence neurogenesis. Understanding what causes this age-related decrease in neurogenesis may be important in assessing the potential utility of potential future neuronal replacement therapies based in the manipulation of endogenous precursors in elderly patients. Although aged rats have dramatically lower levels of neurogenesis than young rats, adrenalectomized aged rats have levels of neurogenesis very similar to those of young adrenalcctomized rats *(67,68)*. These results suggest that it is at least partially increased corticosteroids, which are produced by the adrenal glands, and not a decrease in the number of multipotent precursors, which leads to age-related decreases in neurogcnesis. At least in the hippocampus, multipotent precursors survive with advancing age.

Seizure can also increase hippocampal neurogenesis. However, it appears that seizure-induced neurogenesis may contribute to inappropriate plasticity, highlighting the fact that newly introduced neurons need to be appropriately integrated into the brain in order to have beneficial effects. Chemically or electrically induced seizures induce the proliferation of subgranular zone precursors, the majority of which differentiate into neurons in the granule cell layer *(69,70)*. However, some newborn cells differentiate into granule cell neurons in ectopic locations in the hilus or molecular layers of the hippo-campus and form aberrant connections to the inner molecular layer of the dentate gyrus, in addition to the CA3 pyramidal cell region *(70,71)*. It is hypothesized that these ectopic cells and aberrant connections may contribute to hippocampal kindling *(72,73)*.

Hippocampal cell death or activity-related signals resulting from seizures may modify signals that lead to increased neurogenesis. Induced seizures lead to the degeneration of hippocampal neurons *(74–76)*, which is fol-lowed by neurogenesis *(69,70)*. Excitotoxic or physical lesions of the hippocampal granule cell layer induce precursor cell proliferation within the dentate gyrus and the formation of neurons that have the morphological and immunocytochemical properties of granule cell neurons *(77)*. These results suggest that hippocampal granule cells either inhibit neurogenesis, similar to neurons in the olfactory epithelium, or that they or surrounding neurons produce signals that induce neurogenesis as they die. However, because neurogenesis is also increased by less pathological levels of electrophysiological activity *(78)*, it is also possible that signals induced by electrophysiological activity play a role in seizure-induced hippocampal neurogenesis.

Events occurring in the hippocampus dramatically demonstrate that behavior and environment can have a quite direct influence on the brain's microcircuitry. Animals living in an enriched environment containing toys, running wheels, and social stimulation contain more newborn cells in their hippocampus than control mice living in standard cages *(79)*. Experiments to further assess which aspects of the enriched environment contribute to increased neurogenesis reveal that a large portion of the increase can be attributed to simply exercise via running *(80,81)*. Associative learning tasks that involve the hippocampus also appear to increase neurogenesis *(82)*. Stress, on the other hand, can reduce neurogenesis in both rodents *(83,84)* and primates *(55)*. An intriguing, but completely speculative, idea that has been advanced by some in this field is that the processes mediating these effects on neurogenesis may underlie some of the benefits that physical and social therapies provide for patients with stroke and brain injury.

Some of the molecular mechanisms that mediate behavioral influences on hippocampal neurogenesis have begun to be elucidated. For instance, IGF-1, which increases adult hippocampal neurogenesis *(85)*, is preferentially transported into the brain in animals that are allowed to exercise. Blocking IGF-1 activity in exercising animals reduces hippocampal neurogenesis, suggesting that IGF-1, at least partially, mediates the effects of exercise on neurogenesis *(86)*. Stress increases systemic adrenal steroid levels and reduces hippocampal neurogenesis *(84)*. Adrenalectomy, which reduces adrenal steroids, including corticosteroids, increases hippocampal neurogenesis *(67,87)*, suggesting that adrenal hormones, at least partially, mediate the effects of stress on hippocampal neurogenesis. Intriguingly, some antidepressant medications, which ostensibly reduce stress, also appear to increase neurogenesis *(88,89)*. Together, these results demonstrate that adult neurogenesis can be modified by systemic signals, suggesting that modifying such systemic signals, and not only local ones, may be useful in developing potential future neuronal replacement therapies involving manipulation of endogenous precursors *(90)*.

Adult hippocampal multipotent precursors can adopt a variety of fates in vivo, suggesting that they may be able to appropriately integrate into neuronal microcircuitry outside of the dentate gyrus of the hippocampus. Hippocampal precursors transplanted into neurogenic regions of the brain can differentiate into neurons, whereas precursors transplanted into non-neurogenic regions do not differentiate into neurons at all. Adult rat hippocampal precursors transplanted into the rostral migratory stream migrate to the olfactory bulb and differentiate into a neuronal subtype not found in the hippocampus: tyrosine-hydroxylase-positive neurons *(91)*. However, although adult hippocampal precursors transplanted into the retina can adopt neuronal fates and extend neurites, they do not differentiate into photoreceptors, demonstrating a limitation of their differentiation fate potential *(92,93)*. These findings demonstrate the importance of the local cellular and molecular microenvironment in determining the fate of multipotent precursors. These results also highlight that, although adult hippocampal precursors can adopt a variety of neuronal fates, they may not be able to adopt every neuronal fate.

Some recent correlative evidence suggests that newly generated neurons in the adult hippocampus may participate in some way in hippocampal-dependent memory. Nonspecifically inhibiting hippocampal neurogenesis using a systemic mitotic toxin impairs trace conditioning in a manner not seen in relevant controls, suggesting a role for newly-born neurons in the formation of memories *(94)*. These results, although correlative so far, suggest that adult-born hippocampal neurons integrate functionally into the adult mammalian brain. Ongoing research in multiple laboratories is exploring the precise role they play in the adult hippocampal circuitry.

An interesting, but as yet unproven, hypothesis concerning the role of hippocampal neurogenesis in human depression has been proposed. Jacobs et al. propose that insufficient hippocampal neurogenesis causally underlies depression *(95)*. Consistent with this hypothesis, stress-related glucocorticoids are associated with a decrease in neurogenesis, and increased serotonin levels *(95,96)*, or antidepressants *(89)* are associated with an increase in neurogenesis. However, the hippocampus is generally thought to be involved in memory consolidation and less involved in the generation of mood, suggesting that altered hippocampal neurogenesis may be secondary to depression rather than causative.

CORTICAL NEUROGENESIS

The vast majority of studies investigating potential neurogenesis in the neocortex of the well-studied rodent brain do not report normally occurring adult cortical neurogenesis. Our own results demonstrate a complete absence of constitutively occurring neurogenesis in murine neocortex *(4)*. However, two studies reported low-level constitutively occurring neurogenesis in specific regions of the neocortex of adult primates *(97)* and in the visual cortex of adult rat *(7)*. In ref. *97*, neurogenesis of 2–3 new neurons/mm^3 was reported in the prefrontal, inferior temporal, and posterior parietal cortexes of the adult macaque, but not in the striate cortex, a presumably simpler primary sensory area. Other groups have not yet been able to reproduce these findings in rodents or primates. There exists a single report of neurogenesis in the visual cortex of the adult rat *(7)*, but this study used tritiated thymidine and purely morphological cell-type identification and has not been confirmed by any other group, including our own. It is unclear whether neurogenesis occurs normally in the neocortex of any mammals, but further examination of potential constitutively occurring neurogenesis in classically non-neurogenic regions will be required to definitively assess the potential existence of perhaps extremely low-level neurogenesis.

FUNCTIONAL ADULT NEUROGENESIS OCCURS IN NONMAMMALIAN VERTEBRATES

Functional adult neurogenesis also occurs in many nonmammalian vertebrates. The medial cerebral cortex of lizards, which resembles the dentate gyrus of mammals, undergoes postnatal neurogenesis and can regenerate in response to injury *(98)*. Newts can regenerate their tails, limbs, jaws, ocular tissues, and the neurons that occupy these regions *(99,100)*. Goldfish undergo retinal neurogenesis throughout life *(101)* and, impressively, can regenerate surgically excised portions of their retina in adulthood *(102)*. Although lower animals can undergo quite dramatic regeneration of neural tissue, it is unclear how relevant this is to mammals. It is thought that selective evolutionary pressures have led mammals to lose such abilities during normal life.

Birds, complex vertebrates whose brains are much closer to mammals in complexity, also undergo postnatal neurogenesis. Lesioned postnatal avian retina undergoes some neurogenesis, with the new neurons most likely arising from Muller glia *(103)*. In songbirds, new neurons are constantly added to the high vocal center *(104,105)*, a portion of the brain necessary for the production of learned song *(106–108)*, as well as to specific regions elsewhere in the brain (but not all neuronal populations). It is possible to experimentally manipulate the extent to which new neurons are produced in at least one songbird, the zebra finch, which does not normally seasonally replace HVC–RA projection neurons in the song production network (e.g., as canaries do) *(3)*. Inducing cell death of HVC–RA neurons in zebra finches leads to deterioration in song. Neurogenesis increases following induced cell death, and birds variably recover their ability to produce song coincident with the formation of new projections from area HVC to area RA, suggesting that induced neuronal replacement can restore a learned behavior.

THE LOCATION OF ADULT MAMMALIAN
MULTIPOTENT PRECURSORS

If adult multipotent precursors were limited to the two neurogenic regions of the brain, the olfactory bulb and hippocampal dentate gyrus, it would severely limit the clinical potential of neuronal replacement therapies based on *in situ* manipulation of endogenous precursors. However, adult multipotent precursors are not limited to the olfactory epithelium, anterior SVZ, and hippocampus of the adult brain; they have been cultured in vitro from caudal portions of the SVZ, septum *(2)*, striatum *(2)*, cortex *(109)*, optic nerve *(109)*, spinal cord *(110,111)*, and retina *(112)*. The precursors derived from all of these regions can self-renew and differentiate into neurons, astrocytes, and oligodendrocytes in vitro. It is thought that they normally differentiate only into glia or die in vivo. Cells from each region have differing requirements for their proliferation and differentiation. Precursors derived from septum, striatum, cortex, and optic nerve are reported to require bFGF to proliferate and differentiate into neurons in vitro. There are conflicting reports on whether bFGF is sufficient to culture spinal cord precursors *(110,111)*. Retinal precursors do not require any mitogens in order to divide in vitro, although they do respond to both EGF and bFGF. As with all primary cultures, the particular details of the protocols used can strongly influence the proliferation, differentiation, and viability of the cultured cells, so it is difficult to compare results from different labs. It is estimated that adult multipotent precursors can be found in small-but-significant number in various regions of the brain (e.g., separating cortical neuronal precursors by Percoll gradient yields about 140 multipotent precursors/mg vs 200/mg in the hippocampus *(109)*. Understanding the similarities and differences between the properties of multipotent precursors derived from different regions of the brain will be instrumental in potentially developing neuronal replacement therapies based on the manipulation of endogenous precursors.

Although it is not generally accepted, there are reports that, in addition to the undifferentiated multipotent precursors that are found in various portions of the brain, mature neurons themselves can be induced to divide *(113,114)*. Although it seems unlikely that a neuron could maintain the elaborate neurochemical and morphologic differentiation state of a mature neuron while replicating its DNA and remodeling its nucleus and soma, it is still theoretically possible. Although it is generally accepted that other neural cells, such as astrocytes, can divide, most reports suggest that any attempt by differentiated neurons to re-enter the cell cycle results in aborted cycling and, ultimately, death *(115)*. Significant evidence will need to be presented to convincingly demonstrate that mature neurons in the adult brain are capable of mitosis.

MANIPULATING THE CORTICAL ENVIRONMENT

Endogenous multipotent precursors in the adult brain can divide, migrate, differentiate into neurons, receive afferents, and extend axons to their targets. Multipotent precursors are concentrated in the olfactory epithelium, anterior SVZ, and the dentate gyrus of the hippocampus, but they can be found in lower densities in a number of other regions of the adult brain. In addition, these precursors also have the a broad potential; they can differentiate into at least three different cell types (astroglia, oligodendroglia, and neurons) given an appropriate in vitro or in vivo environment. These properties of endogenous multipotent precursors led us to explore the fate of these precursors in an adult cortical environment that has been manipulated to support neurogenesis.

Our lab has previously shown that cortex undergoing synchronous degeneration of apoptotic projection neurons forms an instructive environment that can guide the differentiation of transplanted immature neurons or neural precursors. Immature neurons or multipotent neural precursors transplanted into targeted cortex migrate selectively to layers of cortex undergoing degeneration of projection neurons *(116–118)*, differentiate into projection neurons *(116–119)*, receive afferent synapses *(116,119)*, express appropriate neurotransmitters and receptors, and re-form appropriate long-distance connections to the original contralateral targets of the degenerating neurons *(118)* in adult murine neocortex. Immature neurons or neural precursors transplanted into control intact or kainic acid-lesioned cortex do not migrate, differentiate into projection neurons, or integrate into cortex. Together, these results suggested to us that cortex undergoing targeted apoptotic degeneration could direct endogenous multipotent precursors to integrate into adult cortical microcircuitry. However, before discussing the fate of endogenous precursors in such an environment, we will review in more detail the results that led us to believe that cortex undergoing targeted apoptosis forms an instructive environment for neurogenesis.

Biophysically targeted apoptotic neurodegeneration produces highly specific cell death of selected projection neurons within defined regions of neocortex (Fig. 1). Importantly, surrounding glia, interneurons, and projection neurons that have not been targeted are not injured by this approach *(116,120)*. Degeneration results from the photoactivation of retrogradely transported nanospheres carrying the chromophore chlorin e_6 *(116,117,121)*. First, chlorin e_6 carrying nanospheres are injected into the axonal terminal

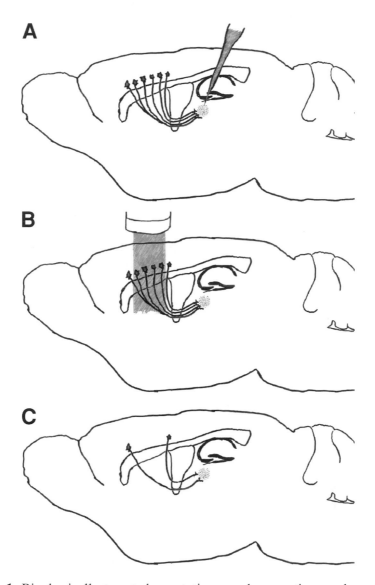

Fig. 1. Biophysically targeted apoptotic neurodegeneration produces highly specific cell death of selected projection neurons within defined regions of neocortex. Degeneration results from the photoactivation of retrogradely transported nanospheres carrying the chromophore chlorin e_6 *(116,117,121)*. (**A**) Nanospheres carrying chlorin e_6 are injected into the axonal terminal fields of the targeted projection neurons. The nanospheres are retrogradely transported to the somata of the projection neurons and stored nontoxically within neuronal lysosomes. (**B**) The projection neurons are then exposed to long-wavelength light, which specifically induces the chlorin e_6 to produce singlet oxygen and induces apoptosis of neurons containing chlorin e_6. (**C**) Because neither the chlorin e_6 nor the long-wavelength light causes degeneration by themselves, only neurons both selectively labeled with the chromophore and located in the controlled light path undergo apoptotic degeneration.

fields of the targeted projection neurons. The nanospheres are retrogradely transported to the soma of the projection neurons and stored nontoxically within neuronal lysosomes. The projection neurons are then exposed to long-wavelength light, which specifically induces the chlorin e_6 to produce singlet oxygen and induces apoptosis of chlorin e_6-containing cells. Chlorin e_6 only produces the reactive oxygen species when excited by near-infrared light. The long-wavelength light used is not absorbed by neural tissue and does not cause damage by either heating or radiation-based effects. Because neither the chlorin e_6 nor the long-wavelength light causes degeneration by themselves, only neurons both selectively labeled with the chromophore and located in the controlled light path undergo apoptotic degeneration.

We have applied this approach in several mammalian and avian CNS systems, effecting 50–99% degeneration (typically, approx 70%) of the targeted populations *(3,4,116,117)*. In the neocortex, we have produced selective degeneration of pyramidal neurons with callosal projections (laminae 2/3 and/or 5 neurons) or corticothalamic projections *(4,116,117,120)*. Targeted neurons die by apoptosis via a cascade of characteristic apoptotic cellular events. Targeted neurons undergo nuclear and internucleosomal DNA fragmentation, undergo heterochromatin condensation, and form apoptotic bodies *(121,122)*. Targeted neurons are TUNEL positive and are phagocytically removed from the brain without inflammation or gliosis. Neuronal death depends on both new protein synthesis and intrinsic endonuclease activity. This relatively synchronous initiation of neuronal apoptosis leads to prolonged alterations in gene expression *(123)*, which may not be induced by the relatively sporadic cell death that occurs in neurodegenerative disease.

The instructive environment that exists in the apoptotically degenerating cortex can alter the differentiation fate of transplanted immature neurons or multipotent precursors. To examine the fate of heterotopically transplanted immature neurons in the cortex undergoing apoptotic degeneration, immature neurons derived from E14 mice, which normally contribute to layer 5 and 6 of cortex, were transplanted into cortex undergoing the targeted apoptosis of layer 2/3 callosal projection neurons. The transplanted E14 immature neurons differentiated into neurons in layer 2/3 and formed contralateral callosal projections, which are typical of layer 2/3 neurons but absent in layer 6 neurons. We also tested the potential of two multipotent precursor cell lines: cerebellar-derived C17.2 precursor cells and hippocampus-derived HiB5 precursor cells, derived from an even earlier stage. Both of these multipotent precursor cell lines can differentiate into neurons and glia when transplanted into the embryonic developing brain and differentiate only into glia (or remain undifferentiated) when transplanted into adult intact or kainic acid–lesioned cortex. C17.2 cells differentiated into neurons at low efficiencies when transplanted into regions of cortex undergoing targeted apoptosis. In contrast, similarly transplanted HiB5 cells, perhaps as a result of the absence of the TrkB receptor, did not differentiate into neurons. These results highlight the fact that although cortex undergoing apoptotic degeneration is instructive for neuronal differentiation and integration, the newly introduced cells must be competent to respond to such environmental signals.

Immature neurons or precursors transplanted near targeted regions of cortex migrated up to 800 μm to occupy degenerating regions of cortex, whereas neurons or precursors transplanted into control mice remain within approx 40 μm of their injection site. Approximately 45% of transplanted E17 neurons underwent directed migration and pyramidal neuron differentiation; the rest remained in the injection site, surviving as small, presumptive interneurons. In living-slice cultures, transplanted immature neurons adopted classic migratory morphologies, showing leading and trailing processes and elongated cell bodies, and underwent saltatory migration. They migrated across intact layers 5 and 6, toward targeted regions of cortex. The immature neurons migrated along axons and host-derived RC-2-positive radial glial cells. Host astrocytes, identified by the expression of a host-specific transgene, dedifferentiated and formed radial glia only in targeted cortex that contained donor cells. Most neurons completed their migration by 2 wk after transplantation and had already begun extending axons toward the corpus callosum.

After migrating, transplanted immature neurons integrated into the microcircuitry of the targeted cortex. Synaptophysin staining and electron microscopy demonstrated that more than 65% of donor neurons accepted afferent input. In addition, transplanted neurons expressed neurotransmitters and neurotransmitter receptors appropriate to the region of cortex into which they were transplanted. Donor neurons expressed the same complement of neurotransmitters (glutamate, aspartate, and GABA) and receptors (kainate-R, AMPA-R, NMDA-R, and GABA-R) as endogenous cortical projection neurons *(124)*. Taken together, these results demonstrate that targeted cortex induces immature neurons to differentiate into neurons that closely resemble the neurons that were induced to degenerate.

Donor neurons or neural precursors transplanted into cortex undergoing apoptotic degeneration re-form specific axonal projections to the original targets of the degenerating projection neurons. Cells transplanted into control intact or kainic acid–lesioned cortex do not form any long-distance projections. To examine the specificity of the donor neuron's long-distance projections, we transplanted immature neurons into cortex undergoing degeneration of transcallosal cortico-cortical layer 2/3 projection neurons. Using a retrograde label, we found that donor cells re-formed specific callosal connections 6–10 mm long to contralateral cortex, the original targets of the degenerating projection neurons. The donor cells did not re-form connections to closer, alternate targets of other nearby populations of S1 neurons, including ipsilateral thalamus, motor cortex, and secondary somatosensory cortex. The efficiency of this anatomic reconnectivity appears to be highly dependent on the developmental stage of the donor neuroblasts or precursors. Forty percent of donor E19 neuroblasts, 21% of donor E17 neuroblasts, 10–15% of donor E14 neuroblasts, and only 1–3% of transplanted multipotent precursors make long-distance projections. The properties of the donor cells are important in determining whether they can respond to the instructive signals produced by degenerating cortex.

To begin to identify the molecular signals that are responsible for the instructive environment produced by cortex undergoing targeted apoptosis of projection neurons, we analyzed the gene expression of candidate neurotrophins, growth factors, and receptors in regions of targeted neuronal death. We compared gene expression to that of intact adult cortex using Northern-blot analysis, *in situ* hybridization, and immunocytochemical analysis. The genes for BDNF, NT-4/5, and NT-3 are upregulated only in degenerating regions of cortex, specifically during the period of projection neuron apoptosis *(123)*. The expression of a variety of other growth factors that are not as developmentally regulated is not altered. These results are in contrast to the less specific, more immediate, and short-lived changes in gene expression observed in response to nonspecific necrotic injuries or seizure-induced injury. Surrounding glia and neurons may change their gene expression in response to activity-dependent changes at their synapses or factors released by degenerating projection neurons. For example, BDNF is upregulated specifically by local interneurons adjacent to degenerating projection neurons. We are currently using polymerase chain reaction (PCR)-based suppression subtraction hybridization and DNA microarray techniques to examine the expression of other known and novel factors that contribute to the instructive environment formed by cortex undergoing targeted apoptosis of projection neurons.

INDUCTION OF NEUROGENESIS
IN THE NEOCORTEX OF ADULT MICE

Based on the above-outlined results, we investigated the fate of endogenous multipotent precursors in cortex undergoing targeted apoptotic degeneration. Although endogenous multipotent precursors exist in the adult brain, including the cortex, neurogenesis does not normally occur in postnatal mouse cortex. By examining the fates of newborn cells in targeted neocortex, an environment that is instructive for neurogenesis by exogenous precursors, we addressed the question of whether the normal absence of endogenous cortical neurogenesis reflects an intrinsic limitation of the endogenous neural precursors' potential or a lack of appropriate microenvironmental signals necessary for neuronal differentiation or survival. In addition, these experiments provide information regarding whether endogenous neural precursors could potentially be manipulated *in situ*, toward future neuronal replacement therapies.

We found that endogenous multipotent precursors, normally located in the adult brain could be induced to differentiate into neurons in the adult mammalian neocortex. We induced synchronous apoptotic degeneration *(116,117)* of corticothalamic neurons in layer 6 of anterior cortex and examined the fates of dividing cells within the cortex using bromo-2-deoxy uridine (BrdU) and markers of progressive neuronal differentiation. BrdU+ newborn cells expressed NeuN, a mature neuronal marker, and survived at least 28 wk. First, subsets of BrdU+ precursors expressed Doublecortin, a protein expressed exclusively in migrating neurons *(125,126)*, and Hu, an early neuronal marker *(127,128)*. We observed no new neurons in control, intact cortex. Retrograde labeling from thalamus demonstrated that newborn BrdU+ neurons can form long-distance corticothalamic connections. Together, these results demonstrate that endogenous neural precursors can be induced *in situ* to differentiate into cortical neurons, survive for many months, and form appropriate long-distance connections in the adult mammalian brain.

We induced synchronous apoptotic degeneration of layer 6 corticothalamic projection neurons via chromophore-targeted neuronal degeneration. We targeted layer 6 corticothalamic projection neurons in the anterior neocortex because they are located close to the subventricular zone, a region that contains a relatively large number of multipotent neural precursors. Chlorin *e*6-conjugated nanospheres were injected into the thalamus in adult female mice, from which they were retrogradely transported to the somata of layer 6 corticothalamic projection neurons. Layer 6 anterior cortex was noninvasively exposed to long-wavelength light via intact dura, inducing apoptosis of the corticothalamic projection neurons in anterior cortex. We treated mice with BrdU for the following 2 wk. After 2, 5, 9, and 28 wk of survival, we examined the phenotype of BrdU+ newborn cells using markers of developing and mature neurons and glia, including Doublecortin,

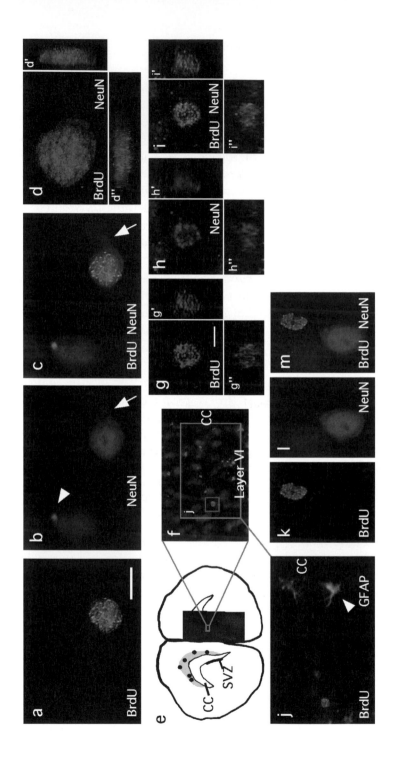

Hu, NeuN, glial fibrillary acidic protein (GFAP), and myelin basic protein (MBP). We examined their axonal projections using the retrograde label FluoroGold.

We found that newborn cells differentiated into mature NeuN-expressing neurons and survived for at least 28 wk. Recently born, BrdU+/NeuN+ neurons were found only in regions of cortex undergoing apoptotic degeneration of layer 6 thalamocortical projection neurons 2–28 wk after inducing degeneration (Fig. 2). To confirm that newborn cells were double-labeled and not merely closely apposed to pre-existing neurons, we imaged them using laser scanning confocal microscopy and produced three-dimensional digital reconstructions. The reconstructions confirmed that these newborn cells were neurons. Some newborn neurons had pyramidal neuron morphology (large, 10–15-μm-diameter somata with apical processes; Figs. 2A–D) characteristic of neurons that give rise to long-distance projections. In order to examine the theoretical possibility that these newborn cells were

---

Fig. 2. *(see opposite page)* Newborn BrdU+ cells can be induced to differentiate into neurons, expressing NeuN, a mature neuronal marker, in regions of cortex undergoing targeted apoptotic degeneration of corticothalamic neurons. **(a)** A large densely stained BrdU+ nucleus (red) 28 wk after induction of apoptosis. Scale bar = 10 μm. **(b)** The BrdU+ neuron (lower right) shows typical NeuN staining, with strong nuclear and weaker cytoplasmic labeling; apical dendrite (arrow). The neuron at left in a different focal plane remains from the original nanosphere targeting for apoptosis; a lysosome containing nanospheres is indicated (arrowhead). No BrdU+/NeuN+ newborn neurons contained nanospheres, demonstrating that they are not original targeted neurons. **(c)** Overlay. **(d)** Confocal images combined to produce a three-dimensional (3D) reconstruction of this newborn neuron. **(d′, d″)** Cell viewed along the *x*-axis and *y*-axis, respectively, demonstrating colocalization of BrdU and NeuN. **(e)** Left: Camera lucida showing location of BrdU+/NeuN+ cells (dots) within the targeted region (gray). Right: A sample newborn neuron in cortical layer VI; corpus callosum (CC); subventricular zone (SVZ). **(f)** Higher-magnification view of layer VI shows a BrdU+ newborn neuron (red box). **(g–i)** Confocal 3D reconstruction of red-boxed region. **(g′, h′, i′)** *x*-axis; **(g″, h″, i″)** *y*-axis. **(g)** BrdU (red), indicating that the cell underwent mitosis during the 2 wk following induction of apoptosis. Scale bar = 5 μm. **(h)** NeuN (green). **(i)** Merged image. BrdU (red) and NeuN (green) labeling are coincident in all three dimensions. **(j)** The BrdU+/NeuN+ neuron (red) is GFAP-negative (blue). A GFAP+ astrocyte (arrowhead). **(k, l, m)** The presence of BrdU+/NeuN– and BrdU–/NeuN+ cells demonstrate that the double-labeling protocol is specific. Image from same section as **(a–d)**. **(k)** BrdU(red) and **(l)** NeuN(green) do not show crossreactivity. **(m)** Overlay. (Reprinted by permission from *Nature* [*see* ref. *4*]. Copyright © 2000 Macmillan Magazines Ltd.)

Table 1
The Difference Between Control and Experimental
Mice is Highly Statistically Significant ($p < 0.0001$,
Two-Tailed t-Test)[a]

Survival (wk)	No. of BrdU+/NeuN+ (cells/mm^3)	No. of experimentals
2	97 ± 69	6
5	43 ± 20	3
28	78 ± 15	2

[a]These results demonstrate that endogenous neural precursors can be induced *in situ* to differentiate into mature neurons and survive for at least many months in the adult murine neocortex.

inappropriately expressing multiple phenotypic markers, we triple-labeled sections with antibodies to GFAP, an astroglial marker, and MBP, an oligodendroglial marker. BrdU+/NeuN+ neurons were negative for GFAP (Fig. 2J) and MBP, confirming that they had specifically differentiated into mature neurons. We found approx 50–100 newborn neurons/mm^3 following induction of apoptosis (*see* Table 1), whereas we found no newborn neurons in the neocortex of control mice at any time ($n = 13$), confirming previous reports that neurogenesis does not occur in the adult rodent brain. These results demonstrate that endogenous neural precursors can be induced *in situ* to differentiate into mature neurons and survive for many months in the adult murine neocortex.

To further understand the source of these newborn neurons, we examined where newly born cells were located in experimental and control animals. After 2 wk of BrdU treatment, newly born, BrdU+ non-neuronal cells were found throughout both experimental and control cortexes. At 2 wk, experimental mice ($n = 6$) had 5400 ± 1500 BrdU+ cells/mm^3 in the experimental regions, not significantly more than control mice ($n = 3$), which had 4100 ± 1700 newborn cells/mm^3. Similarly, aspiration lesions of the cortex have been reported to yield inconsistent increases in BrdU-labeled cells in the subventricular zone *(129)*. Previous studies have established that proliferation normally occurs in the adult cortex; however, under normal circumstances, mitotic cells differentiate into glia, remain undifferentiated, or die.

Some newborn cells located within the cortex of experimental animals appear to originate from precursors located within the cortex itself, whereas a second population, which was not present in control animals, appeared to originate in or near the subventricular zone. At 2 wk, pairs of BrdU+ cells, apparently daughters of the same precursor, were found throughout both control and experimental cortex (Fig. 2I). It is possible that some of these newborn cells are the daughters of cortically located adult multipotent neural precursors that have been described in vitro *(109,130)*. The adult subventricular and ependymal zones, located subjacent to the deep layers of cortex containing degenerating projection neurons, also contain a population of constitutively dividing, multipotent neural precursors *(6,8,131–135)* and appear to be the source of the second population of newborn cells. Cells from the SVZ population, and perhaps both of these populations may contribute to induced cortical neurogenesis.

We investigated the early differentiation and migration of the newborn neurons using a marker of early postmitotic neurons, Doublecortin (Dcx). In experimental mice only, newborn BrdU+/Dcx+ neurons with migratory morphologies appeared to migrate from the SVZ through the corpus callosum (Figs. 3A–D) and into experimental regions of the cortex (Figs. 3E,F). Doublecortin is a microtubule-associated protein that is exclusively expressed in migrating and differentiating neurons, but is not detectable in mature neurons *(125,126)*. Confocal analysis confirmed that these BrdU+ cells express Doublecortin (Figs. 3C,D). No BrdU+/Dcx+ cells were found in the corpus callosum or cortex of control animals. However, they exist in the rostral migratory stream of both controls and experimentals, as previously described *(125,126)*. Newly born BrdU+/Dcx+ neurons within the corpus callosum displayed morphologies characteristic of migrating neurons, whereas newborn Dcx+ neurons located in deep layers of cortex displayed more complex morphologies with apical processes that suggest their further differentiation. These results demonstrate the progressive differentiation of endogenous precursors: first into Dcx+ migratory neuroblasts, then into immature Dcx+ neurons with process extension, then into more mature neurons.

The location of Dcx+ neurons suggests that at least some of the newborn neurons that form in targeted cortex are derived directly from SVZ precursors. However, these data do not rule out the possibility that other precursors contribute to neurogenesis. Newly born cortical neurons could, theoretically, be derived in three different ways. Newly born neurons could arise from endogenous cortical precursors, directly from SVZ precursors, or from endogenous precursors, such as SVZ precursors, that enter the cortex and then divide within the cortex. Neuronal replacement therapies that depend on precursors derived from normally neurogenic portions of the brain could be limited to regions of the brain that are adjacent to these neurogenic regions.

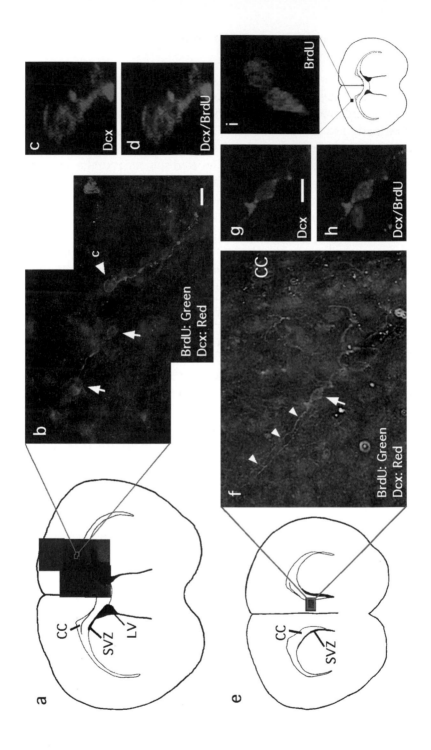

Thus, further understanding the source of the cells that can contribute to induced neurogenesis could be critical for deciding whether endogenous precursors could potentially form the basis of effective neuronal replacement therapies.

The continued differentiation of these newborn neurons was examined using antibodies to Hu, an RNA-binding protein that begins to be expressed in neuronal nuclei and somata soon after differentiation *(127,128)*. BrdU+/Hu+ neurons with large ovoid nuclei were found in the cortex of experimental mice 2 wk after induction of apoptosis (Fig. 4), whereas no BrdU+/Hu+ cells were found in control mice. We used Hoechst 33258, which stains DNA, as a nuclear counterstain that allowed us to confirm that a specific nucleus belonged to a particular cell. The dispersed heterochromatin pattern demonstrated by Hoechst staining coincident with the BrdU+ nuclei is characteristic of neuronal nuclei and further confirms BrdU+/Hu+ double labeling. The expression of Hu further confirms induced cortical neuron differentiation by endogenous neural precursors.

These newborn neurons derived from endogenous precursors can re-form long-distance projections. To examine the identities of the newborn, BrdU+ neurons and examine their potential ability to establish long-distance connections, we injected the retrograde label FluoroGold into the same thalamic sites as the original nanosphere injections. Previous experiments from our lab show that large numbers of embryonic neurons transplanted into regions of adult cortex undergoing apoptotic degeneration of projection neurons can re-form long-distance connections to the original targets of the

---

Fig. 3. *(see opposite page)* Newborn cells express the migratory neuronal marker Doublecortin (Dcx) 2 wk after induction of apoptosis. BrdU+/Dcx+ neurons appear to migrate from the SVZ, through the corpus callosum, and into experimental regions of the cortex. **(a)** Camera lucida showing BrdU+ (green)/Dcx+ (red) neurons within the superficial corpus callosum (CC); Lateral ventricle (LV). **(b)** Fluorescence micrograph of three BrdU+/Dcx+ neurons. Scale bar = 25 µm. **(c,d)** Confocal micrographs confirm that the neuron in (b) (arrowhead) is BrdU+/Dcx+. **(e)** Camera lucida: A BrdU+/Dcx+ newborn neuron within cortex. **(f)** Higher magnification: the differentiating BrdU+/ Dcx+ newborn neuron (arrow) extending processes (arrowheads) in cortex. No BrdU+/Dcx+ cells were found in the CC or cortex of control mice. Dcx staining is specific for migrating neurons *(125,126)*; it does not overlap with A2B5, O4, RIP, MBP, or GFAP staining. **(g, h)** Confocal micrographs of this newborn neuron confirm that it is BrdU+/Dcx+. Scale bar = 10 µm. **(i)** Pairs of BrdU+ non-neuronal cells were found throughout the cortex in experimental and control mice. (Reprinted by permission from *Nature* [*see* ref. *4*] copyright © 2000 Macmillan Magazines Ltd.)

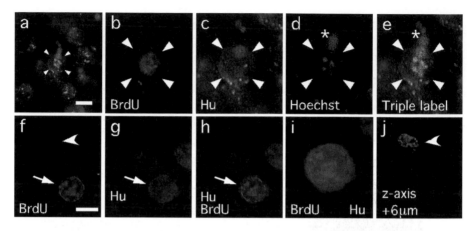

Fig. 4. Newborn BrdU+ cells express the early neuronal marker, Hu, in the experimental cortex. **(a)** A BrdU+/Hu+ newborn neuron (arrowheads) at 2 wk. Surrounding neurons are BrdU-negative. Scale bar = 10 μm. **(b–e)** Higher magnification: **(b)** BrdU (green) labels the nucleus of a newborn neuron. c. Hu (red) labels the BrdU+ newborn cell. **(d)** Hoechst (blue) shows the large dispersed nucleus of the newborn neuron and the compact nucleus of an adjacent glial cell (asterisk). **(e)** Overlay. **(f)** Arrow indicates BrdU+ (green) nucleus of a BrdU+/Hu+ neuron at 28 wk. Arrowhead indicates a BrdU+/Hu– cell in another plane of focus. **(g)** Hu+ soma (red). **(h)** Overlay. **(i)** Confocal confirmation of double labeling. **(j)** A BrdU+/Hu– cell located in a focal plane 6 μm higher demonstrates specificity of staining. (Reprinted by permission from *Nature* [*see* ref. *4*] copyright © 2000 Macmillan Magazines Ltd.)

degenerating neurons *(118)*. At 9 wk, we observed newborn BrdU+ neurons retrogradely labeled with FluoroGold that had large nuclei and large cell bodies denoting pyramidal projection neuron morphology (Fig. 5). These results show that endogenous precursors that differentiate into mature neurons can establish appropriate long-distance corticothalamic connections in the adult brain.

Several lines of evidence reinforce that BrdU+ neurons in these experiments are newborn neurons, not pre-existing neurons artifactually labeled by BrdU due to DNA synthesis during apoptosis, as has been seen in some cell culture experiments. First, we directly controlled for this theoretical possibility, because the photoactive targeting nanospheres we employed also carry a fluorescent label, allowing us to identify all neurons targeted to undergo apoptosis. No BrdU+ cell was a targeted neuron and no targeted neuron was BrdU+. Second, the appearance of BrdU incorporation during

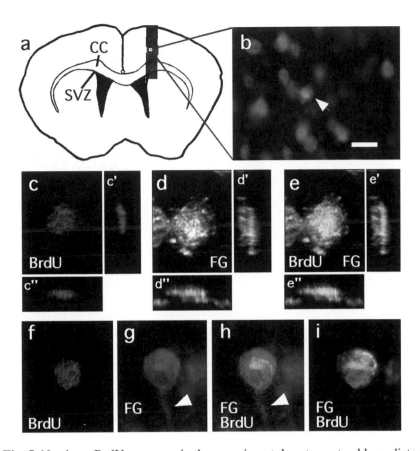

Fig. 5. Newborn BrdU+ neurons in the experimental cortex extend long-distance projections to the thalamus. (**a**) Camera lucida of a BrdU+ (red) / FG+ (white) retrogradely labeled neuron in layer VI of the cortex at 9 wk; corpus callosum (CC). (**b**) Higher magnification shows the newborn corticothalamic neuron (arrow) and FG+/BrdU– original neurons. Scale bar = 20 µm. (**c–e**) Confocal three-dimensional reconstruction of the neuron. (**c′, d′, e′.**) x-Axis; (**c″, d″, e″**) y-axis. (**c**) BrdU+ nucleus, indicating the cell underwent mitosis in the 2 wk following induction of apoptosis. (**d**) FluoroGold+ soma, indicating that this neuron projects to the thalamus. (**e**) Merged image, confirming double labeling in three dimensions. (**f**) BrdU+ (red) nucleus of an adult-born FG+ corticothalamic neuron. (**g**) FG+ (blue) cell body with labeled axon (arrow). (**h**) Overlay. (**i**) Confocal microscopy confirms double labeling. (Reprinted by permission from *Nature* [*see* ref. *4*] copyright © 2000 Macmillan Magazines Ltd.)

DNA repair is that of minor nuclear speckling, not uniform, whole-nucleus labeling, as we observed in these experiments. Third, BrdU+ neurons were present in the neocortex over 6 mo after the end of the BrdU treatment, demonstrating their viability and long-term survival. Fourth, and perhaps arguing most definitively for the identification of BrdU+ neurons as newborn neurons, is the fact that we were able to trace the early developmental progression of these neurons using Doublecortin, an immature migratory neuronal marker. We found BrdU+/Doublecortin+ newborn neurons with migratory morphologies in the corpus callosum and appropriate cortical layer 6 of experimental mice, and we identified BrdU+/Doublecortin+ immature neurons extending early processes in these regions. Furthermore, we have also conducted experiments investigating whether BrdU integrates into dying cells in vivo or whether this is simply a theoretical concern from in vitro studies. We examined apoptotic neurons in early postnatal brain ($n >$ 1000) and neurons undergoing excitotoxic degeneration induced by ibotenic acid in adult animals ($n > 5000$). Absolutely no apoptotic, cleaved caspase-3-positive or dying neurons with pyknotic condensed nuclei integrated BrdU, even when BrdU doses three to eight times higher than normal were used. From this combination of evidence, we conclude that healthy, long-lived neurons with dense BrdU labeling throughout their nuclei can be reliably identified as newborn neurons.

Taken together, our results demonstrate that endogenous neural precursors can be induced to differentiate into neocortical neurons in a layer- and region-specific manner and re-form appropriate corticothalamic connections in regions of adult mammalian neocortex that do not normally undergo neurogenesis. The same microenvironment that supports the migration, neuronal differentiation, and appropriate axonal extension of transplanted neuroblasts and precursors also supports and instructs the neuronal differentiation and axon extension of endogenous precursors.

CONCLUSIONS

Recent research suggests that it may be possible to manipulate endogenous neural precursors *in situ* to undergo neurogenesis in the adult brain, toward future neuronal replacement therapy for neurodegenerative disease and other CNS injury. Multipotent precursors, capable of differentiating into astrocytes, oligodendrocytes, and neurons exist in many portions of the adult brain. These precursors have considerable plasticity, and although they may have limitations in their integration into some host sites *(92)*, they are capable of differentiation into neurons appropriate to a wide variety of recipient regions, when heterotopically transplanted *(91)*. Many adult precursors are capable of migrating long distances, using both tangential and radial forms of migration *(12–14)*. Endogenous adult neural precursors are also capable of extending axons significant distances through the adult brain *(4,50,51)*. In addition, in vitro and in vivo experiments have begun to elucidate the responses of endogenous precursors to both growth factors and behavioral manipulations and are beginning to provide key information toward manipulation of their proliferation and differentiation. Recent experiments from our lab have shown that, under appropriate conditions, endogenous precursors can differentiate into neurons, extend long-distance axonal projections, and survive for long periods of time in regions of the adult brain that do not normally undergo neurogenesis.

Together, these data suggest that neuronal replacement therapies based on manipulation of endogenous precursors may be possible in the future. However, several questions must be answered before neuronal replacement therapies using endogenous precursor become a reality. The multiple signals that are responsible for endogenous precursor division, migration, differentiation, and axon extension must be elucidated in order for such therapies to be developed efficiently. These challenges also exist for neuronal replacement therapies based on transplantation of precursors, because donor cells, whatever their source, must interact with the mature CNS's environment in order to integrate into the brain. In addition, it remains an open question whether potential therapies manipulating endogenous precursors *in situ* would be necessarily limited to portions of the brain near adult neurogenic regions. However, even if multipotent precursors are not located in very high numbers outside of neurogenic regions of the brain, it may be possible to induce them to proliferate from the smaller numbers that are more widely distributed throughout the neuraxis. Research in this field is promising, and we may, one day, be able to induce the brain to heal itself. However, this field is still young and we do not even yet know all of the questions we will need to ask.

ACKNOWLEDGMENTS

This work was partially supported by grants from the NIH (N541590, HD28478, MRRC HD18655), the Alzheimer's Association, the Human Frontiers Science Program, and the National Science foundation to J.D.M. S.S.M. was partially supported by a NIH predoctoral training grant.

REFERENCES

1. Bjorklund, A. and Lindvall, O. (2000) Cell replacement therapies for central nervous system disorders. *Nat. Neurosci.* **3(6),** 537–544.

2. Palmer, T. D., Ray, J., and Gage, F. H. (1995) FGF-2-responsive neuronal progenitors reside in proliferative and quiescent regions of the adult rodent brain. *Mol. Cell. Neurosci.* **6(5),** 474–486.

3. Scharff, C., Kirn, J. R., Grossman, M., Macklis, J. D., and Nottebohm, F. (2000) Targeted neuronal death affects neuronal replacement and vocal behavior in adult songbirds [see comments]. *Neuron* **25(2),** 481–492.

4. Magavi, S. S., Leavitt, B. R., and Macklis, J. D. (2000) Induction of neurogenesis in the neocortex of adult mice [see comments]. *Nature* **405(6789),** 951–955.

5. Altman, J. and Das, G. D. (1965) Autoradiographic and histological evidence of postnatal hippocampal neurogenesis in rats. *J. Comp. Neurol.* **124(3),** 319–335.

6. Altman, J. (1969) Autoradiographic and histological studies of postnatal neurogenesis. IV. Cell proliferation and migration in the anterior forebrain, with special reference to persisting neurogenesis in the olfactory bulb. *J. Comp. Neurol.* **137(4),** 433–457.

7. Kaplan, M. S. (1981) Neurogenesis in the 3-month-old rat visual cortex. *J. Comp. Neurol.* **195(2),** 323–338.

8. Reynolds, B. A. and Weiss, S. (1992) Generation of neurons and astrocytes from isolated cells of the adult mammalian central nervous system. *Science* **255(5052),** 1707–1710.

9. Pincus, D. W., Keyoung, H. M., Harrison-Restelli, C., Goodman, R. R., Fraser, R. A., Edgar, M., et al. (1998) Fibroblast growth factor-2/brain-derived neurotrophic factor-associated maturation of new neurons generated from adult human subependymal cells. *Ann. Neurol.* **43(5),** 576–585.

10. Kirschenbaum, B., Nedergaard, M., Preuss, A., Barami, K., Fraser, R. A., and Goldman, S. A. (1994) In vitro neuronal production and differentiation by precursor cells derived from the adult human forebrain. *Cereb. Cortex* **4(6),** 576–589.

11. Bernier, P. J., Vinet, J., Cossette, M., and Parent, A. (2000) Characterization of the subventricular zone of the adult human brain: evidence for the involvement of Bcl-2. *Neurosci. Res.* **37(1),** 67–78.

12. Lois, C. and Alvarez-Buylla, A. (1994) Long-distance neuronal migration in the adult mammalian brain. *Science* **264(5162),** 1145–1148.

13. Doetsch, F. and Alvarez-Buylla, A. (1996) Network of tangential pathways for neuronal migration in adult mammalian brain. *Proc. Natl. Acad. Sci. USA* **93(25),** 14,895–14,900.

14. Luskin, M. B. and Boone, M. S. (1994) Rate and pattern of migration of lineally-related olfactory bulb interneurons generated postnatally in the subventricular zone of the rat. *Chem. Senses* **19(6),** 695–714.

15. Rousselot, P., Lois, C., and Alvarez-Buylla, A. (1995) Embryonic (PSA) N-CAM reveals chains of migrating neuroblasts between the lateral ventricle and the olfactory bulb of adult mice. *J. Comp. Neurol.* **351(1),** 51–61.

16. Mason, H. A., Ito, S., and Corfas, G. (2001) Extracellular signals that regulate the tangetntial migration of olfactory bulb neuronal precursors: inducers, inhibitors and repellents. *J. Neurosci.* **21(19),** 7654–7663.

17. Hu, H. and Rutishauser, U. (1996) A septum-derived chemorepulsive factor for migrating olfactory interneuron precursors. *Neuron* **16(5),** 933–940.
18. Kirschenbaum, B., Doetsch, F., Lois, C., and Alvarez-Buylla, A. (1999) Adult subventricular zone neuronal precursors continue to proliferate and migrate in the absence of the olfactory bulb. *J. Neurosci.* **19(6),** 2171–2180.
19. Hu, H., Tomasiewicz, H., Magnuson, T., and Rutishauser, U. (1996) The role of polysialic acid in migration of olfactory bulb interneuron precursors in the subventricular zone. *Neuron* **16(4),** 735–743.
20. Gates, M. A., Thomas, L. B., Howard, E. M., Laywell, E. D., Sajin, B., Faissner, A., et al. (1995) Cell and molecular analysis of the developing and adult mouse subventricular zone of the cerebral hemispheres. *J. Comp. Neurol.* **361(2),** 249–266.
21. Garcia-Verdugo, J. M., Doetsch, F., Wichterle, H., Lim, D. A., and Alvarez-Buylla, A. (1998) Architecture and cell types of the adult subventricular zone: in search of the stem cells. *J. Neurobiol.* **36(2),** 234–248.
22. Law, A. K., Pencea, V., Buck, C. R., and Luskin, M. B. (1999) Neurogenesis and neuronal migration in the neonatal rat forebrain anterior subventricular zone do not require GFAP-positive astrocytes. *Dev. Biol.* (Orlando) **216(2),** 622–634.
23. Johansson, C. B., Svensson, M., Wallstedt, L., Janson, A. M., and Frisen, J. (1999) Neural stem cells in the adult human brain. *Exp. Cell Res.* **253(2),** 733–736.
24. Chiasson, B. J., Tropepe, V., Morshead, C. M., and van der Kooy, D. (1999) Adult mammalian forebrain ependymal and subependymal cells demonstrate proliferative potential, but only subependymal cells have neural stem cell characteristics. *J. Neurosci.* **19(11),** 4462–4471.
25. Doetsch, F., Caille, I., Lim, D. A., Garcia-Verdugo, J. M., and Alvarez-Buylla, A. (1999) Subventricular zone astrocytes are neural stem cells in the adult mammalian brain. *Cell* **97(6),** 703–716.
26. Laywell, E. D., Rakic, P., Kukekov, V. G., Holland, E. C., and Steindler, D. A. (2000) Identification of a multipotent astrocytic stem cell in the immature and adult mouse brain. *Proc. Natl. Acad. Sci. USA* **97(25),** 13,883–13,888.
27. Whittemore, S. R., Morassutti, D. J., Walters, W. M., Liu, R. H., and Magnuson, D. S. (1999) Mitogen and substrate differentially affect the lineage restriction of adult rat subventricular zone neural precursor cell populations. *Exp. Cell Res.* **252(1),** 75–95.
28. Gritti, A., Parati, E. A., Cova, L., Frolichsthal, P., Galli, R., Wanke, E., et al. (1996) Multipotential stem cells from the adult mouse brain proliferate and self-renew in response to basic fibroblast growth factor. *J. Neurosci.* **16(3),** 1091–1100.
29. Gritti, A., Cova, L., Parati, E. A., Galli, R., and Vescovi, A. L. (1995) Basic fibroblast growth factor supports the proliferation of epidermal growth factor-generated neuronal precursor cells of the adult mouse CNS. *Neurosci. Lett.* **185(3),** 151–154.
30. Gritti, A., Frolichsthal-Schoeller, P., Galli, R., Parati, E. A., Cova, L., Pagano, S. F., et al. (1999) Epidermal and fibroblast growth factors behave as mitogenic

regulators for a single multipotent stem cell-like population from the subventricular region of the adult mouse forebrain. *J. Neurosci.* **19(9)**, 3287–3297.

31. Gross, R. E., Mehler, M. F., Mabie, P. C., Zang, Z., Santschi, L., and Kessler, J. A. (1996) Bone morphogenetic proteins promote astroglial lineage commitment by mammalian subventricular zone progenitor cells. *Neuron* **17(4)**, 595–606.

32. Williams, B. P., Park, J. K., Alberta, J. A., Muhlebach, S. G., Hwang, G. Y., Roberts, T. M., et al. (1997) A PDGF-regulated immediate early gene response initiates neuronal differentiation in ventricular zone progenitor cells. *Neuron* **18(4)**, 553–562.

33. Arsenijevic, Y. and Weiss, S. (1998) Insulin-like growth factor-I is a differentiation factor for postmitotic CNS stem cell-derived neuronal precursors: distinct actions from those of brain-derived neurotrophic factor. *J. Neurosci.* **18(6)**, 2118–2128.

34. Kirschenbaum, B. and Goldman, S. A. (1995) Brain-derived neurotrophic factor promotes the survival of neurons arising from the adult rat forebrain subependymal zone. *Proc. Natl. Acad. Sci. USA* **92(1)**, 210–214.

35. Goldman, S. A., Kirschenbaum, B., Harrison-Restelli, C., and Thaler, H. T. (1997) Neuronal precursors of the adult rat subependymal zone persist into senescence, with no decline in spatial extent or response to BDNF. *J. Neurobiol.* **32(6)**, 554–566.

36. Ahmed, S., Reynolds, B. A., and Weiss, S. (1995) BDNF enhances the differentiation but not the survival of CNS stem cell-derived neuronal precursors. *J. Neurosci.* **15(8)**, 5765–5778.

37. Craig, C. G., Tropepe, V., Morshead, C. M., Reynolds, B. A., Weiss, S., and van der Kooy, D. (1996) In vivo growth factor expansion of endogenous subependymal neural precursor cell populations in the adult mouse brain. *J. Neurosci.* **16(8)**, 2649–2658.

38. Kuhn, H. G., Winkler, J., Kempermann, G., Thal, L. J., and Gage, F. H. (1997) Epidermal growth factor and fibroblast growth factor-2 have different effects on neural progenitors in the adult rat brain. *J. Neurosci.* **17(15)**, 5820–5829.

39. Wagner, J. P., Black, I. B., and DiCicco-Bloom, E. (1999) Stimulation of neonatal and adult brain neurogenesis by subcutaneous injection of basic fibroblast growth factor. *J. Neurosci.* **19(14)**, 6006–6016.

40. Zigova, T., Pencea, V., Wiegand, S. J., and Luskin, M. B. (1998) Intraventricular administration of BDNF increases the number of newly generated neurons in the adult olfactory bulb. *Mol. Cell. Neurosci.* **11(4)**, 234–245.

41. Lim, D. A., Fishell, G. J., and Alvarez-Buylla, A. (1997) Postnatal mouse subventricular zone neuronal precursors can migrate and differentiate within multiple levels of the developing neuraxis. *Proc. Natl. Acad. Sci. USA* **94(26)**, 14,832–14,836.

42. Herrera, D. G., Garcia-Verdugo, J. M., and Alvarez-Buylla, A. (1999) Adult-derived neural precursors transplanted into multiple regions in the adult brain. *Ann. Neurol.* **46(6)**, 867–877.

43. Clarke, D. L., Johansson, C. B., Wilbertz, J., Veress, B., Nilsson, E., Karlstrom, H., et al. (2000) Generalized potential of adult neural stem cells. [see comments]. *Science* **288(5471)**, 1660–1663.
44. Huard, J. M. and Schwob, J. E. (1995) Cell cycle of globose basal cells in rat olfactory epithelium. *Dev. Dynam.* **203(1)**, 17–26.
45. Caggiano, M., Kauer, J. S., and Hunter, D. D. (1994) Globose basal cells are neuronal progenitors in the olfactory epithelium: a lineage analysis using a replication-incompetent retrovirus. *Neuron* **13(2)**, 339–352.
46. Graziadei, G. A. and Graziadei, P. P. (1979) Neurogenesis and neuron regeneration in the olfactory system of mammals. II. Degeneration and reconstitution of the olfactory sensory neurons after axotomy. *J. Neurocytol.* **8(2)**, 197–213.
47. Carr, V. M. and Farbman, A. I. (1992) Ablation of the olfactory bulb up-regulates the rate of neurogenesis and induces precocious cell death in olfactory epithelium. *Exp. Neurol.* **115(1)**, 55–59.
48. Calof, A. L., Hagiwara, N., Holcomb, J. D., Mumm, J. S., and Shou, J. (1996) Neurogenesis and cell death in olfactory epithelium. *J. Neurobiol.* **30(1)**, 67–81.
49. Samanen, D. W. and Forbes, W. B. (1984) Replication and differentiation of olfactory receptor neurons following axotomy in the adult hamster: a morphometric analysis of postnatal neurogenesis. *J. Comp. Neurol.* **225(2)**, 201–211.
50. Barber, P. C. (1982) Neurogenesis and regeneration in the primary olfactory pathway of mammals. *Bibl. Anat.* **(23)**, 12–25.
51. Crews, L. and Hunter, D. (1994) Neurogenesis in the olfactory epithelium. *Perspect. Dev. Neurobiol.* **2(2)**, 151–161.
52. Calof, A. L., Rim, P. C., Askins, K. J., Mumm, J. S., Gordon, M. K., Iannuzzelli, P., et al. (1998) Factors regulating neurogenesis and programmed cell death in mouse olfactory epithelium. *Ann. NY Acad. Sci.* **855**, 226–229.
53. Pixley, S. K., Dangoria, N. S., Odoms, K. K., and Hastings, L. (1998) Effects of insulin-like growth factor 1 on olfactory neurogenesis in vivo and in vitro. *Ann. NY Acad. Sci.* **855**, 244–247.
54. Newman, M. P., Feron, F., and Mackay-Sim, A. (2000) Growth factor regulation of neurogenesis in adult olfactory epithelium. *Neuroscience* **99(2)**, 343–350.
55. Gould, E., Tanapat, P., McEwen, B. S., Flugge, G., and Fuchs, E. (1998) Proliferation of granule cell precursors in the dentate gyrus of adult monkeys is diminished by stress. *Proc. Natl. Acad. Sci. USA* **95(6)**, 3168–3171.
56. Gould, E., Reeves, A. J., Fallah, M., Tanapat, P., Gross, C. G., and Fuchs, E. (1999) Hippocampal neurogenesis in adult Old World primates. *Proc. Natl. Acad. Sci. USA* **96(9)**, 5263–5267.
57. Kornack, D. R. and Rakic, P. (1999) Continuation of neurogenesis in the hippocampus of the adult macaque monkey. *Proc. Natl. Acad. Sci. USA* **96(10)**, 5768–5773.
58. Eriksson, P. S., Perfilieva, E., Bjork-Eriksson, T., Alborn, A. M., Nordborg, C., Peterson, D. A., et al. (1998) Neurogenesis in the adult human hippocampus [see comments]. *Nat. Med.* **4(11)**, 1313–1317.

59. Stanfield, B. B. and Trice, J. E. (1988) Evidence that granule cells generated in the dentate gyrus of adult rats extend axonal projections. *Exp. Brain Res.* **72(2),** 399–406.
60. Markakis, E. A. and Gage, F. H. (1999) Adult-generated neurons in the dentate gyrus send axonal projections to field CA3 and are surrounded by synaptic vesicles. *J. Comp. Neurol.* **406(4),** 449–460.
61. Hastings, N. B. and Gould, E. (1999) Rapid extension of axons into the CA3 region by adult-generated granule cells [published erratum appears in *J. Comp. Neurol.* 1999 **415(1),** 144]. *J. Comp. Neurol.* **413(1),** 146–154.
62. Gage, F. H., Kempermann, G., Palmer, T. D., Peterson, D. A., and Ray, J. (1998) Multipotent progenitor cells in the adult dentate gyrus. *J. Neurobiol.* **36(2),** 249–266.
63. Lowenstein, D. H. and Arsenault, L. (1996) The effects of growth factors on the survival and differentiation of cultured dentate gyrus neurons. *J. Neurosci.* **16(5),** 1759–1769.
64. Kukekov, V. G., Laywell, E. D., Suslov, O., Davies, K., Scheffler, B., Thomas, L. B., et al. (1999) Multipotent stem/progenitor cells with similar properties arise from two neurogenic regions of adult human brain. *Exp. Neurol.* **156(2),** 333–344.
65. Roy, N. S., Wang, S., Jiang, L., Kang, J., Benraiss, A., Harrison-Restelli, C., et al. (2000) In vitro neurogenesis by progenitor cells isolated from the adult human hippocampus [see comments]. *Nat. Med.* **6(3),** 271–277.
66. Kuhn, H. G., Dickinson-Anson, H., and Gage, F. H. (1996) Neurogenesis in the dentate gyrus of the adult rat: age-related decrease of neuronal progenitor proliferation. *J. Neurosci.* **16(6),** 2027–2033.
67. Cameron, H. A. and McKay, R. D. (1999) Restoring production of hippocampal neurons in old age. *Nat. Neurosci.* **2(10),** 894–897.
68. Montaron, M. F., Petry, K. G., Rodriguez, J. J., Marinelli, M., Aurousseau, C., Rougon, G., et al. (1999) Adrenalectomy increases neurogenesis but not PSA-NCAM expression in aged dentate gyrus. *Eur. J. Neurosci.* **11(4),** 1479–1485.
69. Bengzon, J., Kokaia, Z., Elmer, E., Nanobashvili, A., Kokaia, M., and Lindvall, O. (1997) Apoptosis and proliferation of dentate gyrus neurons after single and intermittent limbic seizures. *Proc. Natl. Acad. Sci. USA* **94(19),** 10,432–10,437.
70. Parent, J. M., Yu, T. W., Leibowitz, R. T., Geschwind, D. H., Sloviter, R. S., and Lowenstein, D. H. (1997) Dentate granule cell neurogenesis is increased by seizures and contributes to aberrant network reorganization in the adult rat hippocampus. *J. Neurosci.* **17(10),** 3727–3738.
71. Huang, L., Cilio, M. R., Silveira, D. C., McCabe, B. K., Sogawa, Y., Stafstrom, C. E., et al. (1999) Long-term effects of neonatal seizures: a behavioral, electrophysiological, and histological study. *Brain Res.* Dev. Brain Res. **118(1–2),** 99–107.
72. Parent, J. M., Janumpalli, S., McNamara, J. O., and Lowenstein, D. H. (1998) Increased dentate granule cell neurogenesis following amygdala kindling in the adult rat. *Neurosci. Lett.* **247(1),** 9–12.

73. Represa, A., Niquet, J., Pollard, H., and Ben-Ari, Y. (1995) Cell death, gliosis, and synaptic remodeling in the hippocampus of epileptic rats. *J. Neurobiol.* **26(3),** 413–425.

74. Magloczky, Z. and Freund, T. F. (1993) Selective neuronal death in the contralateral hippocampus following unilateral kainate injections into the CA3 subfield. *Neuroscience* **56(2),** 317–335.

75. Pollard, H., Charriaut-Marlangue, C., Cantagrel, S., Represa, A., Robain, O., Moreau, J., et al. (1994) Kainate-induced apoptotic cell death in hippocampal neurons. *Neuroscience* **63(1),** 7–18.

76. Whittington, D. L., Woodruff, M. L., and Baisden, R. H. (1989) The time-course of trimethyltin-induced fiber and terminal degeneration in hippocampus. *Neurotoxicol. Teratol.* **11(1),** 21–33.

77. Gould, E. and Tanapat, P. (1997) Lesion-induced proliferation of neuronal progenitors in the dentate gyrus of the adult rat. *Neuroscience* **80(2),** 427–436.

78. Derrick, B. E., York, A. D., and Martinez, J. L., Jr. (2000) Increased granule cell neurogenesis in the adult dentate gyrus following mossy fiber stimulation sufficient to induce long-term potentiation. *Brain Res.* **857(1–2),** 300–307.

79. Kempermann, G., Kuhn, H. G., and Gage, F. H. (1997) More hippocampal neurons in adult mice living in an enriched environment. *Nature* **386(6624),** 493–495.

80. van Praag, H., Kempermann, G., and Gage, F. H. (1999) Running increases cell proliferation and neurogenesis in the adult mouse dentate gyrus [see comments]. *Nat. Neurosci.* **2(3),** 266–270.

81. van Praag, H., Christie, B. R., Sejnowski, T. J., and Gage, F. H. (1999) Running enhances neurogenesis, learning, and long-term potentiation in mice. *Proc. Natl. Acad. Sci. USA* **96(23),** 13,427–13,431.

82. Gould, E., Beylin, A., Tanapat, P., Reeves, A., and Shors, T. J. (1999) Learning enhances adult neurogenesis in the hippocampal formation [see comments]. *Nat. Neurosci.* **2(3),** 260–265.

83. Gould, E., McEwen, B. S., Tanapat, P., Galea, L. A., and Fuchs, E. (1997) Neurogenesis in the dentate gyrus of the adult tree shrew is regulated by psychosocial stress and NMDA receptor activation. *J. Neurosci.* **17(7),** 2492–2498.

84. Tanapat, P., Galea, L. A., and Gould, E. (1998) Stress inhibits the proliferation of granule cell precursors in the developing dentate gyrus. *Int. J. Dev. Neurosci.* **16(3–4),** 235–239.

85. Aberg, M. A., Aberg, N. D., Hedbacker, H., Oscarsson, J., and Eriksson, P. S. (2000) Peripheral infusion of IGF-I selectively induces neurogenesis in the adult rat hippocampus. *J. Neurosci.* **20(8),** 2896–2903.

86. Trejo, J. L., Carro, E., and Torres-Aleman, I. (2001) Circulating insulin-like growth factor I mediates exercise-induced increases in the number on new neurons in the adult hippocampus. *J. Neurosci.* **21(5),** 1628–1634.

87. Gould, E., Cameron, H. A., Daniels, D. C., Woolley, C. S., and McEwen, B. S. (1992) Adrenal hormones suppress cell division in the adult rat dentate gyrus. *J. Neurosci.* **12(9),** 3642–3650.

88. Chen, G., Rajkowska, G., Du, F., Seraji-Bozorgzad, N., and Manji, H. K. (2000) Enhancement of hippocampal neurogenesis by lithium. *J. Neurochem.* **75(4),** 1729–1734.

89. Malberg, J. E., Eisch, A. J., Nestler, E. J., and Duman, R. S. (2000) Chronic antidepressant treatment increases neurogenesis in adult rat hippocampus. *J. Neurosci. (Online)* **20(24),** 9104–9110.

90. Kempermann, G., van Praag, H., and Gage, F. H. (2000) Activity-dependent regulation of neuronal plasticity and self repair. *Prog. Brain Res.* **127,** 35–48.

91. Suhonen, J. O., Peterson, D. A., Ray, J., and Gage, F. H. (1996) Differentiation of adult hippocampus-derived progenitors into olfactory neurons in vivo. *Nature* **383(6601),** 624–627.

92. Young, M. J., Ray, J., Whiteley, S. J., Klassen, H., and Gage, F. H. (2000) Neuronal differentiation and morphological integration of hippocampal progenitor cells transplanted to the retina of immature and mature dystrophic rats. *Mol. Cell. Neurosci.* **16(3),** 197–205.

93. Nishida, A., Takahashi, M., Tanihara, H., Nakano, I., Takahashi, J. B., Mizogu-chi, A., et al. (2000) Incorporation and differentiation of hippocampus-derived neural stem cells transplanted in injured adult rat retina. *Invest. Ophthalmol. Visual Sci.* **41(13),** 4268–4274.

94. Shors, T. J., Miesegas, G., Beylin, A., Zhao, M., Rydel, T., and Gould, E. (2001) Neurogenesis in the adult is involved in the formation of trace memories. *Nature* **410,** 372–376.

95. Jacobs, B. L., Praag, H., and Gage, F. H. (2000) Adult brain neurogenesis and psychiatry: a novel theory of depression. *Mol. Psychiatry* **5(3),** 262–269.

96. Brezun, J. M. and Daszuta, A. (1999) Depletion in serotonin decreases neurogenesis in the dentate gyrus and the subventricular zone of adult rats. *Neuroscience* **89(4),** 999–1002.

97. Gould, E., Reeves, A. J., Graziano, M. S., and Gross, C. G. (1999) Neurogenesis in the neocortex of adult primates. [see comments]. *Science* **286(5439),** 548–552.

98. Lopez-Garcia, C., Molowny, A., Martinez-Guijarro, F. J., Blasco-Ibanez, J. M., Luis de la Iglesia, J. A., Bernabeu, A., et al. (1992) Lesion and regeneration in the medial cerebral cortex of lizards. *Histol. Histopathol.* **7(4),** 725–746.

99. Jones, J. E. and Corwin, J. T. (1993) Replacement of lateral line sensory organs during tail regeneration in salamanders: identification of progenitor cells and analysis of leukocyte activity. *J. Neurosci.* **13(3),** 1022–1034.

100. Brockes, J. P. (1997) Amphibian limb regeneration: rebuilding a complex structure. *Science* **276(5309),** 81–87.

101. Johns, P. R. and Easter, S. S. (1977) Growth of the adult goldfish eye. II. Increase in retinal cell number. *J. Comp. Neurol.* **176(3),** 331–341.

102. Hitchcock, P. F., Lindsey Myhr, K. J., Easter, S. S., Mangione-Smith, R., and Jones, D. D. (1992) Local regeneration in the retina of the goldfish. *J. Neurobiol.* **23(2),** 187–203.

103. Fischer, A. J. and Reh, T. A. (2001) Muller glia are a potential source of neural regeneration in the postnatal chicken retina. *Nat. Neurosci.* **4(3),** 247–252.

104. Nottebohm, F. (1985) Neuronal replacement in adulthood. *Ann. NY Acad. Sci.* **457,** 143–161.
105. Alvarez-Buylla, A., Kirn, J. R., and Nottebohm, F. (1990) Birth of projection neurons in adult avian brain may be related to perceptual or motor learning [published erratum appears in *Science* 1990 **250(4979),** 360]. *Science* **249(4975),** 1444–1446.
106. Nottebohm, F., Stokes, T. M., and Leonard, C. M. (1976) Central control of song in the canary, *Serinus canarius. J. Comp. Neurol.* **165(4),** 457–486.
107. Simpson, H. B. and Vicario, D. S. (1990) Brain pathways for learned and unlearned vocalizations differ in zebra finches. *J. Neurosci.* **10(5),** 1541–1556.
108. Williams, H., Crane, L. A., Hale, T. K., Esposito, M. A., and Nottebohm, F. (1992) Right-side dominance for song control in the zebra finch. *J. Neurobiol.* **23(8),** 1006–1020.
109. Palmer, T. D., Markakis, E. A., Willhoite, A. R., Safar, F., and Gage, F. H. (1999) Fibroblast growth factor-2 activates a latent neurogenic program in neural stem cells from diverse regions of the adult CNS. *J. Neurosci.* **19(19),** 8487–8497.
110. Weiss, S., Dunne, C., Hewson, J., Wohl, C., Wheatley, M., Peterson, A. C., et al. (1996) Multipotent CNS stem cells are present in the adult mammalian spinal cord and ventricular neuroaxis. *J. Neurosci.* **16(23),** 7599–7609.
111. Shihabuddin, L. S., Ray, J., and Gage, F. H. (1997) FGF-2 is sufficient to isolate progenitors found in the adult mammalian spinal cord. *Exp. Neurol.* **148(2),** 577–586.
112. Tropepe, V., Coles, B. L., Chiasson, B. J., Horsford, D. J., Elia, A. J., McInnes, R. R., et al. (2000) Retinal stem cells in the adult mammalian eye. *Science* **287(5460),** 2032–2036.
113. Brewer, G. J. (1999) Regeneration and proliferation of embryonic and adult rat hippocampal neurons in culture. [see comments]. *Exp. Neurol.* **159(1),** 237–247.
114. Gu, W., Brannstrom, T., and Wester, P. (2000) Cortical neurogenesis in adult rats after reversible photothrombotic stroke. *J. Cereb. Blood Flow Metab.* **20(8),** 1166–1173.
115. Yang, Y., Geldmacher, D. S., and Herrup, K. (2001) DNA replication precedes neuronal cell death in Alzheimer's disease. *J. Neurosci.* **21(8),** 2661–2668.
116. Macklis, J. D. (1993) Transplanted neocortical neurons migrate selectively into regions of neuronal degeneration produced by chromophore-targeted laser photolysis. *J. Neurosci.* **13(9),** 3848–3863.
117. Sheen, V. L. and Macklis, J. D. (1995) Targeted neocortical cell death in adult mice guides migration and differentiation of transplanted embryonic neurons. *J. Neurosci.* **15(12),** 8378–8392.
118. Hernit-Grant, C. S. and Macklis, J. D. (1996) Embryonic neurons transplanted to regions of targeted photolytic cell death in adult mouse somatosensory cortex re-form specific callosal projections. *Exp. Neurol.* **139(1),** 131–142.
119. Snyder, E. Y., Yoon, C., Flax, J. D., and Macklis, J. D. (1997) Multipotent neural precursors can differentiate toward replacement of neurons undergoing

targeted apoptotic degeneration in adult mouse neocortex. *Proc. Natl. Acad. Sci. USA* **94(21)**, 11,663–11,668.

120. Madison, R. and Macklis, J. D. (1993) Noninvasively induced degeneration of neocortical pyramidal neurons in vivo: selective targeting by laser activation of retrogradely transported photolytic chromophore. *Exp. Neurol.* **121**, 153–159.

121. Sheen, V. L. and Macklis, J. D. (1994) Apoptotic mechanisms in targeted neuronal cell death by chromophore-activated photolysis. *Exp. Neurol.* **130(1)**, 67–81.

122. Sheen, V. L., Dreyer, E. B., and Macklis, J. D. (1992) Calcium-mediated neuronal degeneration following singlet oxygen production. *NeuroReport* **3(8)**, 705–708.

123. Wang, Y., Sheen, V. L., and Macklis, J. D. (1998) Cortical interneurons upregulate neurotrophins in vivo in response to targeted apoptotic degeneration of neighboring pyramidal neurons. *Exp. Neurol.* **154(2)**, 389–402.

124. Shin, J. J., Fricker-Gates, R. A., Perez, F. A., Leavitt, B. R., Zurakowski, D., and Macklis, J. D. (2000) Transplanted neuroblasts differentiate appropriately into projection neurons with correct neurotransmitter and receptor phenotype in neocortex undergoing targeted projection neuron degeneration. *J. Neurosci. (Online)* **20(19)**, 7404–7416.

125. Gleeson, J. G., Lin, P. T., Flanagan, L. A., and Walsh, C. A. (1999) Doublecortin is a microtubule-associated protein and is expressed widely by migrating neurons. *Neuron* **23(2)**, 257–271.

126. Francis, F., Koulakoff, A., Boucher, D., Chafey, P., Schaar, B., Vinet, M. C., et al. (1999) Doublecortin is a developmentally regulated, microtubule-associated protein expressed in migrating and differentiating neurons. *Neuron* **23(2)**, 247–256.

127. Marusich, M. F., Furneaux, H. M., Henion, P. D., and Weston, J. A. (1994) Hu neuronal proteins are expressed in proliferating neurogenic cells. *J. Neurobiol.* **25(2)**, 143–155.

128. Barami, K., Iversen, K., Furneaux, H., and Goldman, S. A. (1995) Hu protein as an early marker of neuronal phenotypic differentiation by subependymal zone cells of the adult songbird forebrain. *J. Neurobiol.* **28(1)**, 82–101.

129. Szele, F. G. and Chesselet, M. F. (1996) Cortical lesions induce an increase in cell number and PSA-NCAM expression in the subventricular zone of adult rats. *J. Comp. Neurol.* **368(3)**, 439–454.

130. Marmur, R., Mabie, P. C., Gokhan, S., Song, Q., Kessler, J. A., and Mehler, M. F. (1998) Isolation and developmental characterization of cerebral cortical multipotent progenitors. *Dev. Biol. (Orlando)* **204(2)**, 577–591.

131. Richards, L. J., Kilpatrick, T. J., and Bartlett, P. F. (1992) De novo generation of neuronal cells from the adult mouse brain. *Proc. Natl. Acad. Sci. USA* **89(18)**, 8591–8595.

132. Luskin, M. B. (1993) Restricted proliferation and migration of postnatally generated neurons derived from the forebrain subventricular zone. *Neuron* **11(1)**, 173–189.

133. Vescovi, A. L., Reynolds, B. A., Fraser, D. D., and Weiss, S. (1993) bFGF regulates the proliferative fate of unipotent (neuronal) and bipotent (neuronal/astroglial) EGF-generated CNS progenitor cells. *Neuron* **11(5)**, 951–966.

134. Lois, C. and Alvarez-Buylla, A. (1993) Proliferating subventricular zone cells in the adult mammalian forebrain can differentiate into neurons and glia. *Proc. Natl. Acad. Sci. USA* **90(5)**, 2074–2077.

135. Morshead, C. M., Reynolds, B. A., Craig, C. G., McBurney, M. W., Staines, W. A., Morassutti, D., et al. (1994) Neural stem cells in the adult mammalian forebrain: a relatively quiescent subpopulation of subependymal cells. *Neuron* **13(5)**, 1071–1082.

10

Neural Stem Cells in the Adult Hippocampus

Jasodhara Ray and Daniel A. Peterson

INTRODUCTION

It had long been held as dogma that no new neurons are added to the central nervous system after birth. The fixed neuronal population of the adult brain was understood to be necessary to maintain the functional stability of the adult brain and was also taken as an explanation for the lack of endogenous repair within the central nervous system following injury. However, there were early reports of cell proliferation in the adult brain *(1–4)*, and these observations have been validated by the use of new, more powerful labeling techniques over the last decade *(5)*. Nevertheless, under normal conditions, neurogenesis in the adult brain appears to be restricted to the subventricular zone and the hippocampal dentate gyrus. This review summarizes our current understanding regarding neurogenesis in the adult dentate gyrus, the extent to which primitive neural progenitor (stem) cells in the hippocampus can be manipulated in vitro and in vivo, and the implications of hippocampal neurogenesis for brain function and plasticity. Because of the importance of quantitative studies to our understanding of neurogenesis, some methodological and conceptual issues regarding histological quantitation are also addressed.

CHARACTERIZATION OF ADULT HIPPOCAMPAL NEUROGENESIS

It is possible to identify cells throughout the body that are undergoing cell cycle by the systemic administration of nucleotide analog markers. Once these markers are incorporated in the replicating DNA, they can be subsequently localized by histological methods to "birthdate" the cell to the period when the marker was available. Whereas early studies employed ^{3}H-thymidine to label newly generated cells, most recent studies have utilized bromodeoxyuridine (BrdU) for its ability to be detected through immunocytochemical labeling (however, *see* ref. *6*). Thus, newly generated,

From: *Neural Stem Cells for Brain and Spinal Cord Repair*
Edited by: T. Zigova, E. Y. Snyder, and P. R. Sanberg © Humana Press Inc., Totowa, NJ

BrdU-positive cells can be identified in conjunction with other cell phenotype markers using multiple immunofluorescence labeling.

Examining the adult hippocampus at successive intervals following BrdU administration reveals a characteristic progression of BrdU-positive cell location, number, morphology, and terminal differentiation *(7)*. Within the first days after labeling, newly generated cells lie at the border of the dentate gyrus and the hilus in the so-called subgranular zone. Cells within this period typically show compact, elongated nuclei and are often clustered within groups. The subgranular zone is the region of cell proliferation and the clustering of cells suggests that these cells may be daughter cells of a single progenitor, although definitive evidence for this conclusion is still lacking. Within the first week following their generation, these cells begin to disperse and integrate within the dentate granule cell layer. This process suggests a directed migration of these cells and is consistent with their expression of early neuronal markers such as PSA-NCAM and neuron-specific tubulin (TuJ1) *(8–10)*. After approximately 1 wk, most surviving newly generated cells lie fully integrated with the dentate granule cell layer and begin to express mature neuronal markers characteristic of dentate granule cells, such as NeuN and calbindin *(11)*. The majority of the newly generated cells fail to survive beyond their first week.

Based on the descriptive progression of newly generated cells from birth to maturity, neurogenesis in the adult hippocampus has been divided into three components in accordance with the sequence of neurogenesis during central nervous system (CNS) development: proliferation, a short migration, and terminal differentiation. Mounting evidence, described later in this review, supports the view that each of these phases of neurogenesis in the adult hippocampus may be under separate regulation. As a result, many studies now design experimental groups to examine separately cellular proliferation (within days of BrdU administration) or neuronal differentiation (greater than 1 wk following BrdU administration). Unfortunately, the broad term "neurogenesis" is still frequently misapplied to studies examining only cellular proliferation.

ADULT-DERIVED HIPPOCAMPAL STEM CELLS IN VITRO

The temporal and spatial expression of trophic factors plays a critical role in the development of the nervous system and in the support of some differentiated mature neurons *(12)*. In the course of investigating the ability of trophic factor supplementation to maintain long-term primary hippocampal cultures, we observed the unexpected emergence of a population of undifferentiated proliferative cells *(13)*. The proliferative cells could be extracted from late embryonic or adult hippocampus and grown almost indefinitely in culture. Further characterization revealed that, under appropriate culture conditions, these proliferative cells could differentiate to give rise to neurons, astrocytes, and oligodendrocytes *(11,14,15)*. A similar capacity of trophic factors to stimulate a proliferative population of primitive cells in vitro has been reported for adult tissue from the forebrain subventricular zone *(16)*. Many reports since have both confirmed and extended these observations and demonstrated that the adult brain can produce undifferentiated proliferative cells in response to appropriate culture conditions. Recent studies suggest these cells can even be extracted after death *(17–19)*.

The ability of these primitive cells from adult brain to proliferate and subsequently differentiate into distinct cell types suggests an analogy with stem cells of the hematopoietic system *(20)*. There are at least two distinct properties that these primitive neural cells must exhibit in order to be identified as stem cells *(21)*. First, these cells must self-renew; that is, they must reproduce themselves by symmetric cell division to produce daughter cells identical to the parent cell. Second, when one of these daughter cells undergoes asymmetric cell division, the progeny must be capable of producing different cell phenotypes. The original stem cell population of an individual, the embryonic stem cells (ES cells), possesses a totipotency that enables them to eventually produce all cell types in the body. A lesser capacity, termed multipotency, is characteristic of stem cells in the hematopoietic system that, under normal conditions, can produce the various cell types in blood circulation. Based on their ability in vitro to differentiate into various neural lineages, it seems appropriate to refer to these primitive, proliferative cells derived from the adult hippocampus as neural stem cells. However, how well does the ability of these adult-derived cells to differentiate in vitro mimic their potential fate choices in an in vivo environment?

SPECIFICATION OF GRAFTED NEURAL STEM CELL FATE

A number of studies indicate that the local environment is of critical importance in directing the fate of neural stem cells. Cultured neural stem cells, derived from the adult hippocampus, can be grafted back into the hippocampus, where they adopt the appropriate phenotype of mature granule neurons *(11,22)*. The successful differentiation of these cells when placed in their appropriate adult environment suggests that essential survival and differentiation factors continue to be produced by the adult hippocampus. To discriminate between environment-specified differentiation and the endogenous capacity of neural stem cells to determine their own fate, adult hippocampal-derived stem cells were grafted into heterotypic environments.

Placement of these cells into a non-neurogenic region, such as the cerebellum, produced limited glial differentiation and no neuronal differentiation *(22)*. Minimal differentiation was found following engraftment to the mature retina, a non-neurogenic region *(23)*. However, grafting adult hippocampal-derived neural stem cells into developing, injured, or diseased retina improved differentiation and incorporation of grafted cells, suggesting that such normally non-neurogenic environments may once again express developmental cell fate specification signals in response to injury *(23–25)*.

Environmental signals for stem cell survival and differentiation may be region-specific and direct nearby stem cells to differentiate into particular phenotypes. Hippocampal-derived stem cells grafted into the olfactory bulb differentiated into neurons, including differentiation into specific olfactory neurons that were distinct from hippocampal phenotypes *(22)*. Furthermore, the adult hippocampus has also shown an ability to direct the differentiation of nonhippocampal-derived neural stem cells, such as human neural progenitor cells *(26)* and spinal-cord-derived progenitor cells *(27,28)*, into neurons of appropriate hippocampal phenotypes. These reports suggest that diverse neural stem cells are competent to respond to local environmental factors and differentiate into an appropriate phenotype as instructed *(29)*. Furthermore, whereas neurogenic regions of the adult brain express signals appropriate to direct a wide variety of stem cells to adopt a neuronal fate, these signals are either lacking or insufficient in non-neurogenic regions of the adult brain under normal conditions. These findings are encouraging for the potential use of neural stem cells for brain repair in that it may be possible to instruct grafted neural stem cells to adopt the desired differentiation phenotype by altering the local environment to express the necessary differentiation signals or to augment such signals that may be expressed in injured or diseased environments.

ENVIRONMENTAL REGULATION
OF HIPPOCAMPAL NEUROGENESIS

The above-discussed studies demonstrate that the adult hippocampus expresses signals that can instruct the survival and differentiation of grafted neural stem cells. Accumulating evidence suggests that signals expressed by the adult hippocampus also regulate the proliferation of endogenous stem cells and their subsequent survival and phenotypic fate. Interestingly, the regulation of these events is not static, but the rate of neurogenesis in the adult hippocampus fluctuates in response to changes in the environment (Table 1). Even macroenvironmental experiences can produce changes in

Table 1
Some Environmental Factors Affecting Adult
Hippocampal Neurogenesis

Variable/substance	References
Enhancers of neurogenesis	
Enriched environment	*30, 32, 33*
Physical activity	*31, 34*
Caloric restriction	*35*
Vitamin E deficiency	*36*
Modulation of excitatory afferents	*37–39*
Antidepressants	*50*
Seizures	*10, 40–42*
Growth factors	*53–55*
Inhibitors of neurogenesis	
Aging	*43–45*
Stress	*38, 46, 47*
Serotonin depletion	*48, 49*
Opiates	*52*
Mcthamphetamines	*51*

neurogenesis in the adult brain. For example, hippocampal neurogenesis is enhanced by such macroenvironmental alterations as free physical activity and the opportunity to engage in exploratory behavior in an enriched housing environment *(30–34)*. Nutritional status also influences the rate of neurogenesis with both caloric restriction and vitamin E deficiency increasing neurogenesis *(35,36)*. Increased hippocampal neurogenesis can also be produced through the experimental modulation of excitatory hippocampal afferents *(37–39)* and following initiation of status epilepticus and seizure activity *(10,40–42)*.

Environmental changes can also reduce the rate of hippocampal neurogenesis (Table 1). Hippocampal neurogenesis declines with increasing age *(43–45)*. Undergoing psychosocial and/or physiological stress *(38,46,47)* reduces the rate of hippocampal neurogenesis, apparently mediated through elevated levels of circulating glucocorticoids. Serotonin levels, likewise, are implicated in regulating neurogenesis with experimental serotonin depletion producing a decline in neurogenesis *(48,49)*, whereas administration of selective serotonin reuptake inhibitor (SSRI) antidepressants elevate neurogenesis *(50)*. Administration of some drugs of abuse has also been found to suppress the rate of hippocampal neurogenesis *(51,52)*.

The responsiveness of adult neurogenesis to environmental influence suggests that stem cell proliferation and/or differentiation may be under the control of a common mechanism or factor whose availability dictates the rate of neurogenesis. The balance of environmental influences thus may determine the availability of such a factor. For example, the age-related decline in hippocampal neurogenesis can be alleviated by experiencing an enriched environment *(44)* or by lowering corticosteroid levels *(45)*. Candidates for regulatory factors include growth factors, such as basic fibroblast growth factor (FGF-2), and trophic factors, such as brain-derived neurotrophic factor (BDNF), based on the observed proliferative effects of FGF-2 and differentiation effects of BDNF on cultures of neural stem cells from the hippocampus *(11,13,15)*. In addition to a neuroprotective function for growth factors *(56,57)*, a regulatory role for neurogenesis by growth factors is also supported by in vivo evidence. Peripheral administration of FGF-2 (53) and insulinlike growth factor (IGF)-1 *(54)* enhanced neurogenesis in both olfactory bulb and hippocampus. Stress-related suppression of neurogenesis may also act through growth factor expression as glucocorticoids are known to regulate hippocampal FGF-2 and BDNF levels *(58,59)*. However, Yoshimura et al. *(55)* found that the rate of hippocampal neurogenesis in FGF-2 null mutant mice was the same as in wild-type littermates. Increasing the level of FGF-2 by gene delivery more than doubled the rate of neurogenesis in both null mutant and wild-type mice. Therefore, FGF-2 may not be required to maintain neurogenesis, but it may act as a mitogenic factor for neural stem cells.

The neurotrophic factor BDNF may be important for the recruitment of neural stem cells to a neuronal fate. Intraventricular delivery of BDNF increased neuronal differentiation of neural stem cells in the olfactory bulb *(60–62)*. BDNF has also been shown to be important for neuronal recruitment in adult neurogenesis in the avian CNS *(63)*. Elevated BDNF expression has also been correlated with increased neurogenesis in the adult mammalian CNS. Seizure activity increases neurogenesis (Table 1) and is also known to effect expression of BDNF (review in ref. *64*). Increased hippocampal BDNF has been correlated with increased neurogenesis in dietary-restricted animals *(35)*. Finally, BDNF may function in the effect of antidepressants on enhancing neurogenesis (Table 1), given that antidepressants elevate hippocampal BDNF *(65,66)*, particularly in combination with increased activity *(67,68)*. The elevation of dentate BDNF following activation of the 5-HT2A receptor *(69)* may be a mechanism whereby SSRI antidepressants increase neurogenesis.

QUANTITATION OF NEUROGENESIS

As discussed above, the examination of adult neurogenesis in vivo primarily addresses mechanisms of enhancement or suppression of cellular proliferation, survival of newly generated cells, or assessment of their terminal differentiation. Therefore, most studies investigating the regulation of neurogenesis in the adult brain and the functional significance of newly generated neurons require an accurate determination of cell number to test their hypotheses. Accurate quantitation of cell number in tissue sections presents a number of methodological and theoretical difficulties. In addition, the need to simultaneously identify several markers to evaluate the state of phenotypic differentiation in labeled, newly generated cells has raised the criteria for accuracy in histological quantitation.

Counting cells in histological sections based on single-focal-plane (i.e., two dimensional) images are subject to a number of biases and errors that render such simple counts unreliable *(70–72)*. Three-dimensional sampling approaches, such as the optical disector where cells are directly counted in a known volume, allow the investigator to avoid the methodological errors and most biases associated with two-dimensional cell counts *(73–75)*. These three-dimensional counting approaches, known as design-based stereology, have been widely used in the study of in vivo neurogenesis. Nevertheless, design-based stereological procedures can be implemented in a way that can make them vulnerable to a number of quantitative artifacts.

There are at least three considerations for the appropriate use of design-based stereology in the estimation of adult neurogenesis in vivo. First, it is necessary to accurately identify the cell to be counted (known in stereology literature as "The General Requirement"). Although this may seem obvious, the use of BrdU as a marker for newly generated cells presents some difficulty in meeting this requirement (see also ref. *6*). It is unlikely that all proliferating cells in the region being examined are synchronous with regard to their cell cycle and therefore, any given exposure to BrdU will result in variable incorporation among proliferating cells. Subsequent proliferation of labeled cells will produce dilution of the incorporated BrdU, possibly leading to the failure to detect an unknown number of those progeny if the dilution is too great. Finally, the immunohistochemical detection of BrdU may lead to variability of cell identification based on the pretreatment procedures used to allow antibody access to BrdU, the sensitivity of various antibodies, and the level of background relative to the immunoprecipitate. The use of immunofluorescence detection methods produces another potential artifact in that red blood cells may remain in the highly vascularized subgranular proliferative region even after careful perfusion. These red blood cells will autofluoresce and can be mistaken for BrdU-positive cells detected using fluorophore-conjugated antibodies unless careful morphological discrimination is made and nonspecific emission at longer fluorescence wavelengths is examined.

A second consideration for accurate stereological quantitation is the sampling of the entire structure. Investigators may be tempted to sample only the dorsal hippocampus, a legacy from quantitative studies of previous years. However, failure to sample the entire structure with equal probability can produce erroneous results if the volume of the structure changes as a result of the experimental manipulation *(70)*. Use of the optical fractionator method *(73,74)* provides no remedy for this problem, as the systematic random sampling of the fractionator would be valid only for the region actually examined. Given that cellular proliferation is being assessed, making an assumption that the volume of the dentate granule cell layer remains unchanged cannot be justified. Unfortunately, many authors fail to describe the extent of the sampling in their methods and only state the stereological method used without providing the necessary data on the sampling parameters employed in the study.

Even under conditions that enhance proliferation in the adult hippocampus, cell proliferation and/or neurogenesis is still a relatively rare event, with only a small percentage of granule cells showing BrdU labeling. Therefore, a third consideration for accurate quantitation is to assess the amount of variance that can be accepted with systematic sampling of a rare event. Design-based stereology relies on sampling designs that are random but structured, so that the random sampling is systematic throughout the entire structure being counted *(76–79)*. As a result, any cell in the structure should have an equal probability of being sampled in three dimensions. However, not every cell in the tissue is counted, because only a fraction of the total tissue is sampled. For example, the optical fractionator procedure samples a known fraction of the tissue and then the total number of cells is estimated from the number actually counted in this known fraction *(73,74)*. However, how many cells should be counted to achieve a reliable estimate? If too few cells are counted, then the effect of any one cell is disproportionate or, in other words, it may have too great an impact upon the final calculation. Therefore, as with any statistical sampling, the variance decreases and the power increases with larger sample sizes. For counting standard-shaped nuclei, such as BrdU-positive nuclei in the hippocampus, a sampling design that produces between 100 and 200 cells will likely achieve an acceptable level of variance *(78)*. However, because BrdU-positive cells in the hippocampus are a relatively rare occurrence when sampling with a high-resolution objective lens (owing to the smaller linear field of view with high magnification, high-resolution lenses), it may be difficult to obtain over 100 cells unless the sampling density is high, thereby producing many sites to sample. Obtaining a large enough sample of cells should not be a problem when studying neurogenesis in the subventricular zone–olfactory bulb system, where the rate of cell proliferation in much higher than in the hippocampus.

There is a final consideration for the quantitation of neurogenesis deriving from the need to estimate not only the number of newly generated cells but also to assess their fate. Although proliferation can be studied using only a marker against BrdU, it is necessary to simultaneously label for a distinguishing phenotype marker to determine if the cell has become a neuron or glial cell. Fortunately, with the different fluorophores available for conjugation to secondary antibodies, it is possible to perform multiple immunofluorescence staining to address these questions. However, accurate determination of BrdU cell phenotypic identity is complicated by the fact that BrdU staining is nuclear, whereas phenotypic markers are typically cytoplasmic (including much of the expression of the neuronal nuclear marker, NeuN). Given that these labels may be in different cell compartments, a high degree of optical resolution is required to determine if the BrdU-positive nucleus is, indeed, surrounded by the expression of the phenotype marker or if there are really two separate cells in close apposition. For example, an early report of cortical and striatal neurogenesis in the rat where BrdU-positive neurons were observed using standard immunofluorescence *(80)* was later determined to consist of BrdU-positive satellite cells in close apposition to cortical and striatal neurons through a three-dimensional confocal microscopic analysis *(81)*. A similar issue of spatial resolution between BrdU-positive nuclei and phenotype label may underlie a current debate regarding cortical neurogenesis in primate brain *(82–84)*. The resolution of correct identification of BrdU-positive cell phenotype is of critical importance for accurate quantification of neurogenesis in vivo. Until recently, the best approach for accurate quantitation involved sampling the tissue twice: The first sampling estimated the number of BrdU-positive cells and the second sampling of multiple immunofluorescently stained material using confocal microscopy scored BrdU-positive cells encountered in a simplified sampling scheme according to their phenotypic identity. The ratio of BrdU-positive cell phenotype was then applied to the number of BrdU-positive cells to extrapolate the estimate of neurogenesis or gliagenesis. Fortunately, recent advances in computer-assisted stereological sampling and confocal microscopy permit the direct counting of BrdU-positive cells while simultaneously determining their phenotype in the same three-dimensional sampling (75). Although confocal stereology requires access to appropriate instrumentation, it provides for the most reliable and efficient approach to quantifying neurogenesis.

WHAT IS THE FUNCTIONAL ROLE OF NEW HIPPOCAMPAL NEURONS?

Hippocampal neurogenesis continues throughout life and appears to be a regulated process based on the response to environmental and physiological changes. Newly generated dentate granule neurons elaborate processes and extend axons to area CA3, establishing a morphological integration into hippocampal circuitry (85,86). However, despite emerging evidence about regulation of adult neurogenesis, there remains no clear consensus concerning the functional role that these newly generated hippocampal neurons play. Given the role of the hippocampal formation in the processing of learning and memory, it seems plausible that the generation of new granule neurons may be involved in memory formation. Indeed, Shors and Gould have reported in two important studies that hippocampal-dependent memory tasks enhanced hippocampal neurogenesis (87), and blockade of neurogenesis impaired hippocampal-dependent memory formation (88).

Alternatively, hippocampal neurogenesis may serve to maintain a constant granule cell population, and alterations in the rate of neurogenesis may indirectly reflect the loss of mature granule cells. Consistent with this possibility, direct injury to mature cells in the hippocampus generates a proliferative response (89), as does more generalized injury (90–92). These reports may suggest that acceleration of neurogenesis represents a self-repair program for the adult hippocampus. However, rapid generation of new neurons may not promote hippocampal function, but rather result in aberrant reorganization contributing to hippocampal dysfunction, as reported by Parent et al. (10). Therefore, despite the importance of understanding their role, additional evidence will be required to explain the range of function (or dysfunction) supported by newly generated dentate granule neurons.

CONCLUSIONS

Neurogenesis in the adult mammalian hippocampus is an ongoing process throughout life and has generally been accepted to occur even in humans (93,94). Recent work has revealed that neurogenesis is a dynamic, plastic process whose proximal regulation may be through the expression of specific growth factors. However, the specific regulation of the hippocampal environment appears to be orchestrated by the larger context of the animal's overall environment. Although there are some data correlating active neurogenesis with memory formation *(87,88)* and increased neurogenesis with improved behavioral performance *(34)*, it is still not clear if more new neurons are better. Nevertheless, understanding that the mature brain has the capacity, under certain conditions, to support the generation and differentiation of new neurons provides an opportunity to harness this capacity for therapeutic purposes.

Note Added in Proof. Two recent papers have shed additional light upon the regulation and function of neurogenesis in the adult brain. By injecting a retroviral vector expressing GFP into the adult hippocampus to selectively label dividing cells, van Praag et al. *(95)* were able to produce living slice preparations and locate newly generated neurons in the dentate gyrus. New neurons produced 4 wk previously were identified by their GFP expression and displayed a mature dendritic arbor and could be subsequently shown to express mature neuronal markers. Electrophysiological recording of GFP-expressing newly generated neurons demonstrated characteristic neuronal responses including spontaneous and evoked potentials. Importantly, these new neurons can also respond to stimulation of perforant pathway afferents indicating their incorporation in hippocampal circuitry. This report confirms that, at least in the hippocampus, newly generated neurons not only adopt appropriate morphology and express appropriate markers, but also behave as functional neurons. In an elegant series of in vitro experiments, Song et al. *(96)* found that hippocampal astrocytes increased proliferation of cocultured neuronal progenitors derived from the adult hippocampus. Significantly, these hippocampal astrocytes instructed differentiation down a neuronal lineage whereas astrocytes isolated from other CNS regions did not, perhaps explaining the regional specificity of neurogenesis in the adult brain. As neuronal production precedes astrocytic production during development, these data also suggest there is a fundamental switch in the mechanisms regulating the production and differentiation of new neurons in the adult brain.

ACKNOWLEDGMENTS

Dr. Peterson gratefully acknowledges support from the NIH (AG18538 and AG20047) and from the Schweppe Foundation.

REFERENCES

1. Altman, J. (1962) Are neurons formed in the brains of adult mammals? *Science* **135,** 1127–1128.

2. Kaplan, M. S. and Hinds, J. W. (1977) Neurogenesis in the adult rat: electron microscopic analysis of light radioautographs. *Science* **197,** 1092–1094.
3. Bayer, S. A. (1985) Neuron production in the hippocampus and olfactory bulb of the adult rat brain: addition or replacement? *Ann. NY Acad. Sci.* **457,** 163–172.
4. Kaplan, M. S. (2001) Environment complexity stimulates visual cortex neurogenesis: death of a dogma and a research career. *Trends Neurosci.* **24,** 617–620.
5. Gross, C. G. (2000) Neurogenesis in the adult brain: death of a dogma. *Nat. Rev. Neurosci.* **1,** 67–73.
6. Nowakowski, R. S. and Hayes, N. L. (2000) New neurons: extraordinary evidence or extraordinary conclusion? *Science* **288,** 771.
7. Gage, F. H., Kempermann, G., Palmer, T. D., Peterson, D. A., and Ray, J. (1998) Multipotent progenitor cells in the adult dentate gyrus. *J. Neurobiol.* **36,** 249–266.
8. Seki, T. and Arai, Y. (1993) Highly polysialylated neural cell adhesion molecule (NCAM-H) is expressed by newly generated granule cells in the dentate gyrus of the adult rat. *J. Neurosci.* **13,** 2351–2358.
9. Seki, T. and Arai, Y. (1999) Temporal and spacial relationships between PSA–NCAM-expressing, newly generated granule cells, and radial glia-like cells in the adult dentate gyrus. *J. Comp. Neurol.* **410,** 503–513.
10. Parent, J. M., Yu, T. W., Leibowitz, R. T., Geschwind, D. H., Sloviter, R. S., and Lowenstein, D. H. (1997) Dentate granule cell neurogenesis is increased by seizures and contributes to aberrant network reorganization in the adult rat hippocampus. *J. Neurosci.* **17,** 3727–3738.
11. Gage, F. H., Coates, P. W., Palmer, T. D., Kuhn, H. G., Fisher, L. J., Suhonen, J. O., et al. (1995) Survival and differentiation of adult neuronal progenitor cells transplanted to the adult brain. *Proc. Natl. Acad. Sci. USA* **92,** 11,879–11,883.
12. Peterson, D. A. and Gage, F. H. (1999) Trophic factors in experimental models of adult central nervous system injury, in *Neurodegeneration and Age-Related Changes in Structure and Function of Cerebral Cortex* (Jones, E. G., Peters, A., and Morrison, J. H., eds.), pp. 129–173.
13. Ray, J., Peterson, D. A., Schinstine, M., and Gage, F. H. (1993) Proliferation, differentiation and long-term culture of primary hippocampal neurons. *Proc. Natl. Acad. Sci. USA* **90,** 3602–3606.
14. Palmer, T. D., Takahashi, J., and Gage, F. H. (1997) The adult rat hippocampus contains primordial neural stem cells. *Mol. Cell. Neurosci.* **8,** 389–404.
15. Takahashi, J., Palmer, T. D., and Gage, F. H. (1999) Retinoic acid and neurotrophins collaborate to regulate neurogenesis in adult-derived neural stem cell cultures. *J. Neurobiol.* **38,** 65–81.
16. Reynolds, B. A. and Weiss, S. (1992) Generation of neurons and astrocytes from isolated cells of the adult mammalian central nervous system. *Science* **255,** 1707–1710.
17. Laywell, E. D., Kukekov, V. G., and Steindler, D. A. (1999) Multipotent neurospheres can be derived from forebrain subependymal zone and spinal cord of adult mice after protracted postmortem intervals. *Exp. Neurol.* **156,** 430–433.

18. Roisen, F. J., Klueber, K. M., Lu, C. L., Hatcher, L. M., Dozier, A., Shields, C. B., et al. (2001) Adult human olfactory stem cells. *Brain Res.* **890,** 11–22.
19. Palmer, T. D., Schwartz, P. H., Taupin, P., Kaspar, B., Stein, S. A., and Gage, F. H. (2001) Progenitor cells from human brain after death. *Nature* **411,** 42–43.
20. Scheffler, B., Horn, M., Blumcke, I., Laywell, E. D., Coomes, D., Kukekov, V. G., et al. (1999) Marrow-mindedness: a perspective on neuropoiesis. *Trends Neurosci.* **22,** 348–357.
21. Gage, F. H. (2000) Mammalian neural stem cells. *Science* **287,** 1433–1438.
22. Suhonen, J. O., Peterson, D. A., Ray, J., and Gage, F. H. (1996) Differentiation of adult hippocampus-derived progenitors into olfactory neurons in vivo. *Nature* **383,** 624–627.
23. Takahashi, M., Palmer, T. D., Takahashi, J., and Gage, F. H. (1998) Widespread integration and survival of adult-derived neural progenitor cells in the developing optic retina. *Mol. Cell. Neurosci.* **12,** 340–348.
24. Young, M. J., Ray, J., Whiteley, S. J., Klassen, H., and Gage, F. H. (2000) Neuronal differentiation and morphological integration of hippocampal progenitor cells transplanted to the retina of immature and mature dystrophic rats. *Mol. Cell. Neurosci.* **16,** 197–205.
25. Nishida, A., Takahashi, M., Tanihara, H., Nakano, I., Takahashi, J. B., Mizoguchi, A., et al. (2000) Incorporation and differentiation of hippocampus-derived neural stem cells transplanted in injured adult rat retina. *Invest. Ophthalmol. Vis. Sci.* **41,** 4268–4274.
26. Fricker, R. A., Carpenter, M. K., Winkler, C., Greco, C., Gates, M. A., and Bjorklund, A. (1999) Site-specific migration and neuronal differentiation of human neural progenitor cells after transplantation in the adult rat brain. *J. Neurosci.* **19,** 5990–6005.
27. Shihabuddin, L. S., Horner, P. J., Ray, J., and Gage, F. H. (2000) Adult spinal cord stem cells generate neurons after transplantation in the adult dentate gyrus. *J. Neurosci.* **20,** 8727–8735.
28. Thomas, R., Peterson, L. D., and Peterson. D.A. (2001) Region-specific differentiation of grafted neuronal lineage-restricted progenitor cells: hippocampal neurons from spinal cord cells. *Soc. Neurosci. Abstr.* **27,** 791.9.
29. Clarke, D. L., Johansson, C. B., Wilbertz, J., Veress, B., Nilsson, E., Karlstrom, H., et al. (2000) Generalized potential of adult neural stem cells. *Science* **288,** 1660–1663.
30. Kempermann, G., Kuhn, H. G., and Gage, F. H. (1997) More hippocampal neurons in adult mice living in an enriched environment. *Nature* **386,** 493–495.
31. van Praag, H., Kempermann, G., and Gage, F. H. (1999) Running increases cell proliferation and neurogenesis in the adult mouse dentate gyrus. *Nat. Neurosci.* **2,** 266–270.
32. Kempermann, G. and Gage, F. H. (1999) Experience-dependent regulation of adult hippocampal neurogenesis: effects of long-term stimulation and stimulus withdrawal. *Hippocampus* **9,** 321–332.

33. Nilsson, M., Perfilieva, E., Johansson, U., Orwar, O., and Eriksson, P. S. (1999) Enriched environment increases neurogenesis in the adult rat dentate gyrus and improves spatial memory. *J. Neurobiol.* **39,** 569–578.
34. van Praag, H., Christie, B. R., Sejnowski, T. J., and Gage, F. H. (1999) Running enhances neurogenesis, learning, and long-term potentiation in mice. *Proc. Natl. Acad. Sci. USA* **96,** 13,427–13,431.
35. Lee, J., Duan, W., Long, J. M., Ingram, D. K., and Mattson, M. P. (2000) Dietary restriction increases the number of newly generated neural cells, and induces BDNF expression, in the dentate gyrus of rats. *J. Mol. Neurosci.* **15,** 99–108.
36. Ciaroni, S., Cuppini, R., Cecchini, T., Ferri, P., Ambrogini, P., Cuppini, C., et al. (1999) Neurogenesis in the adult rat dentate gyrus is enhanced by vitamin E deficiency. *J. Comp. Neurol.* **411,** 495–502.
37. Cameron, H. A., McEwen, B. S., and Gould, E. (1995) Regulation of adult neurogenesis by excitatory input and NMDA receptor activation in the dentate gyrus. *J. Neurosci.* **15,** 4687–4692.
38. Gould, E., McEwen, B. S., Tanapat, P., Galea, L. A., and Fuchs, E. (1997) Neurogenesis in the dentate gyrus of the adult tree shrew is regulated by psychosocial stress and NMDA receptor activation. *J. Neurosci.* **17,** 2492–2498.
39. Nacher, J., Rosell, D. R., Alonso-Llosa, G., and McEwen, B. S. (2001) NMDA receptor antagonist treatment induces a long-lasting increase in the number of proliferating cells, PSA-NCAM-immunoreactive granule neurons and radial glia in the adult rat dentate gyrus. *Eur. J. Neurosci.* **13,** 512–520.
40. Scott, B. W., Wojtowicz, J. M., and Burnham, W. M. (2000) Neurogenesis in the dentate gyrus of the rat following electroconvulsive shock seizures. *Exp. Neurol.* **165,** 231–236.
41. Madsen, T. M., Treschow, A., Bengzon, J., Bolwig, T. G., Lindvall, O., and Tingstrom, A. (2000) Increased neurogenesis in a model of electroconvulsive therapy. *Biol. Psychiatry* **47,** 1043–1049.
42. Scharfman, H. E., Goodman, J. H., and Sollas, A. L. (2000) Granule-like neurons at the hilar/CA3 border after status epilepticus and their synchrony with area CA3 pyramidal cells: functional implications of seizure-induced neurogenesis. *J. Neurosci.* **20,** 6144–6158.
43. Kuhn, H. G., Dickinson-Anson, H., and Gage, F. H. (1996) Neurogenesis in the dentate gyrus of the adult rat: age-related decrease of neuronal progenitor proliferation. *J. Neurosci.* **16,** 2027–2033.
44. Kempermann, G., Kuhn, H. G., and Gage, F. H. (1998) Experience-induced neurogenesis in the senescent dentate gyrus. *J. Neurosci.* **18,** 3206–3212.
45. Cameron, H. A. and McKay, R. D. (1999) Restoring production of hippocampal neurons in old age. *Nat. Neurosci.* **2,** 894–897.
46. Cameron, H. A. and Gould, E. (1994) Adult neurogenesis is regulated by adrenal steroids in the dentate gyrus. *Neuroscience* **61,** 203–209.
47. Lemaire, V., Koehl, M., Le Moal, M., and Abrous, D. N. (2000) Prenatal stress produces learning deficits associated with an inhibition of neurogenesis in the hippocampus. *Proc. Natl. Acad. Sci. USA* **97,** 11,032–11,037.

48. Brezun, J. M. and Daszuta, A. (1999) Depletion in serotonin decreases neurogenesis in the dentate gyrus and the subventricular zone of adult rats. *Neuroscience* **89,** 999–1002.

49. Brezun, J. M. and Daszuta, A. (2000) Serotonergic reinnervation reverses lesion-induced decreases in PSA-NCAM labeling and proliferation of hippocampal cells in adult rats. *Hippocampus* **10,** 37–46.

50. Malberg, J. E., Eisch, A. J., Nestler, E. J., and Duman, R. S. (2000) Chronic antidepressant treatment increases neurogenesis in adult rat hippocampus. *J. Neurosci.* **20,** 9104–9110.

51. Hildebrandt, K., Teuchert-Noodt, G., and Dawirs, R. R. (1999) A single neonatal dose of methamphetamine suppresses dentate granule cell proliferation in adult gerbils which is restored to control values by acute doses of haloperidol. *J. Neural. Transm.* **106,** 549–558.

52. Eisch, A. J., Barrot, M., Schad, C. A., Self, D. W., and Nestler, E. J. (2000) Opiates inhibit neurogenesis in the adult rat hippocampus. *Proc. Natl. Acad. Sci. USA* **97,** 7579–7584.

53. Wagner, J. P., Black, I. B., and DiCicco-Bloom, E. (1999) Stimulation of neonatal and adult brain neurogenesis by subcutaneous injection of basic fibroblast growth factor. *J. Neurosci.* **19,** 6006–6016.

54. Aberg, M. A., Aberg, N. D., Hedbacker, H., Oscarsson, J., and Eriksson, P. S. (2000) Peripheral infusion of IGF-I selectively induces neurogenesis in the adult rat hippocampus. *J. Neurosci.* **20,** 2896–2903.

55. Yoshimura, S., Takagi, Y., Harada, J., Teramoto, T., Thomas, S. S., Waeber, C., et al. (2001) FGF-2 regulation of neurogenesis in adult hippocampus after brain injury. *Proc. Natl. Acad. Sci. USA* **98,** 5874–5879.

56. Peterson, D. A., Lucidi-Phillipi, C. A., Murphy, D. P., Ray, J., and Gage, F. H. (1996) Fibroblast growth factor-2 protects entorhinal layer II glutamatergic neurons from axotomy-induced death. *J. Neurosci.* **16,** 886–898.

57. Torres-Aleman, I. (2000) Serum growth factors and neuroprotective surveillance. *Mol. Neurobiol.* **21,** 153–160.

58. Hansson, A. C., Cintra, A., Belluardo, N., Sommer, W., Bhatnagar, M., Bader, M., et al. (2000) Gluco- and mineralocorticoid receptor-mediated regulation of neurotrophic factor gene expression in the dorsal hippocampus and the neocortex of the rat. *Eur. J. Neurosci.* **12,** 2918–2934.

59. Hansson, A. C., Sommer, W., Andbjer, B., Bader, M., Ganten, D., and Fuxe, K. (2001) Induction of hippocampal glial cells expressing basic fibroblast growth factor RNA by corticosterone. *NeuroReport* **12,** 141–145.

60. Zigova, T., Pencea, V., Wiegand, S. J., and Luskin, M. B. (1998) Intraventricular administration of BDNF increases the number of newly generated neurons in the adult olfactory bulb. *Mol. Cell. Neurosci.* **11,** 234–245.

61. Benraiss, A., Chmielnicki, E., Lerner, K., Roh, D., and Goldman, S. A. (2001) Adenoviral brain-derived neurotrophic factor induces both neostriatal and olfactory neuronal recruitment from endogenous progenitor cells in the adult forebrain. *J. Neurosci.* **21,** 6718–6731.

62. Pencea, V., Bingaman, K. D., Wiegand, S. J., and Luskin, M. B. (2001) Infusion of brain-derived neurotrophic factor into the lateral ventricle of the adult rat leads to new neurons in the parenchyma of the striatum, septum, thalamus, and hypothalamus. *J. Neurosci.* **21,** 6706–6717.

63. Rasika, S., Alvarez-Buylla, A., and Nottebohm, F. (1999) BDNF mediates the effects of testosterone on the survival of new neurons in an adult brain. *Neuron* **22,** 53–62.

64. Jankowsky, J. L. and Patterson, P. H. (2001) The role of cytokines and growth factors in seizures and their sequelae. *Prog. Neurobiol.* **63,** 125–149.

65. Nibuya, M., Nestler, E. J., and Duman, R. S. (1996) Chronic antidepressant administration increases the expression of cAMP response element binding protein (CREB) in rat hippocampus. *J. Neurosci.* **16,** 2365–2372.

66. Chen, B., Dowlatshahi, D., MacQueen, G. M., Wang, J., and Young, L. T. (2001) Increased hippocampal BDNF immunoreactivity in subjects treated with antidepressant medication. *Biol. Psychiatry* **50,** 260–265.

67. Russo-Neustadt, A. A., Beard, R. C., Huang, Y. M., and Cotman, C. W. (2000) Physical activity and antidepressant treatment potentiate the expression of specific brain-derived neurotrophic factor transcripts in the rat hippocampus. *Neuroscience* **101,** 305–312.

68. Russo-Neustadt, A., Ha, T., Ramirez, R., and Kesslak, J. P. (2001) Physical activity-antidepressant treatment combination: impact on brain-derived neurotrophic factor and behavior in an animal model. *Behav. Brain Res.* **120,** 87–95.

69. Vaidya, V. A., Marek, G. J., Aghajanian, G. K., and Duman, R. S. (1997) 5-HT2A receptor-mediated regulation of brain-derived neurotrophic factor mRNA in the hippocampus and the neocortex. *J. Neurosci.* **17,** 2785–2795.

70. Peterson, D. A., Leppert, J. T., Lee, K. F., and Gage, F. H. (1997) Basal forebrain neuronal loss in mice lacking neurotrophin receptor p75. *Science* **277,** 837–838.

71. Peterson, D. A., Dickinson-Anson, H. A., Leppert, J. T., Lee, K. F., and Gage, F. H. (1999) Central neuronal loss and behavioral impairment in mice lacking neurotrophin receptor p75. *J. Comp. Neurol.* **404,** 1–20.

72. Peterson, D. A. (2002) The use of fluorescent probes in cell counting procedures, in *Quantitative Methods in Neuroscience—A Neuroanatomical Approach* (Evans, S. M., Janson, A. M., and Nyengaard, J. R., eds.), Oxford University Press, Oxford, in press.

73. West, M. J. (1993) New stereological methods for counting neurons. *Neurobiol. Aging* **14,** 275–285.

74. West, M. J. (1993) Regionally specific loss of neurons in the aging numan hippocampus. *Neurobiol. Aging* **14,** 287–293.

75. Peterson, D. A. (1999) Quantitative histology using confocal microscopy: implementation of unbiased stereology procedures. *Methods* **18,** 493–507.

76. Cruz-Orive, L. M. and Weibel, E. W. (1981) Sampling designs for stereology. *J. Microsc.* **122,** 235–257.

77. Gundersen, H. J. G. and Østerby, R. (1981) Optimizing sampling efficiency of stereological studies in biology: or "Do more less well!". *J. Microsc.* **121,** 65–73.

78. Gundersen, H. J. G. and Jensen, E. B. (1987) The efficiency of systematic sampling in stereology and its prediction. *J. Microsc.* **147,** 229–263.
79. Ree, K. (1988) Morphometry of rarely occuring cells, in *Stereology and Morphometry in Electron Microscopy* (Reith, A. and Mayhew, T. M., eds.), Hemisphere, New York, pp. 63–70.
80. Craig, C. G., Tropepe, V., Morshead, C. M., Reynolds, B. A., Weiss, S., and van der Kooy, D. (1996) In vivo growth factor expansion of endogenous subependymal neural precursor cell populations in the adult mouse brain. *J. Neurosci.* **16,** 2649–2658.
81. Kuhn, H. G., Winkler, J., Kempermann, G., Thal, L. J., and Gage, F. H. (1997) Epidermal growth factor and fibroblast growth factor-2 have different effects on neural progenitors in the adult rat brain. *J. Neurosci.* **17,** 5820–5829.
82. Gould, E., Reeves, A. J., Graziano, M. S., and Gross, C. G. (1999) Neurogenesis in the neocortex of adult primates. *Science* **286,** 548–552.
83. Gould, E., Vail, N., Wagers, M., and Gross, C. G. (2001) Adult-generated hippocampal and neocortical neurons in macaques have a transient existence. *Proc. Natl. Acad. Sci. USA* **98,** 10,910–10,917.
84. Kornack, D. R. and Rakic, P. (2001) Cell proliferation without neurogenesis in adult primate neocortex. *Science* **294,** 2127–2130.
85. Stanfield, B. B. and Trice, J. E. (1988) Evidence that granule cells generated in the dentate gyrus of adult rats extend axonal projections. *Exp. Brain Res.* **72,** 399–406.
86. Markakis, E. A. and Gage, F. H. (1999) Adult-generated neurons in the dentate gyrus send axonal projections to field CA3 and are surrounded by synaptic vesicles. *J. Comp. Neurol.* **406,** 449–460.
87. Gould, E., Beylin, A., Tanapat, P., Reeves, A., and Shors, T. J. (1999) Learning enhances adult neurogenesis in the hippocampal formation. *Nat. Neurosci.* **2,** 260–265.
88. Shors, T. J., Miesegaes, G., Beylin, A., Zhao, M., Rydel, T., and Gould, E. (2001) Neurogenesis in the adult is involved in the formation of trace memories. *Nature* **410,** 372–376.
89. Gould, E. and Tanapat, P. (1997) Lesion-induced proliferation of neuronal progenitors in the dentate gyrus of the adult rat. *Neuroscience* **80,** 427–436.
90. Kee, N. J., Preston, E., and Wojtowicz, J. M. (2001) Enhanced neurogenesis after transient global ischemia in the dentate gyrus of the rat. *Exp. Brain Res.* **136,** 313–320.
91. Dash, P. K., Mach, S. A., and Moore, A. N. (2001) Enhanced neurogenesis in the rodent hippocampus following traumatic brain injury. *J. Neurosci. Res.* **63,** 313–319.
92. Jin, K., Minami, M., Lan, J. Q., Mao, X. O., Batteur, S., Simon, R. P., et al. (2001) Neurogenesis in dentate subgranular zone and rostral subventricular zone after focal cerebral ischemia in the rat. *Proc. Natl. Acad. Sci. USA* **98,** 4710–4715.
93. Eriksson, P. S., Perfilieva, E., Bjork-Eriksson, T., Alborn, A. M., Nordborg, C., Peterson, D. A., et al. (1998) Neurogenesis in the adult human hippocampus. *Nat. Med.* **4,** 1313–1317.

94. Kornack, D. R. and Rakic, P. (1999) Continuation of neurogenesis in the hippocampus of the adult macaque monkey. *Proc. Natl. Acad. Sci. USA* **96,** 5768–5773.

95. van Praag, H., Schinder, A. F., Christie, B. R., Toni, N., Palmer, T. D., and Gage, F. H. (2002) Functional neurogenesis in the adult hippocampus. *Nature* **415,** 1030–1034.

96. Song, H., Stevens, C. F., and Gage, F. H. (2002) Astroglia induce neurogenesis from adult neural stem cells. *Nature* **417,** 39–44.

III
STEM/PROGENITOR CELLS
IN REPRESENTATIVE THERAPEUTIC
PARADIGMS FOR THE CNS

11
Global Gene and Cell Replacement Strategies Via Stem Cells

Kook In Park, James J. Palacino, Roseanne Taylor,
Karen S. Aboody, Barbara A. Tate, Vaclav Ourednik,
Jitka Ourednik, Mahesh Lachyankar,
and Evan Y. Snyder

INTRODUCTION

Cell-based therapies such as neural transplantation have, until recently, been reserved for focal or regionally restricted neurologic diseases. These are best exemplified by Parkinson's disease, in which encouraging progress in the use of neural transplantation, especially the grafting of fetal tissue, has been made experimentally *(1,2)* and clinically *(3)*. [Even recent clinical studies that seemed to call into question such efficacy indicated that implanted fetal cells do exert a local impact albeit one that seemed to provoke an "overdose" effect *(4)*]. Donor tissue replaces dopamine via the engraftment and enhanced survival of neurotransmitter-secreting cells within the striatum or by forestalling degeneration of dopaminergic cells within the substantia nigra. However, the pathologic lesions of most neurogenetic diseases—indeed, most neurologic disorders—are usually widely disseminated in the brain and spinal cord and have not typically been regarded as within the purview of neural transplantation. Such diseases include not only the inherited neurodegenerative diseases of the pediatric age group (e.g., the lysosomal storage diseases, the leukodystrophies, inborn errors of metabolism, hypoxic–ischemic encephalopathy) but also such adult maladies as Alzheimer's disease (AD), Huntington's disease (HD), multi-infarct dementia, multiple sclerosis (MS), amyotrophic lateral sclerosis (ALS), and brain tumors (especially glioblastomas). Therapeutic approaches for such "global" problems have typically depended on pharmacologic or genetic interventions; they have been regarded as beyond the purview of cellular-mediated approaches. Cell replacement therapies have largely been

From: *Neural Stem Cells for Brain and Spinal Cord Repair*
Edited by: T. Zigova, E. Y. Snyder, and P. R. Sanberg © Humana Press Inc., Totowa, NJ

limited to transplantation of somatic cells derived from the hematopoietic system administered via bone marrow transplantation (BMT). In the majority of these disorders, such strategies have been unsatisfactory for treating the central nervous system (CNS) component of the disease.

The recognition that neural progenitor or stem cells, or cells which model their behavior, might integrate appropriately *throughout* the mammalian CNS following transplantation (e.g., refs. *5–9*) has unveiled a new role for neural transplantation and a possible strategy for addressing the CNS manifestations of diseases that heretofore had been refractory to intervention. Multipotent neural progenitor or stem cells are operationally defined by their ability to self-renew, to differentiate into cells of all glial and neuronal lineages throughout the neuraxis, and to populate developing or degenerating CNS regions (reviewed in refs. *10–14*). Thus, their use as graft material can be considered analogous to hematopoietic stem-cell-mediated reconstitution and gene transfer.

LIMITATIONS OF VARIOUS GENE TRANSFER AND CELLULAR REPAIR STRATEGIES FOR THE CNS

A number of inherited metabolic diseases, in which a single gene product is deficient, can be partially treated by BMT or enzyme replacement. In many cases, such interventions have been successful in addressing peripheral manifestations but have been disappointing in reversing or forestalling damage to CNS because the blood-brain barrier (BBB) restricts entry of therapeutic molecules from the vascular compartment. BMT also usually involves conditioning regimens such as irradiation, which are deleterious to the developing CNS. Pharmacologic agents for CNS disease administered systemically often have erratic effects, transient efficacy, and undesirable side effects.

The delivery of gene products directly to the CNS might circumvent these problems. Gene transfer may be achieved by the direct delivery of genetic material to the host's own neural tissue. The vectors presently available, however, are sometimes difficult to target *in situ* to the specific neural cell types and regions most in need of correction. For example, retroviral vectors infect only mitotic cells, which are less abundant in the postnatal CNS and often not the cells needing therapy. Although encouraging progress is being made in the use of viral vectors (especially lenti- and adeno-associated virus-based vectors for postmitotic neural tissue), they do not target the widespread lesions characteristic of most neurogenetic diseases. Typically, only CNS cells in a relatively spatially restricted area are corrected. Also, the safety and efficacy in vivo of many of these vectors remains to be established.

Alternatively, gene products may be imported into the host CNS by the implantation of synthetic "pumps"—or genetically modified donor cells that can reside within the CNS to deliver exogenous factors to host brain cells. Donor cells may be chosen for their ability to provide a source of exogenous substances that can diffuse to appropriate targets. Genetically engineered non-neural cells (e.g., fibroblasts) can be used for localized delivery of discrete molecules to the CNS and can be implanted autologously *(15–17)*. However, this approach is limited to correction of disease only in the vicinity of the graft and they lack the ability to incorporate widely into host cytoarchitecture in a functional manner following implantation. Thus, essential circuits may not be reformed and the regulated release of important factors through feedback loops may be missing.

For some substances and impairments, this shortcoming may not be crucial. The replacement of acetylcholine (ACh) to a specific region of denervated neocortex, even in an unregulated fashion, appears to be sufficient for the restoration of the complex cognitive function of spatial navigation in nucleus basalis-lesioned adult rats *(18–20)*. For other substances, such as some neurotrophins, the unregulated, inappropriate, excessive, or ectopic release may be harmful to the host. This has been shown for nerve growth factor (NGF), which improved the performance of aged rats but caused impairment in cognitive tasks in non-aged rats *(15)*. And then there are the surprises. For example, it had been assumed that the local replacement of dopamine to a unilaterally 6-hydroxy-dopamine-lesioned rat striatum would correct asymmetric apomorphine-induced rotary behavior regardless of how it is supplied and in what amounts; dopamine could successfully be replaced by fetal mesencephalon, engineered fibroblasts, or even a synthetic pump *(1–3,21)*. However, when fetal mesencephalic tissue was implanted into the caudate of some actual patients with Parkinson's disease, some who initially responded best to the neurotransmitter replacement actually seemed to go on to develop symptoms of dopamine *excess*, suggesting that cells that better regulate their dopamine expression in vivo may be better suited *(4)*. The choice of vehicle for a particular disease process may, therefore, need to be determined on an individual basis by a better understanding of the pathobiology and replacement needs underlying a particular defect. Of note, downregulation of neural gene expression in engineered *non*-neural cells may leave them "incapacitated." However, donor tissue originating from the brain may sustain expression of neural genes longer. Low levels of normal neural cell products expressed intrinsically by donor neural-derived cells may enhance the therapeutic effects of such engineered cells.

Many neurogenetic diseases are characterized by the degeneration of specific neural cell types or circuits. These losses may be the result of the presence of certain toxins in the milieu, an insufficiency of various trophins in the microenvironment, or to pathologic processes intrinsic to the metabolic deficiency of the diseased cell. Therefore, ideal grafts would not only provide exogenous therapeutic gene products but would also effect repair of damaged host brain by becoming integral components of the host cytoarchitecture and circuitry. Indeed, one of the major deficiencies for most extant gene therapy techniques for neurodegenerative diseases is that they involve inserting *new* genetic information on *old* neural substrates that may have already become dysfunctional or degenerated. The challenge is to create new substrates on which these therapeutic genes can operate. Transplant tissue derived from the CNS may also provide as-yet-unrecognized endogenous neural-specific substances that are beneficial to the host. It has already been recognized that neural stem cells (NSCs) constitutively produce a broad range of peptide neurotrophic factors (e.g., glial-derived neurotrophic factor [GDNF], brain-derived neurotrophic factor [BDNF], NGF, NT-3, NT-4).

Mature or even young neurons derived from the CNS would seem to be the ideal graft material. However, there are restrictions on the types and ages of neurons that successfully survive implantation in a functionally meaningful way for prolonged periods. Primary neurons also have limited usefulness as vehicles for stable gene transfer because their limited mitotic capacity restricts ex vivo transduction by retroviral vectors. Primary fetal neuronal tissue has historically been the most successful donor tissue for CNS grafting and has shown promise for the amelioration of certain neurologic conditions (for review, *see* ref. *22*). However the use of fetal tissue involves significant concerns, including the ready availability of sufficient amounts of suitable disease-free material (as many as 5–15 fetuses have been required for a given patient); ensuring survival of desired cells in tissue, which is typically heterogeneous and contains non-neural cells; augmenting the expression of biological molecules by donor fetal tissue; and limited or very focal integration of the fetal graft into the host brain.

PROPERTIES OF NEURAL PROGENITOR AND STEM CELLS USEFUL FOR THERAPEUTICS

The recognition that neural progenitor and stem cells propagated in culture could be reimplanted into mammalian brain, where they could reintegrate appropriately and stably express foreign genes *(23,24)* made this strategy an attractive alternative for CNS gene therapy and repair (Table 1).

Table 1
Properties of Neural Stem Cells That Make Them Appealing Vehicles for CNS Gene Therapy and Repair

- **Genetic manipulability**
  Progenitor/stem cells easily transduced ex vivo by most viral and nonviral gene transfer methods
- **Facile engraftability following simple implantation procedures**
  From engraftment in germinal zones (as well as into parenchyma) can circumvent BBB unimpeded; no requirement for conditioning regimes (e.g., irradiation as in BMT or opening of BBB)
- **Sustained foreign (therapeutic) gene expression**
  Throughout CNS, from fetus to adult, following technically simple and safe reimplantation procedures; CNS levels rise immediately
- **Potential for normal reintegration into host cytoarchitecture and circuitry**
  Differentiate along *all* CNS cell-type lineages; important for diseases in which neurons and glia both affected (e.g., asphyxia, trauma, MS); not only allows direct, stable, and perhaps, regulated delivery of therapeutic molecules, but also enables replacement of range of dysfunctional neural cells and possible reconstruction of connections and networks.
- **Ability to migrate**
  Particularly within germinal zones, enabling replacement of genes and cells to be directed not only to discrete sites but to widely disseminated lesions as well for diseases of a more global nature; for more focal implants, ability of cells to intermingle with host cells rather than clump at injection track ensures homogeneous distribution of therapeutic molecules throughout target tissue
- **Plasticity**
  Ability to accommodate to region of engraftment and assume array of phenotypes; obviates necessity for obtaining donor cells from many specific CNS regions, or imperative for precise targeting of donor cells during reimplantation, or need for tissue-specific promoters for foreign gene expression
- **Compensatory of transgene nonexpression**
  Low levels of normal neural products expressed intrinsically by progenitor/stem cells (lysosomal enzymes; neurotrophic, matrix, adhesion, and homeodomain molecules; myelin) helps safeguard against transgene inactivation; neural cells may sustain expression of foreign neural genes longer than non-neural vehicles; ability to integrate multiple copies of a transgene into its genome (e.g., following repeated sequential retroviral infection) helps thwart loss of expression; may also provide as-yet-unrecognized beneficial neural-specific substances
- **One stem cell may carry multiple transgenes**
  Following multiple transfection events, one cell can transfer multiple gene products simultaneously

(continued)

Table 1 *(continued)*

- **Minimization of side effects**
  Distribution of gene products restricted to CNS; although proteins may be disseminated by stem cells throughout the brain for diseases of global nature, altering mode of administration can permit cells to selectively integrate in proximity to cells that require given factor without affecting cells for which the molecule might be problematic; conditioning regimes not required prior to transplantation as in BMT
- **Ability to serve as producer cells for the in vivo dissemination of viral vectors**
  May help amplify distribution of virus-mediated genes to large CNS regions and numbers of cells
- **Tropism for and trophism within regions of CNS degeneration**
  When confronted with neurodegenerative environments, stem cells alter their migration and differentiation patterns toward replacement of dying cells; probably a vestigial developmental strategy with therapeutic value
- **Immunotolerance**
  In rodent transplant studies, multiple recipients and mouse strains can integrate the same murine stem cell clone without immunosuppression, suggesting a need for generating very few effective clones (one clone used by many)

These and numerous subsequent studies over the past decade (reviewed in refs. *14* and *25*) represented a unique example of integration by exogenous mammalian CNS tissue that was neither tumorigenic nor of tumor or primary fetal origin. Furthermore, neural progenitors from many regions could be maintained, perpetuated, and passaged in vitro by a number of methods. Examples include the transduction of genes interacting with cell cycle proteins (e.g., v-*myc* and large T-antigen) and by cytokine stimulation (e.g., with epidermal growth factor [EGF] and/or basic fibroblast growth factor [bFGF]). Some of these methods may operate through common cellular mechanisms. This speculation is supported by the interesting observation that many progenitor cell lines behave similarly in their ability to reintegrate into the CNS despite the fact that they were generated by different methods, obtained from various locations, and reimplanted into various CNS regions.

Some of these stem cell lines appear sufficiently plastic to participate in normal CNS development from germinal zones of multiple regions along the rodent neuraxis and at multiple stages of development from embryo to adult *(5,6,26,27)*. This appears to model the in vitro and in vivo behavior of some primary fetal and adult neural cells *(28–31)*, suggesting that insights gleaned from these cells may legitimately reflect the potential of CNS progenitor or stem cells.

Some of the inherent biologic properties of NSCs may circumvent some of the limitations of other techniques for treating metabolic, degenerative, or other widespread lesions in the brain. They are easy to administer (often directly into the cerebral ventricles), are readily engraftable, and circumvent the blood-brain barrier. A preconditioning regime is not required prior to administration (e.g., total-body irradiation), as is required for BMT. One important property of NSCs is their apparent ability to develop into integral cytoarchitectural components of many regions throughout the host brain as neurons, astrocytes, oligodendrocytes, and even incompletely differentiated but quiescent progenitors. Therefore, they may be able to replace a range of missing or dysfunctional neural cell types. A given NSC clone can give rise to multiple cell types within the same region. This is important in the likely situation where return of function may require the reconstitution of the whole milieu of a given region (e.g., not just the neurons but also the glia and support cells required to nurture, detoxify, and/or myelinate the neurons). They appear to respond in vivo to neurogenic signals not only when they occur appropriately during development but even when induced at later stages by certain neurodegenerative processes [e.g., during apoptosis *(32,33)*]. Progenitor or stem cells may be attracted to regions of neurodegeneration *(34,35)*.

Stem cells also appear to accommodate to the region of engraftment, perhaps obviating the necessity for obtaining donor cells from many specific CNS regions or the imperative for precise targeting during reimplantation. The cells can express foreign genes in vivo, which may be delivered and stably expressed *throughout* the host CNS *(5,8,35,36)*. The cells might express certain genes of interest intrinsically (e.g., many neurotrophic factors) or they can be engineered ex vivo to do so because they are readily transduced by gene transfer vectors. These gene products can be delivered to the CNS in a direct, immediate, and sustained manner. Finally, NSCs can migrate and integrate widely throughout the brain when implanted into germinal zones, which may allow them to reconstitute enzyme or cellular deficiencies in a global manner *(8)*. Appealingly, despite their extensive plasticity, NSCs never give rise to cell types inappropriate to the brain (e.g., muscle, bone, teeth) or yield neoplasms.

These attributes of NSCs may provide multiple strategies for treating CNS dysfunction. In early studies employing neural progenitor cell engraftment, the feasibility of transporting foreign genes into the CNS was demonstrated by transplanting the cells into discrete neural regions (cerebellum, cortex, hippocampus, striatum, septum) *(23,32,37,38)*. For example, tyrosine hydroxylase expressed from these cells within the striatum of parkinsonian rats and monkeys could improve motor performance *(39)* and Bcl-2-expression could spare some degenerating striatal neurons *(40)*. NGF expressed near fimbria–fornix lesions in rats appeared to salvage severed septal–hippocampal cholinergic fibers and aid performance on memory tasks *(41,42)*. However, it could be argued that, for focal lesions, grafts of neural progenitors or stem cells offer no advantage over grafts of primary non-neural cells [such as fibroblasts *(17,18)*], which have the added appeal of being easily biopsied from the host into whose brain they will be reimplanted. Indeed, even synthetic pumps could be efficacious *(19,20)*. In contrast, in diseases with *widespread* neuropathology, grafts of neither non-neural cells nor even primary fetal neural tissue would appear to be comparably effective. It is likely that only neural progenitors or stem cells can address this global transplantation challenge.

Many human patients with inherited metabolic neurodegenerative storage diseases are mentally retarded. Mice that model these conditions through the targeted or naturally occurring mutation of identified genes are characterized by widely disseminated lesions that cannot be fully corrected by local CNS grafts. Therefore, to establish *proof-of-principle*, the appealing characteristics of neural progenitors and stem cells were actually first exploited experimentally in mouse models of genetically based neurodegeneration requiring disseminated enzyme and/or neural cell replacement.

PROOF OF CONCEPT: TESTING THE THERAPEUTIC POTENTIAL OF NSC BY ATTEMPTING TREATMENT OF MODEL NEUROGENETIC DEGENERATIVE DISEASES

The first model in which the therapeutic potential of NSCs was actually put to a test was in a neurogenetic lysosomal storage disease, mucopolysaccharidosis type VII (MPS VII) *(5)*. Mice homozygous for a frame-shift mutation in the β-*glucuronidase* gene are devoid of the secreted enzyme β-glucuronidase (GUSB). The enzymatic deficiency results in lysosomal accumulation of undegraded glycosaminoglycans in the brain and other tissues (spleen, liver, kidney, cornea, skeleton), causing a fatal progressive degenerative disorder. The disease in mice is essentially the same as in human patients, who have variable degrees of mental retardation. Disease lesions are present in the mouse brain at birth and reach their full extent by 3 wk of age.

Treatments for MPS VII and most other lysosomal storage disorders are designed to provide a source of normal enzyme for uptake by diseased cells, a process termed "cross-correction" *(43)*. The mannose 6-phosphate receptor, which sorts newly synthesized enzyme to the lysosomal compartment within the cell, is also present on the cell surface and mediates endocytosis of GUSB. Thus, the goal of gene therapy is to engineer donor cells to express the normal GUSB protein for export to other host cells. Stable gene transfer to MPS VII hematopoietic stem cells showed that long-term expression in vivo of even low levels of normal GUSB could produce clinically significant improvements in liver and spleen pathology. However, as for many neurogenetic metabolic diseases, these approaches have generally not been successful in reversing or forestalling damage to the CNS, presumably because of restrictions imposed by the BBB to the sustained entrance of corrective levels of GUSB. Treatment of newborn mice by BMT, prior to maturation of the BBB, did not result in significant transfer of GUSB or donor-derived cells into the CNS and the requisite pretreatment with ablative irradiation damaged the developing CNS. The temporary treatment of injecting very large amounts of purified enzyme intravenously at birth did result in some GUSB reaching the brain and reduced storage, which suggested that progression of the disease could be slowed if normal enzyme could be supplied to the CNS in a sustained, diffuse manner *(44)*. With this information in hand, GUSB-expressing NSCs would seem to be ideally suited for such pathology. Indeed, the engraftment of GUSB-expressing NSCs succeeded in providing a sustained and widespread source of cross-correcting lysosomal enzyme following their integration and differentiation *throughout* brain cytoarchitecture *(5)*.

A rapid intraventricular injection technique was devised for the diffuse engraftment of the NSCs. Injecting the progenitors into the cerebral ventricles presumably allowed them to gain access to most of the subventricular germinal zone (SVZ) as well as to networks of cerebral vasculature, along the surface of which they might also migrate. Recipients survived this rapid, simple procedure without cerebrospinal fluid obstruction or other morbidity. This approach worked equally well in the fetus, where donor stem cells gained access to the ventricular germinal zone (VZ) *(36)*. Donor cells migrated into the parenchyma from the ventricles within 24 h of transplantation and took up residence on the brain side of the BBB as integral members of the CNS cytoarchitecture. This engraftment technique exploited many of the inherent properties of NSCs to integrate into migratory germinal zones and become components of normal structures throughout the host brain, permitting missing gene products to be delivered without disturbing other neurobiological processes (Fig. 1I).

The mouse NSC line used intrinsically expressed high levels of GUSB and was additionally engineered ex vivo by infection with a retroviral vector encoding GUSB to increase the amount of cross-corrective enzyme available

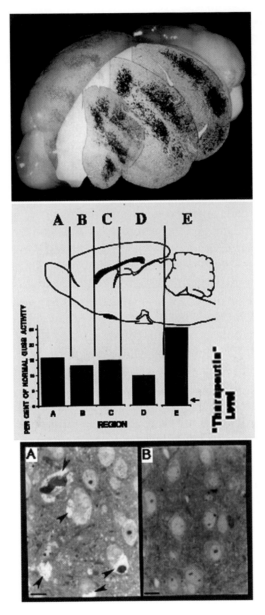

Fig. 1. Widespread engraftment of neural progenitors expressing β-glucuronidase (GUSB) throughout the brain of the mucopolysaccaridosis type VII (MPS VII), GUSB-deficient mouse. (**I**) Brain of a mature MPS mouse after receiving a neonatal intraventricular transplant of C17-2 neural progenitor cells expressing GUSB. (**A**) GUSB staining (red reaction product) of C17-2 cells in culture before transplantation (expressing approx 800 units of GUSB activity/mg protein in vitro). (**B–E**)

to diseased cells from these NSC-derived "enzyme pumps". By injecting the NSCs into the cerebral ventricles of newborn MPS VII mice, engraftment occurred throughout the neuraxis of 94% of the transplanted mutants, resulting in widespread distribution of GUSB in the mutant brain (Fig. 1II). At maturity, donor-derived cells were present as normal constituents of diverse brain regions (Fig. 1I). Intense GUSB activity colocalized with integrated donor cells. All of the engrafted recipients had evidence of corrective levels of GUSB activity throughout the brain with nearly heterozygote levels in some regions. Nearly half of the engrafted brain regions expressed more GUSB than is necessary to reverse pre-existing storage in other tissues. This diffuse GUSB expression along the neuraxis (Fig. 1II) resulted

---

Fig. 1. *(continued)* Identification in vivo of donor-derived C17-2 cells by X-gal histochemical reaction (blue precipitate) for expression of the *lacZ* marker gene. The blue cells can be seen to have engrafted throughout the recipient mutant brain. Representative regions are shown, proceeding clockwise from rostral to caudal: **(B)** olfactory bulbs; **(C)** telencephalon at the level of the striatum (including striatum, corpus callosum, and fronto-parietal cortex); **(D)** telencephalon at the level of the caudal aspect of the hippocampus (including parieto-occipital cortex and hippocampus); and **(E)** posterior telencephalon (including occipital cortex) and midbrain. **(F)** Expression of GUSB (red cells) in a sister section to (E), showing that all GUSB-positive (red) cells colocalized precisely with the distribution of the donor X-gal-positive (blue) cells in (E). Untreated MPS VII mice show no GUSB activity histochemically. Scale bars = 400 μm **(B–E)** and 320 μm **(F)**. **(II)** Distribution of GUSB enzymatic activity throughout brains of MPS VII recipients of C17-2 transplants. Serial sections were collected from throughout the brains of MPS VII transplant recipients and quantitatively assayed for GUSB activity. Serial coronal slices were pooled to reflect the activity present within the regions demarcated in the schematic. The regions were defined by anatomical landmarks in the anterior-to-posterior plane (as per ref. 5) to permit comparison among animals. The mean levels of GUSB enzyme activity for each region ($n = 17$) are presented as the percentage of average normal levels for each region. Untreated MPS VII mice show not GUSB activity biochemically or histochemically. Enzyme activity = 2% of normal is corrective based on data from liver and spleen. **(III)** Decreased lysosomal storage in a treated MPS VII mouse brain at 8 mo of age. **(A)** Extensive vacuolation representing distended lysosomes (arrowheads) in both neurons and glia in the cortex of an 8-mo-old untransplanted control MPS VII mouse. **(B, C)** Decrease in lysosomal storage in the neocortex of an MPS VII mouse treated at birth from a region analogous to the untreated control section (A). The other regions of this animal's brain showed a similar decreased in lysosomal storage compared to untreated age-matched mutants in regions where GUSB was expressed. Scale bars = 21 μm (A, C) and 31 μm (B). (Adapted with permission from ref. *145*.)

in widespread cross-correction of lysosomal storage in both mutant host neurons and glia in all engrafted areas examined, including cerebral cortex, compared to age-matched untreated MPS VII controls (Fig. 1III). Engraftment, expression, and neuropathologic rescue could be detected at least 8 mo posttransplant (the average life-span of an MPS mouse is less than 6 mo), suggesting that this correction was permanent. No tumors or disruption of the CNS cytoarchitecture were ever seen. Subjective observation of cage behavior suggested that the MPS VII recipients were more active and alert than age-matched untreated controls.

The neonatal transplantation of GUSB-producing NSCs produced long-term improvements in the MPS VII mouse brain, as neonatal BMT and peripheral enzyme replacement did for the skeletal and visceral disease. These observations suggested a strategy for CNS gene therapy of a class of neurogenetic diseases that, heretofore, had not been adequately treated. Although MPS VII may be regarded as "uncommon," the broad category of diseases that it models (neurogenetic conditions) afflicts as many as 1 in 1500 children and serves as a model for many adult neurodegenerative processes of genetic origin. (AD, e.g., could broadly fall into this category.) Therapy instituted early in life might arrest CNS disease progression in many inherited metabolic diseases. This may be important for treating conditions causing mental retardation because it is unclear whether permanent alterations in brain function occur early that may not be reversible even if normal metabolism is later restored.

More recent findings suggest that similar treatment of adult recipients might be possible. Even in the adult there are established routes of relatively extensive migration followed by both endogenous and transplanted NSCs. If injected into the ventricles of normal adult mice, NSCs will integrate into the SVZ and migrate long distances [e.g., to the olfactory bulb, where they differentiate into interneurons and occasionally into subcortical parenchyma where they become glia *(26,27,45–47)*. Because reporter genes are successfully expressed by NSCs in these distant locations, they might also be capable of expressing therapeutic enzymes that might cross-correct host cells even in adult brains. Although these migratory paths are present in the adult, the argument could be made that they are still relatively restricted and stereotyped compared to that seen in the fetal or newborn brain. However, as discussed below, in the *degenerating*, *abnormal*, or *injured* adult brain, migration by foreign-gene-expressing NSCs can be extensive and directed specifically to those regions of pathology, a phenomenon observed to date with stroke, head injury, brain tumors, and amyloid plaques.

The therapeutic paradigm described above can be extended to other untreatable neurodegenerative diseases characterized by an absence of gene products and/or the accumulation of toxic metabolites. The recent characterization of a number of mouse models of other neurogenetic diseases, either naturally occurring or generated by targeted mutagenesis, permits this hypothesis to be tested by employing NSC-mediated gene transfer for the replacement of other gene products whose absence produces disease extensively involving the CNS.

For example, engraftable NSCs engineered to overexpress β-hexosaminidase (deficient in Tay–Sachs [TSD]), when implanted into the ventricles of fetal or neonatal mice, could boost enzyme levels by 14–76% above wild type and, in pilot studies in TSD mice *(48)*, bring levels in some affected mice to at least 10% of normal, amounts theoretically sufficient to restore normal ganglioside turnover.

In almost all cases, NSCs, because they are normal cells, constitutively express normal amounts of the particular enzyme in question. The extent to which this amount needs to be augmented may vary from model to model and enzyme to enzyme. Reassuringly, in most inherited metabolic disease, the amount of enzyme required to restore normal metabolism and forestall CNS disease may be quite small. It is significant to note that although the histograms in Fig. 1II illustrate the widespread distribution of a lysosomal enzyme, they could similarly reflect the NSC-mediated distribution of other diffusible (e.g., synthetic enzymes, neurotrophins, viral vectors) and nondiffusible (e.g., myelin, extracellular matrix) factors, as well as the distribution of "replacement" neural cells (see following section). For example, neural progenitors and stem cells have been used for the local expression of NT-3 within the rat spinal cord *(49)*, NGF and BDNF within the septum *(41,42)*, and tyrosine hydroxylase *(39,50)*, and Bcl-2 *(40)* to the striatum. Therefore, studies in genetic diseases helped establish the paradigm of using genetically modified NSCs for the transfer of a range of factors of therapeutic, reconstructive, or developmental interest throughout the CNS, both in mature brains and especially, in developing brains (where the cells actually contribute to organogenesis of multiple CNS structures). Their use as graft material in the brain may be considered analogous to hematopoietic stem-cell-mediated reconstitution and gene transfer to the periphery. They have provided a new role for neural transplantation and a possible strategy for addressing the CNS manifestations of a range of diseases (i.e., those with extensive, disseminated, or global manifestations) that heretofore had been refractory to intervention. These earlier studies helped not only set the stage for the experiments summarized in other chapters of this book, but have also helped to advance the idea that NSCs—as a prototype for stem cells from any solid organ—might aid in reconstructing both the molecules as well as the cells of a maldeveloped or damaged organ.

Replacing More Than Genes: Cells, Myelin, Extracellular Matrix

It was hypothesized, as noted above, that the techniques described for the widespread engraftment of NSCs for gene replacement could be extended to the remediation of other types of diffuse neuropathologies requiring other types of intervention.

Many neurologic diseases, even those that can be ameliorated by replacement of a gene product, are characterized by the degeneration of cells or circuits. These cytoarchitectural components may need to be replaced in a functional manner and be resistant to residual toxic processes. One reassuring insight from the classic fetal transplant literature is that even modest anatomical reconstruction may sometimes have an unexpectedly beneficial functional effect (22). Early experiments with NSC clones in various rodent mutants and injury models have provided evidence that NSCs may be able to replace some degenerated or dysfunctional neural cells. In the *meander tail* (*mea*) mutant mouse, which is characterized by a deficiency of cerebellar granule cell (GC) neurons, NSCs implanted at birth were capable of "repopulating" large portions of the GC-poor internal granular layer (IGL) with neurons (37) [a phenomenon that was subsequently duplicated with human NSCs (6)]. A pivotal observation to emerge from this work, with implications for fundamental stem cell biology, was that cells with the potential for multiple fates will "shift" their differentiation to compensate for a deficiency in a particular cell type. As compared with their fate in normal cerebella, a majority of these donor NSCs in GC-deficient regions pursued a GC phenotype in preference to other potential phenotypes, suggesting a "push" on undifferentiated, multipotent NSCs toward repletion of the "unmet" quota. This work presented a possible developmental mechanism with therapeutic value.

Work in another mutant, the *reeler* (*rl*) mouse, has suggested that NSCs may not only replace developmentally impaired cells but may also help correct certain aspects of abnormal cytoarchitecture, particularly those characterized by deficiencies in extracellular matrix (ECM). The laminar assignment of neurons in *rl* brain is abnormal as a result of a mutation in a gene encoding the secreted ECM molecule, Reelin. NSCs, implanted at birth into the defective developing *rl* CB, appeared not only to replace missing GCs but also to promote a more wild-type-laminated appearance in engrafted regions by "rescuing" aspects of the abnormal migration, positioning, and survival of host neurons most likely by providing molecules (including Reelin) at the cell surface to guide proper histogenesis *(51)*. These findings suggested a possible NSC-based strategy for the treatment of CNS diseases characterized by abnormal cellular migration, lamination, and cytoarchitectural arrangement.

White-Matter Disease as a Model System

As noted above, many neurologic diseases, particularly those of neuro-genetic etiology, are characterized by even more extensive—indeed, "global"—degeneration or dysfunction. Mutants characterized by CNS-wide white-matter disease provided ideal models for testing our hypothesis that NSCs might be useful in neuropathologies requiring widespread neural cell replacement *(8)*. The oligodendroglia of the dysmyelinated *shiverer* (*shi*) mouse are dysfunctional because they lack myelin basic protein (MBP), which is essential for proper myelination. Therapeutic intervention, therefore, requires widespread replacement with oligodendrocytes expressing MBP. NSCs transplanted at birth—employing the same intra-cerebroventricular implantation technique devised for diffuse engraftment of enzyme-expressing NSCs to treat global metabolic lesions—resulted in widespread engraftment throughout the *shi* brain (including within white tracts) (Fig. 2I) with repletion of significant amounts of MBP (Fig. 2II). Accordingly, of the many donor-derived oligodendroglia, a subgroup myelinated up to 52% of host neuronal processes with better compacted myelin. Forty percent of host axons became surrounded by myelin that contained major dense lines (a sign of good compaction) and that had a thickness and periodicity approximating those of normal myelin (Fig. 2III). Some animals experienced symptomatic improvement—a decrease in their symptomatic tremor (Fig. 2IV) *(8)*. Therefore, "global" cell replacement seems feasible for some pathologies if cells with stemlike features are employed. More specifically, the ability of NSCs to generate myelinating cells is important because disordered myelination plays a critical role in many genetic and acquired (injury, infectious) neurodegenerative processes; oligodendroglial pathology is prominent in stroke, spinal cord injury, head trauma, and ischemia and may account for a significant proportion of the neurologic handicap seen in asphyxiated and premature newborns. More broadly, complementation studies in mutants such as those described above, help support an NSC-based approach—whether with exogenous cells or with mobilized endogenous ones—for compensating for neurodevelopmental problems of many etiologies.

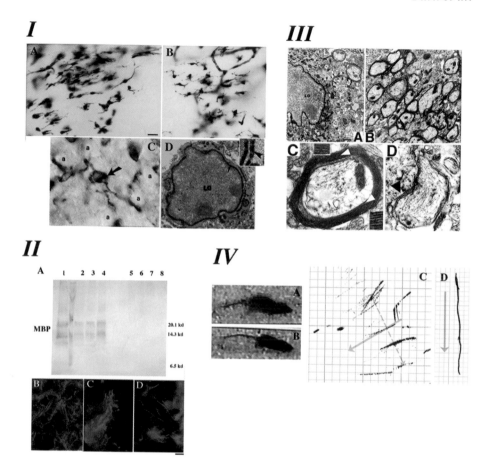

Fig. 2. "Global" cell replacement is feasible via neural stem cell transplantation: evidence from the dysmyelinated *shiverer* (*shi*) mouse brain. (Modified from ref. *8.*) **(I)** NSCs engraft extensively throughout the *shi* dysmyelinated brain, including within white tracts, and differentiate into oligodendrocytes. *lacZ*-expressing, β-galactosidase (β-gal)-producing NSCs were transplanted into newborn *shi* mutants and analyzed systematically at intervals between 2 and 8 wk following engraftment. Coronal sections through the *shi* brain at adulthood demonstrated widely disseminated integration of blue X-gal+ donor-derived cells throughout the neuraxis, similar to the pattern seen in (I) in the MPS VII mutant mouse. Donor-derived cells in the *shi* mouse brain are shown at higher magnification and greater detail in (A–D) (A, B) Donor-derived X-gal+ cells in representative sections through the corpus callosum possessed characteristic oligodendroglial features (small, round, or polygonal cell bodies with multiple fine processes oriented in the direction of the neural fiber tracts). **(C)** Close-up of a representative donor-derived anti-β-gal immunoreactive oligodendrocyte (arrow) extending multiple processes toward and

Fig. 2. *(continued)* beginning to enwrap large adjacent axonal bundles (a) viewed on end in a section through the corpus callosum. That cells such as those in (A–C) [and in II(B–D)] were oligodendroglia was confirmed by the representative electron micrograph in (**D**) (and in III) demonstrating their defining ultrastructural features (see ref. *8*). A donor-derived X-gal-labeled oligodendrocyte (LO) can be distinguished by the electron-dense X-gal precipitate that is typically localized to the nuclear membrane (arrow), ER (arrowhead) and other cytoplasmic organelles. The area indicated by the arrowhead is magnified in the inset to demonstrate the unique crystalline nature of individual precipitate particles. (**II**) Myelin basic protein (MBP) expression in mature transplanted and control brains. (**A**) Western analysis for MBP in whole-brain lysates. The brains of three representative transplanted *shi* mutants (lanes 2–4) expressed MBP at levels close to that of an age-matched unaffected mouse (lane 1, positive control) and significantly greater than the amounts seen in untransplanted (lanes 7 and 8, negative control) or unengrafted (lanes 5 and 6, negative control) age matched *shi* mutants. (Identical total protein amounts were loaded in each lane.) (**B–D**) Immunocytochemical analysis for MBP. (**B**) The brain of a mature unaffected mouse was immunoreactive to an antibody to MBP (revealed with a Texas Red-conjugated secondary antibody). (**C, D**) Age-matched engrafted brains from *shi* mice similarly showed immunoreactivity. Untransplanted *shi* brains lack MBP. Therefore, MBP immunoreactivity has also classically been a marker for normal donor-derived oligodendrocytes (C, D) in transplant paradigms. (**III**) NSC-derived "replacement" oligodendrocytes are capable of myelination of *shi* axons. In regions of MBP-expressing NSC engraftment, *shi* neuronal processes became enwrapped by thick, better compacted myelin. (**A**) At 2 wk posttransplant, a representative donor-derived, labeled oligodendrocyte (LO) (recognized by extensive X-gal precipitate (p) in the nuclear membrane, cytoplasmic organelles, and processes) was extending processes (a representative one is delineated by arrowheads) to host neurites and was beginning to ensheathe them with myelin (m). (**B**) If engrafted *shi* regions, such as that in (**A**), were followed over time (e.g., to 4 wk of age as pictured here), the myelin began to appear healthier, thicker, and better compacted (examples indicated by arrows) than that in age-matched untransplanted control mutants. (**C**) By 6 wk posttransplant, these matured into even thicker wraps; approx 40% of host axons were ensheathed by myelin [a higher power view of a representative axon is illustrated in (C)] that was dramatically thicker and better compacted than that of *shi* myelin (an example of which is shown in (D) (black arrowhead) from an unengrafted region of an otherwise successfully engrafted *shi* brain). In (C), white arrowheads indicate representative regions of myelin that are magnified in the adjacent insets; MDLs are evident. (**IV**) Functional and behavioral assessment of transplanted *shi* mutants and controls. The *shi* mutation is characterized by the onset of tremor and a "shivering gait" by the second to third postnatal week. The degree of motor dysfunction in animals was gauged in two ways: (1) by blindly scoring periods of standardized videotaped cage behavior of experimental and control animals and (2) by measuring the amplitude of tail displacement from the body's rostral–caudal axis (an objective,

Accordingly, the approach employed in the shi mouse has been extended to other demyelinated and oligodendrocyte-deficient mutants. Loss of galactocerebrosidase-α (GALC) enzyme activity results in the accumulation in oligodendrocytes of psychosine, a toxic glycolipid. The resultant disease in humans is *Krabbe's* or *globoid cell leukodystrophy* (GLD). Children afflicted with GLD exhibit inexorable psychomotor deterioration and early death, presumably as a result of dysfunctional and ultimately degenerated oligodendrocytes (that become globular in appearance) with loss of myelin sheaths. The *twitcher* (*twi*) mouse is an authentic model of Krabbe's leukodystrophy. In preliminary studies, an established clone of murine NSCs has been transplanted into the brains of both neonatal *twi* mice and symptomatic *twi* mice (i.e., juvenile/young adults toward the end of their projected life-span). Brains of transplanted animals were examined at both the light and electron microscopic levels. Transplantation of these NSCs into the ventricles of newborn *twi* mice resulted in exuberant engraftment throughout the brains when examined at maturity. The NSCs differentiated extensively into healthy oligodendrocytes that elaborated thick, well-

---

Fig. 2. *(continued)* quantifiable index of tremor). Video freeze-frames of representative unengrafted and successfully engrafted *shi* mice are seen in **(A)** and **(B)**, respectively. The whole-body tremor and ataxic movement observed in the unengrafted symptomatic animal **(A)** causes the frame to blur, a contrast with the well-focused frame of the asymptomatic transplanted *shi* mouse **(B)**. Sixty percent of transplanted mutants evinced nearly normal-appearing behavior as in (B) and attained scores that were not significantly different from normal controls (*see* ref. 8 for details). **(C, D)** depict the manner in which whole-body tremor was mirrored by the amplitude of tail displacement [hatched gray arrow in (C)] measured perpendicularly from a line drawn in the direction of the animal's movement (solid gray arrow, represents the body's long axis). Measurements were made by permitting a mouse, whose tail had been dipped in India ink, to move freely in a straight line on a sheet of graph paper as shown. Large degrees of tremor cause the tail to make widely divergent ink marks away from the midline, representing the body's axis (C). Absence of tremor allows the tail to make long, straight, uninterrupted ink lines on the paper congruent with the body's axis (D). The distance between points of maximal tail displacement from the axis was measured and averaged for transplanted and untransplanted *shi* mutants and for unaffected controls (hatched gray arrow). (C) shows data from a poorly engrafted mutant that did not improve with respect to tremor and (D) reveals lack of tail displacement in a successfully engrafted asymptomatic mutant. Overall, 64% of transplanted *shi* mice examined displayed at least a 50% decrement in the degree of tremor or "shiver." Several showed zero displacement. (*See* ref. 8 for details.)

compacted, normal-appearing myelin. Each NSC-derived oligodendrocyte appeared to remyelinate up to 30–50 host axons throughout the engrafted brain. Unambiguous markers were used to confirm that the observed oligodendrocytes and myelin were, indeed, of exogenous NSC derivation. That oligodendrocytes and myelin were essentially absent from regions where donor cells were not evident further supported the donor origin of the observed oligodendrocytes and myelin. Similar phenomena were observed in the juvenile/young adult transplanted *twi* mice. In these cases, the cells were implanted unilaterally into the parenchyma of the mice. The cells migrated extensively from the site of implantation and resulted in extensive engraftment, albeit more restricted than that seen in newborns. Again, oligodendrocyte differentiation was robust. Interestingly, one could see dramatic examples of individual NSCs migrating from the implantation site and differentiating into oligodendrocytes that were each surrounded by a "halo" of myelinated host axons. (We anticipate that implants bilaterally into the parenchyma as well as into such germinal zones as the SVZ via intraventricular injection will result in even greater degrees of engraftment in the mature animals.)

Interestingly, unlike in the *shi* mouse, cell replacement, even to this extent and magnitude, was nevertheless unable to remediate symptoms or prolong the life of the *twi* mouse. This observation became instructive as to how one might need to treat some complex global neurogenetic diseases (and also how NSCs may be useful in "teasing out" an underlying pathophysiological mechanism). Although engraftment of these unengineered NSCs did not appear to result in a decrease in behavioral symptoms or an increase in life-span when compared to unengrafted controls, the fact remained that exogenous NSCs could survive and differentiate into myelinating oligodendrocytes within the *twi* (i.e., Krabbe) environment. First, this observation helps address a long-standing question regarding the pathophysiology of Krabbe's leukodystrophy; namely, is the observed pathology related to a cell autonomous defect (i.e., a sick cell that dies) or to a toxic environment (one that would theoretically kill not only host cells but any new cells placed into that environment, hence dooming any attempts at cell replacement)? The environment in the *twi* brain was *not* inherently nonpermissive for the implantation of stem cells. Indeed, cell replacement strategies *do* seem to be viable approaches.

One could hypothesize that the absence of GalC not only permits the toxic buildup of psychosine but predisposes Krabbe neural cells (especially oligodendrocytes) to be more vulnerable to that toxicity. To test that hypothesis, in vitro pilot experiments were performed in which the same clone of donor NSCs that had differentiated into oligodendrocytes in the *twi* mouse brain was exposed in culture to increasing concentrations of psychosine, particularly concentrations that would prove inimical to Krabbe cells. These NSCs actually proved to be *resistant* to those toxic effects. As the concentrations were further increased, the NSCs did begin to show some toxicity. However, interestingly, populations of those same NSCs that were engineered (via retroviral transduction) to overexpress human GalC were now no longer affected by that higher concentration and seemed to be even more resistant to even higher psychosine concentrations in vitro. These same engineered GalC-overexpressing NSCs—documented to secrete enough GalC in their culture medium to cross-correct fibroblasts from actual Krabbe patients—were then implanted into the brains of newborn *twi* mice. In these cases, in pilot experiments, some of the animals did seem to have a 2.5-fold to 3-fold increased life-span, GalC levels of 10–20% of wild-type and 25–40% of heterozygote levels, and psychosine levels diminished to at least 30% of predicted. (For example, the brains of two transplanted *twi* mice had psychosine levels of approx 150 pM/mg protein at 100 d of age—an age much longer than the normal life-span for *twi* mice, whereas untreated *twi* animals of 40 d of age usually already have levels of approx 195 pM/mg protein). These experiments with GalC-overexpressing NSC lines warrant greater study and confirmation. It also remains to be determined whether NSCs that have differentiated down various neural lineages retain their psychosine resistance and, if not, whether oligodendrocytes are more vulnerable than neurons or astrocytes. The *in vivo* engraftment studies would suggest that even mature oligodendrocytes remain viable, at least in the environment of the brain—ultimately, probably the most significant result.

Although preliminary, these results already suggest a number of important points regarding the fundamental pathophysiology of Krabbe's disease and the approaches to treating it, particularly with regard to cellular and molecular therapies. First, it does validate the "psychosine hypothesis" of Krabbe's pathogenesis. Second, it suggests that GalC not only permits a buildup of psychosine but endows cells with the capacity to resist its toxicity; that is, an absence of GalC makes them more vulnerable and an increasing amount of GalC makes them more resistant. Third, a component of treatment will be to cross-correct host cells to overexpress GalC and, hence, be more resistant. The effectiveness in treatment appears to be less related to the absolute level of GalC achieved but rather to the diminished level of psychosine accumulated. Fourth, although oligodendrocyte replacement alone is not a sufficient treatment for Krabbe's leukodystrophy (even when extensive), the replacement of both cells *and* of molecules (e.g., NSCs that can both become oligodendrocytes and be pumps for GalC [to antagonize psychosine]) remains a very promising and important basis for multidisciplinary strategies against the disease. In many respects, therefore, the lessons from the behavior of NSCs in this model may help us understand the role and value of such cells in other complex genetically based neurodegenerative processes.

Additional lessons may also be gleaned. For example, it is not entirely unexpected that the present experimental designs did not result in a prolonged life-span in animals with systemwide (i.e., intracranial *and* extracranial) disease (not, in fact, the initial intent of these experiments). Future interventions may call for implantation of NSCs at multiple anatomical locations as well as at *multiple* points in time in the evolution of the disease, beginning, if possible, presymptomatically and continuing after the disease is established. The present interventions, because they were designed to deal solely with the differentiation potential of NSCs, had not yet addressed the spinal cord or cerebellar manifestations or the peripheral nervous system (PNS) aspects of the disease. Interventions that hope to truly be lifesaving will likely need to combine NSC implantation in the brain with strategies to address these other problems (e.g., spinal and cerebellar implants; bone marrow transplantation for the PNS) that *together* may likely prolong the life-span. Most neurological disease is complex in this way and will likely require multifaceted approaches, perhaps with stem cells serving as the "glue" to some extent.

"Homing in" on Pathology from Even Long Distances

As described in Chapter__, in a model of neuron-specific apoptotic degeneration in adult neocortex, transplanted NSCs differentiated specifically into the type of degenerating neuron and partially replaced the lost neuronal population *(32)*. The neuronal phenotype assumed by these NSCs within regions of apoptosis suggested that degeneration created a microenvironment permissive or instructive for a neuronal fate choice, perhaps through reactivation of signals ordinarily available exclusively during embryonic corticogenesis. It is tempting to speculate that the ability of NSCs to pursue alternative differentiation paths in response to certain types of neurodegeneration can be combined with their ability to migrate in order to treat neurologic diseases characterized by widespread neural cell death. Evidence (some published, some preliminary) in various models suggests that this may be feasible: during phases of active neurodegeneration, as-yet-unidentified factors seem to be transiently elaborated to which

NSCs may respond by migrating to degenerating regions and differentiating toward replacement of dying neural cells.

Hypoxic–Ischemic Brain Injury

For example, in pilot studies, when NSCs are transplanted into brains of postnatal mice subjected to unilateral hypoxic–ischemic (HI) brain injury (a model for cerebral palsy), donor-derived cells migrated preferentially to and integrated extensively within the large ischemic areas that typically spanned the injured hemisphere. A subpopulation of donor stem cells, particularly in the penumbra of the infarct, "shifted" their differentiation fate toward neurons and oligodendrocytes, the neural cell types typically damaged following asphyxia/stroke and the cell types least likely to regenerate spontaneously in the postnatal CNS. Furthermore, there appeared to be an optimal window of time following asphyxia (3–7 d) during which signals were elaborated (probably transiently) within the degenerating region and to which NSCs responded with migration and reconstitution of lost neural cells.

Because engrafted donor-derived cells continue to express their *lacZ* reporter gene, it appeared feasible that desired differentiation, neurite outgrowth, and proper connectivity of both host and donor-derived cells might be enhanced if donor NSCs were genetically manipulated ex vivo to express certain trophins, cytokines, or other factors. These could include the overexpression of gene products already within their repertoire or molecules that might serve to recreate a more favorable *milieu* containing some of these immediate repair signals to operate on host cells as well as on the donor cells themselves in an autocrine/paracrine fashion. When, in pilot studies, a subclone of the same murine NSCs was engineered via retroviral transduction to overexpress neurotrophin-3 (NT-3) (known to play a role in inducing neuronal differentiation) was then implanted into asphyxiated mouse brains, the percentage of donor-derived neurons increased from 5% (in the above-described experiments) to 20% in the infarction cavity and up to >80% in the penumbra. Many of the neurons appeared to be variously and appropriately GABA-ergic, glutamatergic, or cholinergic. Donor-derived glia were rare. The engineered NSCs had been documented to produce large amounts of NT-3 in vitro and in vivo, and both the parent NSC clone and its NT-3-producing subclones were determined to express a functional trkC. Hence, it seemed likely that the NSCs were producing a factor that worked in an autocrine/paracrine fashion. Therefore, migratory NSCs may actually be capable of simultaneous gene therapy and cell replacement with the same cells, during the same transplantation procedure, in the same recipient. This approach can likely be applied to other types of genetic and acquired CNS degeneration and injury. A somewhat similar approach has already been described above in the *twi* mouse model of Krabbe's leukodystrophy.

Spinal Cord

That such a tropism and trophism by NSCs for neurodegenerative environments pertains throughout the CNS is suggested by additional examples. One such example was briefly described in Chapter 12 dealing with spinal cord. Induction of segmental α-motorneuron (MN) degeneration in spinal cord (SC) by neonatal sciatic axotomy is a classic experimental model of specific spinal neuron degeneration and SC dysfunction. Normally, there is no spontaneous regeneration of MNs because they are born only in the fetus. If NSCs are implanted—even into the *dorsal* horn—during active degeneration, a significant proportion will engraft, migrate *anteriorly* toward the MN-impoverished ventral horn and as much as 1 mm in the rostral–caudal axis within the ventral horn, and will differentiate into cells that resemble the lost α-MNs. Engrafted NSCs continue to express foreign reporter genes, suggesting that, as in the asphyxiated brain, implantation of genetically engineered NSCs expressing trophic agents, cytokines, or other factors might enhance neuronal differentiation, neurite outgrowth, and proper connectivity. In chronic CNS lesions, nonpermissive signals interfere with engraftment and integration of NSCs. The problem could partially be overcome by employing NSCs specifically engineered to create a more favorable *milieu*. If developmental signals are transiently re-expressed during acute/subacute phases of active degeneration and NSCs respond to these by differentiating into replacement cells, then it might be possible to treat *chronic* lesions by artificially expressing signals that *emulate* the acute phase, "fooling" the NSCs to respond.

Brain Tumors

A dramatic example of using NSCs as gene delivery vehicles that home in on pathology is illustrated by their use against brain tumors. One of the impediments to the treatment of primary human brain tumors (e.g., gliomas) has been the degree to which they expand, infiltrate surrounding tissue, and migrate widely into normal brain, usually rendering them "elusive" to effective resection, irradiation, chemotherapy, or gene therapy. Aboody et al. *(35)* (Fig. 3) demonstrated that migratory murine and human NSCs, when implanted into experimental intracranial gliomas in vivo in adult rodents, distributed themselves quickly and extensively throughout the tumor bed [Fig. 3I(A)] and migrated uniquely in juxtaposition to widely expanding and aggressively advancing tumor cells while continuing to stably express a

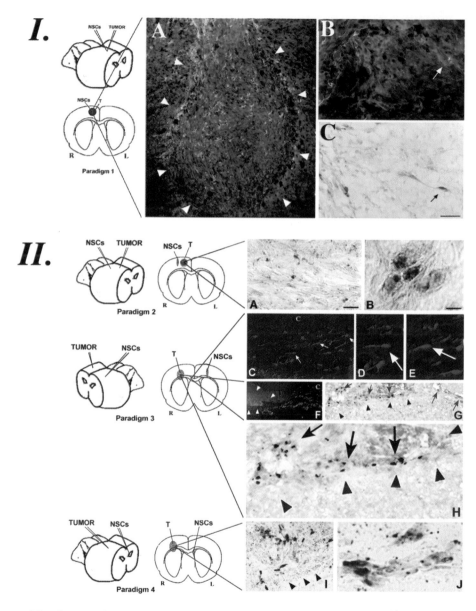

Fig. 3. Neural stem cells (NSCs) display extensive tropism for pathology in the adult brain: evidence from intracranial gliomas. (Modified from ref. *35*.) **(I)** NSCs migrate extensively throughout a brain tumor mass in vivo and "trail" advancing tumor cells. Paradigm 1, in which NSCs are implanted directly into an established experimental intracranial glioblastoma, is illustrated schematically. **(A)** The virulent and aggressively invasive CNS-1 glioblastoma cell line, used to create the tumor, has been labeled ex vivo by transduction (via a retroviral vector) with green

Fig. 3. *(continued)* fluorescent protein (GFP) cDNA, allowing those cells to fluoresce green (enhanced when revealed by an anti-GFP antibody). The NSCs have been stably retrovirally transduced with *lacZ* cDNA to produce β-galactosidase (β-gal), allowing them to be visualized as red (by anti-β-gal immunocytochemistry under fluorescence microscopy) or as blue (by X-gal histochemistry under light microscopy). This panel, processed for double immunofluorescence using an anti-β-gal antibody (NSCs, red) and an anti-GFP antibody (glioblastoma cells, green), shows a section of brain (under low power) from an adult nude mouse sacrificed 10 d after NSC injection into the CNS-1 glioblastoma; arrowheads demarcate approximate edges of the tumor mass where it interfaces with normal tissue. Donor red β-gal+ NSCs can be seen extensively distributed throughout the mass, interspersed among the green tumor cells. This degree of interspersion by NSCs occurs within 48 h following injection. Interestingly, although NSCs have extensively migrated and distributed themselves within the mass, they largely stop at the junction between tumor and normal tissue *except* where a tumor cell is infiltrating normal tissue; then NSCs appear to "follow" the invading tumor cell into surrounding tissue. This "trailing" of individual glioblastoma cells migrating away from the main tumor bed is examined in greater detail in (**B, C**). (**B**) High-power view, under fluorescence microscopy, of single migrating infiltrating GFP+ tumor cells (green) in apposition to β-gal+ NSCs (red) (white arrow). In a similar and perhaps even more impressive demonstration, the section in (**C**) is costained with X-gal (allowing the *lacZ*-expressing NSCs to stain blue, arrow) and with neutral red (allowing the elongated glioblastoma cells to stain dark red). The blue NSC is in direct juxtaposition to a single migrating, invading neutral red+, spindle-shaped tumor cell (arrow), the NSC "riding" the glioma cell in "piggyback" fashion. Scale bar (C) = 60 μm. (**II**) NSCs implanted at various intracranial sites *distant* from main tumor bed migrate through normal adult tissue towards glioblastoma cells. (**A, B**) Same hemisphere (Paradigm 2). Shown here is a section through the tumor from an adult nude mouse 6 d following NSC implantation caudal to tumor. In (**A**) (as per the schematic, a coned down view of a tumor populated as pictured under low-power in I(A); note X-gal+ blue NSCs interspersed among dark neutral red+ tumor cells. (**B**) High-power view of NSCs in juxtaposition to islands of tumor cells. (**C–H**) Contralateral hemisphere (Paradigm 3). (**C–E**) As indicated on the schematic, these panels are views through the corpus callosum (c) where β-gal+ immunopositive NSCs (red cells, arrows) are seen migrating from their site of implantation on one side of the brain toward a tumor on the other. Two representative NSCs indicated by arrows in (C) are viewed at higher magnification in (D) and (E), respectively, to visualize the classic elongated morphology and leading process of a migrating neural progenitor oriented toward its target. In (**F**), β-gal+ NSCs (red) are "homing in" on the GFP+ tumor (green), having migrated from the other hemisphere. In (**G**), and magnified further in (**H**), the X-gal+ blue NSCs (arrows) have now actually entered the neutral red+ tumor (arrowheads) from the opposite hemisphere. (**I, J**) Intraventricular (Paradigm 4). Shown here is a section through the brain tumor of an adult nude mouse 6 d following NSC injection into the contralateral cerebral ventricle. In (**I**), as per the schematic, blue X-gal+ NSCs are foreign gene [Fig. 3I(B,C)]. The NSCs, in effect, "surrounded" the invading tumor border while "chasing down" migrating infiltrating tumor cells. Furthermore, when implanted intracranially at *distant* sites from the tumor bed in adult brain (e.g., into normal tissue [Fig. 3II(A,B)], into the contralateral hemisphere [Fig. 3(II)C-H], or into the cerebral ventricles [Fig. 3II(I,J)]), the donor NSCs will even *cross* from one hemisphere to the *other* and migrate through normal tissue *to* the brain tumor on the *opposite* side (including human glioblastomas), as if specifically "drawn" to and targeting tumor cells. The NSCs appeared to be drawn by either factors elaborated by brain tumors or by the tissue destruction they rendered. NSCs could deliver a bioactive therapeutically relevant molecule—the oncolysis-promoting enzyme cytosine deaminase—such that in vitro (Fig. 4I) and in vivo (Fig. 4II), a dramatic, quantifiable reduction in tumor cell burden resulted. These data suggested the adjunctive use of inherently migratory NSCs as a delivery vehicle for more effectively targeting a wide variety of therapeutic genes and vectors to refractory, migratory, invasive brain tumor cells. More broadly, they suggested that NSC migration can be extensive, even in the adult brain and along nonstereotypical routes, if pathology (as modeled here by tumor) is present.

Amyloid Plaques

More recently, this notion has been extended in preliminary studies in our lab *(52)* to address lesions present in an adult animal model of Alzheimer's disease (AD). AD is a global, progressive neurodegenerative disorder of adulthood with severe cognitive and behavioral abnormalities. The extensive and widespread destruction characteristic of AD makes it difficult to treat with conventional grafting approaches (for the potential replacement of dead or dying neurons) or with traditional gene delivery vectors (for intracranial molecular therapy). Although the signals that stimulate migration of NSCs are not yet identified, inflammatory molecules are among the likely candidates. Tate previously demonstrated that chronically infused human

---

Fig. 3. *(continued)* distributed within the neutral red+ main tumor bed (edge delineated by arrowheads). At higher power in **(J)**, the NSCs are in juxtaposition to migrating islands of red glioblastoma cells. Fibroblast control cells never migrated from their injection site in any paradigm. All X-gal-positivity was corroborated by anti-β-gal immunoreactivity. Scale bars: (A) 20 μm, and applies to (C); (B) 8 μm, and is 14 μm in (D) and (E), 30 μm in (F) and (G), 15 μm in (H), 20 μm in (I), and 15 μm in (J).

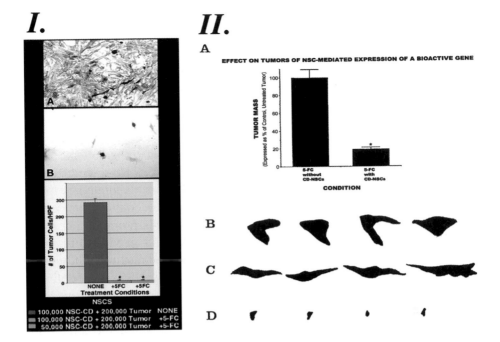

Fig. 4. NSCs can express functional genes within a pathological situation. (Modified from ref. *35*.) **(I)** Bioactive transgene (cytosine deaminase [CD]) remains functional (as assayed by in vitro oncolysis) when expressed within NSCs. CNS-1 glioblastoma cells (red) were cocultured with CD-transduced murine NSCs (CD-NSCs) **(A, B)** (blue). Cocultures unexposed to 5-FC grew healthily and confluent **(A)**, whereas plates exposed to 5-FC showed dramatic loss of tumor cells **(B)**, represented quantitatively by the histograms. ($^*p < 0.001$). The oncolytic effect was identical whether 1×10^5 CD-NSCs or half that number were cocultured with a constant number of tumor cells. (In this paradigm, subconfluent NSCs were still mitotic at the time of 5-FC exposure and, thus, also subject to self-elimination by the generated 5-FU and its toxic metabolites.) **(II)** Expression of a bioactive transgene (CD) delivered by NSCs in vivo as assayed by reduction in tumor mass. The size of an intracranial glioblastoma populated with CD-NSCs in an adult nude mouse treated with 5-FC was compared with that of tumor treated with 5-FC but lacking CD-NSCs. These data, standardized against and expressed as a percentage of a control tumor populated with CD-NSCs receiving no treatment, are presented in the histograms in **(A)**. These measurements were derived from measuring the surface area of tumors (like those in Fig. 3), representative camera lucidas of which are presented in **(B–D)**. Note the large areas of a control non-5-FC-treated tumor containing CD-NSCs **(B)** and a control 5-FC-treated tumor lacking CD-NSCs **(C)** as compared to the dramatically smaller tumor areas of the 5-FC-treated animal who also received CD-NSCs **(D)** (approx 80% reduction as per the histogram in (A), $^*p < 0.001$), suggesting both activity and specificity of the transgene. The lack of effect of 5-FC on tumor mass when no CD-bearing NSCs were within the tumor **(C)** was identical to the effect of CD-NSCs in the tumor without the gene being employed **(B)**.

amyloid will cause an inflammatory response in the rat brain. With that in mind, normal adult male rats were fitted with a cannula placed into one lateral cerebral ventricle through which either (1) a human amyloid peptide, or (2) a control peptide, or (3) vehicle alone were infused for one week via a mini-osmotic pump. A single injection of murine NSCs was placed in the opposite lateral ventricle. After 1 wk, the NSCs were found to have migrated from their site of injection to the opposite hemisphere, where they surrounded areas of amyloid infusion. Although the NSCs colocalized, with activated astrocytes, the NSCs themselves did *not*, surprisingly, differentiate into astrocytes. These preliminary experiments are being extended to transgenic mouse models of AD (the Swiss variety), which is also characterized by diffuse inflammation. Pilot studies similarly suggest a tropism of foreign gene-expressing NSCs for amyloid depositions. NSCs may, therefore, play a role in the delivery of therapeutic molecules in a global degenerative disease like AD. It is not yet known whether they can similarly replace the neural cells that die in that disease.

Reciprocal NSC–Host Interactions and Augmenting This Dynamic

Up to this point, this chapter has dealt primarily with the concept that the *host CNS environment—as it changes over the course of development and aging or as it is altered by injury or degeneration—influences or instructs the stem cell*, including drawing the NSC to areas of degeneration. Indeed, a good deal of effort is being devoted to attempting to use NSCs as "probes" to "interrogate" environments and to help discern which, of the multitude of factors whose expression is altered by injury, are biologically or developmentally significant *from the vantage point of the NSC*.

However, Jitka and V. Ourednik in our lab have accumulated preliminary data to suggest that not only does the host environment "talk" to the NSC but communication *also* occurs in a reciprocal fashion (i.e., the NSC can communicate with and instruct the host CNS). This speculation derives in part from an interesting observation using murine NSCs. In collaboration with Mellitta Schachner, NSCs were implanted *unilaterally* into the mesencephalon of aged mice that, 1 mo previously, had received repetitive systemic administrations of high-dose MPTP, a neurotoxin that produces a persistent loss of tyrosine hydroxylase (TH) expression in the substantia nigra (SN) and striatum, creating a model of Parkinson's disease *(53,54)*. Nigral damage was reversed, both functionally and biochemically, by the exogenous NSCs: the lesioned SNs on *both sides* of the midline presented a gross appearance of virtual reconstitution when compared to untreated lesioned controls. Intriguingly, although a small proportion of the transplanted NSCs did integrate into the SN as TH+ neurons (having spontaneously differentiated into this cell type in the TH-deficient environment), approx 80% of the TH+ cells in the "reconstituted SN" were actually of *host* origin; host cells appeared to have been "rescued"; that is, the re-expression of TH within damaged endogenous dopaminergic neurons of the mesostriatal nuclei of the SN had apparently been facilitated by stem-cell-derived factors.

Teng and Lachyankar in the lab in collaboration with Maragakas, Llado, Rothstein, and Tuszynski (see Chapter 12) have marshaled another example that seems to make a similar point. They observed that the same NSC clone, which, in a motorneuron-depleted spinal ventral horn, is capable of differentiating into motorneurons, will, when cocultured side by side with a spinal cord explant or placed within the scooped out cavity of a dorsal funicultomized rat spinal cord in vivo, cause an exuberant directed outgrowth of host motorneuronal fibers into the NSCs. In other words, the NSCs in these circumstances do not become neurons but rather constitutively exert a previously unrecognized but compelling tropic effect.

Taken together, these findings suggest that the host environment not only influences the fate of NSCs, but, in a heretofore unappreciated reciprocal manner, NSCs offer powerful signals to the *host*. Given the unexpected existence of such *bidirectional* signaling between NSC and host over long distances, NSCs (even in small numbers) may be able to mediate repair by promoting the recovery of damaged *endogenous* neurons even in the absence of donor cell replacement.

Although the mechanism by which NSCs can influence the host remains uncertain, one mode is likely to be the production by NSCs of trophic and tropic agents. We have determined that murine NSCs constitutively produce a broad range of peptide neurotrophic factors (including NT-3, NT-4/5, NGF, BDNF, GDNF), adhesion molecules (e.g., L1), ECMs (e.g., reelin), and lysosomal enzymes. It is quite likely that many of the neuroprotective/trophic effects (e.g., in the MPTP mouse above) and the tropic effects (on spinal motorneurons) are mediated by the constitutive production by NSCs of GDNF.

That efficacy of NSCs within neurodegenerative environments could be mediated by trophic molecules—either those produced at low levels by the injured environment to which NSCs (donor or host) respond or by those produced by the NSCs to which host cells respond—suggests that these effects can be augmented. In other words, if the NSCs produce a small amount constitutively, it seems logical to attempt to enhance that amount by engineering them ex vivo to achieve a larger impact, not only on host cells but also on the donor cells themselves in an autocrine/paracrine fashion.

Some Unexpected Uses of NSCs

The ability of NSCs to migrate, distribute themselves throughout the brain, and disseminate a foreign gene product have prompted the use of these cells in a number of somewhat unconventional areas of investigation.

Generating Models of Neural Disease

Neural stem cells have proven useful for the distribution of "disease-producing" agents throughout normal brains in order to create models of CNS disease where none presently exist *(55,56)*. For example, a model of "retrovirus-induced spongiform myeloencephalopathy" (a histopathology that resembles HIV-induced CNS disease and with a motorneuron involvement that resembles ALS) was created through the diffuse engraftment of genetically engineered NSCs expressing neurovirulent retroviral envelope proteins and replication-competent retroviruses *(55)*. This technique not only created a model of retrovirally-mediated neurodegeneration but permitted a sequential dissection of its underlying pathophysiology by using NSCs variously genetically manipulated to deliver isoforms of molecules and viral strains specific for each stage of the retroviral life cycle. It could be determined that envelope protein, contrary to a popular hypothesis, is not sufficient for the induction of disease nor are early-retroviral-replication events. Rather, late-replication processes, probably within infected host microglia, are actually required. Having generated this model, attempts may now ensue to reverse degeneration using normal or engineered cells. This novel use of NSCs to make a "poor man's transgenic animal" may help generate and test models of other types of neuropathology via the transfer of putatively toxic molecules or viral particles to the CNS. It may help to sequentially dissect the role particular factors may play in complex neuropathologic processes, particularly in instances where transgenic mice have either been unsuccessful, unhelpful, or too cumbersome to generate (particularly when multiple variations of a molecule require testing).

NSCs as Packaging Cells for Retroviral Vectors

Up to this point, the extensive distribution of large quantities of viral vector-mediated therapeutic genes to the brain has been disappointingly limited. Retroviral vectors are typically produced in an incubator by packaging lines that spew infectious particles into the medium. The medium bearing the viral vector particles is then typically collected and injected into the intended target tissue. The region of the target tissue that become genetically altered or improved is usually quite limited to a "halo" around the injection site. It would be interesting, however, if the packaging line itself were not simply fibroblasts in a culture dish but engraftable, migratory, integrated

NSCs within the brain. It seems quite likely that widely engrafted NSCs that have been engineered to package replication-defective retroviral vectors may serve as platforms for the brainwide dissemination of these viral vectors. A stem-cell-based approach has, in fact, been devised and has, indeed, proven successful in magnifying the efficacy of viral-mediated gene delivery throughout the brain *(56–58)*.

HUMAN NSCS RESPOND TO DEVELOPMENTAL CUES, MIGRATE TOWARD PATHOLOGY, REPLACE NEURONS, EXPRESS FOREIGN GENES, AND CROSS-CORRECT METABOLIC DEFECTS

In neuroregeneration research, the relevance of rodent studies to humans is always somewhat of a "leap of faith." The identification of human NSCs *(6,59–63)* contemporaneously with recent evidence for human neurogenesis postnatally *(64)*—both of which mimic rodent phenomena—seemed to vouchsafe conservation of certain neurodevelopmental principles to the human CNS. Stem-cell-based strategies, therefore, merit consideration for human neurodegenerative conditions, both genetic and acquired.

Lines of engraftable human NSCs (hNSCs) have been isolated from normal human fetuses (ideally from the ventricular zone) that appear to be the "human equivalent" of the rodent NSCs described by us and by others (reviewed in refs. *10* and *12*) (albeit with an approximately three times longer cell cycle). Insights into how therapeutic goals might be achieved with such cells have derived directly from observations of murine NSCs. In most studies, hNSCs have mimicked murine NSCs, possessing a range of capabilities to suggest an efficacy in true clinical situations much as rodent NSCs have in animal models. These shared behaviors likely reflect a fundamental stem cell biology that cuts across species.

For example, hNSCs can, like their rodent counterparts, give rise to all three neural lineages in vivo and in vitro and, following transplantation, participate in normal murine CNS development, including migration from germinal zones along migratory streams, to widely disseminated CNS regions *(6,27,60,65)*. They retain a responsiveness to regional and temporal developmental cues to become multiple cell types in these regions (e.g., migrating as far rostral as the olfactory bulbs [at one end of the neuraxis], becoming interneurons; integrating into the cerebellum [at the opposite end of the neuraxis], becoming granule neurons; integrating into white tracts, becoming oligodendroglia and astroglia). Differentiating appropriately into neurons where neurogenesis normally persists and into glia where gliogenesis predominates, they intermix permanently and nondisruptively with host neural progenitors and their progeny. Many of the cells into which these hNSCs differentiate, interestingly, are born not at the developmental stage from which the cells were obtained, but rather at the stage and region of their engraftment, thus affirming their multipotency and responsiveness to environmental cues. Genetically manipulatable ex vivo hNSCs, in their widely disseminated locations, can express a retrovirally transduced transgene, offering promise for future gene therapy applications in a manner similar to what we have demonstrated for rodent NSCs. Secretory products from these hNSCs can cross-correct a prototypical genetic metabolic defect (that of Tay–Sachs disease) in neurons and glia in vitro as effectively as do mouse NSCs. When transplanted into a mouse mutant (*meander tail*) in which a specific neuron type was deficient, donor hNSCs differentiated into that deficient cell type *(6)*, much as murine NSCs could *(37)*, suggesting their potential as well for neural cell replacement and repair. Just as murine NSCs do, hNSCs were drawn to pathology (as modeled by brain tumor) from as far as the opposite side of the adult rodent brain and continued to express foreign genes *(35)*. In a rodent model of asphyxial injury, the hNSCs could, just as described for murine NSCs, yield neurons in the cortex of all neurotransmitter subtypes (including GABAergic, glutamatergic, cholinergic, and even catecholaminergic). In the contused spinal cord of adult rats, hNSCs could robustly yield neurons (including motor neurons) with widely coursing processes (interspersed with donor-derived oligodendrocytes) that could make long-distance connections both rostral and caudal to the lesion, had the ability to conduct cortico-spinal signals, and resulted in apparent functional improvement (see Chapter 14).

Human NSCs in the Nonhuman Primate CNS Interact Appropriately

If insights from rodents may be applicable to humans, then experiments with monkeys should provide a necessary intermediate step for ensuring translation of those insights. To assess whether they might engraft and respect developmental cues in hosts that more closely resemble human and to help assess their potential utility in primate models of human disease, hNSCs were transplanted into normal fetal Old World monkeys via intracerebroventricular injections that allowed the cells access to the VZ. The hNSCs successfully engrafted throughout the primate brain. Following entrance into the VZ, the hNSCs migrated into parenchyma along radial glial fibers and integrated throughout the developing brain, yielding neurons and glia appropriate to the cues for a given cortical lamina (glia in deeper lamina, neurons in superficial lamina) as well as contributing to such secondary germinal zones as the SVZ that serve homeostatic and self-repair functions into adulthood *(9)*. [It bears noting that this intervention was somewhat of a "dress rehearsal" for procedures that could actually be used for *in utero* hNSC-mediated therapy of human neurogenetic disorders. Such prenatal treatments could be directed not only at congenital disorders but also, theoretically, at neurodegenerative diseases that are not expressed until adulthood or middle age but whose antenatal genetic diagnosis is possible (e.g., Huntington's disease).]

Human NSCs in Nonhuman Primate Models of Disease

Experiments have begun involving hNSCs in lesioned nonhuman primates—often the primate equivalent of some of the mouse models described above. Such experiments not only help determine the cells' safety and efficacy in models of human disease, but also provide a more direct understanding of "humanlike" development and the response of human stem cells to "humanlike" neurodegenerative environments. They also permit one to work out the logistics of cell administration in an anatomy more relevant to that of humans.

For example, analysis of the fate and impact of hNSCs in the MPTP-induced model of dopamine depletion and parkinsonism in Old World St. Kitts African green monkeys has begun in collaboration with Gene Redmond. In encouraging pilot studies, hNSCs were implanted into nonimmunosuppressed adult monkeys that, 8 mo previously, had been lesioned with MPTP. After 1 mo, the brains were examined for the presence of hNSC-derived cells in the substantia nigra, for evidence of immune rejection, and for the phenotype of the hNSC-derived cells. Preliminary analysis revealed robust survival of hNSCs in both normal and MPTP-lesioned monkeys without evidence of immunorejection. There appeared to be spontaneous conversion of NSCs to TH+ cells in this dopamine-depleted adult primate mesencephalon; approx 10% of donor-derived cells appeared to express TH. In spite of being injected unilaterally above the right substantia nigra, donor-derived TH+ cells appeared to be distributed *bilaterally* in the region of the mesostriatal nuclei. Some recipient monkeys were assessed in their living state for dopamine activity (via SPECT analysis) and for behavioral response. These highly preliminary studies suggest that NSC-derived cells may not only survive and produce TH in the nigrostriatal system, but may also express dopamine transporter sites. Published and unpublished data indicate that SPECT quantitation of β-CIT uptake correlates both with behavioral measures of parkinsonian symptoms ("Parkscore") and with dopamine concentrations in the striatum (as subsequently measured directly postmortem). Behavioral changes in these pilot animals were consistent with their respective β-CIT uptake on SPECT. Compared to a sham-injected monkey that showed decreased SPECT activity and remained severely parkinsonian, there was improvement in the Parkscore of an NSC-transplanted monkey that also showed an increase bilaterally in SPECT activity as well as more exuberant TH+ fibers that spanned relatively long expanses on postmortem histology. Whether this possible amelioration of a functional dopamine deficit is the direct result of dopamine cell replacement by NSCs or the provision by NSCs of factors that promote survival of host dopaminergic neurons (as described above for the mouse) or a combination of the two remains to be determined. Either mechanism, if reliable, will likely be therapeutically important.

BROADER IMPLICATIONS

With the recent attention paid to embryonic stem cells (ESCs) and their ability to give rise to neural precursors, with the discovery of stem cells in other solid organ systems, and with the possibility that some of these tissue-resident stem cells from other organ systems may enter the CNS (possibly giving rise to neural elements), the question arises as to whether these other variety of stem cells can also address the global demands of neurological diseases or will this still be the unique niche for neural stem cells. Only future research will determine whether such stem cells can meet the gold standard of safety, efficiency, simplicity, and efficacy established by NSCs. Regardless of how that question is resolved by future experiments, it seems likely that NSCs have unveiled a novel therapeutic paradigm that may now be emulated for other organ systems. The existence of stem cells in other organs or the ability of ESCs to give rise to such progenitors suggests that stem-cell-mediated widespread repair of those tissues (e.g., heart, liver, muscle, pancreas) may similarly be possible in much the same way as NSCs seem to do for the CNS. The demand for organ donation may be forestalled, for example. Although approaches to disease have heretofore focused on stopping pathology, stem cells now allow for the complementary approach of "starting over"—replacing defective components with more normal ones. The possibility of doing autografts with stem cells derived from an adult remains controversial, however, particularly for genetic diseases; the cells derived from an individual with a genetic impairment likely already harbor that genetic defect or predisposition and may not be useful or effective.

POTENTIAL THERAPEUTIC STRATEGIES USING NSC

In summary, emerging data suggest that a population of CNS progenitors or stem cells exists whose isolation, expansion, and transplantation may have clinical utility for cell replacement or molecular support therapy for some degenerative, developmental, and acquired insults to the CNS. The use of NSCs as graft material has helped reorient and broaden the scope of neural transplantation as a therapeutic intervention. Most neurologic diseases are not unifocal like parkinsonism but, rather, are characterized by extensive, widespread, multifocal, or even global pathology, often requiring multiple repair strategies. By virtue of their inherent biologic properties, NSCs seem to possess the multiple therapeutic capabilities demanded. At least three therapeutic interventions are possible from the implantation of NSCs into the dysfunctional CNS.

First, transplanted NSCs may differentiate into and replace damaged or degenerated neurons and/or glia. Host cells degenerating not only by cell-intrinsic disease processes (the obvious situation) but also cell-*extrinsic* forces might theoretically be replaced. For example, in some pathologic conditions, host factors may be elaborated within the microenvironment that are inimical to the survival and/or differentiation of donor progenitors. However, donor cells that are not naturally resistant may be engineered ex vivo to be less sensitive to these factors, to secrete substances that might neutralize those factors, or to express trophic agents that might overcome those factors, perhaps allowing "sister" donor progenitors within the graft to differentiate into desired cell types.

In a second category of potential interventions, even those implanted cells that do not differentiate into the desired missing cell types may, nevertheless, serve a therapeutic purpose. Because they are of CNS origin, donor NSCs may intrinsically provide factors—both diffusible (e.g., trophins) and nondiffusible (e.g., "bridges," cell–cell contact signals, myelin, ECM)—that might enable the injured host to regenerate its own lost cells and/or neural circuitry. In the event that such factors are not naturally produced in sufficient quantities, in a third therapeutic strategy, the cells can be genetically engineered ex vivo before transplantation to become resident "factories" for the sustained production of substances known to reactivate, mobilize, and recruit quiescent host progenitor pools, to promote regeneration and further differentiation of immature nerve cells (either endogenous or, in a paracrine/autocrine fashion, donor), to attract ingrowth of host fibers, to forestall degeneration resulting from insufficiency of a trophin, enzyme, or other factor in the milieu, to neutralize toxins already present in the microenvironment, or to allow cells to utilize alternative metabolic pathways.

Because NSCs can incorporate into the host cytoarchitecture, possibly in a functional manner, they may prove to be more than vehicles for passive delivery of substances. The regulated release of certain substances through feedback loops may be reconstituted, as might the reformation of essential circuits. Intriguingly, one multipotent NSC line may, in certain conditions, perform many of above-mentioned functions, even concurrently in one transplant. Transplanting these cells, possibly in combination with other systemic or somatic cell therapies, may be a broadly applicable treatment for a number of neurological diseases.

The field of NSC biology is at a very early stage of development. Many of our suggestions are highly speculative, and much needs to be learned about the properties of such cells. While work is ongoing on the isolation, propagation, and transplantation of human NSCs, many important questions need to be addressed experimentally before it will be possible to use such cells in clinical applications. For instance, what factors ensure the expansion, stability, engraftment, migration, and differentiation of transplanted NSCs? What variables dictate the efficiency of foreign gene expression by engrafted NSCs? What are the fundamental pathophysiological needs in a given disease for reversing progression and/or restoring function. When is the proper time to administer cells? What are the limits of reconstitution in the brain? Do donor-derived cells function normally? It is hoped that the biological properties of human NSCs can be combined with molecular engineering to restore normal functions as well as repair lesions in both inherited and acquired diseases affecting the CNS.

SUMMARY

Several methods have been proposed to apply gene transfer to the CNS to ameliorate neurological defects. One promising possibility is the use of NSCs. These cells can intrinsically secrete missing or therapeutic gene products or can be genetically engineered ex vivo to do so. NSCs may provide a strategy for long-term treatment of the CNS manifestations of a number of neurogenetic diseases. NSCs are capable of differentiating along the three major CNS cell-type lineages. Following transplantation, they can integrate seamlessly within structures throughout the host CNS, assuming the morphological, immunocytochemical, and ultrastructural characteristics appropriate for cells of that region. Although it is not yet known whether these NSC-derived cells function normally in these contexts, they show no evidence of disturbing neurobiological processes in the normal brain and may contribute to improvement in some models. NSCs can also be easily genetically manipulated ex vivo to overexpress potentially therapeutic genes. By exploiting their basic biologic properties, these NSCs may be able to deliver these gene products in a sustained, direct, perhaps regulated fashion throughout the CNS. Furthermore, whereas they may disseminate these proteins throughout the brain, they nevertheless restrict that distribution to *only* the CNS. Thus, these vehicles may overcome many of the limitations of viral and non-neural cellular vectors, as well as pharmacologic and genetic interventions. The feasibility of this NSC-based strategy has been demonstrated in a number of animal models. NSCs of human derivation seem to have conserved many of these appealing qualities.

The profound effect a degenerative environment may have on the fate of the NSC may particularly pay potential therapeutic dividends when attempting to treat widely disseminated, multifocal, or global pathology in the CNS. During phases of active neurodegeneration, as-yet-unidentified factors seem to be transiently elaborated to which NSCs may respond by migrating to affected regions and differentiating toward replacement of dying neural cells. In other words, NSCs have the capacity to respond to neurogenic cues in vivo not only during their normal developmental expression but even, we believe, when "reactivated" at later stages, for example, by certain types of cell death in the "postdevelopmental" CNS. Viewed another way, neurodegeneration appears to "alter" the CNS "landscape" such that NSCs, when confronted with degenerating environments, appear impelled constitutively to recapitulate their developmental programs—even in the adult—and "attempt" to "recreate" or "repopulate" abnormal regions. The CNS may, indeed, be "attempting"—often futilely—to "reconstruct" itself. To the extent that degeneration might influence the fate of NSCs, it might actually draw a this powerful gene delivery vehicle toward it.

In short, the inherent biologic properties of NSCs may elevate them beyond simply serving as an alternative to fetal tissue in transplantation paradigms or being just another vector in gene therapy paradigms. NSCs in many ways have helped to reorient and broaden the paradigmatic scope of neural transplantation as a therapeutic intervention. Although it has traditionally been reserved for neuropathologies that are localized anatomically (e.g., Parkinson's disease), the use of NSCs, because of their ability to engraft within germinal zones and to migrate and integrate widely, may permit this strategy for widespread enzyme and/or cell replacement to tackle conditions conventionally consigned solely to pharmacologic, genetic, and BMT interventions.

ACKNOWLEDGMENTS

Some of the work described here was supported in part by grants to E.Y.S. from the National Institute of Neurologic Diseases and Stroke, March of Dimes, Project ALS, Brain Tumor Society, Hunter's Hope, Canavan

Research Fund, Late Onset Tay Sachs Foundation, A-T Children's Project, International Organization for Glutaric Acidemia, and Parkinson's Action Network/Michael J. Fox Foundation.

REFERENCES

1. Nikkah, G., Cunningham, M. G., Cenci, M. A., McKay, R. D., and Bjorklund, A. (1995) Dopaminergic microtransplants into the substantia nigra of neonatal rats with bilateral 6-OHDA lesion. I. Evidence for anatomical reconstruction of the nigrostriatal pathway. *J. Neurosci.* **15(5),** 3548–3561.

2. Nikkah, G., Cunningham, M. G., McKay, R., and Bjorklund, A. (1995) Dopaminergic microtransplants into the substantia nigra of neonatal rats with bilateral 6-OHDA lesions. II. Transplant-induced behavioral recovery. *J. Neurosci.* **15(5),** 3562–3570.

3. Kordower, J. H., Freeman, T. B., Snow, B. J., Vingerhoets, F. J., Mufson, E. J., Sanberg, P. R., et al. (1995) Neuropathological evidence of graft survival and striatal reinervation after the transplantation of fetal mesencephalic tisssue in a patient with Parkinson's disease. *N. Engl. J. Med.* **332(17),** 1118–1124.

4. Freed, C. R., Greene, P. E., Breeze, R. E., Tasai, W. Y., DuMoucehl, W., Kao, R., et al. (2001) Transplantation of embryonic dopamine neurons for severe Parkinson's disease. *N. Engl. J. Med.* **344,** 763–765.

5. Snyder, E. Y., Taylor, R. M., and Wolfe, J. H. (1995) Neural progenitor cell engraftment corrects lysosomal storage throughout the MPS VII mouse brain. *Nature* **374,** 367–370.

6. Flax, J. D., Aurora, S., Yang, C., Simonin, C., Wills, A. M., Billinghurst. L., et al. (1998) Engraftable human neural stem cells respond to developmental cues, replace neurons, and express foreign genes. *Nat. Biotechnol.* **16,** 1033–1039.

7. Brüstle, O., et al. (1998) Chimeric brains generated by intraventricular transplantation of fetal human brain cells into embryonic rats. *Nat. Biotechnol.* **11,** 1040–1049.

8. Yandava, B., Billinghurst, L., and Snyder, E. (1999) "Global" cell replacement is feasible via neural stem cell transplantation: evidence from the dysmyelinated shiverer mouse brain. *Proc. Natl. Acad. Sci. USA* **96,** 7029–7034.

9. Ourednik, V., Ourednik, J., Flax, J. D., Zawada, M., Hutt, C., Yang, C. L., et al. (2001) Segregation of human neural stem cells in the developing primate forebrain. *Science* **293,** 1820–1829.

10. McKay, R. (1997) Stem cells in the central nervous system. *Science* **276,** 66–71.

11. Fisher, L. J. (1997) Neural precursor cells: applications for the study and repair of the central nervous system. *Neurobiol. Dis.* **4,** 1–22.

12. Gage, F. H. (2000) Mammalian neural stem cells. *Science* **287,** 1433–1438.

13. Alvarez-Buylla, A. and Temple, S. (1998) Neural stem cells. *J. Neurobiol.* **36,** 105–314.

14. Vescovi, A. L. and Snyder, E. Y. (1999) Establishment and properties of neural stem cell clones: plasticity in vitro and in vivo. *Brain Pathol.* **9,** 569–598.

15. Chen, K. S. and Gage, F. H. (1995) Somatic gene transfer of NGF to the aged brain: behavioral and morphologic amelioration. *J. Neurosci.* **15(4)**, 2819–2825.
16. Tuzynski, M. H. and Gage, F. H. (1995) Bridging grafts and transient nerve growth factor infusions promote long-term central nervous system neuronal rescue and partial functional recovery. *PNAS* **92**, 4621–4625.
17. Grill, R., Murai, K., Blesch, A., Gage, F. H., and Tuszynski, M. H. (1997) Cellular delivery of neurotrophin-3 promotes corticospinal axonal growth and partial functional recovery after spinal cord injury. *J. Neurosci.* **17**, 5560–5572.
18. Winkler, J., Suhr, S. T., Gage, F. H., Thal, L. J., and Fisher, L. J. (1995) Essential role of neocortical acetylcholine in spatial memory. *Nature* **375**, 484–487.
19. Emerich, D. F., Winn, S. R., Harper, J., Hammang, J. P., Baetge, E. E., and Kordower, J. H. (1994) Implants of polymer-encapsulated human NGF-secreting cells in the nonhuman primate: rescue and sprouting of degenerating cholinergic basal forebrain neurons. *J. Comp. Neurol.* **349**, 148–164.
20. Kordower, J. H., Winn, S. R., Liu, Y.-T., Mufson, E. J., Sladek, J. R., Hammang, J. P., et al. (1994) The aged basal forebrain: rescue and sprouting of axotomized basal forebrain neurons after grafts of encapsulated cells secreting human nerve growth factor. *PNAS* **91**, 10,898–10,902.
21. Fisher, L. J., Jinnah, H. A., Kale, L. C., Higgins, G. A., and Gage, F. H. (1991) Survival and function of intrastriatally grafted primary fibroblasts genetically modified to produce L-dopa. *Neuron* **6**, 371–380.
22. Dunnett, S. B. and Bjorklund, A. (2000) *Functional Neural Transplantation* Raven, New York.
23. Snyder, E. Y., Deitcher, D. L., Walsh, C., Arnold-Aldea, S., Hartwieg, E. A., and Cepko, C. L. (1992) Multipotent neural cell lines can engraft and participate in development of mouse cerebellum. *Cell* **68**, 33–55.
24. Renfranz, P. J., Cunningham, M. G., and McKay, R. D. G. (1991) Region-specific differentiation of the hippocampal stem cell line HiB5 upon implantation into the developing mammalian brain. *Cell* **66**, 713–719.
25. Whittemore, S. R. and Snyder, E. Y. (1996) The biologic relevance and functional potential of central nervous system-derived cell lines. *Mol. Neurobiol.* **12**, 13–38.
26. Suhonen, J. O., Peterson, D. A., Ray, J., and Gage, F. H. (1996) Differentiation of adult hippocampus-derived progenitors into olfactory neurons in vivo. *Nature* **383**, 624–627.
27. Fricker, R. A., Carpenter, M. K., Winkler, C., Greco, C., Gates, M. A., and Bjorklund, A. (1999) Site-specific migration and neuronal differentiation of human neural progenitor cells after transplantation in the adult rat brain. *J. Neurosci.* **19(14)**, 5990–6005.
28. Park, K. I., Jensen, F. E., Stieg, P. E., and Snyder, E. Y. (1998) Hypoxic–ischemic (HI) injury may direct the proliferation, migation, and differentiation of endogenous neural progenitors. *Soc. Neurosci. Abstr.* **24**, 1310.
29. Magavi, S. S., Leavitt, B. R., and Macklis, J. D. (2000) Induction of neurogenesis in the neocortex of adult mice. *Nature* **405**, 951–955.

30. Gould, E., Reeves, A. J., Graziano, M. S., and Gross, C. G. (1999) *Science* **286,** 548–552.

31. Weiss, S., Reynolds, B. A., Vescovi, A. L., Morshead, C., and Van der Kooy, D. (1996) Is there a neural stem cell in the mammalian forebrain? *Trends Neurosci.* **19,** 387–393.

32. Snyder, E. Y., Yoon, C., Flax, J. D., and Macklis, J. D. (1997) Multipotent neural precursors can differentiate toward replacement of neurons undergoing targeted apoptotic degeneration in adult mouse neocortex. *Proc. Natl. Acad. Sci. USA* **94(21),** 11,663–11,668.

33. Doering, L. and Snyder, E. Y. (2000) Cholinergic expression by a neural stem cell line grafted to the adult medial septum/diagonal band complex. *J. Neurosci. Res.* **61,** 597–604.

34. Park, K. I., Liu, S., Flax, J. D., Nissim, S., Stieg, P. E., and Snyder, E. Y. (1999) Transplantation of neural progenitor and stem-like cells: developmental insights may suggest new therapies for spinal cord and other CNS dysfunction. *J. Neurotrauma* **16(8),** 675–687.

35. Aboody, K. S., Brown, A., Rainov, N. G., Bower, K. A., Liu, S., Yang, W., et al. (2000) Neural stem cells display extensive tropism for pathology in the adult brain: evidence from intracranial gliomas. *Proc. Natl. Acad. Sci. USA* **97,** 12,846–12,851.

36. Lacorazza, H. D., Flax, J. D., Snyder, E. Y., and Jendoubi, M. (1996) Expression of human β-hexosaminidase α-subunit gene (the gene defect of Tay–Sachs disease) in mouse brains upon engraftment of transduced progenitor cells. *Nat. Med.* **4,** 424–429.

37. Rosario, C. M., Yandava, B. D., Kosaras, B., Zurakowski, D., Sidman, R. L., and Snyder, E. Y. (1997) Differentiation of engrafted multipotent neural progenitors towards replacement of missing granule neurons in meander tail cerebellum may help determine the locus of mutant gene action. *Development* **124,** 4213–4224.

38. Gage, F. H., Coates, P. W., Palmer, T. D., Kuhn, H. G., Fisher, L. J., Suhonen, J. O., et al. (1996) Survival and differentiation of adult neuronal progenitor cells transplanted to the adult brain. *PNAS*, in press.

39. Anton, R., Kordower, J. H., Maidment, N. T., Manaster, J. S., Kane, D. J., Rabizadeh, S., et al. (1994) Neural-targeted gene therapy for rodent and primate hemiparkinsonism. *Exp. Neurol.* **127,** 207–218.

40. Anton, R., Kordower, J. H., Kane, D. J., Markahma, C. H., and Bredesen, D. E. (1995) Neural transplantation of cells expressing the anti-apoptotic gene bcl-2. *Cell Transplant.* **4,** 49–54.

41. Martinez-Serrano, A., Lundberg, C., Horellou, P., Fischer, W., Bentlage, C., Campbell, K., et al. (1995) CNS-derived neural progeniutor cells for gene transfer of nerve growth factor to the adult rat brain: complete rescue of axotomized cholinergic neurons after transplantation into the septum. *J. Neurosci.* **15,** 5668–5680.

42. Martinez-Serrano, A., Fischer, W., and Bjorklund, A. (1995) Reversal of age-dependent cognitive impairments and cholinergic neuron atrophy by

NGF-secreting neural progenitors grafted to the basal forebrain. *Neuron* **15,** 473–484.

43. Neufeld, E. F. and Fratantoni, J. C. (1970) Inborn errors of mucopolysaccharide metabolism. *Science* **169,** 141–146.

44. Sands, M. S., Vogler, C., Kyle, J. W., Grubb, J. H., Levy, B., Galvin, N., et al. (1994) Enzyme replacement therapy for murine mucopolysaccharidosis Type VII. *J. Clin. Invest.* **93(6),** 2324–2331.

45. Snyder, E. Y. (1998) Neural stem-like cells: developmental lessons with therapeutic potential. *The Neuroscientist* **4(6),** 408–425.

46. Lois, C. and Alvarez-Buylla, A. (1993) Proliferating subventricular zone cells in the adult mammalian forebrain can differentiate into neurons and glia. *Proc. Natl. Acad. Sci. USA* **90,** 2074–2077.

47. Lois, C. and Alvarez-Buylla, A. (1994) Long distance neuronal migration in the adult mammalian brain. *Science* **264,** 1145–1148.

48. Yamanaka, S., Johnson, M. D., Grinberg, A., Westphal, H., Crawley, J. N., Taniike, M., et al. (1994) Targeted disruption of the HexA gene results in mice with biochemical and pathologic features of Tay-Sachs disease. *Proc. Natl. Acad. Sci. USA* **91,** 9975–9979.

49. Himes, B. T., Liu, Y., Solowska, J. M., Snyder, E. Y., Fischer, I., and Tessler, A. (2001) Transplants of cells genetically modified to express neurotrophin-3 rescue axotomized Clarke's nucleus neurons after spinal cord hemisection in adult rats. *J. Neurosci. Res.* **65,** 549–564.

50. Sabaate, O., Horellou, P., Vigne, E., Colin, P., Perricaudaet, M., Buc-Caron, M.-H., et al. (1995) Transplantation to the rat brain of human neural progenitors that were genetically modified using adenoviruses. *Nat. Genet.* **9,** 256–260.

51. Auguste et al. (1997).

52. Tate, B. A., Werzanski, D., Marciniack, A., and Snyder, E. Y. (2000) Migration of neural stem cells to Alzheimer-like lesions in an animal model of AD. *Soc. Neurosci. Abstr.* **26,** 496.

53. Ourednik, J., Ourednik, V., Lynch, W. P., Snyder, E. Y., and Schachner, M. (1999) Massive regeneration of substantia nigra neurons in aged parkinsonian mice after transplantation of neural stem cells overexpressing L1. *Soc. Neurosci. Abstr.* **25,** 1310.

54. Ourednik, V., Ourednik, J., Park, K. I., and Snyder, E. Y. (1999) Neural stem cells: a versatile tool for cell replacement and gene therapy in the CNS. *Clin. Genet.* **46,** 267–278.

55. Lynch, W. P., Snyder, E. Y., Qualtierre, L., Portis, J. L., and Sharpe, A. H. (1996) Neither neurovirulent retroviral envelope protein nor viral particles are sufficient for the induction of acute spongiform neurodegeneration: evidence from engineered neural progenitor-derived chimeric mouse brains. *J. Virol.* **70,** 8896–8907.

56. Poulsen, D. J., Favara, C., Snyder, E. Y., Portis J, and Chesebro, B. (1999) Increased neurovirulence of polytropic mouse retroviruses delivered by inoculation of brain with infected neural progenitor cells. *Virology* **263,** 23–29.

57. Lynch, W. P., Sharpe, A. H., and Snyder, E. Y. (1999) Neural stem cells as engraftable packaging lines optimize viral vector-mediated gene delivery to the CNS: evidence from studying retroviral env-related neurodegeneration. *J. Virol.* **73,** 6841–6851.

58. Herrlinger, U., Woiciechowski, C., Aboody, K. S., Jacobs, A. H., Rainov, N. G., Snyder, E. Y., et al. (2000) Neural stem cells for delivery of replication–conditional HSV-1 vectors to intracerebral gliomas. *Mol. Ther.* **1,** 347–357.

59. Villa, A., Snyder, E. Y., Vescovi, A., and Martinez-Serrano, A. (2000) Establishment and properties of a growth factor-dependent, perpetual neural stem cell line from the human CNS. *Exp. Neurol.* **161,** 67–84.

60. Vescovi, A. L., et. al. (1999) Isolation and cloning of multipotential stem cells from the embryonic human CNS and establishment of transplantable human neural stem cell lines by epigenetic stimulation. *Exp. Neurol.* **156,** 71–83.

61. Uchida, N., et al. (2000) *Proc. Natl. Acad. Sci. USA* **97,** 14,720–14,725.

62. Roy, N. S., et al. (2000) *Nat. Med.* **6,** 271–277.

63. Pincus, D. W., Keyoung, H. M., Harrison-Restelli, C., et al. (1998). FGF2/BDNF-associated maturation of new neurons generated from adult human subependymal cells. *Ann. Neurol.* **43,** 576–585.

64. Erickson et al. (1998).

65. Rubio, F. J,, Bueno, C., Villa, A., Navarro, B., and Martinez-Serrano, A. (2000) Genetically perpetuated human neural stem cells engraft and differentiate into the adult mammalian brain. *Mol. Cell. Neurosci.* **16,** 1–13.

SUGGESTED READINGS

Akerud, P., Canals, J. M., Snyder, E. Y., and Arenas, E. (2001) Neuroprotection through delivery of GDNF by neural stem cells in a mouse model of Parkinson's Disease. *J. Neurosci.*, in press.

Billinghurst, L. L., Taylor, R. M., and Snyder, E. Y. (1998) Remyelination: cellular and gene therapy, *Semin. Pediatr. Neurol.* **5,** 211–228.

Brustle, O., Maskos, U., and McKay, R. D. G. (1995) Host-guided migration allows targeted introduction of neurons into the embryonic brain. *Neuron* **15,** 1275–1285.

Campell, K., Olsson, M., and Bjorklund, A. (1995) Regional incorporation and site-specific differentiation of strial precursors transplanted to the embryonic forebrain ventricle. *Neuron* **15,** 1259–1273.

Davis, A. A. and Temple, S. (1994) A self-renewing multipotential stem cell in embryonic rat cerebral cortex. *Nature* **372,** 263–266.

Dunnett, S. B. (1995) Functional repair of striatal systems by neural transplants: evidence for circuit reconstruction. *Behav. Brain Res.* **66,** 133–142.

Fishell, G., Mason, C. A., and Hatten, M. E. (1993) Dispersion of neural progenitors within the germinal zones of the forebrain. *Nature* **362,** 636–638.

Gritti, A., Parati, E. A., Cova, L., Frolichsthal, P., Galli, R., Wanke, E., et al. (1996) Multipotential stem cells from the adult mouse brain proliferate and self-renew in response to basic fibroblast growth factor. *J. Neurosci.* **16(3),** 1091–1100.

Johnson, E. M. (1995) Possible role of neuronal apoptosis in Alzheimer's disease. *Neurobiol. Aging* S188–S189.

Kilpatrick, T., Richards, L. J., and Bartlett, P. F. (1995) The regulation of neural precursor cells within the mammalian brain. *Mol. Cell. Neurosci.* **6,** 2–15.

Liu, Y., Himes, B. T., Solowska, J., Moul, J., Chow, S. Y., Park, K. I., et al. (1999) Intraspinal delivery of neurotrophin-3 using neural stem cells genetically modified by recombinant retrovirus. *Exp. Neurol.* **158,** 9–26.

Loo, D. T., Copani, A., Pike, C. J., Whittemore, E. R., Walencewicz, A. J., and Cotman, C. W. (1993) Apoptosis is induced by β-amyloid in cultured central nervous system neurons. *PNAS* **90,** 7951–7955.

Lu, P., Jones, L., Park, K. I., Snyder E. Y., and Tuszynski, M. (2000) Neural stem cells secrete BDNF and GDNF, and promote axonal growth after spinal cord injury. *Soc. Neurosci. Abstr.* **26,** 332.

Luskin, M. B. (1993) Restricted proliferation and migration of postnatally generated neurons derived from the forebrain subventricular zone. *Neuron* **11,** 173–189.

Morshead, C. M., Reynolds, B. A., Craig, C. G., McBurney, M. W., Staines, W. A., Morassutti, D., et al. (1994) Neural stem cell in the adult mammalian forebrain: a relatively quiescent subpopulation of subependymal cells. *Neuron* **13,** 1071–1082.

Onnifer, S. M., Whittemore, S. R., and Holets, V. R. (1993) Variable morphological differentiation of a raphe-derived neuronal cell line following transplantation into the adult rat CNS. *Exp. Neurol.* **122,** 130–142.

Ray, J. and Gage, F. H. (1994) Spinal cord neuroblasts proliferate in response to basic fibroblast growth factor. *J. Neurosci.* **14(6),** 3548–3564.

Reynolds, B. A. and Weiss, S. (1992) Generation of neurons and astrocytes from isolated cells of the adult mammalian central nervous system. *Science* **27,** 1707–1710.

Rubio, F. J., Kokai, Z., del Arco, A., Garcia-Simon, M. I., Snyder, E. Y., Lindvall, O., et al. (1999) BDNF gene transfer to the mammalian brain using CNS-derived neural precursors. *Gene Ther.* **6,** 1851–1866.

Sango, K., Yamanaka, S., Hoffmann, A., Okuda, Y., Grinberg, A., Westphal, H., et al. (1995) Mouse models of Tay–Sachs and Sandhoff diseases differ in nuerologic phenotype and ganglioside metabolism. *Nat. Genet.* **11,** 170–176.

Shihabuddin, L., Horner, P., Ray, J., and Gage, F. (2000) Adult spinal cord stem cells generate neurons after transplantation in the adult dentate gyrus. *J. Neurosci.* **20(23),** 8727–8735.

Shihabuddin, L. S., Hetz, J. A., Holets, V. R., and Whittemore, S. R. (1995) The adult CNS retains the potential to direct region-specific differentiation of a transplanted neuronal precursor cell line. *J. Neurosci.* **10,** 6666–6678.

Shihabuddin, L. S., Ray, J., and Gage, F. H. (1997) FGF-2 is sufficient to isolate progenitors found in the adult mammalian spinal cord. *Exp. Neurol.* **148(2),** 577–586.

Svendsen, C. N., Caldwell, M. A., Shen, J., ter Borg, M. G., Rosser, A. E., and Tyers, P. (1997) Long-term survival of human central nervous system progenitor cells transplanted into a rat model of Parkinson's disease. *Exp. Neurol.* **148,** 135–146.

Walsh, C. and Cepko, C. L. (1993) Dispersion of neural progenitors within the germinal zones of the forebrain. *Nature* **362,** 632–635.

Williams, B. P. and Price, J. (1995) Evidence for multiple precursor cell types in the embryonic rat cerebral cortex. *Neuron* **14,** 1181–1188.

Winn, S. R., Tresco, P. A., Zielinski, B., Greene, L. A., Jaeger, C. B., and Aebischer, P. (1991) Behavioral recovery following intrastriatal implantation of microencapsulated PC12 cells. *Exp. Neurol.* **113,** 322–329.

12
Neural Stem Cells
in and from the Spinal Cord

Yang D. Teng, Mary Katherine White, Erin Lavik, Shaoxiang Liu, Mahesh Lachyankar, Kook In Park, and Evan Y. Snyder

NEUROGENESIS IN THE "POSTDEVELOPMENTAL" SPINAL CORD?

It had long been held that neurogenesis in the mammalian central nervous system (CNS) was largely completed by birth. This opinion held sway until the early 1960s, when cellular proliferation was discovered in the subventricular zone (SVZ) (1) and neurogenesis was observed to occur in the olfactory bulb (2,3). The significance of these studies went unexplored until recently, when the ability to extract, grow, and reimplant neural stem cells (NSCs) into the brain forced a re-examination of the intrinsic capacity of the cerebrum to marshal endogenous stem cell pools (4–6). Similarly for the spinal cord, Adrian and Walker (7), employing ^{3}H-thymidine, a mitotic marker, labeled a small population of cells lining the central canal of the intact adult rat. Because these cells had only a limited life-span, however, their role was deemed to be trivial. The significance of this observation for spinal cord dysfunction and repair has also recently been revisited in light of the possible existence of NSCs within the spinal cord.

The use of bromodeoxyuridine (BrdU) a nucleotide analog that, like thymidine, is incorporated into dividing cells but which can be detected much less laboriously, has facilitated examination of proliferative neural cells *in situ*. Echoing the results of Adrian and Walker, latter day investigators have observed BrdU-positive cells within the adult spinal cord. Horner et al. (8) employed BrdU to label proliferating cells in the cord and follow their distribution and differentiation fate. They observed frequent cell division throughout the entire adult rat cord, particularly in the white-matter tracts. Most of the labeled cells colocalized with markers of immature glial

From: *Neural Stem Cells for Brain and Spinal Cord Repair*
Edited by: T. Zigova, E. Y. Snyder, and P. R. Sanberg © Humana Press Inc., Totowa, NJ

cells. After 4 wk, 10% of these cells expressed mature astrocyte and oligodendroglial markers. None differentiated into neurons, as assessed by their expression of either the mature neuronal marker, NeuN, or even the immature neuronal marker, βIII-tubulin. Interestingly, in contrast to what is observed during development where spinogenesis emerges from a centralized proliferative germinal zone lining the central canal, in the intact adult cord, dividing cells were located primarily in the outer circumference, a region limited to glia and fiber tracts (Fig. 1). Johansson et al. *(5)* did report rapidly proliferative cells in the ependyma lining the central canal following spinal cord injury, an observation that held significance for these authors because they hypothesized that the ependyma throughout the CNS harbors NSCs in adulthood. These spinal ependymal cells were reported to migrate outward radially from the center. However, the differentiated into astrocytes and participated in scar formation; no neurogenesis was observed *in situ.*

Although studies such as those above suggested that neurogenesis did not occur in the adult spinal cord in vivo, they contrasted with observations made on proliferative cells that were isolated from the adult spinal cord and expanded ex vivo in a tissue culture dish *(10,11).* When basic fibroblast growth factor (bFGF) was applied to dissociated neural cultures derived from the cervical, thoracic, lumbar, and sacral regions of the adult rat spinal cord, cells began to propagate, which, upon differentiation, could yield neurons in addition to glia, suggestive of multipotent progenitors or stemlike cells *(11).* Interestingly, when the cells were transplanted back into the adult spinal cord, a homotopic but non-neurogenic site (based on the BrdU and thymidine studies reported above), they differentiated into glial, but not neuronal, cell types. When cells from the same clone were transplanted into the adult hippocampus, a heterotopic but neurogenic site, the cells that integrated into the granular layer yielded neurons. These data suggest that the fate of adult neural progenitors from even non-neurogenic regions, such as the spinal cord, is not predefined or lineage restricted by their site of origin but, rather, is significantly determined by exogenous signals in their tissue environment *(12).* In general, it is believed that NSCs, regardless of their site of origin in the CNS (e.g., cerebellum, hippocampus, SVZ, cortex, or spinal cord), can properly respond to regional and developmental stage-specific environmental cues and differentiate into tissue-appropriate neurons in vivo *(13).*

SHIFTING FATES

What the above studies left unanswered was whether the normal regional and developmentally appropriate cues present within a given CNS structure

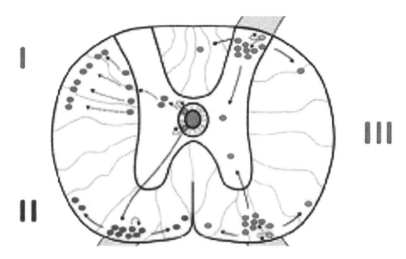

Fig. 1. Models of endogenous stem cell proliferation and migration in the intact adult spinal cord (as adapted from ref. *8*). Three models illustrate how dividing stem cells (solid circles) may give rise to progenitors (open circles) that migrate and proliferate (arrowed curves). Model I represents early postnatal gliogenesis. Horner et al. (*8*) demonstrated that dividing cells in the adult are located primarily in the outer circumference of the cord, suggesting that stem cell activity in the intact adult rat spinal cord—represented by Models II and III—give rise to glial or undifferentiated neural cells. This model does not represent what occurs when injury causes the death of motor neurons in the anterior horn gray matter. In that case, neurogenic signals appear to be re-expressed and NSCs may give rise to new neurons (*9*).

can be shifted from their normal trajectory by certain events or conditions. For example, can the non-neurogenic "postdevelopmental" spinal cord be "turned into" a neurogenic structure? Indeed, might this actually happen spontaneously and constitutively during certain types of pathological processes? One way to begin to answer this question might be to investigate the response of NSCs to pathological conditions within the cord (e.g., trauma, degenerative diseases such as amyotrophic lateral sclerosis [ALS] and multiple sclerosis [MS], and neurogenetic disorders such sphingolipid storage diseases) (*13*). In an early experiment designed to begin approaching this question for the CNS in general, a clone of murine NSCs was implanted into an adult mouse neocortex in which a circumscribed region was induced to undergo the selective and targeted apoptosis of layer II/III pyramidal neurons (*14*). These NSCs integrated into the regions of selective neuronal death; about 15% differentiated into neurons with many characteristics of the degenerated pyramidal neurons despite the fact that the developmental window for neuron birth—an event of *fetal* corticogenesis—had long since passed. The exogenous NSC-derived neurons extended axons and dendrites, received afferent synaptic contacts *(14)*, and projected axons to their normal target region on the contralateral side of the brain *(15)*. In intact control adult neocortices, NSCs engrafted in a similarly robust manner but yielded only non-neurons, the developmentally anticipated fate. Similarly, NSCs implanted into a necrotic, nonapoptotic lesion (induced by kainic acid) yielded non-neurons, suggesting that pathology can alter the CNS landscape to support neurogenesis even in non-neurogenic environments, but that the mode of neuronal death must be of a particular type, probably invoking molecular events associated with apoptosis. Wang et al. *(16)* observed that interneurons surrounding or actually serving as the synaptic neighbors of those apoptotic pyramidal neurons upregulate expression of a specific set of developmental signaling molecules, possibly representing the signals to which the exogenous NSCs are responding.

These early experiments in the cortex might explain observations similar to those cited above in which exogenous spinal-cord-derived NSCs yielded only non-neurons when implanted into the intact *(12)* or even lesioned *(5)* spinal cord. It has been well established that excitotoxicity (i.e., toxicity engendered by glutamate, aspartate, and kainic acid) is largely responsible for the secondary cell death observed after spinal cord injury *(17,18)*. It is likely that the microenvironmental alterations produced by synchronous *apoptotic* neuronal degeneration, but not by necrotic cell death, are both necessary and sufficient to permit multipotent NSCs to undergo a *neuronal* differentiation that ordinarily occurs only during embryonic corticogenesis *(14)*. Indeed, in the spinal cord, if one promotes the selective apoptotic degeneration of α-motor neurons (e.g., via sciatic axotomy), implanted NSCs will migrate to the motor neuron-deficient regions of the anterior horn and differentiate into motor neurons, suggesting that, in the cord as well, certain types of degeneration can "convert" a non-neurogenic region into one that supports neurogenesis *(9)*.

The question, of course, arises whether endogenous NSCs might occur spontaneously in the cord following degenerative processes. Magavi et al. *(19)* observed that synchronous apoptotic degeneration of adult murine corticothalamic neurons in the cerebrum (similar to that employed in ref. *14*) promoted the appearance of new neurons in the appropriate cortical layer even without transplantation, presumably the contribution of endogenous neural progenitors or stem cells. Park et al. *(20)* showed that such endogenous progenitors may not require special induction, but that ostensibly "hard-wired" developmental patterns are nevertheless spontaneously and constitutively "shifted" to effect cell replacement in response to such common and clinically-relevant degenerative processes as hypoxic–ischemic (HI) brain injury. By using two independent methods to label proliferative endogenous neural stem/progenitor cells in the SVZ—(1) a retroviral vector transducing the *lacZ* reporter gene and (2) BrdU—these authors tracked SVZ cells that entered the cell cycle at the time of HI in order to elucidate the brain's *intrinsic spontaneous* response to this severe prototypical cerebral insult, particularly as it involves this secondary germinal zone and endogenous progenitor pool. Unilateral HI increased the number of newly born cells in the SVZ. Interestingly, while *expanding* the periventricular population ipsilateral to the lesion, injury *reduced* the total number of newly born cells reaching the ipsilateral olfactory bulb, the stereotypical developmentally programmed destination of migratory SVZ cells in the intact brain. Rather, these newly proliferative cells appeared to be directed spontaneously from this migratory pattern toward the region of infarction. Indeed, HI appeared to increase significantly and diffusely the number of newly born cells throughout such classically non-neurogenic regions as the neocortex and striatum by the altered migration of SVZ cells and/or by the stimulation of local NPCs. Progenitors were even recruited from the contralateral ostensibly intact side. Many of these newly born cells differentiated into neurons (c-*fos*-active, decorated by synapsin-immunoreactivity) and oligodendrocytes (the two neural cell types most jeopardized by HI). These findings suggested that destructive injury—at least in the cerebrum—can spontaneously promote the widespread establishment of a significant number of new neurons, even in non-neurogenic CNS regions, suggesting that the brain invokes constitutive mechanisms of self-repair by mobilizing and even altering the stereotypical migratory/developmental patterns of an endogenous pool of NPCs without the need for specific induction.

In other preliminary studies focusing on the spinal cord, White et al. *(21)* have begun to explore the possibility that a neurodegenerative disease, like ALS, in which there is massive apoptotic motor neuron death in the spinal cord and brain stem, might actually be providing environmental cues supportive of neurogenesis. The SOD1^{G93A} transgenic mouse emulates one of the human familial forms of ALS and remains the best extant animal model for that disease. In pilot studies, SOD1^{G93A} transgenic mice were pulsed on 12 successive days with BrdU either at a presymptomatic stage (approx 60 d old) or at a later stage (approx 98 d old), along with age-matched wild-type littermate controls. At 4 wk following the last BrdU injection, very few BrdU+ cells were found in the spinal cord in the early symptomatic and wild-type control groups. Similarly, animals with a gradual progression of symptoms to later stages had results similar to that in wild-type controls. However, in rapidly progressive late-stage animals (in fact, those that reached the end stage so rapidly that they typically succumbed prior to the 4-wk postinjection goal), large numbers of BrdU+ cells were detected. In these late-stage/early-death animals, in addition to the many BrdU-labeled cells that colocalized with glial fibrillary acidic protein (GFAP), a marker for astrocytes, a small proportion of BrdU+ cells colocalized with βIII-tubulin, a marker for immature neurons. These data suggest that neurodegenerative diseases at times create an environment that promotes or supports neurogenesis and that there may be a threshold injury level that must be reached within a relatively short period in order to produce a signal sufficient for neurogenesis. Although such a dynamic (as observed in cerebrum and possibly spinal cord) is obviously ineffective in restoring full function in the most devastating of injuries and degenerative processes, its presence suggests that their numbers may be augmented by providing exogenous neural stem/progenitor cells and/or that supplying exogenous trophic/tropic factors during key postinjury temporal windows may better recruit and mobilize internal stem/progenitor cell pools.

This strategy has been attempted for spinal cord in at least three experimental situations in our laboratory, as detailed in the section below.

A ROLE FOR EXOGENOUS STEM CELLS IN THE ABNORMAL SPINAL CORD?

In preliminary studies, NSCs have been transplanted into lumbar cords of symptomatic adult SOD1^{G93A} mice (targeting the central canal or surrounding parenchyma). The impact of these NSCs was examined on the coordinated hind limb function of two groups of SOD1^{G93A} mice that were at different stages of disease progression (i.e., early–mid symptomatic (89–98 d old) vs end stage (near-death) (117 d old). It was hypothesized that this intervention would have an impact on motor function (although it was not designed specifically to affect the multiple pathological processes contributing to death). A single focal intraparenchymal, intracentral, and/or pericentral canal implantation of NSCs into the cord of symptomatic SOD1^{G93A} mice appeared to result in robust and extensive integration and migration of donor-derived cells across many segments of the lumbar cord, with many cells migrating into the ventral horn and differentiating into cells with the following morphological characteristics: large (approx 50–70 µm) polygonal perikarya bearing large nuclei, prominent nucleoli, and multiple processes. By Golgi stain criteria, such cells in the spinal anterior horn gray matter would be categorized as motor neurons, identical to similar host cells in the vicinity and to wild-type controls. Transplantation of NSCs into the lumbar spinal cord significantly improved rotorod performance of early symptomatic SOD1^{G93A} mice relative to those transplanted with control, non-neural cells. Although rotorod is the key behavioral test for assessing SOD1 mouse function, if translated to the Basso, Beattie, and Bresnahan (BBB) scale of locomotor function used in spinal cord injury *(22)*, these animals achieved an impressive 14–16 out of a possible 21; 0 indicates no hind limb function and 21 means normal locomotion. Animals attaining a BBB score of 16 can perform body-weight bearing, front and hind limb coordinated walking with plantar stepping, and a nearly normal paw position. (Although not the focus of this experiment, SOD1^{G93A} mice that received a single lumbar segment NSC transplantation did live 9 d longer on average when compared with those receiving control cells, at a weakly significant level.)

Therefore, it appeared that an NSC-mediated intervention could significantly delay progression of functional deficits in an anatomical distribution attributable to the region of cell engraftment. Interest, however, focused on the possible mechanisms underlying this effect. Although it appeared that this clone of NSCs had the ability to differentiate into cells with the morphological and immunocytochemical properties of neurons in the anterior horn, their number and degree of maturity did not seem adequate to account for the extent of functional improvement observed. Although neuronal replacement might underlie some degree of improvement, our focus shifted to the large number (indeed, the majority) of clearly donor NSC-derived *non*-neuronal cells in close proximity to such neurons that might theoretically provide trophic or neuroprotective support for both host and donor-derived motor neurons. Indeed, it was determined via enzyme-linked immunosorbent assay (ELISA) that this clone of NSCs constitutively produced a broad range of peptide neurotrophic and neurite outgrowth-promoting factors that functioned appropriately in a bioassay (the promotion of motor neuron outgrowth from a spinal explant). Of the various factors, GDNF was of greatest interest. Indeed, GDNF antisense and a soluble GDNF receptor scavenging GDNF were sufficient to blunt neurite outgrowth in this bioassay. Of additional interest was the observation that when the clonal NSC progeny existed in a *non*-neuronal (undifferentiated or glial) state, GDNF expression was robust but GDNF receptor (ret) expression was not present; yet, when induced to differentiate into neurons, GDNF production diminished, giving way to expression of a functional ret-receptor. This suggested a developmental mechanism by which a single "mother"

NSC will give rise to progeny that, in a symbiotic fashion, provide reciprocal support for each other. Although, teleologically, GDNF, for example, may be "intended" by *non*-neuronal progeny for support of their juxtaposed *neuronal* clonal members, this factor likely has a broad sphere of influence, including support and/or neuroprotection of host motor neurons. Hence, functional improvement might result from both the relatively small number of donor-derived neurons and, more significantly, from the large number of donor-derived non-neuronal cells. That GDNF seems to be expressed by donor-derived cells in vivo is suggested by preliminary immunocytochemical analysis. Accordingly, host SOD1 motor neurons appear (in preliminary analyses) to survive in greater numbers for longer periods of time in NSC implanted mice than in nonimplanted mice. Thus, early-onset SOD1^{G93A} mice that received NSC transplantation may have more anterior horn motor neurons spared relative to end-stage SOD1^{G93A} mice. That the amount of GDNF expressed by these NSCs does, indeed, have significance in vivo is supported by observations in the in vivo correlate of the in vitro bioassay in which the same NSCs are implanted as a graft into the dorsal funiculized rat spinal cord *(23)*. These cells promote the ingrowth of choline acetyltransferase (ChAT)-positive neuronal fibers into the graft from the host. Taken together, the above-summarized preliminary data suggest an additional hypothesis: NSC transplantation may protect spinal motor neurons of the SOD1^{G93A} mouse via the production of neurotrophic and/or neuroprotective factors (including the reduction of excitotoxicity). In this preliminary experiment, one injection in one region at one time-point was performed in all animals to assess most cleanly their impact on the function of a defined region of spinal cord. Future studies (and ultimate interventions) will likely include multiple NSC injections into multiple foci over successive time points within the SOD1^{G93A} spinal cord to determine if such a strategy would further increase the observed beneficial effects (including into near-death end-stage animals).

The salubrious effect of engrafted murine NSCs on somatomotor function of SOD1^{G93} transgenic mice has also served to suggest a little recognized mechanism by which cellular therapies, in general, may play a role for a range of neurological disorders. Although the degenerating host environment seems to instruct an implanted exogenous stem cell to alter its differentiation fate toward that of a lost cell (as in the sciatic axotomy model in which the abnormal host CNS prompted donor NSCs to become motor neurons), implanted NSC may also—in a reciprocal manner—alter or influence their host. Such a situation is most likely playing a pivotal role in a second experimental model we have been exploring in the spinal cord. NSCs

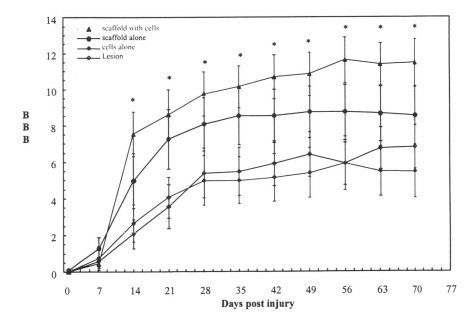

Fig. 2. Functional impact of NSCs seeded onto a biodegradable synthetic scaffold and implanted into the injury site of a hemisectioned adult rat spinal cord. The BBB score is a behavioral evaluation system that examines expanded local motor function of hind limbs after middle or lower thoracic spinal cord injury *(22)*. It employs a 0 to 21 scale, where 0 indicates no hind limb function and 21 represents normal locomotion. BBB open-field walking scores are shown for the four experimental groups on the hemisectioned side of the rat. The rate of improvement for the NSC-seeded scaffold group (triangles) was significantly higher than the rate for cells-alone ($p < 0.001$) (diamond) or lesion-control (circle) groups ($p = 0.004$) (two-way repeated measures of analysis of variance [ANOVA]). Overall, the scaffold-with-cells group showed significant improvement in open-field locomotion as compared to the cells-alone ($p = 0.006$) and lesion-control groups ($p = 0.007$) for all time-points from 14 d on. (ANOVA, Bonferroni post-hoc analysis).

were seeded onto a unique biodegradable synthetic scaffold (composed of polyglycolic acid and polylysine) and implanted into the injury site of adult rats subjected to a hemisection at spinal levels T9–10 *(26)*. Control animals received scaffolds alone, NSCs alone, or no implant. Hind-limb deficits were evaluated by a battery of behavioral tests on a weekly basis for at least 10 wk. At 70 d postimplant, treated animals exhibited coordinated weight-bearing stepping, whereas the lesion-control group only exhibited movement of two or three joints (Fig. 2). The level of functional improvement resulting from treatment with the NSC-seeded-scaffold persisted for at least 1 yr (at which time the study was terminated). Histology, immunocytochemistry (including for GAP-43 immunostaining), and tract tracing suggested that the long-term reduction in functional deficits resulted not from cell replacement by the NSCs but, rather, from their role in providing trophic support, reducing scar formation, increasing the amount of preserved host tissue by mitigating secondary cell death, and, possibly, by promoting host fiber regrowth and regeneration of damaged host tissue.

This is not to say that NSC-derived neurons cannot play a role in the restoration of function following spinal cord injury or in helping to promote reconnectivity. In preliminary studies, NSCs were transplanted into the contused region of an adult rat thoracic spinal cord. Donor-derived cells were identified in vivo by two independent means: expression of a stable lipophillic fluorescent dye and, most reliably, by species-specific markers, the NSCs having been of human origin. The NSCs engrafted well, migrated as far as one to two adjacent cord segments from their point of implantation, and differentiated into neurons (as well as glia). Some donor-derived neurons expressed ChAT, suggesting their differentiation toward motor neurons. Tract tracing suggested that some donor-derived neurons extended neurites at least four spinal segments caudal to the transplant site and extended processes as far rostral as the sensorimotor cortex. Accordingly, transplanted rats showed electrophysiological improvements in cortically recorded somatosensory and motor-evoked potentials and demonstrated hind-limb weight support and coordination on gait analysis (BBB of 14). These preliminary observations suggested that NSCs may yield neurons that can, in fact, become integrated into cortico-spinal and intraspinal circuitry, offering a possible substrate for functional improvement directed from higher centers.

McDonald and colleagues in independent experiments *(24,25)* have shown that neural progenitors derived from embryonic stem cells can mediate functional improvement after spinal cord injury by differentiating into oligodendrocytes that remyelinate host axons and seem, by that mechanism, to promote better conductivity.

SUMMARY AND CONCLUSIONS

In summary, we believe that endogenous stem/progenitor cells may naturally participate in some element of repair following injury or degeneration in the spinal cord as they appear to do in the brain. Although exogenous NSCs transplanted into the "non-neurogenic" normal adult spinal cord will not differentiate into neurons—an appropriate developmental response—in a degenerative milieu, particularly one characterized by apoptotic cell death, NSCs do appear to respond to "developmental" neurogenic signals that seem to be re-expressed in the "postdevelopmental" cord, if only transiently. Some of these neurons may integrate functionally. Stem cells may also differentiate into myelin-producing cells that promote an element of functional improvement by replacing degenerated host oligodendrocytes and myelinating host neuronal processes. However, transplanted NSCs may also exert a significant therapeutic impact even when not yielding replacement cells. They may do this (even when undifferentiated or differentiated into astrocytes) by inherently providing factors that, for example, neuroprotect, provide trophic support, forestall secondary injury, promote axonal regrowth and tissue regeneration, mobilize endogenous progenitors, and reduce scarring. The basic biology governing stem cell behavior must be further explored in order to develop new therapeutic strategies for neurodegenerative diseases, including those affecting the spinal cord.

REFERENCES

1. Sidman, R. L., Miale, I. L., and Feder, N. (1959) Cell proliferation and migration in the primitive ependymal zone: an autoradiographic study of histogenesis in the nervous system. *Exp. Neurol.* **1**, 322–333.
2. Altman, J. (1963) Autoradiographic investigation of cell proliferation in the brains of rats and cats. *Anat. Rec.* **145**, 573–591.
3. Altman, J. and Das, G. D. (1967) Postnatal neurogenesis in the guinea-pig. *Nature* **214(93)**, 1098–1101.
4. Lois, C. and Alvarez-Buylla, A. (1994) Long-distance neuronal migration in the adult mammalian brain. *Science* **264**, 1145–1148.
5. Johansson, C., Momma, S., Clarke, D., Risling, M., Lendahl, U., and Frisen, J. (1999) Identification of a neural stem cell in the adult mammalian central nervous system. *Cell* **96**, 25–34.
6. Doetsch, F., Caille, I., Lim, D., Garcia-Verdugo, J. M., and Alvarez-Buylla, A. (1999) Subventricular zone astrocytes are neural stem cells in the adult mammalian brain. *Cell* **97**, 703–716.
7. Adrian, E. K., Jr. and Walker, B. E. (1962) Incorporation of thymidine–H 3 by cells in normal and injured mouse spinal cord. *J. Neuropathol. Exp. Neurol.* **21**, 597–609.
8. Horner, P., Power, A., Kempermann, G., Kuhn, H., Palmer, T., Winkler, J., et al. (2000) Proliferation and differentiation of progenitor cells throughout the intact adult rat spinal cord. *J. Neurosci.* **20(6)**, 2218–2228.
9. Park, K. I., Liu, S., Flax, J. D., Nissim, S., Stieg, P. E., and Snyder, E. Y. (1999) Transplantation of neural progenitor & stem-like cells: developmental insights may suggest new therapies for spinal cord and other CNS dysfunction. *J. Neurotrauma* **16(8)**, 675–687.

10. Weiss, S., Dunne, C., Hewson, J., Wohl, C., Wheatley, M., Peterson, A. C., et al. (1996) Multipotent CNS stem cells are present in the adult mammalian spinal cord. *J. Neurosci.* **16,** 7599–7609.

11. Shihabuddin, L. S., Ray, J., and Gage, F. H. (1997) FGF-2 is sufficient to isolate progenitors found in the adult mammalian spinal cord. *Exp. Neurol.* **148(2),** 577–586.

12. Shihabuddin, L., Horner, P., Ray, J., and Gage, F. (2000) Adult spinal cord stem cells generate neurons after transplantation in the adult dentate gyrus. *J. Neurosci.* **20(23),** 8727–8735.

13. Snyder, E. Y. (1998) Neural stem-like cells: developmental lessons with therapeutic potential. *The Neuroscientist* **4(6),** 408–425.

14. Snyder, E. Y., Yoon, C., Flax, J. D., and Macklis, J. D. (1997) Multipotent neural precursors can differentiate toward replacement of neurons undergoing targeted apoptotic degeneration in adult mouse neocortex. *Proc. Natl. Acad. Sci. USA* **94(21),** 11,663–11,668.

15. Leavitt, B. R., Canales, M., Snyder, E. Y., and Macklis, J. D. (1996) Multipotent neural precursors transplanted to regions of targeted neuronal apoptosis in adult mouse somatosensory cortex differentiate into neurons & reform callosal projections. *Soc. Neurosci. Abstr.* **22,** 505.

16. Wang, Y., Sheen, V. L., and Macklis, J. D. (1998) Cortical interneurons upregulate neurotrophins in vivo in response to targeted apoptotic degeneration of neighboring pyramidal neurons. *Exp. Neurol.* **154,** 389–402.

17. Wrathall, J. R., Choiniere, D., and Teng, Y. D. (1994) Dose-dependent reduction of tissue loss and functional impairment after spinal cord trauma with the AMPA/kainate antagonist NBQX. *J. Neurosci.* **14,** 6598–6607.

18. Teng, Y. D., Mocchetti, I., Taveira-DaSilva, A. M., Gillis, R. A., and Wrathall, J. R. (1999) Basic fibroblast growth factor increases long-term survival of spinal cord motor neurons and improves respiratory function after experimental spinal cord injury. *J. Neurosci.* **19,** 7037–7047.

19. Magavi, S., Leavitt, B., and Macklis, J. (2000) Induction of neurogenesis in the neocortex of adult mice. *Nature* **405,** 951–955.

20. Park, K. I., Jensen, F. E., Stieg, P. E., and Snyder, E. Y. (1998) Hypoxic–ischemic injury may direct the proliferation, migation, and differentiation of endogenous neural progenitors. *Soc. Neurosci. Abstr.* **24,** 1310.

21. White, M. K. C. (2001) Potential for neurogenesis in amyotrophic lateral sclerosis. Senior thesis for bachelor of arts, Princeton University, Princeton, NJ.

22. Basso, D. M., Beattie, M. S., and Bresnahan, J. C. (1995) A sensitive and reliable locomotor rating scale for open field testing in rats. *J. Neurotrauma* **12,** 1–21.

23. Lu, P., Jones, L., Park, K. I., Snyder E. Y., and Tuszynski, M. (2000) Neural stem cells secrete BDNF & GDNF, & promote axonal growth after spinal cord injury. *Soc. Neurosci. Abstr.* **26,** 332.

24. McDonald, J. W., et al. (1999) Transplanted embryonic stem cells survive, differentiate and promote recovery in injured rat spinal cord. *Nat. Med.* **5,** 1410–1412.

25. Liu, S., et al. (2000) Embryonic stem cells differentiate into oligodendrocytes and myelinate in culture and after spinal cord transplantation. *Proc. Natl. Acad. Sci. USA* **97,** 6126–6131.

26. Teng, Y. D., Lavik, E., Qu, X., Park, K. I., Ourednik, J., Langer, R., et al. (2000) Transplantation of neural stem cells seeded in biodegradable polymer scaffold ameliorates long-term functional deficits resulting from spinal cord hemisection in adult rats. American Society of Neural Transplantation and Repair; Meeting, Vol 7, p. 61.

13
Stem Cells for Spinal Cord Injury

Paul Lu, Evan Y. Snyder, and Mark H. Tuszynski

INTRODUCTION

Spinal cord injury (SCI) frequently results in the devastating loss of such neurological functions as mobility, sensation, and autonomic control below the level of the lesion. A number of regenerative approaches to possibly regain these functions have been examined in SCI models. Recent progress in isolating and culturing multipotent neural stem cells (NSCs) and neural progenitor cells from developing or even adult CNS tissue (*1–8*) has allowed neural stem cells to emerge as an intriguing and potentially useful means of promoting neural repair.

In this chapter, we will discuss briefly the fundamental problems associated with SCI, the putative mechanisms responsible for the failure of CNS regeneration, and some possible strategies for promoting regeneration. We will then focus on NSCs and progenitors. Possible sources of NSCs and progenitors for spinal cord transplantation will be discussed, together with the characterization and manipulation of these cells. Next, we will review findings from studies that have transplanted NSCs and progenitors into the injured spinal cord. Finally, the use of NSCs as vehicles for gene delivery in SCI will be discussed.

SPINAL CORD INJURY

Functional loss after SCI results from the initial direct damage of the cord and a secondary cascade of inflammation and excitotoxic damage that considerably worsens the extent of damage and cell loss (for reviews, *see* refs. *9* and *10*). SCI not only directly damages neural cell bodies at the site of injury but also, and more importantly, disrupts descending and ascending axonal pathways that traverse the injury site. The clinical consequences of the injury are often permanent loss of sensory, motor, and autonomic function. This permanent loss occurs because the adult mammalian CNS

From: *Neural Stem Cells for Brain and Spinal Cord Repair*
Edited by: T. Zigova, E. Y. Snyder, and P. R. Sanberg © Humana Press Inc., Totowa, NJ

is unable to regenerate severed axons. When axons within the CNS are transected, they exhibit an initial minimal growth response (sprouting) *(11)*. Although sprouting can result in some functional recovery (e.g., *see* ref. *12*), extensive axonal regeneration does not occur in the CNS. The poor regenerative capability of the CNS contrasts to that of many other mammalian tissues, including peripheral nerve, which exhibit greater regenerative potential.

Several factors account for the failure of the CNS to regenerate: (1) the lack of available neurotrophic factors and substrate molecules to which growing neurites can respond and attach, (2) the presence of axonal growth inhibitor molecules *(13–15)*, (3) the formation of a glial scar as a physical impediment to axonal regrowth *(10,16,17)*, and (4) changes in gene expression in injured neurons, such as a failure to upregulate growth-associated proteins. In contrast, injuries to peripheral nerves are frequently followed by efficient removal of cellular debris, upregulation of genes correlated with axonal growth, the production of growth factors, the secretion of an extracellular matrix through which injured axons can attach and extend, and mechanical guidance to appropriate targets via connective tissue and basal lamina sheaths in the peripheral nerve. Thus, experimental strategies to enhance CNS regeneration typically provide one or more of these growth-promoting elements to the injured CNS. With such approaches, some degree of enhancement of CNS axonal growth and, in some cases, partial functional recovery have been described *(18–30)*. The remainder of this chapter will focus on the use of neural stem cells as another promising strategy for enhancing CNS repair.

NEURAL STEM CELLS AND PROGENITORS

Neural stem cells and neural progenitor cells can be isolated from various regions of CNS, including embryonic and adult spinal cord *(31–38)*. In addition, NSCs and neural progenitor cells can be derived from more primitive cells, such as embryonic stem (ES) cells, that have the capacity to generate NSCs and stem cells of other tissues *(7,39,40)*. It is unknown whether stem cells from different regions of the CNS, including the spinal cord, behave differently in vitro and in vivo when implanted into intact or lesioned spinal cord. As we learn more about the basic biology of NSCs and neural progenitor cells and the pathophysiology of SCI, the optimal cell type(s) for spinal cord repair will be determined.

Some NSCs and neural progenitors can be maintained and propagated in vitro for extended periods of time without losing their multipotentiality. This is achieved through either genetic or epigenetic manipulation of stem cells to induce cell proliferation and to inhibit cell differentiation in culture. Genetic manipulation usually involves transduction and chromosomal integration of a propagating gene, such as v-*myc* or SV40 large T-antigen (for review, *see* refs. *41–44*). Several NSC lines have been immortalized and generated from different regions of embryonic or neonatal rodent brain *(45–51)*. For studies of spinal cord injury, immortalized NSCs and progenitor cells can be readily expanded and cultured and, thus, provide a potentially limitless source for transplantation after SCI. In addition, these cells can be maintained and genetically transduced as a cell line in culture, allowing expression of therapeutic genes.

For epigenetic manipulation of NSCs or progenitors, high concentrations of mitogens, such as fibroblast growth factor-2 (bFGF) or epidermal growth factor (EGF), are added to either a defined or supplemented medium to maintain stem cell proliferation *(7,43)*. Using this approach, NSCs and neural-restricted precursors have been successfully isolated from embryonic *(32,33,35)* and adult spinal cord *(31,37,52)* as well as from cultured ES cells *(39,40,53)*. For spinal cord applications, epigenetically manipulated NSCs and progenitors derived from the spinal cord could reconstitute damaged tissue. For example, neuronal-restricted precursors reportedly generate multiple neuronal phenotypes in culture, which could hypothetically reconstruct gray matter of the spinal cord *(33,40)*. Glial-restricted precursors could hypothetically also be induced to differentiate into astrocytes or oligodendrocytes, which, in turn, might reconstitute white matter *(35,36)* or provide a cellular source for remyelinating injured axons *(39,53)*.

TRANSPLANTATION OF NSC AND PROGENITORS TO THE INJURED SPINAL CORD

In this section, we will review recent studies implanting NSCs and progenitors in animal models of spinal cord injury. These models include the chemical or X-irradiated demyelination of the spinal cord, where various types of NSCs and progenitors, especially glial progenitors, are grafted to test their ability to differentiate into the various cell types and to myelinate host axons (for reviews, *see* refs. *54* and *55*). In addition, a genetic model of demyelination has been established in mice and rats. Also, traumatic injury models have been used to examine the effects of NSC or progenitor cell transplantation on axonal regeneration and neural cell replacement. We will discuss several studies of neuronal replacement strategies for SCI by graft of NSCs or progenitors. Finally, we will focus on our work of implanting NSCs and progenitors in lesioned spinal cord to promote host axonal regeneration.

Transplantation of NSCs and Progenitors
into Demyelinated Spinal Cord

Different types of NSCs and progenitors have been transplanted into demyelinated spinal cord. These cells include NSCs, glial progenitors, oligodendrocyte progenitors, and ES cells. Studies show that transplanted NSCs and progenitors, particularly oligodendrocyte progenitors, can myelinate host axons in a variety of animal models (39,53,55).

NSC Transplantation

To determine whether NSCs can differentiate into oligodendrocytes and myelinate host axons in myelin-deficient rat spinal cord, Hammang et al. (56) isolated a population of multipotent NSCs from embryonic murine brain. The NSCs can be maintained in an undifferentiated state in vitro in the presence of EGF and can be induced to differentiate into neurons, astrocytes, and oligodendrocytes by removing growth factor and adding serum. A myelin-deficient (md) mutant rat spinal cord was used as a demyelination model, characterized by reduction of oligodendrocyte number and a concomitant absence of CNS myelin. When the undifferentiated NSCs were transplanted into the mutant myelin-deficient rat spinal cord, these cells preferentially differentiated into oligodendrocytes and extensively myelinated host axons. This study suggests that multipotent NSCs can respond to relevant signals from mutant spinal cord and differentiate into an oligodendrocyte lineage. The ability of NSCs to differentiate into a specific lineage in vivo underscores the potential for these cells to repair injured spinal cord.

In another experiment, EGF-responsive NSCs were committed to the oligodendroglial lineage by exposure to B104 cell-conditioned medium in vitro before transplantation (57). The oligodendroglial progenitors formed cell clumps called oligospheres. The differentiation of multipotent NSCs to oligodendroglial progenitors is associated with the expression of two cytokine receptors: platelet-derived growth factor-α (PDGF) and basic FGF. However, EGF receptor was downregulated. When oligospheres were transplanted into myelin-deficient mutant rat spinal cord, transplanted cells expressed myelin-basic protein, myelin-associated glycoprotein, and myelin oligodendrocyte glycoprotein, and the cells myelinated host axons. Furthermore, this study found that more myelin was formed in vivo from grafted oligodendroglial progenitors than grafted multipotent NSCs, suggesting that lineage-restricted progenitors may be a useful approach for remyelinating the spinal cord.

Glial Progenitor Transplantation

Because more remyelination was observed using cell preparations enriched in glial progenitors *(57)*, attempts have been made to identify an optimal source of progenitors (as opposed to stem cells) that are capable of remyelination. One possible source is the CG-4 cell, an oligodendrocyte type-2 progenitor cell line *(58,59)*. Studies have shown that CG4 cells survive in vivo grafting and migrate extensively following transplantation into the X-irradiated spinal cord. In addition, CG4 cells enter regions of demyelination, reportedly remyelinating host axons. In contrast, grafted CG4 cells did not survive when grafted into the intact spinal cord. These findings suggest that the environment of the demyelinated adult CNS is unique in supporting survival and phenotypic differentiation of glial progenitors *(59)*.

Because neural progenitors expressing the embryonic polysialylated form of neural cell adhesion molecule (PSA-NCAM) have been reported to generate mostly oligodendrocytes and astrocytes in vitro, Keirstead et al. *(60)* isolated and cultured PSA-NCAM neural progenitors by immunoselection from a mixture of cells obtained from neonatal rat brain. They first chemically demyelinated the adult rat spinal cord, then suppressed remyelination using X-irradiation. Following grafting of PSA-NCAM-immunoreactive neural progenitors into this model, remyelination of host axons was reported.

Neural progenitors can also be derived from the subventricular zone (SVZ). Studies show that neonatal SVZ is a rich source of multipotent neural progenitors *(61–63)*. Smith and Blakemore *(64)* reported that neural progenitors isolated from neonatal pig SVZ and exposed to B104-conditioned medium differentiated into oligodendrocytes and remyelinated the spinal cord.

Embryonic Stem Cell Transplantation

Neural stem cells and progenitors have also been derived from ES cells. Recently, two studies reported that ES-cell-derived oligodendrocyte progenitors myelinate host axons after transplantation into demyelinated spinal cord *(39,53)*. In one study *(39)*, ES cells were induced to differentiate into glial progenitors through sequential propagation using medium containing different growth factors. Then, glial progenitors were grafted into the spinal cord of 1-wk-old myelin-deficient rats. Two weeks after transplantation, ES-cell-derived progenitors reportedly differentiated into myelinating oligodendrocytes and astrocytes and generated myelin sheaths.

In addition, grafted myelin-forming progenitors were not restricted to the implant site but had spread in both longitudinal and horizontal directions.

In a second study *(53)*, ES cells were induced to differentiate into NSCs by exposure to retinoic acid. NSCs cultured in neurobasal medium or SATO-defined medium supplemented with serum resulted in production of a mixed culture of neurons, astrocytes, and oligodendrocytes. To enrich cultures in oligodendrocytes and their progenitors, the authors mechanically separated oligodendrocyte progenitors from other types of cell and cultured them as "oligospheres." When these "oligospheres" were transplanted into the dorsal columns of adult rat spinal cord 3 d after chemical demyelination, many ES-cell-derived progenitors survived and differentiated primarily into oligodendrocytes that reportedly remyelinated host axons. In addition, when the ES-cells-derived oligospheres were transplanted into the spinal cord of myelin-deficient *shiverer* (*shi/shi*) mutant mice, these cells became oligodendrocytes that expressed myelin basic protein and myelinated host axons *(53)*. These results demonstrate that ES cells can be programmed to commit to a specific neural cell lineage, the oligodendrocyte, and remyelinate host axons of the adult spinal cord.

NSC Grafts and Neuronal Replacement for SCI

Transplantation of NSCs and neural progenitors at a site of spinal cord injury could potentially replace damaged or lost host neurons and reconstitute local circuitry. In the past decade, NSCs and progenitors have been developed from various sources of CNS tissues and from ES cells. One such cell line, RN33B, has been developed from embryonic rat raphe nuclei immortalized with the temperature-sensitive mutant of the SV40 large T-antigen (tsT-ag) *(65)*. This cell line is mitotically active at 33°C, but becomes quiescent at ≥37°C. In vitro, RN33B NSCs are neuronal restricted, differentiating into cells that express neuron-specific antigens such as neuron-specific enolase (NSE) and neurofilament, as well as glutamate. When these cells were transplanted into different areas of the adult or neonatal CNS, they reportedly formed mature phenotypes that were not observed in vitro, indicating that the further differentiation of these cells was directed by local signals *(66,67)*.

To determine the extent of RN33B NSC differentiation potential in the injured spinal cord, the cells were transplanted into either normal, compressed, or transected spinal cords *(68)*. In the intact spinal cord, the majority of grafted RN33B cells differentiated into neuronal cells with bipolar morphologies 2 wk after transplantation. Some grafted cells in spinal cord gray matter adopted a multipolar morphology that, according to the authors, resembled α-motor neurons; however, expression of specific motor neurons' phenotypic markers (e.g., choline acetyltransferase) was not documented. When the same RN33B NSCs were grafted into lesioned spinal cords, cells with bipolar rather than multipolar morphology were observed. This result suggests that multipolar cell differentiation might be mediated by local signals expressed in the intact spinal cord but not in the injured spinal cord. Although grafted RN33B NSCs show morphological differentiation toward neurons, immunocytochemistry revealed that grafted cells did not express either the neuronal marker, choline acetyltransferase (ChAT), or the astrocytic marker, glial fibrillary acidic protein (GFAP) *(68)*. The limited differentiation of RN33B NSCs in vivo is probably the result of a lack of appropriate signals from the adult spinal cord.

In a later study, a member of this same group of investigators attempted to develop a NSC line from human fetal CNS tissue *(5)*. However, immortalizing human CNS cells using similar approaches resulted in significant chromosomal aberrations. In another approach, freshly isolated human NSCs were cultured in the presence of EGF and bFGF and could be induced to differentiate into neurons and astrocytes at early passages. However, at later passages, cells became restricted to the astrocyte lineage. Further study is needed to find an approach that will maintain these cells as undifferentiated cells and to examine their differentiation potential in intact and injured spinal cord.

In addition to NSC transplantation, mouse ES cells have been transplanted into rat spinal cord after contusion injury *(27)*. The grafted stem cells survived at least 5 wk, migrated at least 8 mm from the injection site, and differentiated into astrocytes, oligodendrocytes, and neurons. Injured rats in this study also showed partial functional recovery. A subsequent study from this group showed that some grafted ES cells remyelinated host axons, suggesting this as a potential mechanism for functional recovery *(53)*.

For neuronal replacement in the injured spinal cord, another interesting cell type is the neuronal-restricted-precursor (NRP) cell, originally derived from embryonic rat spinal cord *(33,34)*. NRPs can also be isolated from mouse neural tube and from cultured ES cells *(40)*. Undifferentiated NRPs express embryonic NCAM and a subset of immature neuronal markers, are mitotically active and electrically immature. When fully mature neurons, including motoneurons, were developed from NRPs, these cells were postmitotic, process bearing, and expressed two mature neuronal markers: neurofilament-M and synaptophysin. These cells also synthesized and responded to various neurotransmitter molecules. Because of these characteristics, NRPs also have the potential to either replace lost neurons in the injured spinal or to support host axonal regeneration, although, to date, there have been no published reports on the use of these cells in models of SCI.

NSC Grafts and Axonal Regeneration

The preceding experiments have primarily focused upon the hypothesis that neural stem cells can be useful for host axonal *remyelination* and host neuronal *replacement* after injury to the nervous system. Our own laboratory has examined in vivo effects of neural stem cells in a slightly different context, by testing that hypothesis that stem cells can promote *host axonal regeneration* after SCI. These experiments have used a prototypical NSC line, C17.2, from the laboratory of Evan Snyder. C17.2 cells were originally derived from the external germinal layer of neonatal mouse cerebellum and immortalized with the propagation gene, v-*myc* (46,69). As a NSC line, C17.2 is able to participate in the normal development of the CNS (46,70,71). At all developmental stages, transplanted C17.2 NSCs reportedly give rise to variety of neurons and glia. For example, C17.2 NSCs can differentiate into pyramidal and nonpyramidal neurons when transplanted into embryonic mouse brain during normal corticogenesis (72). However, C17.2 NSCs either differentiate into glia or become quiescent progenitors when transplanted into the postnatal neocortex during normal gliogenesis. In addition, grafting C17.2 NSCs in models of abnormal development and adult injury of the CNS can correct these abnormalities and regenerate lost neurons at the injury site (71,73–75). For example, C17.2 NSCs can differentiate into neurons that replace apoptotic host neurons in adult mouse neocortex (74).

To examine in vivo effects of C17.2 NSCs on *host* axonal regeneration, adult Fischer 344 rats underwent lesions of the C3 cervical spinal cord using a microwire knife lesion. This focal lesion completely transects the dorsal columns of the spinal cord bilaterally, thereby disrupting both the descending motor corticospinal projection and the ascending dorsal column proprioceptive pathways (76). Immediately after performing the lesion, rats received injections of 300,000 C17.2 NSCs into the lesion site in a volume of 3 µL. The grafted cells were prelabeled by genetically modifying them to express the reporter gene, green fluorescent protein (GFP).

Two and four weeks after in vivo grafting to the spinal cord lesion site, C17.2 cells survived well (Fig. 1) and completely filled the lesion site. Grafts were well vascularized but did not form tumors. Grafted NSCs migrated only relatively short distances of up to 5 mm from the lesion/injection site. Examination of the differentiation state of the grafted cells 2 wk after in vivo grafting indicated that the cells remained undifferentiated, expressing

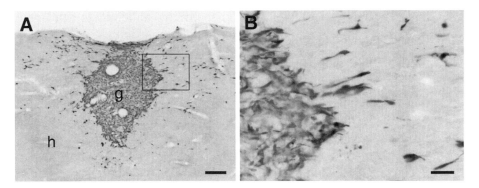

Fig. 1. In vivo morphology of stem cell graft. (**A**) GFP-expressing C17.2 NSC graft in a C3 wireknife lesion in a sagittal spinal cord section. (**B**) High magnification of (A), demonstrating morphology of grafted cells bearing short processes. g, graft; h, host; scale bars = 177 μm (A) and 35 μm (B).

neither neuronal or glial markers in the spinal cord lesion site. Labeling for the mature neuronal markers, Neu N and neurofilament, as well as the mature glial markers, GFAP (for astrocytes) and adenomatous polyposis coli tumor suppressor protein (for oligodendrocytes), was not present in the transplanted stem cells. The presence of labeling for the early neural cell line marker, nestin, indicated that the grafted cells persisted in an undifferentiated state in vivo after the relatively short time of 2 wk.

Despite the absence of stem cell differentiation, or possibly *because* of it (see below), specific classes of host axons grew extensively within the grafts placed in the lesion cavity. Specifically, neurofilament-labeled axons penetrated C17.2 cell grafts in threefold greater numbers than control fibroblast grafts (Fig. 2). Labeling for specific axonal markers indicated that cerulo-spinal axons labeled with tyrosine hydroxylase, motor axons labeled with ChAT, and sensory axons labeled with calcitonin gene-related peptide (CGRP) penetrated C17.2 grafts to a significantly greater extent that control grafts in the lesion site (Figs. 3A,B). These findings indicate that NSCs can indeed support host axonal regeneration. It is highly unlikely that these axons originated from the NSCs for three reasons: (1) there was no evidence that the NSC somata differentiated and produced any of these neurotransmitters, (2) GFP labeling did not disclose the presence of extensive neuritic processes emerging from the NSCs in vivo, and (3) axons could be observed penetrating the grafts directly from the host spinal cord.

We then measured neurotrophin production from NSC cultures in vitro prior to grafting to examine potential mechanisms whereby NSCs might

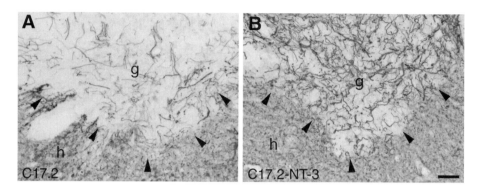

Fig. 2. Axonal penetration into C17.2 and C17.2–NT-3 NSC grafts. **(A)** Neurofilament (NF) immunolabeling in a sagittal spinal cord section showing penetration of axons into stem cell graft. **(B)** Significantly increased numbers of NF-labeled axons penetrate C17.2–NT-3 NSC grafts. g, graft; h, host; arrowheads indicate host/graft interface; scale bar = 71 μm.

be promoting host axonal growth. Specific enzyme-linked immunosorbent assays (ELISAs) were performed on conditioned medium collected from C17.2 NSCs, and nerve growth factor (NGF), brain-derived neurotrophic factor (BDNF), neurotrophin-3 (NT-3), and GDNF protein levels were measured *(77)*. Whereas primary fibroblasts expressed no detectable levels of these growth factors, conditioned medium from C17.2 NSCs contained significant quantities of NGF (7.5 ± 2.5 pg/10^6 cells/d), BDNF (7.1 ± 0.1 pg/10^6 cells/d), and GDNF (70 ± 1 pg/10^6 cells/d). This pattern of growth factor production precisely correlates with the observed growth of axons into the respective graft types (e.g., sensory axons into NGF-producing grafts, motor axons into GDNF-producing grafts). Thus, a likely mechanism of stem-cell-induced axonal growth after SCI is the production of growth factors. As noted above, it may be the case that stem cells remaining in an undifferentiated state may produce more growth factors than committed-fate cells and may, therefore, be superior tools for promoting host axonal regeneration; this possibility is being examined.

Thus, these findings indicate that neural stem cells can provide both permissive substrates and secrete neurotrophic factors to promote the regeneration of *host* axons.

NEURAL STEM CELLS AND SPINAL CORD GENE THERAPY

In a further set of studies, we examined the hypothesis that genetic modification of stem cells to produce greater quantities of nervous system

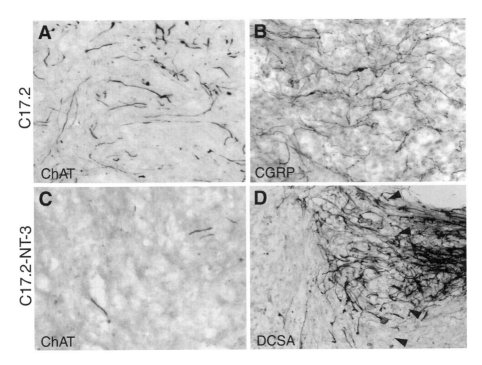

Fig. 3. Immunolabeling of different axonal phenotypes penetrating C17.2 and C17.2–NT-3 NSC grafts. **(A)** ChAT-labeled local spinal motor axons extensively penetrate C17.2 NSC graft. **(B)** CGRP-labeled local spinal sensory axons robustly penetrate C17.2 NSC graft. **(C)** However, fewer ChAT-labeled local spinal motor axons penetrate NT-3 transduced C17.2 NSC graft. **(D)** NT-3 transduced C17.2 NSC graft is penetrated by cholera toxin B subunit-labeled dorsal column sensory axons (DCSA), which readily cross the host/graft interface (arrowheads). Scale bars = 35 µm in (A–C) and 71 µm in (D).

growth factors would further augment their ability to support the growth of injured host spinal cord axons. In particular, because the experiments reviewed above indicated that stem cells produce no detectable quantities of NT-3, we used retroviral vectors *(22,78)* to genetically modify C17.2 cells to produce and secrete human NT-3. Previously, we and others reported that NT-3 promotes the growth of spinal cord corticospinal axons *(18,22)*. Large, myelinated dorsal root ganglion axons are also NT-3-sensitive *(79,80)*. Thus, after genetically modifying C17.2 cells to secrete NT-3, they were implanted into dorsal column lesion sites at the C3 level; effects on host corticospinal and dorsal column sensory axons were examined 2 wk later.

In vitro prior to grafting, NT-3 transduced cells secreted 12 ng NT-3/10^6 cells/d, whereas untransduced C17.2 NSCs expressed no NT-3. Interestingly, transduction of C17.2 NSCs to express NT-3 also induced changes in production of other growth factors by the stem cells. NT-3-transduced stem cells no longer secreted detectable quantities of GDNF, but they secreted significantly greater quantities of NGF and BDNF. Transduction of C17.2 cells to produce NT-3 also altered their morphology and differentiation state in vitro: C17.2 cells assumed a more pyramidal morphology and extended processes, and the majority (approx 60%) of cells expressed a neuronal marker, microtubule-associated protein (MAP)-2, (see also ref. *81*), whereas only 3–4% expressed the astroglial marker, GFAP. Thus, in vitro transduction of C17.2 cells to express NT-3 changed their state of differentiation and their natural pattern of growth factor production.

Cells were then harvested and grafted in vivo at the C3 level. Two weeks later, cells survived well and were densely penetrated by host axons. Quantification revealed that there was a fourfold increase of NF-labeled axons within NT-3-secreting C17.2 NSC grafts compared to untransduced C17.2 NSCs (Fig. 2). This result indicated that overexpression of NT-3 further stimulates the axonal growth into the grafts and is consistent with a previous report using NT-3 transduced fibroblast grafts in injured adult spinal cord *(82)*. Immunolabeling to identify specific axonal classes penetrating grafts demonstrated a significant increase in the penetration of dorsal column sensory axons (detected by transport of cholera toxin B subunit) into the lesion site (Fig. 3D). A previous study showed that dorsal column axonal regeneration can be induced by infusion of NT-3 *(80)*. Significant penetration by CGRP-labeled sensory axons were also observed in NT-3-secreting C17.2 cell grafts. Interestingly, a *reduction* in growth of ChAT-labeled motor axons into grafts was observed (Fig. 3C), correlating with observed elimination of GDNF production by NT-3-transduced C17.2 cells. Augmented growth of lesioned corticospinal axons was not detected either into the lesion site or in host spinal cord surrounding the lesion, although the relatively brief postlesion time of 2 wk may have been insufficient to reflect an effect on this system.

Thus, these findings indicate that the differentiation state and growth-promoting properties of neural stem cells may be altered by transduction to secrete other growth factors. These properties may be useful in designing optimal strategies to support neural repair.

CONCLUSIONS AND FUTURE PROSPECTS

This chapter indicates that neural stem cells have the potential to enhance regeneration after SCI via two distinct mechanisms: by replacing cells that are lost as a consequence of injury, such as host oligodendrocytes, and by directly promoting host axonal regeneration through the lesion site. Stem cells can support axonal regeneration both by providing a cellular and extracellular scaffold to which host axons can attach and extend through the lesion site and by secreting growth factors. Further, as indicated earlier, the regeneration potential of stem cells can be further manipulated by gene therapy.

Future strategies using stem cell approaches can explore several potential questions: (1) Can lineage-restricted cells, such as NRPs *(33,34,40)* or glial-restricted precursors *(34–36)*, impart specific properties to sites of SCI to augment regeneration? For example, NRPs could repopulate neurons lost at the site of injury, whereas glial-restricted precursors possessing properties of radial glia might support axonal growth or guidance. (2) Can stem cell strategies be combined with other approaches, such as myelin neutralization *(25,29,83,84)* or degradation of the extracellular matrix *(85)*, to enhance regeneration? (3) Can stem cells be isolated readily from the adult spinal cord, bone marrow, or other sources to provide an autologous cell source for spinal cord regeneration *(31,37,86–93)*? Studies in progress are exploring these intriguing possibilities.

ACKNOWLEDGMENTS

We are grateful to Mr. Casey Cox for editorial assistance. This work was supported by the Brodie Lockard Foundation, the Veterans Administration, and the NIH (R01 NS37083).

REFERENCES

1. Snyder, E. Y., Park, K. I., Flax, J. D., Liu, S., Rosario, C. M., Yandava, B. D., et al.(1997) Potential of neural "stem-like" cells for gene therapy and repair of the degenerating central nervous system. *Adv. Neurol.* **72,** 121–132.
2. Ourednik, V., Ourednik, J., Park, K. I., and Snyder, E. Y. (1999) Neural stem cells—a versatile tool for cell replacement and gene therapy in the central nervous system. *Clin. Genet.* **56,** 267–278.
3. Ray, J., Palmer, T. D., Shihabuddin, L. S., and Gage, F. H. (1999) The use of neural progenitor cells for therapy in the CNS disorders, in *CNS Regeneration: Basic Science and Clinical Advances* (Tuszynski, M. H. and Kordower, J. H., eds.). Academic, San Diego, CA, pp. 183–202.
4. Shihabuddin, L. S., Palmer, T. D., and Gage, F. H. (1999) The search for neural progenitor cells: prospects for the therapy of neurodegenerative disease. *Mol. Med. Today* **5,** 474–480.
5. Whittemore, S. R. (1999) Neuronal replacement strategies for spinal cord injury. *J. Neurotrauma* **16,** 667–673.
6. Björklund, A. and Lindvall, O. (2000) Cell replacement therapies for central nervous system disorders. *Nat. Neurosci.* **3,** 537–544.

7. Gage, F. H. (2000) Mammalian neural stem cells. *Science* **287,** 1433–1438.
8. Lee, J. C., Mayer-Proschel, M., and Rao, M. S. (2000) Gliogenesis in the central nervous system. *Glia* **30,** 105–121.
9. Schwab, M. E. and Bartholdi, D. (1996) Degeneration and regeneration of axons in the lesioned spinal cord. *Physiol. Rev.* **76,** 319–370.
10. Giménez y Ribotta, M. and Privat, A. (1998) Biological interventions for spinal cord injury. *Curr. Opin. Neurol.* **11,** 647–654.
11. Ramon y Cajal, S. (1928/1991) *Degeneration and Regeneration of the Nerve System.* Hafner, New York.
12. Weidner, N., Ner, A., Salimi, N., and Tuszynski, M. H. (2001) Spontaneous corticospinal axonal plasticity and functional recovery after adult central nervous system injury. *Proc. Natl. Acad. Sci. USA* **98,** 3513–3518.
13. Nieto-Sampedro, M. (1999) Neurite outgrowth inhibitors in gliotic tissue. *Adv. Exp. Med. Biol.* **468,** 207–224.
14. Chen, M. S, Huber, A. B., van der Haar, M. E., Frank, M., Schnell, L., Spillmann, A. A., et al. (2000) NoGo-A is a myelin-associated neurite outgrowth inhibitor and an antigen for monoclonal antibody IN-1. *Nature* **403,** 434–439.
15. GrandPré, T., Nakamura, F., Vartanian, T., and Strittmatter, S. M. (2000) Identification of the Nogo inhibitor of axon regeneration as a Reticulon protein. *Nature* **403,** 439–444.
16. Fawcett, J. W. and Asher, R. A. (1999) The glial scar and central nervous system repair. *Brain Res. Bull.* **49,** 377–391.
17. Fitch, M. T., Doller, C., Combs, C. K., Landreth, G. E., and Silver, J. (1999) Cellular and molecular mechanisms of glial scarring and progressive cavitation: in vivo and in vitro analysis of inflammation-induced secondary injury after CNS trauma. *J. Neurosci.* **19,** 8182–8198.
18. Schnell, L., Schneider, R., Kolbeck, R., Barde, Y. A., and Schwab, M. E. (1994) Neurotrophin-3 enhances sprouting of corticospinal tract during development and after adult spinal cord lesion. *Nature* **367,** 170–173.
19. Xu, X. M., Guénard, V., Kleitman, N., Aebischer, P., and Bunge, M. B. (1995) A combination of BDNF and NT-3 promotes supraspinal axonal regeneration into Schwann cell grafts in adult rat thoracic spinal cord. *Exp. Neurol.* **134,** 261–272.
20. Cheng, H., Cao, Y., and Olson, L. (1996) Spinal cord repair in adult paraplegic rats: partial restoration of hind limb function. *Science* **273,** 510–513.
21. Tuszynski, M. H., Gabriel, K., Gage, F. H., Suhr, S., Meyer, S., and Rosetti, A. (1996) Nerve growth factor delivery by gene transfer induces differential outgrowth of sensory, motor, and noradrenergic neurites after adult spinal cord injury. *Exp. Neurol.* **137,** 157–173.
22. Grill, R., Murai, K., Blesch, A., Gage, F. H., and Tuszynski, M. H. (1997) Cellular delivery of neurotrophin-3 promotes corticospinal axonal growth and partial functional recovery after spinal cord injury. *J. Neurosci.* **17,** 5560–5572.
23. Li, Y., Field, P. M., and Raisman, G. (1997) Repair of adult rat corticospinal tract by transplants of olfactory ensheathing cells. *Science* **277,** 2000–2002.

24. Rapalino, O., Lazarov-Spiegler, O., Agranov, E., Velan, G. J., Yoles, E., Fraidakis, M., et al. (1998) Implantation of stimulated homologous macrophages results in partial recovery of paraplegic rats. *Nat. Med.* **4,** 814–821.
25. Huang, D. W., McKerracher, L., Braun, P.E., and David, S. (1999) A therapeutic vaccine approach to stimulate axon regeneration in the adult mammalian spinal cord. *Neuron* **24,** 639–647.
26. Liu, Y., Kim, D., Himes, B. T., Chow, S. Y., Schallert, T., Murray, M., et al. (1999) Transplants of fibroblasts genetically modified to express BDNF promote regeneration of adult rat rubrospinal axons and recovery of forelimb function. *J. Neurosci.* **19,** 4370–4387.
27. McDonald, J. W., Liu, X. Z., Qu, Y., Liu, S., Mickey, S. K., Turetsky, D., et al. (1999) Transplanted embryonic stem cells survive, differentiate and promote recovery in injured rat spinal cord. *Nat. Med.* **5,** 1410–1412.
28. Weidner, N., Blesch, A., Grill, R. J., and Tuszynski, M. H. (1999) Nerve growth factor-hypersecreting Schwann cell grafts augment and guide spinal cord axonal growth and remyelinate central nervous system axons in a phenotypically appropriate manner that correlates with expression of L1. *J. Comp. Neurol.* **413,** 495–506.
29. Brösamle, C., Huber, A. B., Fiedler, M., Skerra, A., and Schwab, M. E. (2000) Regeneration of lesioned corticospinal tract fibers in the adult rat induced by a recombinant, humanized IN-1 antibody fragment. *J. Neurosci.* **20,** 8061–8068.
30. Bomze, H. M., Bulsara, K. R., Iskandar, B. J., Caroni, P., and Skene, J. H. P. (2001) Spinal axons regeneration evoked by replacing two growth cone proteins in adult neurons. *Nat. Neurosci.* **4,** 38–43.
31. Weiss, S., Dunne, C., Hewson, J., Wohl, C., Wheatley, M., Peterson, A. C., et al. (1996) Multipotent CNS stem cells are present in the adult mammalian spinal cord and ventricular neuroaxis. *J. Neurosci.* **16,** 7599–7609.
32. Kalyani, A., Hobson, K., and Rao, M. S. (1997) Neuroepithelial stem cells from the embryonic spinal cord: isolation, characterization, and clonal analysis. *Dev. Biol.* **186,** 202–223.
33. Kalyani, A. J., Piper, D., Mujtaba, T., Lucero, M. T., and Rao, M. S. (1998) Spinal cord neuronal precursors generate multiple neuronal phenotypes in culture. *J. Neurosci.* **18,** 7856–7868.
34. Mayer-Proschel, M., Kalyani, A. J., Mujtaba, T., and Rao, M. S. (1997) Isolation of lineage-restricted neuronal precursors from multipotent neuroepithelial stem cells. *Neuron* **19,** 773–785.
35. Rao, M. S. and Mayer-Proschel, M. (1997) Glial-restricted precursors are derived from multipotent neuroepithelial stem cells. *Dev. Biol.* **188,** 48–63.
36. Rao, M. S., Noble, M., and Mayer-Proschel, M. (1998) A tripotential glial precursor cell is present in the developing spinal cord. *Proc. Natl. Acad. Sci. USA* **95,** 3996–4001.
37. Shihabuddin, L. S., Ray, J., and Gage, F. H. (1997) FGF-2 is sufficient to isolate progenitors found in the adult mammalian spinal cord. *Exp. Neurol.* **148,** 577–586.

38. Quinn, S. M., Walters, W. M., Vescovi, A. L., and Whittemore, S. R. (1999) Lineage restriction of neuroepithelial precursor cells from fetal human spinal cord. *J Neurosci. Res.* **57,** 590–602.
39. Brüstle, O., Jones, K. N., Learish, R. D., Karram, K., Choudhary, K., Wiestler, O. D., et al. (1999) Embryonic stem cell-derived glial precursors: a source of myelinating transplants. *Science* **285,** 754–756.
40. Mujtaba, T., Piper, D. R., Kalyani, A., Groves, A. K., Lucero, M. T., and Rao, M. S. (1999) Lineage-restricted neural precursors can be isolated from both the mouse neural tube and cultured ES cells. *Dev. Biol.* **214,** 113–127.
41. Martínez-Serrano, A. and Björklund, A. (1997) Immortalized neural progenitor cells for CNS gene transfer and repair. *Trends Neurosci.* **20,** 530–538.
42. Onifer, S. M., Cannon, A. B., and Whittemore, S. R. (1997) Potential of immortalized neural progenitor cells to replace lost adult central nervous system neurons. *Transplant. Proc.* **29,** 2221–2223.
43. Martínez-Serrano, A. and Snyder, E. Y. (1999) Neural stem cell lines for CNS repair, in *CNS Regeneration: Basic Science and Clinical Advances* (Tuszynski, M. H. and Kordower, J. H., eds.), Academic, San Diego, pp. 203–250.
44. Gray, J. A., Grigoryan, G., Virley, D., Patel, S., Sinden, J. D., and Hodges, H. (2000) Conditionally immortalized, multipotential and multifunctional neural stem cell lines as an approach to clinical transplantation. *Cell Transplant.* **9,** 153–168.
45. Renfranz, P. J., Cunningham, M. G., and McKay, R. D. (1991) Region-specific differentiation of the hippocampal stem cell line HiB5 upon implantation into the developing mammalian brain. *Cell* **66,** 713–729.
46. Snyder, E. Y., Deitcher, D. L., Walsh, C., Arnold-Aldea, S., Hartwieg, E. A., and Cepko, C. L. (1992) Multipotent neural cell lines can engraft and participate in development of mouse cerebellum. *Cell* **68,** 33–51.
47. Onifer, S. M., Whittemore, S. R., and Holets, V. R. (1993) Variable morphological differentiation of a raphé-derived neuronal cell line following transplantation into the adult rat CNS. *Exp. Neurol.* **122,** 130–142.
48. Cattaneo, E., Magrassi, L., Butti, G., Santi, L., Giavazzi, A., and Pezzotta, S. (1994) A short term analysis of the behaviour of conditionally immortalized neuronal progenitors and primary neuroepithelial cells implanted into the fetal rat brain. *Brain Res. Dev. Brain Res.* **83,** 197–208.
49. Cattaneo, E. and Conti, L. (1998) Generation and characterization of embryonic striatal conditionally immortalized ST14A cells. *J. Neurosci. Res.* **53,** 223–234.
50. Hoshimaru, M., Ray, J., Sah, D. W., and Gage, F. H. (1996) Differentiation of the immortalized adult neuronal progenitor cell line HC2S2 into neurons by regulatable suppression of the v-myc oncogene. *Proc. Natl. Acad. Sci. USA* **93,** 1518–1523.
51. Rao, M. S. and Anderson, D. J. (1997) Immortalization and controlled in vitro differentiation of murine multipotent neural crest stem cells. *J. Neurobiol.* **32,** 722–746.
52. Laywell, E. D., Kukekov, V. G., and Steindler, D. A. (1999) Multipotent neurospheres can be derived from forebrain subependymal zone and spinal

cord of adult mice after protracted postmortem intervals. *Exp. Neurol.* **156,** 430–433.

53. Liu, S., Qu, Y., Stewart, T. J., Howard, M. J., Chakrabortty, S., Holekamp, T. F., et al. (2000) Embryonic stem cells differentiate into oligodendrocytes and myelinate in culture and after spinal cord transplantation. *Proc. Natl. Acad. Sci. USA* **97,** 6126–6131.

54. Franklin, R. J. and Blakemore, W. F. (1997) Transplanting oligodendrocyte progenitors into the adult CNS. *J. Anat.* **190,** 23–33.

55. Blakemore, W. F. and Franklin, R. J. (2000) Transplantation options for therapeutic central nervous system remyelination. *Cell Transplant.* **9,** 289–294.

56. Hammang, J. P., Archer, D. R., and Duncan, I. D. (1997) Myelination following transplantation of EGF-responsive neural stem cells into a myelin-deficient environment. *Exp. Neurol.* **147,** 84–95.

57. Zhang, S., Lundberg, C., Lipsitz, D., O'Connor, L. T., and Duncan, I. D. (1998) Generation of oligodendroglial progenitors from neural stem cells. *J. Neurocytol.* **27,** 475–489.

58. Franklin, R. J., Bayley, S. A., Milner, R., Ffrench-Constant, C., and Blakemore, W. F. (1995) Differentiation of the O-2A progenitor cell line CG-4 into oligodendrocytes and astrocytes following transplantation into glia-deficient areas of CNS white matter. *Glia* **13,** 39–44.

59. Franklin, R. J., Bayley, S. A., and Blakemore, W. F. (1996) Transplanted CG4 cells (an oligodendrocyte progenitor cell line) survive, migrate, and contribute to repair of areas of demyelination in X-irradiated and damaged spinal cord but not in normal spinal cord. *Exp. Neurol.* **137,** 263–276.

60. Keirstead, H. S., Ben-Hur, T., Rogister, B., O'Leary, M. T., Dubois-Dalcq, M., and Blakemore, W. F. (1999) Polysialylated neural cell adhesion molecule-positive CNS precursors generate both oligodendrocytes and Schwann cells to remyelinate the CNS after transplantation. *J. Neurosci.* **19,** 7529–7536.

61. Luskin, M. B., Zigova, T., Soteres, B. J., and Stewart, R. R. (1997) Neuronal progenitor cells derived from the anterior subventricular zone of the neonatal rat forebrain continue to proliferate in vitro and express a neuronal phenotype. *Mol. Cell. Neurosci.* **8,** 351–366.

62. Morshead, C. M., Craig, C. G., and van der Kooy, D. (1998) In vivo clonal analyses reveal the properties of endogenous neural stem cell proliferation in the adult mammalian forebrain. *Development* **125,** 2251–2261.

63. Chiasson, B. J., Tropepe, V., Morshead, C. M., and van der Kooy, D. (1999) Adult mammalian forebrain ependymal and subependymal cells demonstrate proliferative potential, but only subependymal cells have neural stem cell characteristics. *J. Neurosci.* **19,** 4462–4471.

64. Smith, P. M. and Blakemore, W. F. (2000) Porcine neural progenitors require commitment to the oligodendrocyte lineage prior to transplantation in order to achieve significant remyelination of demyelinated lesions in the adult CNS. *Eur. J. Neurosci.* **12,** 2414–2424.

65. White, L. A. and Whittemore, S. R. (1992) Immortalization of raphe neurons: an approach to neuronal function in vitro and in vivo. *J. Chem. Neuroanat.* **5,** 327–330.

66. Shihabuddin, L. S., Hertz, J. A., Holets, V. R., and Whittemore, S. R. (1995) The adult CNS retains the potential to direct region-specific differentiation of a transplanted neuronal precursor cell line. *J. Neurosci.* **15,** 6666–6678.
67. Shihabuddin, L. S., Brunschwig, J. P., Holets, V. R., Bunge, M. B., and Whittemore, S. R. (1996) Induction of mature neuronal properties in immortalized neuronal precursor cells following grafting into the neonatal CNS. *J. Neurocytol.* **25,** 101–111.
68. Onifer, S. M., Cannon, A. B., and Whittemore, S. R. (1997) Altered differentiation of CNS neural progenitor cells after transplantation into the injured adult rat spinal cord. *Cell Transplant.* **6,** 327–338.
69. Ryder, E. F., Snyder, E. Y., and Cepko, C. L. (1990) Establishment and characterization of multipotent neural cell lines using retrovirus vector-mediated oncogene transfer. *J. Neurobiol.* **21,** 356–375.
70. Taylor, R. M. and Snyder, E. Y. (1997) Widespread engraftment of neural progenitor and stem-like cells throughout the mouse brain. *Transplant. Proc.* **29,** 845–847.
71. Yandava, B. D., Billinghurst, L. L., and Snyder, E. Y. (1999) "Global" cell replacement is feasible via neural stem cell transplantation: evidence from the dysmyelinated shiverer mouse brain. *Proc. Natl. Acad. Sci. USA* **96,** 7029–7034.
72. Park, K. I., Liu, S., Flax, J. D., Nissim, S., Stieg, P. E., and Snyder, E. Y. (1999) Transplantation of neural progenitor and stem cells: developmental insights may suggest new therapies for spinal cord and other CNS dysfunction. *J. Neurotrauma* **16,** 675–687.
73. Snyder, E. Y., Taylor, R. M., and Wolfe, J. H. (1995) Neural progenitor cell engraftment corrects lysosomal storage throughout the MPS VII mouse brain. *Nature* **374,** 367–370.
74. Snyder, E. Y., Yoon, C., Flax, J. D., and Macklis, J. D. (1997) Multipotent neural precursors can differentiate toward replacement of neurons undergoing targeted apoptotic degeneration in adult mouse neocortex. *Proc. Natl. Acad. Sci. USA* **94,** 11,663–11,668.
75. Rosario, C. M., Yandava, B. D., Kosaras, B., Zurakowski, D., Sidman, R. L., and Snyder, E. Y. (1997) Differentiation of engrafted multipotent neural progenitors towards replacement of missing granule neurons in meander tail cerebellum may help determine the locus of mutant gene action. *Development* **124,** 4213–4224.
76. Weidner, N., Grill, R. J., and Tuszynski, M. H. (1999) Elimination of basal lamina and the collagen "scar" after spinal cord injury fails to augment corticospinal tract regeneration. *Exp. Neurol.* **160,** 40–50.
77. Conner, J. M. and Varon, S. (1996) Characterization of antibodies to nerve growth factor: assay-dependent variability in the cross-reactivity with other neurotrophins. *J. Neurosci. Methods* **65,** 93–99.
78. Blesch, A., Uy, H. S., Grill, R. J., Cheng, J. G., Patterson, P. H., and Tuszynski, M. H. (1999) Leukemia inhibitory factor augments neurotrophin expression and corticospinal axon growth after adult CNS injury. *J. Neurosci.* **19,** 3556–3566.

79. Fariñas, I., Yoshida, C. K., Backus, C., and Reichardt, L. F. (1996) Lack of neurotrophin-3 results in death of spinal sensory neurons and premature differentiation of their precursors. *Neuron* **17,** 1065–1078.

80. Bradbury, E. J., Khemani, S., Von, R., et al. (1999) NT-3 promotes growth of lesioned adult rat sensory axons ascending in the dorsal columns of the spinal cord. *Eur. J. Neurosci.* **11,** 3873–3883.

81. Liu, Y., Himes, B. T., Solowska, J., Moul, J., Chow, S. Y., Park, K. I., et al. (1999) Intraspinal delivery of neurotrophin-3 using neural stem cells genetically modified by recombinant retrovirus. *Exp. Neurol.* **158,** 9–26.

82. McTigue, D. M., Horner, P. J., Stokes, B. T., and Gage, F. H. (1998) Neurotrophin-3 and brain-derived neurotrophic factor induce oligodendrocyte proliferation and myelination of regenerating axons in the contused adult rat spinal cord. *J. Neurosci.* **18,** 5354–5365.

83. Bregman, B. S., Kunkel-Bagden, E., Schnell, L., Dai, H. N., Gao, D., and Schwab, M. E. (1995) Recovery from spinal cord injury mediated by antibodies to neurite growth inhibitors. *Nature* **378,** 498–501.

84. Filbin, M. T. (2000) Axon regeneration: vaccinating against spinal cord injury. *Curr. Biol.* **10,** R100–R1033.

85. Moon, L. D., Brecknell, J. E., Franklin, R. J., Dunnett, S. B., and Fawcett, J. W. (2000) Robust regeneration of CNS axons through a track depleted of CNS glia. *Exp. Neurol.* **161,** 49–66.

86. Johansson, C. B., Momma, S., Clarke, D. L., Risling, M., Lendahl, U., and Frisén, J. (1999) Identification of a neural stem cell in the adult mammalian central nervous system. *Cell* **96,** 25–34.

87. Liu, R. H., Morassutti, D. J., Whittemore, S. R., Sosnowski, J.S., and Magnuson, D. S. (1999) Electrophysiological properties of mitogen-expanded adult rat spinal cord and subventricular zone neural precursor cells. *Exp. Neurol.* **158,** 143–154.

88. Brazelton, T. R., Rossi, F. M. V., Keshet, G. I., and Blau, H. M. (2000) From marrow to brain: expression of neuronal phenotypes in adult mice. *Science* **290,** 1775–1779.

89. Horner, P. J., Power, A. E., Kempermann, G., Kuhn, H. G., Palmer, T. D., Winkler, J., et al. (2000) Proliferation and differentiation of progenitor cells throughout the intact adult rat spinal cord. *J. Neurosci.* **20,** 2218–2228.

90. Mezey, É., Chandross, K. J., Harta, G., Maki, R.A., and McKercher, S. R. (2000) Turning blood into brain: cells bearing neuronal antigens generated in vivo from bone marrow. *Science* **290,** 1779–1782.

91. Sanchez-Ramos, J., Song, S., Cardozo-Pelaez, F., Hazzi, C., Stedeford, T., Willing, A., et al. (2000) Adult bone marrow stromal cells differentiate into neural cells in vitro. *Exp. Neurol.* **164,** 247–256.

92. Woodbury, D., Schwarz, E. J., Prockop, D. J., and Black, I. B. (2000) Adult rat and human bone marrow stromal cells differentiate into neurons. *J. Neurosci. Res.* **61,** 364–370.

93. Deng, W., Obrocka, M., Fischer, I., and Prockop, D. J. (2001) In vitro differentiation of human marrow stromal cells into early progenitors of neural cells by conditions that increase intracellular cyclic AMP. *Biochem. Biophys. Res. Commun.* **282,** 148–152.

14

Spinal Ischemia-Induced Paraplegia

A Potential Therapeutical Role of Spinally Grafted Neural Precursors and Human hNT Neurons

Martin Marsala, Osamu Kakinohana, Tony L. Yaksh, and Joho Tokumine

SPINAL ISCHEMIA AND ALTERATIONS IN FUNCTION

Transient spinal cord ischemia and subsequent loss of neurological function (spastic or flaccid paraplegia) represents a serious complication associated with transient aortic cross-clamp (as used in repair of aortic aneurysm). In clinical studies, it has been demonstrated that the incidence of paraparesis or developed spastic or flaccid paraplegia range between 12% and 40% in patients with extensive thoracoabdominal aortic aneurysm repair *(1,2)*. Although a spontaneous recovery of function was noted in a fraction of patients, who displayed motor dysfunction early after reflow, in a majority of cases this deficit is irreversible. In accordance with these data, experimental studies using monkey, cat, dog, rabbit, or rat spinal ischemia models show that aortic occlusion will lead to a comparable dysfunction including transient motor weakness or permanent spastic or flaccid paraplegia *(2–8)*.

Although the precise mechanism leading to spinal neuronal degeneration after ischemic insult is not certain, the neuronal populations affected by spinal ischemia are well defined. Thus, in animals with *spastic* paraplegia, a selective degeneration of small and medium-sized interneurons, in the lumbosacral segments, predominantly in the lamina VII occurs as soon as 30 min after ischemia *(9)*. Immunohistochemical staining of these neurons for glutamate decarboxylase (GAD) or tissue analysis for GABA (γ-aminobutyric acid) or glycine content *(3)* shows that these are likely inhibitory interneurons. In contrast, as defined by light microscopy or retrograde labeling (Fluorogold), the α-motoneuron pool in the ventral horn shows long-term survival (up to 2–3 mo) *(10)*. These observations are further supported by electrophysiological analysis that demonstrates an increase in

From: *Neural Stem Cells for Brain and Spinal Cord Repair*
Edited by: T. Zigova, E. Y. Snyder, and P. R. Sanberg © Humana Press Inc., Totowa, NJ

the monosynaptic reflex and near-complete loss in spinal polysynaptic activity *(6,11,12)*. This profile of damage accounts for the extensive spasticity observed in these animals.

CHARACTERIZATION OF A PRECLINICAL RODENT MODEL OF SPINAL ISCHEMIA

In our laboratory, we have developed and characterized in rat, a relatively noninvasive aortic occlusion model *(8)*. In this model, spinal ischemia is induced by transient balloon occlusion of the descending thoracic aorta and a moderate systemic hypotension (40 mm Hg). Using this model, we have demonstrated that 10 min of aortic occlusion consistently leads to a development of spastic paraplegia, whereas after ischemic intervals longer than 10 min, flaccidity is seen. Consistent with the previous report on several animals models of spinal ischemic injury, histopathological analysis of lumbar spinal cords in animals with spasticity showed a selective loss of small inhibitory neurons localized in lumbosacral segments between laminae V and VII (Figs. 1A,B).

DISTINCTIONS BETWEEN SPINAL ISCHEMIA AND PHYSICAL TRAUMA

All spinal injuries are not the same. An important distinction between spinal *trauma* and *ischemia* is that trauma-induced paraplegia is typically characterized by destruction of long descending tracts. To characterize the changes in the descending motor tracts in rats with ischemic spastic paraplegia, we have employed anterograde (biotinylated dextran; BD) and retrograde tracers (Fluorogold; FG). At 2–3 wk after ischemia, in animals with developed spasticity, anterograde tracer (BD) was stereotaxically injected into motor cortex or nucleus ruber. Retrograde tracer (FG) was applied directly on the lateral surface of the lumbar spinal cord using FG-soaked gelfoam. Subsequent analysis at 1–2 wk after applying the tracers showed a near-normal labeling of rubral neurons (Figs. 1E,F) and continuing labeling of cortico-spinal and rubrospinal fibers and terminals in previously ischemic lumbosacral regions (Figs. 1G,H,I). These data show that in contrast to trauma-induced paraplegia, there is a continuing presence of long descending tracts in perinecrotic regions after ischemic spinal injury *(10)*. To further characterize the functional status of the persisting descending motor tracts, we have used motor-evoked potentials (MEP) elicited by transcranial stimulation (20 V, 200 µs) and recorded from the soleus muscle (Fig. 1J). In control animals, a clear muscle response was recorded after a single stimulus. In animals with spasticity, a 350% increase in the amplitude of MEP was

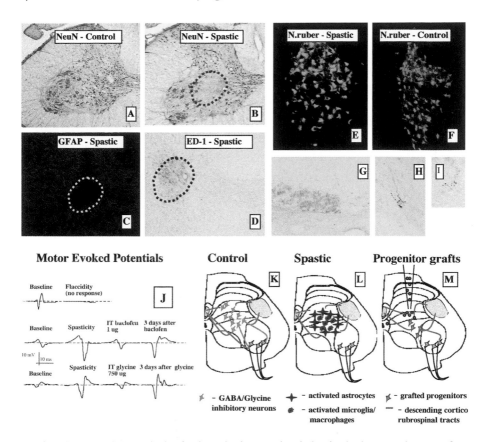

Fig. 1. Spinal morphological and electrophysiological changes in rat after development of ischemic spastic paraplegia. In comparison to control nonischemic animal (**A**), animals with spasticity show a selective loss of small and medium-sized interneurons localized between laminae V and VII in lumbosacral segments (**B**; NeuN staining). In the same regions a clear activation of astrocytes in the perinecrotic region (**C**) and infiltration of activated macrophages/microglia in the core of the necrosis (**D**) is also observed. In contrast to traumatic spinal injury, animals with ischemic spasticity showed near-normal retrograde labeling of nucleus ruber (**E**, **F**; fluorogold tracer) and corticospinal and rubrospinal terminals in previously ischemic lumbosacral segments (**G–I**; biotinylated dextran; 10,000 molecular weight). Consistent with the loss of local inhibition recording of motor-evoked potential (MEP) (**J**) from the soleus muscle (middle and lower trace) showed on average 350% increase in the amplitude of MEP in animals with spastic paraplegia. Intrathecal administration of baclofen (1 µg; GABA$_B$ receptor agonist middle trace) or glycine (750 µg; lower trace) transiently reversed the increase in MEP amplitude. In animals with flaccid paraplegia, no MEP was recorded. Therapeutic strategy in the cell replacement therapy is based on a region-specific grafting of neural precursors or postmitotic neurons into previously ischemic spinal cord segments (L2–L5) (**K,L,M**).

measured on average. In these animals, an intrathecal (IT) injection of baclofen (1 µg; $GABA_B$ receptor agonist) or glycine (750 µg) leads to transient and near-complete recovery in the amplitude of MEP. In animals with flaccidity, no motor response was recorded (consistent with the loss of ventral α-motoneurons). Similarly, a significant increase in monosynaptic reflex *(6)* or H-reflex *(13)* after induction of ischemic spasticity in rat and rabbit spinal ischemia model was described. As in our study, the increase in H-reflex was effectively blocked by intrathecal glycine administration *(13)*.

These data jointly suggest that the mechanism of ischemic *spasticity* is specifically related to the loss of segmental inhibition primarily mediated by GABA and glycine-ergic inhibitory neurons, with a continuing presence of descending motor systems, functional rubro/cortico α-motoneuron synapses, and functional motor neurons.

STRATEGY FOR CELL REPLACEMENT THERAPY AFTER SPINAL ISCHEMIC INJURY

As discussed, depending on the duration of the ischemic interval, transient spinal ischemia can lead to a selective loss of inhibitory neurons and spasticity or, after longer ischemic intervals, to a widespread neuronal loss and resulting flaccidity. The qualitative profile (i.e., spasticity or flaccidity) of the resulting postischemic damage can be reliably defined by neurological assessment and quantified by motor- and spinal-cord-evoked potentials. Unfortunately, at present there is no effective therapy available that would lead to a significant and permanent relief of spasticity or eventually partial or complete recovery of motor function. As demonstrated, because the neuronal loss resulting from previous spinal ischemia is permanent and irreversible, the cell replacement therapy (i.e., a direct spinal implantation of neural precursors or postmitotic neurons into lumbosacral segments) appears to be a rational therapeutic approach, which can potentially lead to functionally useful neurological improvement.

As a model, the ischemic spasticity has merit over trauma. With flaccid paraplegia, functional recovery would require replacement of motor neurons and their projections and a redevelopment of neuromuscular connectivity. In contrast, in spastic paraplegia, a targeted replacement of local interneurons in areas of previous neuronal loss (i.e., intermediate gray matter in LS segments) would seek to restore local segmental inhibition (Figs. 1K–M). Preferential differentiation of grafted cells into a GABA and glycine-ergic phenotype and development of functional synapses between persisting α-motoneurons and descending motor tract terminals would represent a potentially achievable "rewiring" in animals with ischemic spasticity.

An important question relates to the nature of the cellular replacement. Possibilities include neural precursors or postmitotic neurons. However, at present, there is accumulating evidence that implantation of neural precursors or postmitotic neurons into previously *injured* spinal cord is associated with only limited neuronal differentiation and maturation of grafted cells. The presence of local inflammatory changes appears to be the primary limiting factor in a successful functional incorporation and maturation of the grafted cells. The scope of the present chapter is to summarize these findings and outline possible future directions in the cell replacement therapy after spinal cord ichemia-induced paraplegia.

LACK OF NEURONAL PROLIFERATION AFTER SPINAL CORD ISCHEMIA

In recent years, it was established that the birth of new neurons in the hippocampus continues in adult rodents *(14,15)*. The neuronal nature of the newly generated cells has been confirmed by either immunostaining with neuron-specific markers Neuronal N (NcuN), neuron-specific enolase (NSE), or granule cell marker calbindin D28k *(16,17)* or by electron-optical analysis *(18)* and retrograde labeling of neurites *(19)*. It has also been demonstrated that the proliferation of neuronal progenitors in adult animals is under the effective positive control of environmental signals typically associated with learning *(20–23)* and is specifically mediated by qualitatively different molecular modulators, including growth factors, hormones, or excitatory activation *(17,24,25)*.

At present, there is also evidence that an accelerated proliferation of neural progenitors followed by neuronal differentiation occurs after an exposure of neuronal tissue to different pathological stressors associated with neuronal degeneration (such as ischemia or traumatic injury). Thus, after transient global cerebral ischemia (2–10 min) in gerbils, there is a significant increase in cell birth in the dentate subgranular zone, with the peak of proliferation seen at 11 d *(26)*. The time-dependent differentiation of these cells into neurons (as evidenced by NeuN, calbindin D28k, or microtubule-associated protein [MAP]-2 immunoreactivity) was also seen with the peak of neuronal differentiation at 40 d after ischemia. Whether such a proliferation and neuronal differentiation are also followed by functional synaptogenesis and improvement of behavioral function is not known. More recently, similar increases in neuronal proliferation in the dentate gyrus after cerebral ischemia in mice were reported *(27)*.

Neuronal systems may, however, vary with respect to regeneration potential. Thus, in the spinal cord, we were not able to demonstrate proliferation of new neurons after transient *injurious* or *noninjurious* intervals of spinal ischemia. Instead, a significant astrocytic proliferation (as evidenced by BrdU [5-bromo-2′-deoxyuridine] and GFAP [glial fibrillary acidic protein] colocalization) was seen in lumbosacral segments (i.e., in the area of targeted ischemia) *(28)*. This observation is in agreement with other studies which demonstrated that after spinal *traumatic* injury, there is a clear astrocytic proliferation at the site of injury but with a lack of neuronal proliferation *(29)*. Although the explanation for the differences between the responsiveness of brain and spinal neural progenitors to local injury and their capacity to proliferate and differentiate into neurons is not clear, it may reflect a preferential astrocyte differentiation of proliferating precursors resulting from continuing inflammation at the site of the graft. Several cytokines have been demonstrated to effectively stimulate non-neuronal differentiation from noncommitted neural precursors (see following section). Alternatively, it may reflect a limited neurotrophic activity and the synthesis of the trophic factors, which would favor non-neuronal differentiation.

CNS ISCHEMIC INJURY IS ASSOCIATED WITH MICROGLIAL AND MACROPHAGE ACTIVATION

In recent years, there has been an increasing interest in identifying the role of local inflammatory response in (1) the evolution of the secondary brain and spinal cord injury and (2) modulation of differentiation and maturation of noncommitted precursors in the areas of inflammation. It has been demonstrated that after transient spinal ischemia there is a significant accumulation of mononuclear phagocytes in the previously ischemic spinal segments in rabbit *(30)*. Accumulation of mononuclear phagocytes is observed between 24 and 48 h after ischemia and corresponds with a deterioration in neurological function. Inhibition of phagocytic and secretory function in mononuclear phagocytes after chloroquine and colchicine treatment is associated with improved neurological outcome and reduced neuronal degeneration *(30)*. Similarly, massive infiltration of traumatically injured spinal cord by macrophages was demonstrated *(31–33)*. The peak of macrophage infiltration after traumatic spinal injury is seen between 2 and 7 d; however, it is still present at 6 wk. Using the previously described spinal ischemia model in rat, we have demonstrated a similar time-course of macrophage/microglial activation in spinal ischemic segments. Immunostaining with ED1 antibody, which specifically labels activated (transformed) microglia and macrophages revealed the peak of ED1 immunoreactivity between 7 and 14 d after ischemia, but with continuing presence at 6 wk after the insult (Fig. 1D). In parallel, a massive activation and proliferation of astrocytes in the perinecrotic region was observed (Fig. 1C). Comparable macrophage/microglial and astrocyte activation in focal cerebral ischemia models was reported *(34–37)*.

Prolonged infiltration of spinal cord and brain tissue with macrophages and activated microglia and resulting secretory activity of these cells in the local neuronal milieu appears to be important for several reasons: (1) Activation of microglia/macrophages is associated with an increased release of several proinflammatory and anti-inflammatory cytokines including interleukin (IL)-1β, tumor necrosis factor (TNF)-α, Il-6, and leukemia inhibitory factor (LIF) in the affected regions *(31,38–42)*. (2) Increased synthesis of IL1-β, IL-6, TNF-α, and bFGF may contribute to stimulation and proliferation of astrocyte as well as to the progressive neuronal degeneration in the penumbral regions *(43,44)*. In the focal cerebral ischemia model, the expansion of the infarct can be effectively suppressed by IL-1 receptor antagonism *(45)*; (3) LIF acting through its specific receptor and by a direct LIFR-coupled signaling pathway (JAK–STAT pathway) is a potent inducer of astrocyte differentiation from neural precursors in vitro *(46,47)*. The LIF-mediated effect on astrocyte proliferation/differentiation is potentiated by several other factors such as activin *(48)* or bone morphogenetic protein *(46,49)*. Upregulation of some of these factors such as activin after traumatic hippocampal injury has been described recently *(50)*.

The above data demonstrate that after spinal traumatic or ischemic injury, there is a continuing presence of local inflammatory processes and corresponding increased parenchymal synthesis of several cytokines, which may persist for several weeks after injury. Increased activity of these factors appears to provide an environment, which favor astrocyte proliferation and differentiation from noncommitted neural precursors.

In a recent experiment, we have examined the effect of the treatment with macrophage-derived media on the differentiation of *spinal* neural precursors in vitro. Macrophages were harvested from the rat peritoneal cavity *(51)* at 5 d after intraperitoneal administration of thioglycolate medium. After harvesting, macrophages were activated using lipopolysacharide (20 μg/mL) for 3 d and the medium was collected. Spinal neural precursors plated on lysine coated cover slips were then treated with increasing concentrations (1%, 5%, 10%, or 20%) of macrophage-conditioned media (MCM) in neurobasal medium plus Dulbecco's modified Eagle's medium (DMEM) (2:3) for 48 h. At 48 h, the presence of GFAP and MAP-2 (neuronal marker) immunoreactivity was examined using immunofluorescence. In comparison to control, treatment of neural precursors with MCM lead to a significant induction of astrocyte differentiation, with the most potent effect seen after

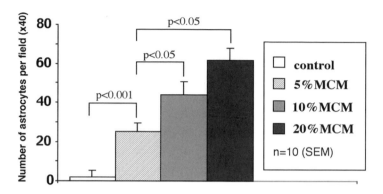

Fig. 2. Treatment of noncommitted spinal neural precursors with macrophage-conditioned media potentiates astrocyte differentiation in vitro. Spinal neural precursors were plated on lysine-coated glass chamber slides and treated with increasing concentrations of lypopolysacharide-conditioned macrophage medium (5%, 10%, 20%). At 48 h, cells were stained with GFAP antibody. A concentration-dependent increase in the number of counted astrocytes (40× optical field) was seen.

treatment with 10–20% of MCM (Fig. 2). These data suggest that the release of cytokines from activated macrophages at the site of the cell graft can play an important modulatory role in the induction of astrocyte differentiation from noncommitted neural precursors. However, at present, the relative contribution of several proinflammatory or anti-inflammatory cytokines known to be released from activated macrophages/microglia and from astrocytes (i.e., LIF, TNF-α, IL-1β; IL-10, IL-6) in this process is not defined.

MODULATION OF NEURONAL DIFFERENTIATION AND MATURATION BY LOCAL INFLAMMATION

In previous studies, it has been shown in *control* animals that intracerebral implantation of neural precursors is followed by a successful neuronal differentiation when progenitors are implanted into specific brain regions *(52)*. It has also been demonstrated that the transmitter and morphological phenotype of these neurons develop similar profiles as found in the surrounding host neurons *(53–57)*.

In contrast, only a relatively limited neuronal differentiation (8%) of spinally implanted neural progenitors has been reported when these cells are implanted into previously injured spinal cord *(58)*. In addition, in the same study, only limited signs of neuronal maturation as evidenced by axo-dendritic outgrowth was seen. Comparable observations have been made after the spinal implantation of immortalized neuronally *committed* progenitor cells (RN33B) into previously transected rat spinal cord *(59)*. In this study, it was shown that the implanted cells develop only a bipolar morphology, an observation typical of immature neurons. In contrast, spinal implantation of the same cells in control animals resulted in a clear neuronal maturation with the development of multipolar neuronal morphology. Similar limited neuronal differentiation of neural precursors after implantation into the dorsal hippocampus was reported. In this study, the differentiation of implanted cells into neurons was only evident in the *intact* region of the granule cell layer, but not in the areas adjacent to the injury site caused by the implantation needle. In the injured region, the majority of implanted cells showed differentiation into oligocytes or astrocytes *(52)*.

Using the previously described spinal ischemia model in rat, we have recently studied the survival and phenotypic differentiation of spinal neural precursors when grafted spinally 7–15 d after spinal ischemia (i.e., during the period of continuing microglial activation and macrophages infiltration at the site of the cell graft).

Spinal neural precursors were isolated from the spinal cords of embryonic day 14 of Sprague-Dawley rats and expanded with bFGF and EGF (20 ng/mL). To label proliferating progenitors cells before in vivo implant, cells were labeled with BrdU (5 μM/24h), fluorescent label PKH-26 or transfected using pLRNL retroviral construct encoding the lac-Z or Tau-GFP gene. To characterize the lineage potential of naive or labeled precursors, cells were treated with (1) FGF (1–10 ng/mL) or (2) 5% fetal bovine serum (FBS) for 2–4 d and then immunostained with neuronal and non-neuronal markers (MAP-2, TUJ1, GFAP, O4) as well as with specific growth factors receptors (GFRα1, TrkB, TrkC) antibodies. Similarly as demonstrated in previous studies, treatment with 5% serum led to a significant astrocyte induction *(60)*, whereas treatment with 0.1 ng of bFGF led to a virtually pure neuronal differentiation *(61)*. The majority of terminally differentiated neurons also showed GAD immunoreactivity. In addition, a consistent presence of growth factor receptors in proliferating or terminally differentiated neurons was detected (Fig. 3).

To implant neural precursors, laminectomy of L2–L5 vertebra was performed and animals received a total of 5–10 injections (0.5 µL each; 40,000–80,00 cells per injection) on each side of the spinal cord (left and right) evenly distributed between exposed L2–L5 segments and targeted into central gray matter (laminae V–VII). For injections, a pulled glass micropipet (diameter of the tip 100–150 µm) was used.

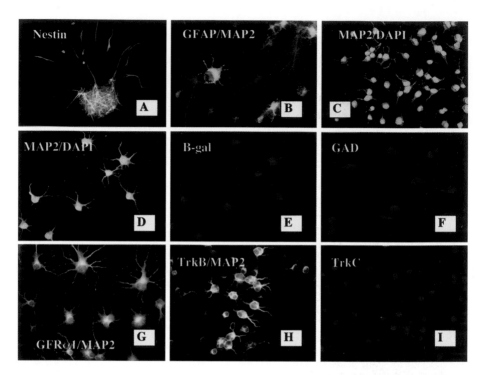

Fig. 3. Differentiation potential of spinal neural precursors in vitro. Spinal cord progenitors (SCP) isolated from E14 lumbar spinal cord and cultivated with EGF plus bFGF proliferated as single cells or showed formation of microspheres and consistent immunoreactivity for Nestin **(A)**. Treatment of SCP with 5% serum led to a significant increase in astrocyte differentiation **(B**; red), whereas treatment with a low concentration of bFGF (1 ng/mL) yielded near completely pure neuronal differentiation **(C, D)**. Differentiation characteristics were not affected by retroviral lac-Z transfection **(D, E)**. The majority of immature or differentiated cells also displayed GAD immunoreactivity **(F)** and the presence of growth factor receptors, including GFRα1, TrkB, and TrkC **(G, H, I)**.

Histopathological analysis at 3 mo after implantation revealed a relatively low survival rate of implanted progenitors. Three to six percent of the implanted cells remained at this time. The majority of the implanted cells were seen in the vicinity of the injection tract. Qualitative and quantitative histopathological/confocal analysis of the surviving cells revealed that fewer than 1% of the surviving cells showed expression of neuronal markers (MAP-2 or neurofilament H) (Figs. 4A–D) and with very limited process outgrowth.

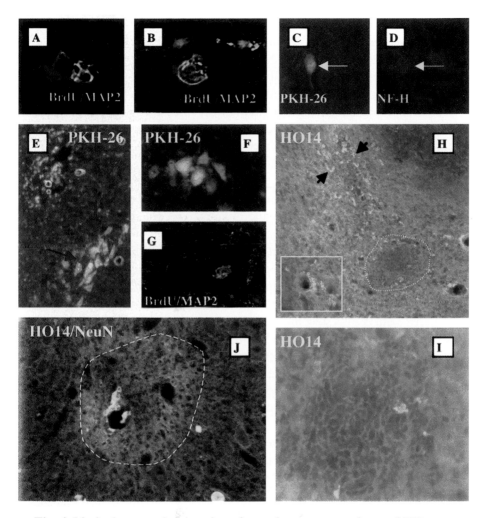

Fig. 4. Limited neuronal maturation of neural precursors or human hNT neurons after grafting into previously ischemic spinal cord. Spinal neural precursors labeled with BrdU or PKH-26 that colocalize with neuronal marker MAP-2 or NF-H showed only limited maturation and axodendritic growth at 3–4 mo after grafting **(A, B, C, D)**. A number of grafted cells, which displayed neuronlike morphology, however, were negative for neuronal markers (MAP-2, TUJ1, or NF-H) were also identified **(E, F, G)**. Similar, limited maturation of human postmitotic hNT neurons in animals treated with cyclosporine A (10 mg/kg/d; ip) was detected. This was evidenced by a weak neurofilament (HO14) staining, restricted into neuronal bodies **(H)**. Only in some neurons localized at the periphery of the graft was colocalization with NeuN and HO14 seen **(J)**. In the core of the graft, a high density of hNT neurons that showed near-complete HO14 negativity were still identified at 3 mo after transplantation **(I)**.

In a number of sections, clearly labeled implanted cells were detected that displayed typical neuronal morphology, but lacked expression of any of the neuronal markers (MAP-2, TUJ1, or NF). These "shadow" neurons were typically organized in small groups consisting of 5–15 cells that were localized at the periphery of the implant (Fig. 4E,F,G).

As a second cell line, postmitotic human hNT neurons were used for spinal grafting in combination with immunosuppressive therapy (cyclosporine A; 10 mg/kg/d). hNT neurons are derived from NT2/D1 clone originally from embryonal teratocarcinoma *(62)*. Following retinoic acid treatment, these cells differentiate into pure postmitotic neurons *(63)*. Using electrophysiological and reverse transcription–polymerase chain reaction (RT-PCR) analysis, the presence of functional GABA and glutamate receptors *(64,65)* as well as L- and N-type calcium channels has been demonstrated in hNT neurons *(66)*. In addition, it has been shown that hNT neurons transcribe the glutamic acid decarboxylase *p67* gene, suggesting their potential to develop an inhibitory phenotype *(67)*. In vivo transplantation studies using hNT neurons reveal that intrastriatal implants of hNT neurons show a long-term (1–3 mo) and about 15% survival of the implanted neurons in control animals or after middle cerebral artery occlusion in cyclosporine A-immunosuppressed rat *(68,69)*. Similarly, long-term survival and extensive hNT neurite outgrowth after implantation into nude mice spinal cord has been demonstrated *(70)*. More recently, hNT cells showed comparable survival when transplanted into previously contused cervical cord in cyclosporine A-immunosuppressed rats *(71,72)*.

After spinal grafting, the presence and the maturation of hNT neurons were analyzed using human-specific antibodies (HO14-axonal neurofilament; MOC-1 neuronal cell-adhesion molecule). At 3 mo, a number of MOC-1-positive neurons were identified that were typically organized in small circular formations or as solitary neurons distributed in mediolateral regions between laminae II and IX. Subsequent staining with HO14-axonal neurofilament antibody revealed immmunoreactivity in neuronal bodies and not axons, suggesting continuing immature stage of grafted neurons (Fig. 4H,I). Only occasional NeuN was observed in grafted hNT neurons at the periphery of the graft (Fig. 4J).

These data demonstrate that even fully *postmitotic neurons* when implanted into previously ischemized spinal cord show only limited maturation and axonal growth when combined with cyclosporine A treatment. A comparable lack of a cycloporine A-mediated neurotrophic effect on axonal regeneration in the transected sciatic nerve in the rat was reported *(73)*.

The reasons for the limited neuronal maturation and axodendritic out-growth of implanted cells is not clear; however, the presence of inhibitory molecules that suppress or block neurite outgrowth after injury has been proposed *(74,75)*. As we have demonstrated, there is an extensive macro-phage and astrocyte activation and proliferation in the previously ischemic spinal cord segments. It was reported that reactive astrocytes can synthesize both (1) molecules that can support neurite growth such as laminin *(76,77)*, PSA-NCAM *(78–80)*, and growth factors (glial-derived neurotrohic factor [GDNF], nerve growth factor [NGF], brain-derived neurotrophic factor [BDNF], and ciliary neurotrophic factor [CNTF]) *(81–84)* and (2) molecules that are potent inhibitors of axodendritic growth. Among inhibitory sub-stances proteoglycans (such as neurocan), thy-1, tenascin, and interferon-δ have been proposed *(85–91)*. It has also been demonstrated that astrocyte synthesis of some of these inhibitory molecules such as neurocan is effec-tively potentiated by transforming growth factor (TGF)-β and by EGF *(85)* (i.e., factors that are synthesized by activated macrophages/microglia *(42,92)*. In addition, a potent (80%) inhibitory effect of leukemia inhibitory factor and ciliary neurotrophic factor on dendritic growth in sympathetic neurons in vitro was reported *(93)*. Both of these factors can be synthesized by activated astrocytes and macrophages *(41,42)*.

These data jointly suggest that a limited neuronal differentiation and maturation of neural precursors or postmitotic hNT neurons when implanted into previously injured spinal cord can be the result of an increased synthesis of molecules that directly or indirectly potentiate a non-neuronal dif-ferentiation and inhibit axodendritic sprouting. The primary source of these molecules appear to be activated astrocytes with a positive stimulatory control from activated macrophages/microglia residing in the same regions (Fig. 5). A recent study shows that the disappearance of macrophages from the site of injury precedes axonal regeneration after Purkinje cell axotomy in vivo *(94)*. In the same study, it was demonstrated that the onset of axonal regeneration is accompanied by a massive astrogliosis and astrocyte-derived expression of PSA-NCAM (i.e., a permissive substratum for neurite outgrowth). This would indicate that the disappearance of macrophages from the site of the cell graft is likely beneficial because of the loss of their stimulatory effect on the synthesis of inhibitory factors from local astrocytes. Conversely, as discussed, a relative increase in the activity of astrocyte-derived neurotrophic molecules such as PSA-NCAM, laminin, and several growth factors may result in an increased axodendritic sprouting. From this perspective, a delayed spinal grafting of neural precursors or

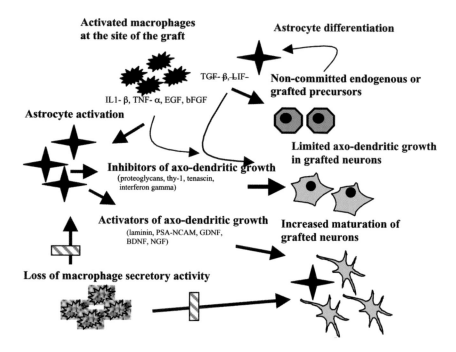

Fig. 5. Modulation of differentiation/maturation of neural precursors by local inflammation after in vivo grafting. Presence of activated *macrophages* at the site of the graft and resulting release of several cytokines and growth factors potentiate astrocyte differentiation from grafted noncommitted precursors, triggers host-derived astrocyte activation/proliferation, and directly inhibits dendrites' growth from grafted cells. Activation of astrocytes by TNF-α and EGF promotes *astrocyte*-derived synthesis of inhibitors of axo-dendritic growth, including proteoglycans, thy-1, tenascin and interferon-δ. Loss of macrophage secretory activity can lead to increased astrocyte migration into the implant and astrocyte-derived synthesis of molecules (laminin, PSA-NCAM, growth factors GDNF, BDNF, NGF) that stimulate and promote axodendritic growth in grafted neurons.

postmitotic neurons (i.e., after the period of acute inflammation resulting from the ischemic insult) may result in more accelerated maturation of grafted cells. Alternatively, pharmacologic manipulations, which are effective in suppressing the secretory activity of local macrophages, could provide similar effects (see the following section).

FK-506 POTENTIATES MATURATION OF NEURAL PRECURSORS AND HNT NEURONS

As discussed, one of the mechanisms of limited neuronal maturation of terminally differentiated neural precursors or postmitotic hNT neurons when implanted into ischemic spinal cord may involve (1) a limited neurotrophic support at the site of the graft and/or (2) release of several inhibitory molecules, which suppress neuronal maturation and axodendritic growth. These data indicate that pharmacological manipulation, which has the ability to promote local neurotrophic activity or ameliorate the synthesis of these inhibitory molecules, should be effective in promoting the maturation and axodendritic growth in grafted cells. With this respect, it has recently been demonstrated that systemic treatment with FK-506 (tacrolimus, Fujisawa) significantly potentiates regeneration of peripheral nerve axons after transection *(95,96)*. Similar neurotrophic effects as evidenced by neurite outgrowth in PC12 cells in culture have been reported *(97)*.

In recent experiments, we tested the effect of FK-506 treatment on the survival, maturation, and axodendritic outgrowth of rat spinal neural precursors or human hNT neurons when implanted spinally in animals with ischemic spasticity *(98,99)*. Three days before cell grafting, treatment with FK-506 (1 mg/kg/d; ip) was initiated and continued for a duration of 2 wk to 4 mo survival.

In comparison with previous data demonstrating limited neuronal maturation of grafted spinal precursors or hNT neurons at 3–4 mo survival, a qualitatively and quantitatively different picture was seen after FK-506 treatment (Fig. 6).

After grafting of hNT neurons, clear signs of the initial neuronal maturation were seen as soon as 3 wk after implantation and with near-complete morphologically defined maturation detected at 3–4 mo after grafting. These signs were expressed as the presence of numerous HO14-positive axons growing from the graft in a distance of more than 1500 μm. The number of HO14-positive axons terminating in the vicinity of persisting host α-motoneurons was also identified. Expression or colocalization of other markers typical for mature neurons, including NSE, NeuN, or MAP-2, was also identified in the implanted cells. In addition, a dense synaptophysin immunoreactivity within and at the periphery of the implant was detected. A number of hNT neurons and their axons were found to colocalize with GABA. Initiation of the maturation of grafted hNT neurons as well as axonal growth was found to be accompanied by a massive astrocyte ingrowth into the graft and astrocyte-derived GDNF expression.

Using spinal neural precursors labeled with the *lacZ* gene and combined with FK-506 treatment, we have seen comparable rapid maturation of grafted cells as soon as at 2 wk after grafting (Fig. 6N,O,P). This was expressed as a number of β-gal-positive cells, which displayed neuron-like morphology with multiple processes. Costaining of these sections with MAP-2 antibody

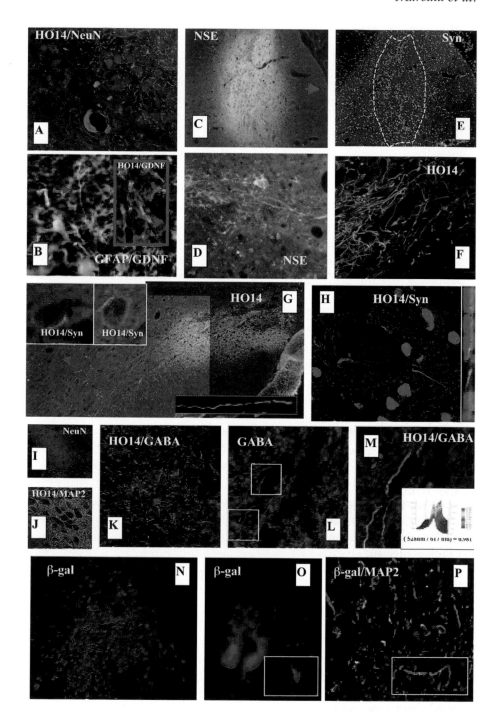

coupled with confocal analysis showed that numerous β-gal-positive cells colocalized with MAP-2.

MECHANISM OF FK-506 NEUROTROPHIC EFFECT

The mechanisms through which FK-506 exerts its neurotrophic effect appear to be complex and may involve a direct or indirect effect on grafted neurons, activated microglia/macrophages, and astrocytes.

First, FK-506 potentiates neurite outgrowth in PC12 cells in culture. This effect depends on the presence of NGF in the cultivating medium *(97)*. In another study, using hippocampal neurons cocultivated with cortical astrocytes, the FK-506-mediated effect on neurite growth was found to be *independent* of the presence of NGF in medium *(96)*. As reported, astrocytes constitutively express a number of growth factors, including GDNF *(81)*, BDNF *(82)*, CNTF *(83)*, or NGF *(84)*, suggesting that the release of these neurotrophins in neuron–astrocyte coculture can potentiate FK-506 neurotrophic activity.

Second, it has been shown that human or mouse astrocytes express several chemotactic cytokine receptors, including IL-1β, TNF-α, C5aE, IL-8R, and FMLPR *(100–102)*. As discussed previously, activation of some of these receptors (IL-1β, TNF-α) by the microglia/macrophage-synthesized cytokines will increase astrogliosis in neonatal and adult animals after trauma *(38,43)* and increase astrocytic synthesis of several other proinflammatory cytokines such as IL-6 and IL-8 *(102,103)*. In vivo or in vitro treatment with

---

Fig. 6. *(see opposite page)* A potent neurotrophic effect of FK-506 on maturation of hNT neurons and neural precursors after grafting into ischemic spinal cord. Human hNT neurons grafted into a previously ischemic spinal cord in combination with FK-506 treatment (1 mg/kg/d; ip) showed rapid maturation (3 wk) as evidenced by NeuN (red) expression and axonal (green) outgrowth from the graft **(A)**. Proliferation and migration of GFAP/GDNF positive astrocytes into implant was also observed at the same time point **(B)**. At 3 mo, an advanced stages of neuronal maturation expressed as NSE **(C, D)** NeuN **(I)**, and MAP-2 **(J)** immunoreactivity were detected. A well-developed synaptophysin immunoreactivity within and at the periphery of the graft were also seen **(E)**. In comparison with 3 wk, an extensive (1000–2000 μm) axonal growth from the graft was observed **(D, F, G)**. Many of hNT neuron-derived axons terminated in the vicinity of the persisting host α-motoneurons **(G insets, H)**. Numerous hNT neurons and their axons showed colocalization with GABA-like immunoreactivity **(K, L, M)**. Comparable rapid maturation of lac-Z-labeled spinal neural precursors at 2 wk after grafting was detected. Numerous labeled cells with extensive axodendritic processes that colocalized with MAP2 immunoreactivity were identified **(N, O, P)**.

anti-inflammatory cytokines such as IL-4 and IL-10 effectively suppresses activation of astrocytes *(39)* in adult mice after cortitectomy or as measured by astrocytic IL-6 production in cultured mouse astrocyte after IL-1β stimulation *(102)*.

Third, a recent in vivo study demonstrated that oral treatment with FK-506 (3.2 mg/kg/d) selectively suppressed tissue IL-10 mRNA and plasma IL-10 protein concentration in a rat heart/kidney allograft transplantation model *(104)*. In the same experimental setting, cyclosporine A treatment had *no* effect on IL-10 mRNA or plasma protein levels.

Taken together, these data suggest that the selective suppression of IL-10 production seen after FK-506 treatment may result in increased astrocyte activation and proliferation at the implantation site. Increased synthesis and release of several growth factors from activated astrocytes, including GDNF, can then potentiate maturation as well as neurite outgrowth of implanted neurons. In agreement with this hypothesis, it has been demonstrated that cultivation of hNT cells on an astrocyte monolayer is required for neuronal maturation, axonal growth, and synapse formation *(105)*. Similarly, it has been shown that in vitro treatment of dopamine neurons with GDNF has a potent neurotrophic effect on neurite outgrowth *(106)*. Whether or not FK-506 treatment also has similar potentiative effect on the synthesis of other astrocyte-derived factors, which promote axodendritic sprouting, such as laminin or PSA-NCAM is not known at present.

Finally, it has been demonstrated that after injury, ischemia, or inflammation, there is an increased *neuronal* expression of several cytokines (TNF-α, IL-1β) with chemotactic activity *(107–109)*. These data suggest that an increased expression of these chemotactic factors in the *implanted* cells can also stimulate astrocyte proliferation and migration toward implanted cells. This is supported by our observation of the active ingrowth of astrocyte processes into hNT grafts. However, at present, a differential effect of cyclosporine A versus FK-506 treatment on the synthesis of several proinflammatory and inflammatory cytokines as well as growth factors in grafted, partially injured cells is not defined yet.

As shown in FK-506-treated animals, a robust axonal growth was seen as soon as 3 wk after grafting, with a number of axons terminating in the vicinity of persisting α-motoneurons. However, in these studies, there was no evidence of target (neuron)-driven axonal growth toward host neurons and a number of axons growing in the peripheral white matter close to the implant or along the ventrally localized α-motoneuronal axons forming the ventral root were also identified. Whether or not these axons form functional synapses with the host neurons is not defined at present. However, previous studies have demonstrated that after implantation of hNT neurons into nude mouse spinal cord, there was extensive axodendritic outgrowth and the development of synapselike structure, as confirmed by Syn/HO14 colocalization *(70)*. Similarly, in the present study, a clear immunoreactivity for synaptophysin was seen within and at the periphery of the implant. These data suggest that hNT neuron-derived axons have a potential to form functional synapses within the host neurons.

As demonstrated, a comparable neurotrophic effect of FK-506 treatment on the maturation of spinal neural *precursors* was seen. These data show that FK-506 treatment has the potential to increase the maturation of not only grafted *postmitotic* neurons but also noncommitted spinal precursors. Whether or not FK-506 will exert similar neurotrophic effects on other premitotic or postmitotic cell lines after grafting into previously injured brain or spinal cord remains to be determined.

CONCLUSION AND FUTURE DIRECTION

As discussed, previous studies have demonstrated that implantation of uncommitted neural precursors or fully postmitotic neurons is associated with only limited survival and maturation of implanted cells when grafted into injured spinal cord. In our studies, we have seen a comparable limited maturation of grafted spinal neural precursors or human hNT neurons when implanted into a previously *ischemic* spinal cord. Subsequent studies have demonstrated that by using immunosuppressive drug treatment with FK-506, it is possible to achieve a rapid and much greater maturation and axonal growth from implanted precursors and hNT neurons. This neurotrophic effect appears to be mediated primarily through (1) increased astrocyte proliferation and migration into the graft and (2) increased astrocyte-derived GDNF synthesis. Although the mechanism of the FK-506-mediated effect is not clear, it may involve modulation of cytokine synthesis in the host astrocytes, microglia/macrophages, and/or grafted cells.

The ability to isolate, propagate, and differentiate spinal neural precursors as well as human hNT neurons into a well-defined phenotype provides a clear advantage. The facts that both cell lines transcribe the GAD gene and that hNT neurons express GABA immunoreactivity at 3–4 mo after grafting suggest that these cell lines are optimal candidates for spinal grafting. An optimal use of this approach would appear to be in the attempt to modulate ischemic spasticity/motor dysfunction after spinal ischemic injury or other spinal injury-related spasticity. Although the therapeutic potency of spinally grafted neural precursors or hNT neurons is not defined at present, future studies using multiple spinal grafts targeted into the regions of defined interneuronal loss will be needed to characterize its effect in the modulation of spasticity and possible improvement of motor function. In addition, as demonstrated, at present we have no evidence of targeted axonal growth from grafted cells toward host neurons. Identifying the molecules with chemotactic/neurotrophic activity, which would promote the development of functional synapses between grafted cells and persisting host neurons, will represent one of the key elements in a successful functional incorporation of grafted cells.

ACKNOWLEDGMENTS

This work was supported by grants NS32794 and NS40386 (M.M.)

REFERENCES

1. Picone, A. L., Green, R. M., Ricotta, J. R., May, A. G., and DeWeese, J. A. (1986) Spinal cord ischemia following operations on the abdominal aorta. *J. Vasc. Surg.* **3,** 94–103.
2. Svensson, L. G., Von Ritter, C. M., Groeneveld, H. T., Rickards, E. S., Hunter, S. J., Robinson, M. F., et al. (1986) Cross-clamping of the thoracic aorta. Influence of aortic shunts, laminectomy, papaverine, calcium channel blocker, allopurinol, and superoxide dismutase on spinal cord blood flow and paraplegia in baboons. *Ann. Surg.* **204,** 38–47.
3. Homma, S., Suzuki, T., Murayama, S., and Otsuka, M. (1979) Amino acid and substance P contents in spinal cord of cats with experimental hind-limb rigidity produced by occlusion of spinal cord blood supply. *J. Neurochem.* **32,** 691–698.
4. Marsala, J., Sulla, I., Santa, M., Marsala, M., Zacharias, L., and Radonak, J. (1991) Mapping of the canine lumbosacral spinal cord neurons by Nauta method at the end of the early phase of paraplegia induced by ischemia and reperfusion. *Neuroscience* **45,** 479–494.
5. Zivin, J. A. and DeGirolami, U. (1980) Spinal cord infarction: a highly reproducible stroke model. *Stroke* **11,** 200–202.
6. Matsushita, A. and Smith, C. M. (1970) Spinal cord function in postischemic rigidity in the rat. *Brain Res.* **19,** 395–410.
7. Marsala, M., Danielisová, V., Chavko, M., Hornáková, A., and Marsala, J. (1989) Improvement of energy state and basic modifications of neuropathological damage in rabbits as a result of graded postischemic spinal cord reoxygenation. *Exp. Neurol.* **105,** 93–103.
8. Taira, Y. and Marsala, M. (1996) Effect of proximal arterial perfusion pressure on function, spinal cord blood flow, and histopathologic changes after increasing intervals of aortic occlusion in the rat. *Stroke* **27,** 1850–1858.
9. Marsala, J., Sulla, I., Santa, M., Marsala, M., Mechírová, E., and Jalc, P. (1989) Early neurohistopathological changes of canine lumbosacral spinal cord segments in ischemia–reperfusion-induced paraplegia. *Neurosci. Lett.* **106,** 83–88.

10. Kakinohana, O., Marsala, M., Cizkova, D., Yang, L., Marsala, J., and Yaksh, T. (1999) Persistent labeling of spinal A-motoneurons and rubrospinal neurons after spinal ischemia-induced paraplegia. *Soc. Neurosci. Abstr.* **25(1–2),** 828.
11. Marsala, M., Vanicky, I., Galik, J., Radonak, J., Kundrat, I., and Marsala, J. (1993) Panmyelic epidural cooling protects against ischemic spinal cord damage. *J. Surg. Res.* **55,** 21–31.
12. Cheng, M. K., Robertson, C., Grossman, R. G., Foltz, R., and Williams, V. (1984) Neurological outcome correlated with spinal evoked potentials in a spinal cord ischemia model. *J. Neurosurg.* **60,** 786–795.
13. Simpson, R. K., Jr., Gondo, M., Robertson, C. S., and Goodman, J. C. (1995) The influence of glycine and related compounds on spinal cord injury-induced spasticity. *Neurochem. Res.* **20,** 1203–1210.
14. Altman, J. and Das, G. D. (1965) Autoradiographic and histological evidence of postnatal hippocampal neurogenesis in rats. *J. Comp. Neurol.* **124,** 319–335.
15. Altman, J. and Das, G. D. (1967) Postnatal neurogenesis in the guinea-pig. *Nature* **214,** 1098–1101.
16. Cameron, H. A., Woolley, C. S., McEwen, B. S., and Gould, E. (1993) Differentiation of newly born neurons and glia in the dentate gyrus of the adult rat. *Neuroscience* **56,** 337–344.
17. Kuhn, H. G., Dickinson-Anson, H., and Gage, F. H. (1996) Neurogenesis in the dentate gyrus of the adult rat: age-related decrease of neuronal progenitor proliferation. *J. Neurosci.* **16,** 2027–2033.
18. Kaplan, M. S. and Hinds, J. W. (1977) Neurogenesis in the adult rat: electron microscopic analysis of light radioautographs. *Science* **197,** 1092–1094.
19. Stanfield, B. B. and Trice, J. E. (1988) Evidence that granule cells generated in the dentate gyrus of adult rats extend axonal projections. *Exp. Brain Res.* **72,** 399–406.
20. Goldman, S. A. and Nottebohm, F. (1983) Neuronal production, migration, and differentiation in a vocal control nucleus of the adult female canary brain. *Proc. Natl. Acad. Sci. USA* **80,** 2390–2394.
21. Barnea, A. and Nottebohm, F. (1994) Seasonal recruitment of hippocampal neurons in adult free-ranging black-capped chickadees. *Proc. Natl. Acad. Sci. USA* **91,** 11,217–11,221.
22. Kempermann, G., Kuhn, H. G., and Gage, F. H. (1997) More hippocampal neurons in adult mice living in an enriched environment. *Nature* **386,** 493–495.
23. Kempermann, G., Kuhn, H. G., and Gage, F. H. (1997) Genetic influence on neurogenesis in the dentate gyrus of adult mice. *Proc. Natl. Acad. Sci. USA* **94,** 10,409–10,414.
24. McEwen, B. S. (1996) Gonadal and adrenal steroids regulate neurochemical and structural plasticity of the hippocampus via cellular mechanisms involving NMDA receptors. *Cell. Mol. Neurobiol.* **16,** 103–116.
25. Craig, C. G., Tropepe, V., Morshead, C. M., Reynolds, B. A., Weiss, S., and van der Kooy, D. (1996) In vivo growth factor expansion of endogenous subependymal neural precursor cell populations in the adult mouse brain. *J. Neurosci.* **16,** 2649–2658.

26. Liu, J., Solway, K., Messing, R. O., and Sharp, F. R. (1998) Increased neurogenesis in the dentate gyrus after transient global ischemia in gerbils. *J. Neurosci.* **18,** 7768–7778.

27. Takagi, Y., Nozaki, K., Takahashi, J., Yodoi, J., Ishikawa, M., and Hashimoto, N. (1999) Proliferation of neuronal precursor cells in the dentate gyrus is accelerated after transient forebrain ischemia in mice. *Brain Res.* **831,** 283–287.

28. Marsala, M., Kakinohana, O., Cizkova, D., Ragin, M., Yang, L., Marsala, J., et al. (1999) Lack of neuronal proliferation after injurious or non-injurious intervals of spinal ischemia in rat. International Symposium on Neuronal Regeneration, p. 40.

29. Johansson, C. B., Momma, S., Clarke, D. L., Risling, M., Lendahl, U., and Frisén, J. (1999) Identification of a neural stem cell in the adult mammalian central nervous system. *Cell* **96,** 25–34.

30. Giulian, D. and Robertson, C. (1990) Inhibition of mononuclear phagocytes reduces ischemic injury in the spinal cord. *Ann. Neurol.* **27,** 33–42.

31. Leskovar, A., Moriarty, L. J., Turek, J. J., Schoenlein, I. A., and Borgens, R. B. (2000) The macrophage in acute neural injury: changes in cell numbers over time and levels of cytokine production in mammalian central and peripheral nervous systems. *J. Exp. Biol.* **203(Pt. 12),** 1783–1795.

32. Blight, A. R. (1992) Macrophages and inflammatory damage in spinal cord injury. *J. Neurotrauma* **9(Suppl. 1),** S83–S91.

33. Giulian, D., Chen, J., Ingeman, J. E., George, J. K., and Noponen, M. (1989) The role of mononuclear phagocytes in wound healing after traumatic injury to adult mammalian brain. *J. Neurosci.* **9,** 4416–4429.

34. Kochanek, P. M. and Hallenbeck, J. M. (1992) Polymorphonuclear leukocytes and monocytes/macrophages in the pathogenesis of cerebral ischemia and stroke. *Stroke* **23,** 1367–1379.

35. Gregersen, R., Lambertsen, K., and Finsen, B. (2000) Microglia and macrophages are the major source of tumor necrosis factor in permanent middle cerebral artery occlusion in mice. *J. Cereb. Blood Flow Metab.* **20,** 53–65.

36. Norenberg, M. D. (1994) Astrocyte responses to CNS injury. *J. Neuropathol. Exp. Neurol.* **53,** 213–220.

37. Mabuchi, T., Kitagawa, K., Ohtsuki, T., Kuwabara, K., Yagita, Y., Yanagihara, T., et al. (2000) Contribution of microglia/macrophages to expansion of infarction and response of oligodendrocytes after focal cerebral ischemia in rats. *Stroke* **31,** 1735–1743.

38. Giulian, D. and Lachman, L. B. (1985) Interleukin-1 stimulation of astroglial proliferation after brain injury. *Science* **228,** 497–499.

39. Balasingam, V. and Yong, V. W. (1996) Attenuation of astroglial reactivity by interleukin-10. *J. Neurosci.* **16,** 2945–2955.

40. Banner, L. R., Moayeri, N. N., and Patterson, P. H. (1997) Leukemia inhibitory factor is expressed in astrocytes following cortical brain injury. *Exp. Neurol.* **147,** 1–9.

41. Suzuki, S., Tanaka, K., Nogawa, S., Ito, D., Dembo, T., Kosakai, A., et al. (2000) Immunohistochemical detection of leukemia inhibitory factor after focal cerebral ischemia in rats. *J. Cereb. Blood Flow Metab.* **20,** 661–668.

42. Robertson, T. A., Maley, M. A., Grounds, M. D., and Papadimitriou, J. M. (1993) The role of macrophages in skeletal muscle regeneration with particular reference to chemotaxis. *Exp. Cell Res.* **207,** 321–331.
43. Balasingam, V., Tejada-Berges, T., Wright, E., Bouckova, R., and Yong, V. W. (1994) Reactive astrogliosis in the neonatal mouse brain and its modulation by cytokines. *J. Neurosci.* **14,** 846–856.
44. Yamasaki, Y., Suzuki, T., Yamaya, H., Matsuura, N., Onodera, H., and Kogure, K. (1992) Possible involvement of interleukin-1 in ischemic brain edema formation. *Neurosci. Lett.* **142,** 45–47.
45. Loddick, S. A. and Rothwell, N. J. (1996) Neuroprotective effects of human recombinant interleukin-1 receptor antagonist in focal cerebral ischaemia in the rat. *J. Cereb. Blood Flow Metab.* **16,** 932–940.
46. Koblar, S. A., Turnley, A. M., Classon, B. J., Reid, K. L., Ware, C. B., Cheema, S. S., et al. (1998) Neural precursor differentiation into astrocytes requires signaling through the leukemia inhibitory factor receptor. *Proc. Natl. Acad. Sci. USA* **95,** 3178–3181.
47. Bonni, A., Sun, Y., Nadal-Vicens, M., Bhatt, A., Frank, D. A., Rozovsky, I., et al. (1997) Regulation of gliogenesis in the central nervous system by the JAK–STAT signaling pathway. *Science* **278,** 477–483.
48. Satoh, M., Sugino, H., and Yoshida, T. (2000) Activin promotes astrocytic differentiation of a multipotent neural stem cell line and an astrocyte progenitor cell line from murine central nervous system. *Neurosci. Lett.* **284,** 143–146.
49. Gross, R. E., Mehler, M. F., Mabie, P. C., Zang, Z., Santschi, L., and Kessler, J. A. (1996) Bone morphogenetic proteins promote astroglial lineage commitment by mammalian subventricular zone progenitor cells. *Neuron* **17,** 595–606.
50. Tretter, Y. P., Munz, B., Hubner, G., ten Bruggencate, G., Werner, S., and Alzheimer, C. (1996) Strong induction of activin expression after hippocampal lesion. *NeuroReport* **7,** 1819–1823.
51. Miller, C., Tsatas, O., and David, S. (1994) Dibutyryl cAMP, interleukin-1 beta, and macrophage conditioned medium enhance the ability of astrocytes to promote neurite growth. *J. Neurosci. Res.* **38,** 56–63.
52. Gage, F. H., Coates, P. W., Palmer, T. D., Kuhn, H. G., Fisher, L. J., Suhonen, J. O., et al. (1995) Survival and differentiation of adult neuronal progenitor cells transplanted to the adult brain. *Proc. Natl. Acad. Sci. USA* **92,** 11,879–11,883.
53. Fricker, R. A., Carpenter, M. K., Winkler, C., Greco, C., Gates, M. A., and Björklund, A. (1999) Site-specific migration and neuronal differentiation of human neural progenitor cells after transplantation in the adult rat brain. *J. Neurosci.* **19,** 5990–6005.
54. Campbell, K., Olsson, M., and Björklund, A. (1995) Regional incorporation and site-specific differentiation of striatal precursors transplanted to the embryonic forebrain ventricle. *Neuron* **15,** 1259–1273.
55. Suhonen, J. O., Peterson, D. A., Ray, J., and Gage, F. H. (1996) Differentiation of adult hippocampus-derived progenitors into olfactory neurons in vivo. *Nature* **383,** 624–627.
56. Svendsen, C. N., Clarke, D. J., Rosser, A. E., and Dunnett, S. B. (1996) Survival and differentiation of rat and human epidermal growth factor-responsive

precursor cells following grafting into the lesioned adult central nervous system. *Exp. Neurol.* **137,** 376–388.

57. Zigova, T., Betarbet, R., Soteres, B. J., Brock, S., Bakay, R. A., and Luskin, M. B. (1996) A comparison of the patterns of migration and the destinations of homotopically transplanted neonatal subventricular zone cells and heterotopically transplanted telencephalic ventricular zone cells. *Dev. Biol.* **173,** 459–474.

58. McDonald, J. W., Liu, X. Z., Qu, Y., Liu, S., Mickey, S. K., Turetsky, D., et al. (1999) Transplanted embryonic stem cells survive, differentiate and promote recovery in injured rat spinal cord. *Nat. Med.* **5,** 1410–1412.

59. Onifer, S. M., Cannon, A. B., and Whittemore, S. R. (1997) Altered differentiation of CNS neural progenitor cells after transplantation into the injured adult rat spinal cord. *Cell Transplant.* **6,** 327–338.

60. Loo, D. T., Althoen, M. C., and Cotman, C. W. (1995) Differentiation of serum-free mouse embryo cells into astrocytes is accompanied by induction of glutamine synthetase activity. *J. Neurosci. Res.* **42,** 184–191.

61. Qian, X., Davis, A. A., Goderie, S. K., and Temple, S. (1997) FGF2 concentration regulates the generation of neurons and glia from multipotent cortical stem cells. *Neuron* **18,** 81–93.

62. Andrews, P. W., Damjanov, I., Simon, D., Banting, G. S., Carlin, C., Dracopoli, N. C., et al. (1984) Pluripotent embryonal carcinoma clones derived from the human teratocarcinoma cell line Tera-2. Differentiation in vivo and in vitro. *Lab. Invest.* **50,** 147–162.

63. Lee, V. M. and Andrews, P. W. (1986) Differentiation of NTERA-2 clonal human embryonal carcinoma cells into neurons involves the induction of all three neurofilament proteins. *J. Neurosci.* **6,** 514–521.

64. Paterlini, M., Valerio, A., Baruzzi, F., Memo, M., and Spano, P. F. (1998) Opposing regulation of tau protein levels by ionotropic and metabotropic glutamate receptors in human NT2 neurons. *Neurosci. Lett.* **243,** 77–80.

65. Neelands, T. R., Greenfield, L. J., Jr., Zhang, J., Turner, R. S., and Macdonald, R. L. (1998) GABAA receptor pharmacology and subtype mRNA expression in human neuronal NT2-N cells. *J. Neurosci.* **18,** 4993–5007.

66. Gao, Z. Y., Xu, G., Stwora-Wojczyk, M. M., Matschinsky, F. M., Lee, V. M., and Wolf, B. A. (1998) Retinoic acid induction of calcium channel expression in human NT2N neurons. *Biochem. Biophys. Res. Commun.* **247,** 407–413.

67. Yoshioka, A., Yudkoff, M., and Pleasure, D. (1997) Expression of glutamic acid decarboxylase during human neuronal differentiation: studies using the NTera-2 culture system. *Brain Res.* **767,** 333–339.

68. Borlongan, C. V., Saporta, S., Poulos, S. G., Othberg, A., and Sanberg, P. R. (1998) Viability and survival of hNT neurons determine degree of functional recovery in grafted ischemic rats. *NeuroReport* **9,** 2837–2842.

69. Saporta, S., Willing, A. E., Zigova, T., Daadi, M. M., and Sanberg, P. R. (2001) Comparison of calcium-binding proteins expressed in cultured hNT neurons and hNT neurons transplanted into the rat striatum. *Exp. Neurol.* **167,** 252–259.

70. Hartley, R. S., Trojanowski, J. Q., and Lee, V. M. (1999) Differential effects of spinal cord gray and white matter on process outgrowth from grafted human NTERA2 neurons (NT2N, hNT). *J. Comp. Neurol.* **415,** 404–418.

71. Velardo, M. J., O'Steen, B. E., McGrogan, M., and Reier, P. (2000) HNT cells and transplantation repair of rat cervical spinal cord contusions. *Exp. Neurol.* **164(2),** 454.

72. Velardo, M., O'Steen, B., McGrogan, M., and Reier, P. (2000) Survival, maturation and innervation of hNT cell transplants in the contused spinal cord of the adult rat. *Soc. Neurosci. Abstr.* **26,** 1104.

73. Wang, M. S., Zeleny-Pooley, M., and Gold, B. G. (1997) Comparative dose-dependence study of FK506 and cyclosporin A on the rate of axonal regeneration in the rat sciatic nerve. *J. Pharmacol. Exp. Ther.* **282,** 1084–1093.

74. Schwab, M. E., Kapfhammer, J. P., and Bandtlow, C. E. (1993) Inhibitors of neurite growth. *Annu. Rev. Neurosci.* **16,** 565–595.

75. McKeon, R. J., Schreiber, R. C., Rudge, J. S., and Silver, J. (1991) Reduction of neurite outgrowth in a model of glial scarring following CNS injury is correlated with the expression of inhibitory molecules on reactive astrocytes. *J. Neurosci.* **11,** 3398–3411.

76. Liesi, P. (1985) Laminin-immunoreactive glia distinguish regenerative adult CNS systems from non-regenerative ones. *EMBO J.* **4,** 2505–2511.

77. Frisén, J., Haegerstrand, A., Risling, M., Fried, K., Johansson, C. B., Hammarberg, H., et al. (1995) Spinal axons in central nervous system scar tissue are closely related to laminin-immunoreactive astrocytes. *Neuroscience* **65,** 293–304.

78. Le Gal La Salle, G., Rougon, G., and Valin, A. (1992) The embryonic form of neural cell surface molecule (E-NCAM) in the rat hippocampus and its reexpression on glial cells following kainic acid-induced status epilepticus. *J. Neurosci.* **12,** 872–882.

79. Alonso, G. and Privat, A. (1993) Reactive astrocytes involved in the formation of lesional scars differ in the mediobasal hypothalamus and in other forebrain regions. *J. Neurosci. Res.* **34,** 523–538.

80. Oumesmar, B. N., Vignais, L., Duhamel-Clérin, E., Avellana-Adalid, V., Rougon, G., and Baron Van Evercooren, A. (1995) Expression of the highly polysialylated neural cell adhesion molecule during postnatal myelination and following chemically induced demyelination of the adult mouse spinal cord. *Eur. J. Neurosci.* **7,** 480–491.

81. Moretto, G., Walker, D. G., Lanteri, P., Taioli, F., Zaffagnini, S., Xu, R. Y., et al. (1996) Expression and regulation of glial-cell-line-derived neurotrophic factor (GDNF) mRNA in human astrocytes in vitro. *Cell Tissue Res.* **286,** 257–262.

82. Alderson, R. F., Curtis, R., Alterman, A. L., Lindsay, R. M., and DiStefano, P. S. (2000) Truncated TrkB mediates the endocytosis and release of BDNF and neurotrophin-4/5 by rat astrocytes and Schwann cells in vitro. *Brain Res.* **871,** 210–222.

83. Kamiguchi, H., Yoshida, K., Sagoh, M., Sasaki, H., Inaba, M., Wakamoto, H., et al. (1995) Release of ciliary neurotrophic factor from cultured astrocytes and its modulation by cytokines. *Neurochem. Res.* **20,** 1187–1193.

84. Boutros, T., Croze, E., and Yong, V. W. (1997) Interferon-beta is a potent promoter of nerve growth factor production by astrocytes. *J. Neurochem.* **69,** 939–946.

85. Asher, R. A., Morgenstern, D. A., Fidler, P. S., Adcock, K. H., Oohira, A., Braistead, J. E., et al. (2000) Neurocan is upregulated in injured brain and in cytokine-treated astrocytes. *J. Neurosci.* **20,** 2427–2438.

86. Rudge, J. S. and Silver, J. (1990) Inhibition of neurite outgrowth on astroglial scars in vitro. *J. Neurosci.* **10,** 3594–3603.

87. Snow, D. M., Lemmon, V., Carrino, D. A., Caplan, A. I., and Silver, J. (1990) Sulfated proteoglycans in astroglial barriers inhibit neurite outgrowth in vitro. *Exp. Neurol.* **109,** 111–130.

88. McKeon, R. J., Schreiber, R. C., Rudge, J. S., and Silver, J. (1991) Reduction of neurite outgrowth in a model of glial scarring following CNS injury is correlated with the expression of inhibitory molecules on reactive astrocytes. *J. Neurosci.* **11,** 3398–3411.

89. Tiveron, M. C., Barboni, E., Pliego Rivero, F. B., Gormley, A. M., Seeley, P. J., Grosveld, F., et al. (1992) Selective inhibition of neurite outgrowth on mature astrocytes by Thy-1 glycoprotein. *Nature* **355,** 745–748.

90. Davies, S. J., Fitch, M. T., Memberg, S. P., Hall, A. K., Raisman, G., and Silver, J. (1997) Regeneration of adult axons in white matter tracts of the central nervous system. *Nature* **390,** 680–683.

91. Fok-Seang, J., DiProspero, N. A., Meiners, S., Muir, E., and Fawcett, J. W. (1998) Cytokine-induced changes in the ability of astrocytes to support migration of oligodendrocyte precursors and axon growth. *Eur. J. Neurosci.* **10,** 2400–2415.

92. Nolte, C., Kirchhoff, F., and Kettenmann, H. (1997) Epidermal growth factor is a motility factor for microglial cells in vitro: evidence for EGF receptor expression. *Eur. J. Neurosci.* **9,** 1690–1698.

93. Guo, X., Chandrasekaran, V., Lein, P., Kaplan, P. L., and Higgins, D. (1999) Leukemia inhibitory factor and ciliary neurotrophic factor cause dendritic retraction in cultured rat sympathetic neurons. *J. Neurosci.* **19,** 2113–2121.

94. Dusart, I., Morel, M. P., Wehrlé, R., and Sotelo, C. (1999) Late axonal sprouting of injured Purkinje cells and its temporal correlation with permissive changes in the glial scar. *J. Comp. Neurol.* **408,** 399–418.

95. Gold, B. G. (1999) FK506 and the role of the immunophilin FKBP-52 in nerve regeneration. *Drug Metab. Rev.* **31,** 649–663.

96. Gold, B. G. (1997) FK506 and the role of immunophilins in nerve regeneration. *Mol. Neurobiol.* **15,** 285–306.

97. Snyder, S. H. and Sabatini, D. M. (1995) Immunophilins and the nervous system. *Nat. Med.* **1,** 32–37.

98. Marsala, M., Kakinohana, O., Tokumine, J., Yang, L., Velardo, M., Cizkova, D. et al. (2001) A potent, FK-506-mediated maturation and axonal sprouting in hNT neurons implanted spinally in rats with ischemic paraplegia. *Am. Soc. Neural Transplant. Repair* **8,** 74.

99. Kakinohana, O., Marsala, M., Tokumine, J., Cizkova, D., and Ragin, M. (2001) FK-506 treatment potentiate neuronal differentiation from spinal stem cells implanted spinally in rats with ischemia-induced paraplegia. *Am. Soc. Neural Transplant. Repair* **8,** 73.

100. Lacy, M., Jones, J., Whittemore, S. R., Haviland, D. L., Wetsel, R. A., and Barnum, S. R. (1995) Expression of the receptors for the C5a anaphylatoxin, interleukin-8 and FMLP by human astrocytes and microglia. *J. Neuroimmunol.* **61,** 71–78.

101. Pita, I., Jelaso, A. M., and Ide, C. F. (1999) IL-1beta increases intracellular calcium through an IL-1 type 1 receptor mediated mechanism in C6 astrocytic cells. *Intl. J. Dev. Neurosci.* **17,** 813–820.

102. Pousset, F., Cremona, S., Dantzer, R., Kelley, K., and Parnet, P. (1999) Interleukin-4 and interleukin-10 regulate IL1-beta induced mouse primary astrocyte activation: a comparative study. *Glia* **26,** 12–21.

103. Ehrlich, L. C., Hu, S., Sheng, W. S., Sutton, R. L., Rockswold, G. L., Peterson, P. K., et al. (1998) Cytokine regulation of human microglial cell IL-8 production. *J. Immunol.* **160,** 1944–1948.

104. Jiang, H., Yang, X. F., Soriano, R., Fujitsu, T., and Kobayashi, M. (1999) Inhibition of IL-10 by FK 506 may be responsible for overcoming ongoing allograft rejection in the rat. *Transplant. Proc.* **31,** 1203–1205.

105. Hartley, R. S., Margulis, M., Fishman, P. S., Lee, V. M., and Tang, C. M. (1999) Functional synapses are formed between human NTera2 (NT2N, hNT) neurons grown on astrocytes. *J. Comp. Neurol.* **407,** 1–10.

106. Costantini, L. C. and Isacson, O. (2000) Immunophilin ligands and GDNF enhance neurite branching or elongation from developing dopamine neurons in culture. *Exp. Neurol.* **164,** 60–70.

107. Knoblach, S. M., Fan, L., and Faden, A. I. (1999) Early neuronal expression of tumor necrosis factor-alpha after experimental brain injury contributes to neurological impairment. *J. Neuroimmunol.* **95,** 115–125.

108. Sairanen, T. R., Lindsberg, P. J., Brenner, M., and Sirén, A. L. (1997) Global forebrain ischemia results in differential cellular expression of interleukin-1beta (IL-1beta) and its receptor at mRNA and protein level. *J. Cereb. Blood Flow Metab.* **17,** 1107–1120.

109. Schlüter, D., Kaefer, N., Hof, H., Wiestler, O. D., and Deckert-Schlüter, M. (1997) Expression pattern and cellular origin of cytokines in the normal and toxoplasma gondii-infected murine brain. *Am. J. Pathol.* **150,** 1021–1035.

15
Utilization of Marrow Stromal Cells for Gene Transfer into the CNS

S. Ausim Azizi, Emily J. Schwarz, Darwin Prockop, Guillermo Alexander, Katherine A. Mortati, and Barbara Krynska

INTRODUCTION

In the past several decades, neurotransplantation has been investigated for the study of development of the normal nervous system and possible functional restoration and repair of diseased and damaged nervous tissue *(1,2)*. In patients with Parkinson's disease, transplantation of dopamine-producing human and pig fetal tissues has resulted in varying degrees of clinical improvements, which correlated with graft survival and anatomical integration of the grafted fetal cells into the host brain *(3–5)*, albeit, the control of dopamine release remains an issue *(5)*. In animal studies, implantation of neural stem cells and partially differentiated embryonic stem cells into lesioned brain and spinal cord resulted in more rapid behavioral improvement in experimental models of spinal cord injury, stroke, and neurotrauma *(6–11)*. Although implantation of embryonic stem cells can be a promising method of cell therapy, obtaining fetal and embryonic tissues has presented major logistical, ethical, and immunological barriers *(12,13)*. Thus, autologous cells from the bone marrow, marrow stromal cells, can be an attractive alternative source of tissue for grafting and treatment of neurological diseases *(11)*.

For therapy of neurological diseases through cell and gene therapy, donor cells should be (1) easily available, (2) capable of rapid expansion in culture, (3) immunologically inert, (4) capable of long-term survival and integration in the host brain, and (5) amenable to stable transfection/transduction and long-term expression of exogenous genes. A variety of donor cells, including fibroblasts, endothelial cells, and astrocytes *(14–17)*, has been tested with varying degrees of success. In several studies by different investigators, fibroblasts and astrocytes were engineered to elaborate tyrosine hydroxylase

From: *Neural Stem Cells for Brain and Spinal Cord Repair*
Edited by: T. Zigova, E. Y. Snyder, and P. R. Sanberg © Humana Press Inc., Totowa, NJ

(15–17) and, after implantation, were noted to reverse the apomorphine-induced rotatory behavior in the rodent model of Parkinson's disease *(16–17)*. Trophic factors are known to play an important role in maintaining native and grafted dopaminergic neurons *(18,19)*. Fibroblasts and astrocytes engineered to produce these factors have been cografted along with the dopaminergic neurons or alone. Reportedly, the grafted dopaminergic neurons survive longer and native cells become more robust, with extensive arborization of their processes *(20–23)*. However, controlled delivery of gene products to the appropriate areas of the brain still remains a major obstacle.

Although fibroblasts and, to some extent, astrocytes fit some of the above-outlined criteria for transplantation, each suffers from a unique drawback. For example, after implantation into the brain, fibroblasts continue to produce collagen and the area of implantation undergoes gliosis *(24)*, whereas autologous astrocytes and endothelial cells are difficult to obtain. Therefore, efforts continue to identify other sources of neural and non-neural donor tissues. One such source is the bone marrow. We believe that nonhematopoietic multipotential stemlike cells obtained from the patient's own bone marrow can be genetically engineered to produce therapeutic molecules, can integrate and migrate into damaged areas, and can be used for rebuilding diseased brains *(25)*.

CHARACTERISTICS OF MARROW STROMAL CELLS

Bone marrow, in addition to hematopoietic precursors, contains cells that can be considered stem cells of nonhematopoietic tissues. These cells are part of a heterogeneous population that has been referred to as nonhematopoietic mesenchymal stem cells, because of their ability to differentiate into cells that are defined as mesenchymal. They are also known as marrow stromal cells (MSCs), because they appear to arise from the stromal structures of the marrow *(26)*. A German pathologist Cohnheim first suggested the existence of stem cells for non-hematopoietic tissue in bone marrow in 1867 *(27)*. He discovered that, in addition to the inflammatory cells, other cells—collagen-producing fibroblasts—that participated in wound healing and tissue repair came from blood. The implication was that blood and, by extension, bone marrow may participate in tissue repair by actually contributing the construction material. A century later, Friedenstein *(28)* observed that a subpopulation of cells from the bone marrow could grow in culture and differentiate into colonies that resemble bone or cartilage. Indeed, the natural tendency of human MSCs, grown in complete media with serum, is to differentiate into osteoblasts *(26)*.

A number of in vitro studies have shown that by manipulating the culture media and growing these cells under different conditions, the marrow stromal cells can be induced to differentiate into bone, fat, cartilage, skeletal muscle, cardiac muscle, and, perhaps, neural cells *(29,30)*.

Transplantation of marrow stromal cells into an organism and their integration into the host organs serves as a stringent measure of both their normalcy and a test of their potential to produce appropriate differentiated progeny. However, incorporation of these cells into the central nervous system (CNS) and assessment of their true functional integration may be more difficult. Mere expression of neural cell markers is not a proof of either their anatomical or functional integration.

In a series of experiments, systemic infusion of MSCs into irradiated mice and rats was followed by the appearance of progeny of donor cells in a variety of nonhematopoietic tissues *(29–32)*. Ferrari et al. *(33)*, upon transplanting genetically marked bone marrow into immune-deficient mice, observed homing of bone-marrow-derived cells to damaged muscle and differentiation of these cells into myocytes. Similarly, intravenous infusion of whole bone marrow into irradiated *mdx* mice, the animal model of Duchenne's muscular dystrophy, resulted in restoration of expression of dystrophin in these mice *(34)*. More recently, it was reported that infusion of labeled bone marrow cells into irradiated mice resulted in infiltration of cells of bone marrow origin into the brain. Bone marrow cells integrated into the brain and expressed neural cell markers *(31,32)*. However, whether these cells differentiate into functional neurons and glia still remains to be proven. Thus, it appears that at least a subpopulation of MSCs behave similarly to embryonic stem cells in that they are capable of differentiating into a variety of cells and tissues and are part of a vestigial system for renewal of a large number of nonhematopoietic cells and tissues.

To determine the possibility of survival and integration of MSCs in the brain, we implanted human MSCs into the striatum of adult rats (10,000 cells/μL; 10 μL per animal). Five to 180 d later, brain sections were examined for the presence of the donor cells. Initially, about 20% of the cells had survived and engrafted. There was no evidence of an inflammatory response or rejection. The cells had migrated from the injection site along the known pathways for the migration of neural stem cells to several areas of the brain *(25)*. These results demonstrate that human marrow stromal cells infused into rat brain can engraft, migrate, and survive in a manner similar to native brain cells (e.g., astrocytes) *(25)*. A decline in the number of implanted MSCs was noted after 30 d, more than likely because human cells were implanted into rat brains. However, compared to other types of cell *(3,35)*, the number of surviving MSCs in the brain is many-fold higher. This was confirmed by immunostaining of rat brain sections, where it was demonstrated that grafted cells express human-specific cell markers *(25)*. The sections did not stain at any time with either alizarin red (a stain for osteoblasts) or oil red-O (stain for adipocytes). Of particular interest was the lack of a vigorous immune reaction by the host *(25)*. This is significant for the use of MSCs in gene therapy, as it shows that MSCs, infused in low concentrations (<10,000 cells/μL) lose their natural tendency to differentiate into osteoblasts or adipocytes *(25)*. Moreover, they integrate into xenogenic host brain and survive for long periods of time, important characteristics of cells used for ex vivo gene therapy.

ISOLATION AND GROWTH OF RAT AND HUMAN MSC

Marrow stromal cells are obtained from small bone marrow aspirates and are separated from the hematopoietic cells by virtue of their adherence to the culture-dish plastic. The cultures are practically devoid of cells that are of interest in the field of bone marrow transplantation. They comprise a heterogeneous population of small and large cells. Precisely which groups of cells within this population are multipotential, easily expandable, and amenable to gene engineering is still under investigation. However, consistency in culturing methods for isolation and expansion of these cells can allow us to predict their behavior in vitro and in vivo. Here, we describe observations on the growth of these cells and methods for isolation and expansion of these cells used in our laboratory.

The rat bone marrow was harvested from femurs and tibias of adult rats. The bone marrow was flushed out under sterile conditions and plated in 15-cm^2 dishes of media containing minimal essential medium with α modification (α-MEM), 20% fetal bovine serum (FBS), 2 m*M* L-glutamine, and 1% penicillin/streptomycin. The MSCs were isolated by their adherence to the culture-dish plastic. The nonadherent cells, presumably the hematopoietic components of the marrow, were removed after 48 h and fresh medium was added. After the cells had grown to near confluence, they were harvested with trypsin. For each passage, the cells were split in 1:5 ratios and were grown to confluence in α-MEM containing 10% FBS.

The human bone marrow samples were obtained from the iliac crest of healthy volunteers by needle aspiration. The bone marrow samples were diluted in 1:3 ratios with sterile phosphate-buffered saline (PBS) and the mononuclear cells were isolated using a Ficoll–Paque density gradient. Nucleated marrow cells were cultured in α-MEM containing 20% FCS, 2 m*M* L-glutamine, and 1% penicillin/streptomycin for 2 d. Subsequently, the nonadherent hematopoetic cells were removed by discarding the media, cultures were rinsed with PBS, fresh media was added, and the adherent cells were allowed to grow for an additional 10 d. At 70% confluence, cells were harvested and frozen for later use, or replated after 1:4 ratio dilution in α-MEM with 10% FCS. For expansion, some of the large colonies composed of mainly fibroblastlike small cells were harvested and recultured (see below).

Close observations of the cultures revealed that small cells have a tendency to grow well in the vicinity of large flat cells surrounding them in a ringlike fashion (Fig. 1D). We believe that the large flat cells comprise a feeder layer similar to those needed by embryonic stem cells *(36)*. The large flat cells are likely essential for growth and proliferation of small cells, similar to the natural hematopoietic "nurse" cells or stromal cells of the bone marrow *(26)*.

When plated at a density of 20,000 cells/well in a six-well plate, both human and rat MSCs formed spherical aggregates similar to the cell aggregates formed by the embryonic stem cell and neural stem cells (neurospheres) *(37)*. These aggregates consisted of small proliferating cells (Figs. 1B,E). The aggregates were lifted by ring cloning and were further expanded in culture (Figs. 1C,F).

EXPANSION OF HUMAN MSC

Given the current methods, the process of gene engineering is notoriously inefficient and needs large numbers of cells. Therefore, expansion of cells prior to definitive differentiation is a critical characteristic for gene engineering. Early passage cells can be expanded by low-density plating *(11)*. Isolated MSCs yield a heterogeneous population with primarily two types of morphology: large, flat cells and smaller elongated cells *(25)*. Large colonies containing small cells were lifted by either ring cloning or using 5-mm disks of trypsin-soaked filter paper. The cell suspensions were diluted and then plated at very low density (3–5 cells/cm^2). After 10–12 d, the dishes were stained with cresyl violet for colony counting. Selected colonies were harvested and the cells within each colony were counted using a hemocytometer, then diluted and replated again at very low densities. This was repeated for four to five passages, using cells from multiple donors. At each passage, the colonies and cells in selected colonies were counted (*see* Table 1). It appears that the smaller cells retain the capacity to generate single-cell colonies over several passages and are likely enriched with multipotential stem cells. Moreover, the MSC cultures display a lag phase of 5–7 d, a log phase of rapid growth of about 5 d, and then a stationary phase. During the log phase, the cells doubled an average of twice every 24 h *(38)*. These cultures are highly proliferative and could be expanded to up to 2000-fold over a period of 10 d *(11)*.

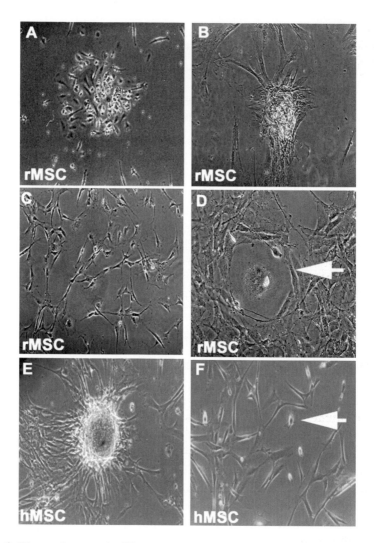

Fig. 1. Photomicrographs illustrate our observations on rat and human marrow stromal cell growth characteristics. **(A,B)** Aggregates of rat MSCs in culture are similar to cell aggregates formed by other stem cells (see text). **(C)** The rat marrow stromal cell aggregates dissociated and replated. **(D)** In cultures of MSCs, small cells tend to grow around and over the large flat cells, which are presumably "nurse" cells. **(E)** A spherical aggregate of cells formed by human MSCs cultured at 20,000 cells per well. **(F)** Photomicrograph demonstrates small proliferative group of the human MSCs after dissociation of cell aggregate in E.

Table 1
**Expansion Potential of MSCs, When Cells Plated
at Low Density (1–4 cells/cm²)**

Days (passage)	No. of cells/colony[a]	No. of colonies/100 plated cells	Estimated no. of cells from a single marrow aspirate
14 (P2)	N/A[b]	12 ± 5	5×10^5
28 (P3)	25,000 ± 8,000	18 ± 7	15×10^8
40 (P4)	30,000 ± 7,000	7 ± 3	6.5×10^{11}
52 (P5)	23,000 ± 6,500	11 ± 6	16.5×10^{14}

[a] Mean number of cells in 12 colonies from 3 different donors.
[b] At the initial passage, colonies were not isolated.
Note: The table demonstrates the estimated number of cells after low-density plating. The cells were counted using a hemocytometer and the mean number was calculated from 12 samples. The next column demonstrates the colony-forming efficiency of the samples. The last column shows the estimated number of cells, given that all cells from each passage were to be plated at low density in a bioreactor.

The ability to rapidly expand human MSCs in culture will be of obvious importance in using these cells for cell and gene therapy. Under the conditions developed here, one bone marrow aspirate obtained under local anesthesia can generate about 10^{12} cells in 6–7 wk, a number that approaches the total number of cells in the human body (*see* Table 1). Colter and colleagues (*38*) using FACS analysis have shown that when cells from low density plated cultures are assayed for size and granularity, three populations of cells can be identified: (1) large cells with medium content of granularity, which are likely mature stromal cells; (2) small agranular cells (13% of population); (3) small and granular cells (30% of population). The later two groups recycle in different phases of cell cycle (i.e., the small granular cells replenish the population of agranular cells).

ENGINEERING MSC TO EXPRESS TRANSGENES

Marrow stromal cells are of particular interest for gene therapy because they are obtained from patient's own bone marrow, and, contrary to other adult stemlike cells, they can be maintained and expanded in unlimited numbers in vitro. Furthermore, MSCs can be genetically transduced and clonally expanded. Such cell lines are attractive tools for targeted gene delivery. The characteristics of MSCs important for gene engineering are their survivability, lack of toxicity, and, perhaps, persistent expression of the introduced transgenes.

Parkinson's disease, because of deficit in a single neurotransmitter, has emerged as an attractive and exploitable model for ex vivo gene therapy. We chose this model to explore the potential of rMSCs for gene transfer into the CNS. Rat MSC were genetically engineered to synthesize L-DOPA by transduction of the cells with retroviruses containing cDNAs encoding tyrosine hydroxylase (TH) and GTP cyclohydrolase I (GC). The initial vectors that we have used were based on Maloney murine leukemia virus (MMLV)-derived plasmids. In LNCX-TH vector, human tyrosine hydroxylase type-2 (hTH2) cDNA *(39)* was inserted in LNCX *(40)* downstream of a cytomegalovirus (CMV) promoter. The vector pΔGHCGC *(15)* contained the rat GTP cyclohydrolase I (GTPCH I) cDNA *(41)* under the control of the CMV promoter in pΔGHC. The plasmids also contain the genes for aminoglycoside phosphotransferase (LNCX-TH) and hygromycin-B-phosphotransferase (pΔGHCGC) as selection markers. Retroviruses were produced following calcium phosphate transfection of LNCX-TH and pΔGHCGC plasmids separately into PT67 amphotropic packaging cells and selected in either G418 or hygromycin B. For the doubly transduced rMSCs, the cells were first infected with the retrovirus-expressing GC, selected for resistance to hygromycin B and expanded in the culture. In the next step, cells were infected with TH-expressing retrovirus and selected for resistance to both hygromycin B and G418. These transduced cells were used for phenotype rescue in the rat model of Parkinson's disease *(42)*.

In a different set of experiments, in order to alleviate the problems associated with using two separate retroviruses in a two-stage procedure, we fabricated a 3.3-kb "central construct" composed of the TH cDNA and the GC cDNA separated by an internal ribosome entry site from the encephalomyocarditis virus. This central construct can be shuffled into various retroviral and non-retroviral expression vectors to obtain L-DOPA-producing MSCs. Also, the central construct can be placed downstream of promoters for ubiquitous "housekeeping" genes. We have used a promoter from the murine phosphoglycerate kinase-1 (PGK) gene that has been used extensively in embryonic stem cells to prepare transgenic mice.

Expression of TH was evaluated by immunostaining of the transduced MSCs samples using antibodies to TH. The enzymatic activity of TH in these cells was measured through conversion of tyrosine to L-DOPA. After determining appropriate controls, L-tyrosine was added to transduced cultures to a final concentration of 50 μM and assayed for L-DOPA synthesis in HBSS. Samples were analyzed at several intervals by electrochemical detection high-pressure liquid chromatography. The concentration of L-DOPA and its metabolites were determined by comparison against known standards *(42)*.

We used a limited dilution method to isolate clones of high L-DOPA-producing cells. Doubly-transduced clones of rat MSCs, engineered to produce tyrosine hydroxylase and GTP cyclohydrolase, were cultured in 10% FCS using the limited dilution method. Statistically, 0.33 cells were plated per well of 96-well plates. Plates were examined microscopically for colony formation. Cells were subcloned after 20 d. Immunocytochemical staining of established L-DOPA-producing clones revealed TH expression in every cell. Three clones that were high producers of L-DOPA were further expanded by plating at intermediate density cultured to 80% confluence and then harvested for implantation into the rat model of Parkinson's disease.

IMPLANTATION OF MSC INTO THE RAT MODEL OF PARKINSON'S DISEASE

It has been observed that stem cells implanted in the developing brain react appropriately to local signals and integrate into the host brain *(43)*. Moreover, in damaged brains, multipotential neural precursors were shown to migrate to the areas of damage, where they replaced depleted cells *(44,45)*. Therefore, stemlike cells from bone marrow could perhaps be used to repair injured brain either from degenerative conditions or trauma and stroke. In a series of experiments, we implanted wild-type MSCs into the striatum of the 6-hydroxydopamine (6-OHDA)-treated rat brains, a rat model of Parkinson's disease. In these animals the nigro-striatal pathway is unilaterally lesioned and the degree of parkinsonism is quantified by counting the number of turns in response to apomorphin treatment. Apomorphine-induced rotation of 6-OHDA-lesioned animals is an accepted measurement of parkinsonism in the rat model of this disease. There was no alteration in the rotational activity following implantation of wild-type MSCs in the 6-OHDA-treated animals. These data indicated the need to genetically engineer MSCs in order to produce dopamine or its precursor, 3,4 -dihydroxyphenylalanine (L-DOPA) to induce phenotype rescue of the rat model of Parkinson's disease.

Marrow stromal cell cultures were transduced (see previous section) with human tyrosine hydroxylase type 2 and rat GTP cyclohydrolase I, the gene encoding for the TH cofactor tetrahydrobiopterin, without which tyrosine hydroxylase remains nonfunctional *(42)*. Transduced cells synthesized L-DOPA in vitro and maintained their multipotentiality after transduction *(42)*. These cells were then grafted into the ipsilateral corpus striatum of 6-OHDA-lesioned rats. Experiments were then conducted to determine whether the implanted MSCs would continue to synthesize and secrete L-DOPA in the rat brain. The level of L-DOPA production was assessed by sampling through a microdialysis probe. The samples were analyzed using a high-pressure liquid chromatography system. The data demonstrated that the levels of L-DOPA were comparable to or higher than that obtained by others in similar experiments employing fibroblasts or astrocytes *(15,17)*. Furthermore, we observed a significant reduction in the apomorphine-induced rotation after implantation of genetically engineered MSCs in the ipsilateral striatum of parkinsonian rats *(42)*.

Gene expression and phenotype rescue persisted for up to 2 wk, then the expression of trans-gene from the integrated retroviruses ceased, even though expression of the gene was stable in culture. These results were similar to others using the same retroviral vector with the TH gene in different types of cell *(16,17)*. Similarly, the phenotype rescue was short-lived, and by d 14, the decrease in rotation returned to the control level *(42)*. Unfortunately, given the current technology, stable and regulated expression of trans-genes remains an issue. New strategies for gene-engineering MSCs are currently being explored to overcome the short-lived transgene expression.

CONCLUSIONS

It has traditionally been considered to be axiomatic that the adult primate brain is structurally stable, except for minor exceptions, and that neurogenesis and synapse formation occurs only during development. Recently, a number of investigators have shown that neurogenesis can occur not only in the adult rodents but also in the adult human and primate brains *(46,47)*. Moreover, it has been shown in monkeys *(48)* that after ischemic damage to the motor cortex, rehabilitative training can reshape the adjacent intact cortex. This indicates that the adult primate brain is not necessarily a static structure but, within limits, has the potential for turnover and repair. We now know that implantation of embryonic stem cells into the adult and developing nervous system can result in the integration of these cells in developing and adult brain, with differentiation of these cells into oligodendroglia *(49,50)* and likely neurons *(51)*. Given this background, regeneration and repair in the nervous system no longer appears implausible. One way that this can be implemented is through implantation and/or infusion of stem cells and/or genetically engineered cells into the damaged CNS. Obtaining embryonic and fetal tissues, however, is laden with major logistical and ethical problems.

Multipotential cells from the adult bone marrow appear to fulfill some of the criteria for the stem cells and appear to be ideal for gene engineering. Because marrow stromal cells can be obtained from small-volume aspirates

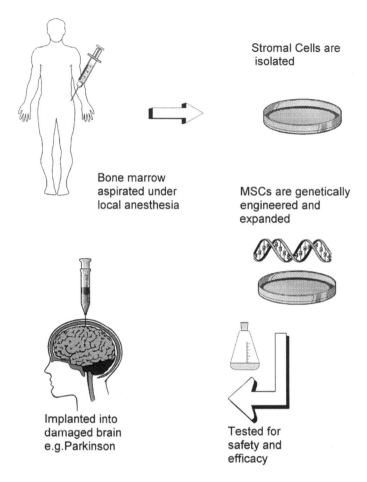

Fig. 2. Diagram depicting the future general strategy for using multipotential cells from bone marrow in the treatment of neurologic diseases.

of the iliac crest, they represent an easily accessible and replenishable source of autologous cells for transplantation into the brain and spinal cord, as well as other tissues. They have the ability to integrate, migrate, and survive in the central nervous system, and they may differentiate into brain cells. They can be isolated from the patient's own bone marrow, circumventing rejection and immune suppression associated with transplantation of other types of cells. These cells are, thus, worth considering as the cells of choice for rebuilding the nervous system. Figure 2 illustrates such a strategy for treatment of neurological diseases. Research carried out over the next few years will help to tell us whether this strategy will be successful.

REFERENCES

1. Bjorklund, A. (1991) Neural transplantation—an experimental tool with clinical possibilities. *TINS* **14,** 319–322.
2. Bjorklund, A. (1993) Better cells for brain repair. *Science* **276,** 66–71.
3. Kordower, J. H., Freeman, T. B., Snow, B. J., Vingerhoets, F. J., and Mufson, E. J. (1995) Neuropathological evidence of graft survival and striatal reinnervation after the transplantation of fetal mesencephalic tissue in a patient with Parkinson's disease. *N. Engl. J. Med.* **332,** 1118–1124.
4. Piccini, P., Brooks, D. J., Bjorklund, A., Gunn, R. N., Grasby, P. M., et al. (1999) Dopamine release from nigral transplants visualized in vivo in a Parkinson's patient. *Nat. Neurosci.* **2,** 1137–1140.
5. Freed, C. R., Greene, P. E., Breeze, R. E., Tsai, W.-Y., DuMouchel, W., Kao, R., et al. (2001) Transplantation of embryonic dopamine neurons for severe Parkinson's disease. *N. Engl. J. Med.* **344(10),** 710–719.
6. Brustle, O., Jones, K., Learish, R., Karram, K., Choudhary, K., et al. (1999) Embryonic stem cell-derived glial precursors: a source of myelinated transplants. *Science* **285,** 754–756.
7. McDonald, J., Liu, X.-Z., Qu, Y., Liu, S., Mickey, S., et al. (1999) Transplanted embryonic stem cells survive, differentiate and promote recovery in injured rat spinal cord. *Nat. Med.* **5,** 1410–1412.
8. Hurlbert, M. S., Gianani, R. I., Hutt, C., Freed, C. R., and Kaddis, F. G. (1999) Neural transplantation of hNT neurons for Huntington's disease. *Cell Transplant.* **8(1),** 143–151.
9. Nishino, H. and Borlongan, C. V. (2000) Restoration of function by neural transplantation in the ischemic brain. *Prog. Brain Res.* **127,** 461–476.
10. Bjorklund, A. and Lindvall, O. (2000) Cell replacement therapies for central nervous system disorders. *Nat. Neurosci.* **3(6),** 537–544.
11. Azizi, S. A. (2000) Exploiting nonneural cells to rebuild the nervous system: from bone marrow to brain. *The Neuroscientist* **6(5),** 353–361.
12. Turner, D. A. and Kearney, W. (1993) Scientific and ethical concerns in neural fetal tissue transplants. *Neurosurgery* **33,** 1031–1037.
13. Rosenstein, J. M. (1995) Why do neural transplants survive? An examination of some metabolic pathophysiological considerations in neural transplantation. *Exp. Neurol.* **133,** 1–6.
14. Snyder, E. (1997) The use of non-neural cells for gene delivery. *Neurobiol. Dis.* **4,** 69–102.
15. Bencsics, C., Wachtel, S. R., Milstien, S., Hatakeyama, K., Becker, J. B., et al. (1996) Double transduction with GTP cyclohydrolase and tyrosine hydroxylase is necessary for spontaneous synthesis of L-DOPA by primary fibroblasts. *J. Neurosci.* **16,** 4449–4456.

16. Horellou, P., Brundin, P., Kalen, P., Mallet, J., and Bjorklund, A. (1990) In vivo release of dopa and dopamine from genetically engineered cells grafted to the denervated rat striatum. *Neuron* **5,** 393–402.

17. Ljungberg, C. M., Stern, G., and Wilkin, G. P. (1999) Survival of genetically engineered, adult derived rat astrocytes grafted into the 6-hydroxy dopamine lesioned adult rat striatum. *Brain Res.* **816(1),** 29–37.

18. Takayama, H., Ray, J., Raymon, H. K., Baird, A., Hogg, J., et al. (1995) Basic fibroblast growth factor increases dopaminergic graft survival and function in a rat model of Parkinson's disease. *Nat. Med.* **1,** 53–58.

19. Lucidi-Phillipi, C. A., Gage, F. H., Shults, C. W., Jones, K. R., Reichardt, L. F., et al. (1995) Brain-derived neurotrophic factor-transduced fibroblasts: production of BDNF and effects of grafting to the adult rat brain. *J. Comp. Neurol.* **354,** 361–376.

20. Choi-Lundberg, D., Lin, Q., Chang, Y., Chiang, Y., Hay, C., et al. (1997) Dopaminergic neurons protected from degeneration by GDNF gene therapy. *Science* **275,** 838–841.

21. Choi-Lundberg, D. L., et al. (1998) Behavioral and cellular protection of rat dopaminergic neurons elicited by an adenoviral vector encoding glial cell line-derived neurotrophic factor. *Exp. Neurol.* **154,** 261–275.

22. Connor, B., et al. (1999) Differential effects of glial cell line-derived neurotrophic factor (GDNF) in the striatum and substantia nigra of the aged Parkinsonian rat. *Gene Ther.* **6,** 1936–1951.

23. Connor, B., Kozlowski, D. A., Unnerstall, J. R., Elsworth, J. D., Tillerson, J. L., Schallert, T., et al. (2001) Glial cell line-derived neurotrophic factor (gdnf) gene delivery protects dopaminergic terminals from degeneration. *Exp. Neurol.* **169,** 83–95.

24. Kawaja, M. and Gage, F. (1992) Morphological and neurochemical features of cultured primary skin fibroblasts of Fischer 344 rats following striatal implantation. *J. Comp. Neurol.* **317,** 102–116.

25. Azizi, S. A., Stokes, D., Augelli, B. J., DiGirolamo, C., and Prockop, D. J. (1998) Engraftment and migration of human bone marrow stromal cells implanted in the brains of albino rats—similarities to astrocyte grafts. *Proc. Natl. Acad. Sci. USA* **95,** 3908–3913.

26. Prockop, D. (1997) Marrow stromal cells as stem cells for nonhematopoietic tissues. *Science* **276,** 71–74.

27. Cohnheim, J. (1867) *Arch. Pathol. Anat. Physiol. Clin. Med.* **40,** 1.

28. Friedenstein, A. J., Gorskaja, U., and Kulaagina, N. N. (1976) Fibroblast precursors in normal and irradiated mouse hematopoietic organs. *Exp. Hematol.* **4,** 267–274.

29. Pittenger, M., Mackay, A., Beck, S., Jaiswal, R., Douglas, R., et al. (1999) Multilineage potential of adult human mesenchymal stem cells. *Science* **284,** 143–147.

30. Periera, R. F., O'Hara, M. D., Laptev, A. V., Halford, K. W., Pollard, M. D., et al. (1998) Marrow stromal cells as a source of progenitor cells for nonhe-

matopoietic tissues in transgenic mice with a phenotype of osteogenesis imperfecta. *Proc. Natl. Acad. Sci. USA* **95,** 1142–1147.

31. Mezey, E., Chandross, K. J., Harta, G., Maki, R. A., and McKercher, S. R. (2000) Turning blood into brain: cells bearing neuronal antigens generated in vivo from bone marrow. *Science* **290(5497),** 1779–1782.

32. Brazelton, T. R., Rossi, F. M., Keshet, G. I., and Blau, H. M. (2000) From marrow to brain: expression of neuronal phenotypes in adult mice. *Science* **290(5497),** 1775–1779.

33. Ferrari, G., Cusella-DeAngelis, C., Coletta, M., Paolucci, E., Stornaiuolo, A., Cossu, G., et al. (1998) Muscle regeneration by bone marrow-derived myogenic progenitors. *Science* **279,** 1528–1530.

34. Gussoni, E., Soneoka, Y., Strickland, C., Buzney, E., Khan, M., et al. (1999) Dystrophin expression in the mdx mouse restored by stem cell transplantation. *Nature* **401,** 390–394.

35. Galpern, W. R., Burns, L. H., Deacon, T. W., Dinsmore, J., and Isacson, O. (1996) Xenotransplantation of porcine fetal ventral mesencephalon in a rat model of Parkinson's disease: functional recovery and graft morphology. *Exp. Neurol.* **140,** 1–13.

36. Brook, F. A. and Gardner, R. L. (1997) The origin and efficient derivation of embryonic stem cells in the mouse. *Proc. Natl. Acad. Sci. USA* **94(11),** 5709–5712.

37. Laywell, E. D., Kukekov, V. G., and Steindler, D. A. Multipotent neurospheres can be derived from forebrain subependymal zone and spinal cord of adult mice after protracted postmortem intervals. *Exp. Neurol.* **156(2),** 430–433.

38. Colter, D., Class, R., DiGirolamo, C., and Prockop, D. (2000) Rapid expansion of recycling stem cells in cultures of plastic-adherent cells from human bone marrow. *Proc. Natl. Acad. Sci. USA* **97,** 2313–3218.

39. Ginns, E. I., Rehavi, M., Martin, B. M., Weller, M., O'Malley, K. L., et al. (1998) Expression of human tyrosine hydroxylase cDNA in invertebrate cells using a baculovirus vector. *J. Biol. Chem.* **263(15),** 7406–7410.

40. Miller, A. D. and Rosman, G. J. (1989) Improved retroviral vectors for gene transfer and expression. *Biotechniques* **7(9),** 980–982, 984–986, 989–990.

41. Hatakeyama, K., Inoue, Y., Harada, T., and Kagamiyama, H. (1991) Cloning and sequencing of cDNA encoding rat GTP cyclohydrolase I. The first enzyme of the tetrahydrobiopterin biosynthetic pathway. *J. Biol. Chem.* **266(2),** 765–769.

42. Schwarz, E. J., Guillermo, A. M., Prockop, D. J., and Azizi, S. A. (1999) Multipotential marrow stromal cells transduced to produce L-DOPA: engraftment in a rat model of Parkinson's disease. *Hum. Gene Ther.* **10,** 2539–2549.

43. Snyder, E. (1998) Neural stem-like cells: developmental lessons with therapeutic potential. *The Neuroscientist* **4,** 408–425.

44. Snyder, E., Yoon, C., Flax, J., and Macklis, J. (1997) Multipotent neural precursors can differentiate toward replacement of neurons undergoing targeted apoptotic degeneration in adult mouse neocortex. *Proc. Natl. Acad. Sci. USA* **94,** 11,663–11,668.

45. Yandava, B., Billinghurst, L., and Snyder, E. (1999) "Global" cell replacement is feasible via neural stem cell transplantation: evidence from the dysmyelinated shiverer mouse brain. *Proc. Natl. Acad. Sci. USA* **96,** 7029–7034.
46. Eriksson, P., Perfilieva, E., Bjork-Eriksson, T., Alborn, A.-M., Nordborg, C., Peterson, D. A., et al. (1998) Neurogenesis in the adult human hippocampus. *Nat. Med.* **4,** 1313.
47. Gould, E., Reeves, A., Graziano, S., and Gross, C. (1999) Neurogenesis in the neocortex of adult primates. *Science* **286,** 548–552.
48. Nudo, R. J., Wise, B. M., SiFuentes, F., and Milliken, G. W. (1996) Neural substrates for the effects of rehabilitative training on motor recovery after ischemic infarct. *Science* **272,** 1791–1794.
49. Zhang, S.-C., Ge, B., and Duncan, I. (1999) Adult brain retains the potential to generate oligodendroglial progenitors with extensive myelination capacity. *Proc. Natl. Acad. Sci. USA* **96,** 4089–4094.
50. Nait-Oumesmar, B., Decker, L., Lachapelle, F., Avellana-Adalid, V., Bachelin, C., et al. (1999) Progenitor cells of the adult mouse subventricular zone proliferate, migrate and differentiate into oligodendrocytes after demyelination. *Eur. J. Neurosci.* **11,** 4357–4366.
51. Snyder, E., Yandava, B., Pan, Z., Yoon, C., and Macklis, J. (1993) Immortalized postnatally-derived cerebellar progenitors can engraft and participate in development of multiple structures at multiple stages along mouse neuroaxis. *Soc. Neurosci. Abstr.* **19,** 613.

16
Preclinical Basis for Use of NT2N Cells in Neural Transplantation Therapy

Cesario V. Borlongan and Paul R. Sanberg

INTRODUCTION

This chapter will review laboratory experiments using human neuroteratocarcinoma cells (hNT cells, also called NT2N cells and LBS neurons) in animal models of neurological disorders. Because NT2N cells are originally derived from adult human cancerous tumors engineered to become neuronlike cells, their use for transplantation therapy eliminates logistical problems associated with the use of fetal and embryonic stem cells. This chapter will focus on functional recovery in stroke animals that received NT2N cell grafts. In light of recent studies demonstrating expression of trophic factors by NT2N cells, such trophic factor property of the cells is presented here as a plausible mechanism underlying the behavioral effects of transplanted NT2N cells. An overview on the clinical report of NT2N cell grafts in stroke patients will also be provided.

NT2N CELLS: FROM CANCEROUS CELLS TO POSTMITOTIC NEURONS

There is a concerted effort among basic scientists and clinicians to find a strategy for the delivery of exogenous proteins into the central nervous system (CNS). A major research interest in recent years has evolved around the use of cellular and gene therapy for neurological disorders such as Parkinson's disease, Huntington's disease, and stroke. Specifically, investigations have been geared toward exploiting a transfectable and transplantable cellular "platform" that can serve as an "on-site" or local delivery system for gene products of therapeutic value (1–3). A range of neuroprotective strategies, including neurotrophic factor treatment and cell transplantation techniques, has shown potential benefits for treatment of neurological disorders.

From: *Neural Stem Cells for Brain and Spinal Cord Repair*
Edited by: T. Zigova, E. Y. Snyder, and P. R. Sanberg © Humana Press Inc., Totowa, NJ

Cellular transplantation reached the clinical arena 15 yr ago. Laboratory studies in animal models of Parkinson's disease (PD) and Huntington's disease (HD) have demonstrated that cell replacement via transplantation surgery has potential therapeutic effects (2,4). A variety of cells have been used as graft source for neural transplantation, such as fetal neurons, neuronal stem and progenitor cells, cells engineered to secrete neurotransmitters or neurotrophic factors (e.g., immortalized cell lines, fibroblasts, astrocytes), para-neuronal cells, which naturally synthesize neuronal substances and/or have neuronlike properties, and bridging grafts that assist in the physical reconstruction of a lost axonal pathway. In general, these cells, when intra-cerebrally transplanted, have been shown to reconstruct partially the neuronal circuitry and form functional synapses (5–9). Fetal cells remain the most widely studied graft source for transplantation. Unfortunately, there are many logistical issues that hinder the use of primary fetal cells in the clinic. Although clinical transplantation procedures have been performed in over 300 PD patients and recently initiated in HD patients, transplanted patients have been observed to display, at best, moderate improvement owing in part to the variable viability of the grafts (2,10,11). Of note, the survival rate of transplanted neural tissue is low, whereas non-neuronal cells show better survival rates (12–15), suggesting that non-neuronal cells may be an alternative graft source. However, transplantation of non-neuronal cells, such as immortalized cells, may form tumors (16). A major limiting factor in graft survival is the host immune response, and the use of cells that can possibly circumvent immunosurveillance, such as autologous cells (i.e., the transplant recipient's own adrenal cells) or stem cells, may limit graft rejection (17–23). Thus, for these and several other reasons, a search for a nonprimary fetal graft source has become a main research endeavor in cell transplantation.

Cell transplantation, neurotrophic factor treatment, and gene therapy appear to complement each other in optimizing their therapeutic benefits. For example, to achieve a sustained therapeutic effect from the application of a gene product to affected regions of the CNS, a continuous secretion of the gene product may be necessary and this could be accomplished by the transplantation of cells genetically engineered to express the therapeutic protein of interest. Although such prolonged release of neuronal surviving promoting proteins may be achieved with transplantation of primary fetal cells, the use of cell lines would be associated with less controversy, the cell line is always readily available, the cells can be maintained in culture indefinitely, and because of their clonal nature, they are uniform and well defined. However, one disadvantage of using genetically engineered cell lines is the unpredictability of expression of the transgene. Even under optimal conditions, transgene expression is not likely to last as long as the life of the host into which the cells are transplanted. Although the treatment of some disorders may not require the continued presence of growth factors, other disorders, especially chronic neurological diseases, are likely to benefit from the continued presence of growth factors for the life-span of the patient. Accordingly, there is considerable interest in establishing and characterizing neuronal cell lines, especially in identifying a human neuronal cell line that might serve as a platform for gene therapy in the CNS. The human embryonal carcinoma cell line appears to possess many of the cellular properties required for an efficacious graft source; the cell line is transfectable and capable of differentiating into postmitotic neuronlike cells (NT2N cells) following treatment with retinoic acid (RA) in vitro *(24–26)*. Thus, NT2N cells can be considered as suitable cellular vehicles for the delivery of potentially therapeutic, exogenous proteins into the CNS for the treatment of neurological disorders *(1)*.

The parent cells of NT2N cells (called NT2) are an embryonal carcinoma cell line derived from a human teratocarcinoma *(24–26)*. During the 6-wk RA treatment period, NT2 cells, which share many characteristics of neuroepithelial precursor cells, undergo significant changes, including the loss of neuroepithelial markers and the appearance of neuronal markers *(27,28)*. Additional enrichment steps, such as exposure to mitotic inhibitors, result in the production of >99% pure populations of NT2N neurons that are terminally differentiated *(29)*. NT2N neurons exhibit outgrowth processes and establish functional synapses. Compared with terminally differentiated postmitotic, embryonic neurons, mature NT2N neurons are virtually indistinguishable; they do not divide and they maintain a neuronal phenotype *(28)*. Compared with previously described germ-cell tumor lines *(27,28)*, NT2 cells have been shown to be unique because they do not give rise to progeny committed to other well-defined neural or non-neural lineages in response to RA or any other differentiating agent *(27,28)*. Accordingly, the NT2 cells have been regarded as in vitro equivalents of CNS neuronal progenitor cells *(1,30)*. Because both NT2 cells and NT2N neurons can be genetically engineered, allowing expression of a gene product of interest in vitro, and possibly in vivo, this offers in-depth studies on the cellular and molecular biology of neurons *(1,30)*. More importantly, accumulating evidence has suggested that both NT2 and NT2N cells can be exploited as possible alternative graft sources for transplantation therapy in CNS disorders.

NT2 AND NT2N CELL GRAFT SURVIVAL IN NORMAL ANIMALS: INFLUENCE OF HOST CNS MICROENVIRONMENT

Trojanowski and Lee and their colleagues performed a series of studies demonstrating that purified NT2N cells can survive, mature, and integrate with the host nervous system following transplantation into the CNS of rodents *(24–26)*. These observations paved the way for examining the biology of human neuronal cells in an in vivo CNS environment, which otherwise may not be fully investigated in an in vitro setting. The availability of NT2N neurons offers many advantages over the use of human fetal neurons; for example, NT2N neurons appear to have better graft survival (15%), low tissue variability, and high degree of host reinnervation *(1,25,26)*. Compared with limited data on grafted mitotic stem cells, none of the NT2N cell grafts have reverted to a neoplastic state despite posttransplantation intervals of more than 1 yr *(24–26)*. Accordingly, these transplantation studies suggest that grafted NT2N cells may provide a platform for developing cell replacement therapies for CNS disorders.

STRIATUM (CAUDOPUTAMEN): THE MOST CONDUCIVE MICROENVIRONMENT FOR TRANSPLANTATION

Although it has been established that NT2N cells attained features of fully differentiated neurons following treatment with RA and mitotic inhibitors and that they do not revert to the neoplastic state after transplantation, a few scientists have raised concerns on the possibility that the "mitotic capacity" of NT2N neurons can be stimulated by the host microenvironment. To this end, Trojanowski and colleagues *(24–26)* have addressed the issue of effects of microenvironment on NT2 cell transplantation. To explore the neoplastic potential of grafted NT2 cells, they grafted these cells into different regions of the brains of subacute combined immunodeficient (SCID) mice and nude mice *(24–26)*. These studies indicated that the anatomical site into which the NT2 cells were implanted significantly influenced the survival, proliferation, and differentiation of NT2 cells. For example, the NT2 cells continued to proliferate and undergo an apoptoticlike cell death, but they exhibited a very limited capacity to differentiate into neurons following implantation into the subarachnoid space and superficial neocortex. In addition to these two sites, NT2 cell grafts in the lateral ventricles, liver, and muscle rapidly progressed into bulky, lethal tumors within 10 wk after transplantation. In contrast, transplantation of the NT2 cells into the caudoputamen of SCID mice stopped proliferating, and showed no evidence of necrosis or apoptosis after more than 20 wk posttransplantation. These results clearly indicate that transplantation of NT2 cells into the caudoputamen cease proliferating, differentiate into neuronlike cells, and do not form lethal tumors, whereas transplantation of these cells in other brain regions can form lethal tumors *(24–26)*. Furthermore, neuronal phenotypic markers demonstrate that the majority of NT2 cell grafts in the caudoputamen have differentiated into postmitotic immature neuronlike cells. Thus, these observations support the notion that the host environment, in this case the mouse caudoputamen, may promote signal molecules or other "cues" (i.e., cell–cell contacts) capable of regulating the proliferation, death, and differentiation of human NT2 cells. These studies highlight the importance of the host microenvironment in regulating NT2 cells' mitotic ability following transplantation into specific brain areas.

The importance of identifying a conducive microenvironment for cell transplantation in neurological disorders is exemplified in a disease characterized by widespread damage or ongoing neurodegeneration, such as that seen in stroke. In stroke animal models, the reported NT2N cell graft survival rate of 15% is a bit higher, but this is still low considering that ischemic stroke is not limited to a specific population of cells (e.g., depleted dopaminergic cells in Parkinson's disease). Furthermore, a stroke is characterized by necrotic infarction (within the ischemic core), as well as an ongoing degeneration of cells (along the ischemic penumbra); therefore, the ischemic core or the ischemic penumbra does not offer a conducive environment for the grafted cells. In general, the brain damage that accompanies stroke encompasses many cell populations and brain structures. In such a case, a much higher number of cells, with high viability and increased survivability, must be transplanted in the ischemic regions. Thus, neural transplantation may require multiple brain targets to repair the damaged neuroanatomical circuitry. However, the transplant recipient's trauma associated with multiple insertions of the transplant needle as well as the difficulty in transplanting into deep brain sites may hinder such a technique. Considering also that many cell populations are destroyed in stroke, different types of donor cell may need to be transplanted. Alternatively, the retinoic acid-naive NT2 cells may possess some "multipotent" properties, such as those attributed to neural stem cells, and these features may be potentially advantageous for transplantation in stroke. Indeed, Snyder and colleagues *(31)* have demonstrated that transplanted human neural stem cells can mature into the phenotype of cells that are undergoing cell death in brains of animals that were introduced to neuronal injury. In the end, the pluripotent features of NT2 cells and the highly differentiated neuronallike characteristics of NT2N cells may be exploited to examine their efficacy as graft source for neural transplantation. For example, the possibility exists that the choice for using NT2 or NT2N cells may be dictated by the target disease, in that a devastating and global brain disorder may benefit from the putative multipotent NT2 cell grafts, whereas a site-specific and focal brain disorder may appeal to the differentiated NT2N cell grafts.

NT2N CELL GRAFT-MEDIATED FUNCTIONAL RECOVERY IN STROKE ANIMALS

We pioneered the transplantation of NT2N neurons in the ischemic rat *(32)*. Because the rodent model of MCA occlusion mimics several motor abnormalities seen in clinical cerebral ischemia, we used this model to investigate treatment strategies for stroke. The study was designed to examine the potential benefits of neural transplantation of fetal rat striatal cells or human neurons derived from a clonal embryonal carcinoma cell line to correct the abnormalities associated with cerebral ischemia. We reported that ischemia-induced behavioral dysfunctions were ameliorated by the neural grafts as early as 1 mo posttransplantation. The novel finding of the study is that transplantation of human neurons induced a significantly more robust recovery than fetal rat striatal grafts. These observations indicate the logistical and ethical concerns inherent with the use of fetal striatal cells for transplantation therapy can be avoided by exploiting cell-line-derived human neurons as alternative graft sources.

Pretransplant and Posttransplant NT2N Cell Viability as a Function of Behavioral Recovery

We demonstrated the direct evidence that the survival of the NT2N neurons correlated with the functional recovery of transplanted animals *(33)*. We reported that the viability and survival of NT2N neurons are critical factors when considering transplantation of NT2N neurons in animals with ischemic stroke. At the pretransplantation period, viability counts revealed a variable range of 52–95% viable neurons. Within-subject comparisons of pretransplantation cell viability and subsequent behavioral changes in transplanted animals revealed that a high cell viability just prior to transplantation surgery correlated highly with a robust and sustained functional improvement in the transplant recipients. Furthermore, histological analysis of grafted brains revealed a positive correlation between the number of surviving NT2N neurons and the degree of functional recovery. The same factors of pretransplantation viability and posttransplantation survival of grafted cells have been reported in fetal tissue transplantation.

Cryopreserved NT2N Cell Grafts Attained Similar Positive Effects as Freshly Cultured NT2N Cell Grafts

In another report, cryopreservation (a method of storing cells for use at a later time-point) of NT2N neurons did not produce any significant deleterious effects on the viability of the cells prior to as well as after transplantation in stroke animals *(33)*. This sustained viability of NT2N neurons following cryopreservation is unprecedented. In many studies using cryopreservation of fetal cells, there was a significant loss of viable cells, rendering these cells nontransplantable. Such efficacious cryopreservation of NT2N neurons would, therefore, allow shipment of the cells to remote hospitals where a patient in need of the transplant resides.

NT2N Cell Graft Dose-Response Behavioral Effects

In another study, we examined the optimal dosage of NT2N neurons that needs to be transplanted to promote functional recovery in ischemic animals *(34)*. Ischemic animals that received 40×10^3, 80×10^3, or 160×10^3 NT2N neurons demonstrated a dose-dependent improvement in performance of both the passive avoidance and elevated body swing tests. At the conclusion of behavioral testing, NT2N neurons were identified in brain sections with antibodies to human cells. Grafts of 80×10^3 or 160×10^3 NT2N neurons demonstrated a 12–15% survival of NT2N neurons in the graft, whereas grafts of 40×10^3 NT2N neurons demonstrated a 5% survival. More importantly, the ischemic animals that received 80×10^3 or 160×10^3 NT2N neurons produced a significantly better amelioration of behavioral deficits produced by ischemic damage compared to those that received lower dosages of NT2N neurons. One can extrapolate from these observations that the greater the stroke-induced brain damage, the more cells may need to be transplanted. In stroke patients with localized basal ganglia stroke, about 2 million cells were transplanted. The required dosage of NT2N neurons should be viewed in terms of cell viability at pretransplantation and posttransplantation, as well as the extent of brain damage.

The Need for Immunosuppression as Adjunct Therapy to NT2N Cell Grafts

We showed that dysfunction in a passive avoidance learning and memory task and in asymmetrical motor behavior was significantly corrected in stroke animals that received transplants of NT2N cells into the ischemic striatum with immunosuppression treatment of cyclosporine-A (CsA) *(32)*. Although all of the NT2N-cell-transplanted animals demonstrated recovery of function, the immunosuppressed animals transplanted with NT2N cells maintained their behavioral recovery for over 6 mo posttransplantation, but the recovery in the nonimmunosuppressed animals transplanted with NT2N cells began to diminish by about 2 mo posttransplantation. Nonetheless, the magnitude of the behavioral recovery produced by NT2N cell grafts in rats that did not receive CsA was greater than that seen in animals with ischemia-induced brain injury followed by injections of rat fetal cerebellar cells or medium alone, suggesting that NT2N cell grafts promote behavioral effects at early time periods posttransplantation, even in the absence of immunosuppression. Moreover, histological analysis revealed surviving NT2N cells in the brains of immunosuppressed transplanted animals but not in those nonimmunosuppressed transplanted animals. Notwithstanding, the need for chronic immunosuppressive therapy as an adjunct to the transplantation of human NT2N cells in rats appears necessary in order to obtain optimal and sustained functional improvement as well as prolonged graft survival. The near absence of visible grafts in the nonimmunosuppressed animals transplanted with NT2N cells results from the immunological rejection of these grafts, as described earlier *(35)*. Nonetheless, these animals were significantly improved compared to control animals at 6 mo posttransplantation, but more impaired than the immunosuppressed animals that received NT2N cells. These data suggest that "trophic" effects of the transplanted NT2N cells are sustained for a prolonged period of time posttransplantation in animals that were not immunosuppressed. Nevertheless, immunosuppression with CsA enhanced the survival of grafted NT2N cells, and this is consistent with a previous study *(35)*. Further, we detected surviving NT2N cells in immunosuppressed animals showing robust functional recovery for over 6 mo posttransplantation. Significantly, there was no evidence that transplanted NT2N cells, with or without immunosuppression, had any deleterious effects on the host brain consistent with previous reports *(24–26)*. In future clinical trials of neural transplantation of NT2N cells, however, chronic immunosuppression, which is normally required for successful xenografting, may not be necessary because these cells are derived from human embryonal cells. Indeed, in some clinical trials of human fetal cell transplantation for PD, the absence of immunosuppression did not deleteriously affect the survival of fetal cell grafts and their ability to produce clinical improvement *(36,37)*. In addition, preliminary data suggest that NT2N neurons may have immunosuppressive properties *(38)*. Thus, long-term systemic immunosuppression may not be necessary in humans. In contrast, because recent studies have indicated that immunosuppressants may have trophic factor effects *(2)*, combining immunosuppression with

NT2N cell transplantation may reveal more robust graft survival, as well as behavioral effects.

NT2N CELL GRAFTS IN OTHER NEUROLOGICAL DISORDERS

NT2N cells have been transplanted also in animal models of Huntington's disease (HD), Parkinson's disease, and spinal cord injury. Freed and colleagues *(39)* compared the efficacy of fetal striatal tissue transplants with NT2N cell grafts in a rodent model of HD. In vitro, they found that purified NT2N neurons have a biochemical phenotype similar to that of human fetal striatal tissue; they both express mRNAs for glutamic acid decarboxylase, choline acetyltransferase, and the D1 and D2 dopamine receptors. Adult rats that received NT2N cell grafts or fetal striatal tissue grafts into their unilaterally lesioned striata displayed significant reductions in methamphetamine-induced circling behavior, as well as partial recovery in skilled use of the paw compared to control sham animals. Thus, NT2N cell grafts have been demonstrated as efficacious as fetal striatal tissue grafts in an animal model of HD.

Mendez and colleagues *(40)* examined the effects of NT2N cell grafts in the substantia nigra (SN) and striatum of the rat model for PD. Unilateral 6-hydroxydopamine-lesioned rats were grafted with one of three NT2N neuronal products; NT2N neurons, NT2N-DA neurons, or lithium chloride (LiCl)-pretreated NT2N-DA neurons. The use of LiCl was based on our recent study demonstrating that LiCl treatment enhances the expression of TH in NT2N neurons *(41)*. The results noted by Mendez and colleagues *(40)* support the potentiating effects of LiCl to induce NT2N neurons to express TH, in that no TH-immunoreactive (THir) neurons were observed in any animals with NT2N neuronal grafts, 43% of animals with NT2N-DA neuronal grafts displayed THir neurons, and 100% of animals with LiCl pretreated NT2N-DA neuronal grafts exhibited THir neurons. However, the observed TH immunoreactivity in NT2N cell grafts was not sufficient to produce significant functional recovery. Nonetheless, this study suggests that NT2N neurons can be treated with LiCl or some other chemicals that can enhance TH expression in these cells, thereby facilitating the cells' utility for transplantation studies in PD.

Trojanowski and Lee examined transplantation of NT2N neurons into the spinal cord of athymic nude mice *(42)*. Grafted NT2N neurons were observed to remain at the implantation site and to integrate with the host. Each graft displayed similar phenotypic features regardless of location or host age at implantation, characterized by morphologic and molecular phenotype of mature neurons. Furthermore, there was host oligodendrocyte ensheathment of NT2N neuronal processes. As noted earlier, the host microenvironment may influence the differentiation of grafted NT2N cells. In this case, the pattern of NT2N cell outgrowth processes was affected by the microenvironment, in that differential morphologies of graft processes extended into white matter versus gray matter. For example, NT2N processes reached long distances (>2 cm) within white matter, whereas NT2N processes located within gray matter had shorter trajectories. Although these results support the view of host microenvironment–NT2N cell graft interaction, these observations may be further advanced to indicate the possibility of transplanting NT2N neurons in a model of spinal cord injury. Indeed, this was the focus of the recent study by Trojanowski and Lee *(30)*.

Finally, McIntosh and colleagues *(43)* grafted NT2N cells into the brains of immunocompetent rats following lateral fluid–percussion brain injury to determine the long-term survivability of NT2N cell grafts in cortices damaged by traumatic brain injury (TBI) and the therapeutic effect of NT2N neurons on cognitive and motor deficits. Twenty-four hours after TBI, rats were stereotactically implanted with NT2N cells into the peri-injured or control cerebral cortex. Although surviving NT2N cell grafts were noted in the peri-injured cortex at 2 and 4 wk posttransplantation, there was no significant improvement in motor or cognitive function. Additional studies are needed to modulate the functional effects of transplanted NT2N neurons in the TBI model.

NT2N CELLS AND TROPHIC FACTORS

Another promising neuronal rescue strategy is treatment with neurotrophic factors. Neurotrophic factors play critical roles in cell survival and proliferation, differentiation, biochemical function, and morphological plasticity. Trophic influences are important during development and in the adult, in both the intact animal and during injury and degenerative events. Indeed, endogenous levels of neurotrophic factors are often found to increase in response to neuronal injury, suggesting a physiological regenerative response (for review, see refs. *4, 44,* and *45*). Many neurotrophic and growth factors are localized to the nigrostriatal system, suggesting a physiological role in the development, maintenance, and/or recovery of these neurons and potentially a role in these neurological disorders. Indeed, many of the factors localized to the nigrostriatal system have neuroprotective actions when administered exogenously in in vivo animal models of PD, HD, and stroke.

The observed functional improvement in our stroke animals was initially ascribed to neurotrophic effects of NT2N cells to the injured area *(32)*.

However, there was no direct evidence in this study that neuroprotection was indeed a function of NT2N neuronal grafts. The first suggestion that NT2N neurons could have neuroprotective activity was reported recently *(46)*, showing that NT2N neurons are positive for mRNA of glial-cell-line-derived neurotrophic factor (GDNF). Because GDNF has been shown as neuroprotective for animal models of neurodegenerative disorders, this view prompted us to investigate the neuroprotective effects of NT2N neuronal grafts. Previous studies using NT2N neurons were all directed toward recovery of function from brain insults, and there has never been a clear demonstration that NT2N cells could promote neuroprotection.

A neurotrophic effect is the more likely explanation for the observed behavioral recovery of stroke animals that received the NT2N cell grafts because of the limited ability of NT2N cells to produce TH, although precursor NT2 cells cultured for 2, 3, 4, and 5 wk with caudatoputamen (CP) extracts produced the following respective percentages of TH+ cells: 2–5%, 8–16%, 28–34%, and 36–42% *(25)*. In our earlier study with stroke animals *(32)*, although we observed a similar pattern of behavioral recovery in animals that received NT2N cells and those that received fetal striatal transplants, the NT2N-transplanted animals showed a more robust recovery at 1 mo posttransplantation. This effect of the NT2N cell grafts also was evident in the nonimmunosuppressed animals transplanted with NT2N cells. Because there has been no reported evidence that neural transplants replace lost host brain tissue at this early time-point, the observed functional effects may be the result of the release of trophic factors from the grafted NT2N cells. Accordingly, the NT2N cell grafts may secrete trophic factors that ameliorated ischemia-induced deficits. Of note, the effective dose of transplanted NT2N cells (an average of 7.8×10^4 and 2.3×10^4 cells contained in each injection for fresh and cryopreserved NT2N cells, respectively) shown to produce functional recovery was 10 times less than the transplanted striatal cells (about 8×10^5 striatal cells).

Recently, we have demonstrated for the first time that intranigral transplantation of NT2N neurons 5 min after a unilateral 6-OHDA lesion of the nigrostriatal system promotes nearly complete protection against dopaminergic depletion on the lesioned NT2N-transplanted side, when evaluated histologically 1 mo after transplantation *(47)*. Furthermore, lesioned animals that received NT2N neurons immediately after lesioning also demonstrated no behavioral deficits, as opposed to lesioned animals receiving only vehicle. The present study indicates that it is feasible to deliver GDNF into the central nervous system by transplantation of GDNF-secreting NT2N neurons. This novel delivery method can provide the opportunity for long-term expression of these (or other) potent neurotrophic factors in PD as well as other neurodegenerative disorders.

In many preclinical and clinical studies of neural transplantation, the use of neurotrophic factors has been shown to significantly enhance the survival rate of the grafted cells. Another potentially beneficial technique that can serve as an adjunct to neural transplantation is the use of neurotrophic factors. The protective effects of neurotrophic factor treatment against brain disorders are supported by accumulating evidence showing the survival-promoting effects of intracerebral infusion of GDNF on the ischemic brain. GDNF has been suggested as the most potent trophic factor for dopaminergic neurons. In vitro and in vivo examination in Parkinson's disease models have demonstrated neuroprotective properties of GDNF. The direct infusion of neurotrophic factors alone or their use as a transplant facilitator (either by pretreating donor cells or coadministration during and after neural transplantation therapy) has proven efficacious in animal models of Parkinson's disease. Promising results have been reported in the use of neural transplantation therapy for stroke; for example, using GDNF-secreting fetal kidney cells *(48,49)*. However, the preclinical and clinical data, so far, indicate efficacy of neural transplantation for treating a stroke that is characterized by focal and localized brain damage. In order to extend such a method to a stroke that has produced significant damage to multiple brain areas may require the use of a multipotential cell type combined with adjunctive neurotrophic factor treatment.

Additional Speculative Mechanisms for NT2N Cell Graft-Mediated CNS Recovery: Maturation into Striatallike Neurons and Dopamine Secretion

In a few stroke animals that received NT2N cell grafts, we noted immuno-stained fibers (for human neural cell-adhesion molecule) that extended from the NT2N grafts *(32)*. In vitro studies have shown that NT2N cells exhibit several hallmark features of normal postmitotic neurons (i.e., a highly asymmetric morphology, a very extended axon, and numerous elongated dendrites) *(24–26)*. Further, axonlike processes have been shown to extend for several millimeters from the grafted NT2N cells *(27)*. Thus, a possible mechanism underlying the observed behavioral recovery produced by NT2N cell grafts is that NT2N cells might have replaced the degenerated host brain cells at the later posttransplantation period. It has been shown that after transplantation into nude mice, NT2N cells can integrate and change phenotype into neurons similar to the target neurons (i.e., striatal neurons) *(24–27)*.

Another speculative mechanism of NT2N cell graft-induced recovery from ischemia is that because NT2N cells can potentially become striatallike neurons, they may be capable of secreting neurochemicals or even performing functions of lost striatal cells of the host brain. Indeed, NT2N cells can be stimulated through application of neurotrophic factors (i.e., acidic fibroblast growth factor) and activating factors (e.g., catecholamines, forskolin) to express the rate-limiting enzyme in catecholamine biosynthesis, tyrosine hydroxylase (TH) *(50,51)*. Of note, in vitro studies have shown that the percentages of TH-positive neuronlike cells in the NT2N cells treated with RA cocultured with striatal extracts exceeded by greater than 10-fold the percentage of TH-positive cells that were induced in sister cultures exposed to retinoic acid alone *(24–26)*. As mentioned earlier, the behavioral dysfunctions we noted in our studies *(33,52)* are dopamine-mediated behaviors. Accordingly, the possibility that NT2N cell grafts can be induced by the host microenvironment (i.e., remaining host striatal neurons or the whole host striatum itself) to secrete dopamine would greatly contribute to amelioration of ischemia-induced behavioral deficits. Of note, an in vitro study demonstrated that NT2N cells respond positively to putative neurotrophic factors secreted by an immortalized human fetal astrocyte cell line *(53)*. In addition, it has been shown that the microenvironment of the adult mouse striatum appears to have the potential ability to induce grafted NT2N cells to differentiate progressively into fully mature, adult CNS neurons *(24–26)*. Further studies are warranted to address this issue of NT2N cell graft-mediated functional effects.

TOWARD CLINICAL TRANSPLANTATION OF NT2N CELLS

The series of preclinical studies that we conducted in an animal model of stroke, as described earlier, provided a twofold justification for initiating neural transplantation of these human-derived NT2N neurons in stroke patients. First, it offers the possibility of reversing motor symptoms associated with a stable stroke because the transplantation in the animals exposed to ischemia was performed at 1 mo post-MCA surgery. Second, whereas the transplantation of human neurons appeared to be well tolerated by the immunosuppressed rats, these cells, being human derived, may not require immunosuppression of the stroke patients. Because adverse effects of general immunosuppression have been documented, being able to circumvent immunosuppression would be an advantage in using these cells. Futhermore, because these cells are cultured, examination of the cells for possible infectious diseases can be performed well ahead of the scheduled transplant surgery and, therefore, a more efficient transplantation protocol can be achieved with the use of these cells compared to using fetal cells. The observation of successful implantation of human clonal neurons into rat brains supports the notion that grafting of cell lines may ultimately replace the use of fetal cells in the clinic. While avoiding the ethical concerns on using fetal cells, transplantation of clonal cells, such as NT2N neurons, also allows a logistical advantage of conducting neural transplantation in a wider therapeutic window following stroke.

Because the success of treating cerebral ischemia depends highly on the timing of intervention, the ready availability of clone cells as a graft source would significantly reduce the time between the ischemic event and the therapeutic intervention. Nonetheless, the robust recovery of animals, with a stable stroke, following transplantation of NT2N neurons, suggests the possibility of treating stroke patients even with a long delay after a stroke episode.

NT2N cells have been approved by the Food and Drug Administration for Phase I human transplantation to evaluate this therapy in the treatment of patients with stable stroke. The cellular attributes of NT2N cells, described earlier, make them suitable for implantation in stroke. Kondziolka and colleagues *(54)* transplanted NT2N cells in patients with basal ganglia stroke and fixed motor deficits, including 12 patients (aged 44–75 yr) with an infarct of 6 mo to 6 yr (stable for at least 2 mo). Serial evaluations (12–18 mo) showed no adverse cell-related serologic or imaging-defined effects. The total European Stroke Scale score improved in six patients (3–10 points), with a mean improvement of 2.9 points in all patients ($p = 0.046$). Six of 11 positron-emitting tomographic (PET) scans at 6 mo showed improved fluorodeoxyglucose uptake at the implant site. These results suggest that transplantation of NT2N cells is feasible in patients with motor infarction.

The significant results, thus far, from this limited clinical trial is that there were no serious adverse events, PET indicated grafted cells that appear to remain viable for 1 yr in some instances, and there was a suggestion that a few patients experienced a measurable improvement in neurologic function. The long-term goal of NT2N cell transplantation in stroke is to restore neurologic function, and this can only be verified by monitoring the patients over the years. Additional concerns that need to be closely examined during this posttransplantation period include the possibility of malignant transformation because the cells were derived from human embryonic carcinoma. However, there was no evidence of this during the 1-yr follow-up examination. The next step will be to provide an efficacy trial, and here the possibility of dose-related toxicity needs to be studied.

Reservations have been raised on the benefits from a small number of NT2N cells transplanted into the large stroke infarct. Although these grafted cells may not readily make proper connections among the denervated targets, it is possible that there is a fair degree of redundancy within the CNS, in that despite the small number of transplanted cells forming the correct connections with the host brain tissue, even rudimentary connections will be of benefit *(55)*. Finally, one needs to emphasize that the preceding clinical trial is an open-label study, which was not designed to prove efficacy. Clearly, to determine whether NT2N cell transplantation is effective in ameliorating the functional abnormalities produced by stroke or other neurological disorders will require carefully designed laboratory studies and limited clinical trials.

ACKNOWLEDGMENTS

The authors wish to thank their colleagues and associates, and Layton Bioscience, Inc., whose support and research collaborations advanced NT2N cells as efficacious graft source for neural transplantation therapy; their studies are core to this review paper. Ms. Christina E. Clark provided excellent technical assistance in the final preparation of the manuscript.

REFERENCES

1. Trojanowski, J. Q., Kleppner, S. R., Hartley, R. S., Miyazono, M., Fraser, N. W., Kesari, S., et al. (1997) Transfectable and transplantable postmitotic human neurons: a potential "platform" for gene therapy of nervous system diseases. *Exp. Neurol.* **144,** 92–97.
2. Borlongan, C. V., Sanberg, P. R., and Freeman, T. B. (1999) Neural transplantation for neurodegenerative disorders. *Lancet* **353,** SI29–SI30.
3. Nishino, H. and Borlongan, C. V. (2000) Restoration of function by neural transplantation in the ischemic brain. *Prog. Brain Res.* **127,** 461–476.
4. Alexi, T., Borlongan, C. V., Faull, R. L., Williams, C. E., Clark, R. G., Gluckman, P. D., et al. (2000) Neuroprotective strategies for basal ganglia degeneration: Parkinson's and Huntington's diseases. *Prog. Neurobiol.* **60,** 409–470.
5. Freund, T. F., Bolam, J. P., Bjorklund, A., Steveni, U., Dunnett, S. B., Powell, J. F., et al. (1985) Efferent synaptic connections of grafted dopaminergic neurons reinnervating the host neostriatum: a tyrosine hydroxylase immunocytochemical study. *J. Neurosci.* **5,** 603–616.
6. Mahalik, T. J., Finger, T. E., Stromberg, I., and Olson, L. (1985) Substantia nigra transplants into denervated striatum of the rat: ultrastructure of graft and host interconnections. *J. Comp. Neurol.* **240,** 60–70.
7. Faull, R. L. M., Waldvogel, H. J., Nicholson, L. F. B., Williams, M. N., and Dragunow, M. (1995) Huntington's disease and neural transplantation: GABAA

receptor changes in the basal ganglia in Huntington's disease, in the human brain and in the quinolinic acid lesioned rat model of the disease following fetal neuron transplants, in *Neurotransmitters in the Human Brain* (Tracey, D. J., ed.), Plenum, New York, pp. 173–197.

8. Hoffer, B. and Olson, L. (1997) Treatment strategies for neurodegenerative diseases based on trophic factors and cell transplantation techniques. *J. Neural Transm.* **49,** 1–10.

9. Isacson, O., Costantini, L., Schumacher, J. M., Cicchetti, F., Chung, S., and Kim, K. (2001) Cell implantation therapies for Parkinson's disease using neural stem, transgenic or xenogeneic donor cells. *Parkinsonism Relat. Disord.* **7,** 205–212.

10. Freeman, T. B., Cicchetti, F., Hauser, R. A., Deacon, T. W., Li, S. J., Hersch, S. M., et al. (2000) Transplanted fetal striatum in Huntington's disease: phenotypic development and lack of pathology. *Proc. Natl. Acad. Sci. USA* **5, 97(25),** 13,877–13,882.

11. Kordower, J. H. and Sortwell, C. E. (2000) Neuropathology of fetal nigra transplants for Parkinson's disease. *Prog. Brain Res.* **127,** 333–344.

12. Bjorklund, A. (1992) Dopaminergic transplants in experimental parkinsonism: cellular mechanisms of graft-induced functional recovery. *Curr. Opin. Neurobiol.* **2,** 683–689.

13. Mahalik, T. J., Hahn, W. E., Clayton, G. H., and Owens, G. P. (1994) Programmed cell death in developing grafts of fetal substantia nigra. *Exp. Neurol.* **129,** 27–36.

14. Nikkhah, G., Bentlage, C., Cunningham, M. G., and Bjorklund, A. (1994) Intranigral fetal dopamine grafts induce behavioral compensation in the rat Parkinson model. *J. Neurosci.* **14,** 3449–3461.

15. Nikkhah, G. M., Cunningham, G., Jodicke, A., Knappe, U., and Bjorklund, A. (1994) Improved graft survival and striatal reinnervation by microtransplantation of fetal nigral cell suspensions in the rat Parkinson model. *Brain Res.* **633,** 133–143.

16. Horellou, P., Marlier, L., Privat, A., Darchen, F., Scherman, D., Henry, J. P., et al. (1990) Exogenous expression of L-dopa and dopamine in various cell lines following transfer of rat and human tyrosine hydroxylase cDNA: grafting in an animal model of Parkinson's disease. *Prog. Brain Res.* **82,** 23–32.

17. Freed, W. J., Adinol, A. M., Laskin, J. D., and Geller, H. M. (1989) Transplantation of B16/C3 melanoma cells into the brains of rats and mice. *Brain Res.* **485,** 349–362.

18. Kawaja, M. D., Fagan, A. M., Firestein, B. L., and Gage, F. H. (1991) Intracerebral grafting of cultured autologous skin fibroblasts into the rat striatum: an assessment of graft size and ultrastructure. *J. Comp. Neurol.* **307,** 695–706.

19. Nishino, H., Hashitani, T., and Kumazaki, M. (1990) Phenotypic plasticity of grafted catecholaminergic cells in the dopamine-depleted caudate nucleus in the rat. *Neurosci. Res.* **13(Suppl.),** S54–S60.

20. Schueler, S. B., Ortega, J. D., Sagen, J., and Kordower, J. H. (1993) Robust survival of isolated bovine adrenal chromaffin cells following intrastriatal

transplantation: a novel hypothesis of adrenal graft viability. *J. Neurosci.* **13,** 4496–4510.

21. Emerich, D. F., Winn, S. R., Christenson, L., Palmatier, M. A., Gentile, F. T., and Sanberg, P. R. (1992) A novel approach to neural transplantation in Parkinson's disease: use of polymer-encapsulated cell therapy. *Neurosci. Biobehav. Rev.* **16,** 437–447.

22. Studer, L., Tabar, V., and McKay, R. D. (1998) Transplantation of expanded mesencephalic precursors leads to recovery in parkinsonian rats. *Nat. Neurosci.* **1,** 290–295.

23. Svendsen, C. N., Caldwell, M. A., Shen, J., ter Borg, M. G., Rosser, A. E., Tyers, P., et al. (1997) Long-term survival of human central nervous system progenitor cells transplanted into a rat model of Parkinson's disease. *Exp. Neurol.* **148,** 135–146.

24. Kleppner, S. R., Robinson, K. A., Trojanowski, J. Q., and Lee, V. M. (1995) Transplanted human neurons derived from a teratocarcinoma cell line (NTera-2) mature, integrate, and survive for over 1 year in the nude mouse brain. *J. Comp. Neurol.* **357,** 618–632.

25. Miyazono, M., Nowell, P. C., Finan, J. L., Lee, V. M., and Trojanowski, J. Q. (1996) Long-term integration and neuronal differentiation of human embryonal carcinoma cells (NTera-2) transplanted into the caudoputamen of nude mice. *J. Comp. Neurol.* **376,** 603–613.

26. Miyazono, M., Lee, V. M., and Trojanowski, J. Q. (1995) Proliferation, cell death, and neuronal differentiation in transplanted human embryonal carcinoma (NTera2) cells depend on the graft site in nude and severe combined immunodeficient mice. *Lab. Invest.* **73,** 273–283.

27. Pleasure, S. J. and Lee, V. M. (1993) NTera 2 cells: a human cell line which displays characteristics expected of a human committed neuronal progenitor cell. *J. Neurosci. Res.* **35,** 585–602.

28. Pleasure, S. J., Page, C., and Lee, V. M. (1992) Pure, postmitotic, polarized human neurons derived from NTera 2 cells provide a system for expressing exogenous proteins in terminally differentiated neurons. *J. Neurosci.* **12,** 1802–1815.

29. Lee, V. M. and Andrews, P. W. (1986) Differentiation of NTERA-2 clonal human embryonal carcinoma cells into neurons involves the induction of all three neurofilament proteins. *J. Neurosci.* **6,** 514–521.

30. Lee, V. M., Hartley, R. S., and Trojanowski, J. Q. (2000) Neurobiology of human neurons (NT2N) grafted into mouse spinal cord: implications for improving therapy of spinal cord injury. *Prog. Brain Res.* **128,** 299–307.

31. Flax, J. D., Aurora, S., Yang, C., Simonin, C., Wills, A. M., Billinghurst, L. L., et al. (1998) Engraftable human neural stem cells respond to developmental cues, replace neurons, and express foreign genes. *Nat. Biotechnol.* **16,** 1033–1039.

32. Borlongan, C. V., Tajima, Y., Trojanowski, J. Q., Lee, V. M., and Sanberg, P. R. (1998) Transplantation of cryopreserved human embryonal carcinoma-derived neurons (NT2N cells) promotes functional recovery in ischemic rats. *Exp. Neurol.* **149,** 310–321.

33. Borlongan, C. V., Tajima, Y., Trojanowski, J. Q., Lee, V. M., and Sanberg, P. R. (1998) Cerebral ischemia and CNS transplantation: differential effects of grafted fetal rat striatal cells and human neurons derived from a clonal cell line. *NeuroReport* **9,** 3703–2709.

34. Saporta, S., Borlongan, C. V., and Sanberg, P. R. (1999) Neural transplantation of human neuroteratocarcinoma (hNT) neurons into ischemic rats. A quantitative dose-response analysis of cell survival and behavioral recovery. *Neuroscience* **91,** 519–525.

35. Trojanowski, J. Q., Mantione, J. R., Lee, J. H., Seid, D. P., You, T., Inge, L. J., et al. (1993) Neurons derived from a teratocarcinoma cell line establish molecular and structural polarity following transplantation into the rodent brain. *Exp. Neurol.* **122,** 283–294.

36. Henderson, B. T. H., Clough, C. G., Hughes, R. C., Hitchcock, E. R., and Kenny, B. G. (1991) Implantation of human ventral mesencephalon to the right caudate nucleus in advanced Parkinson's disease. *Arch. Neurol.* **48,** 822–827.

37. Freed, C. R., Breeze, R. E., Rosenberg, N. L., Schneck, S. A., Kriek, E., Qi, J.-X., et al. (1992) Survival of implanted fetal dopamine cells and neurologic improvement 12 to 46 months after transplantation for Parkinson's disease. *N. Engl. J. Med.* **327,** 1549–1555.

38. Sarnacki, P. G., Engelman, R. W., Chang, Y., Day, N. K., Good, R. A., and Sanberg, P. R. (1998) Immunosuppression by hNT neurons and supernatant. *Exp. Neurol.* **153,** 386.

39. Hurlbert, M. S., Gianani, R. I., Hutt, C., Freed, C. R., and Kaddis, F. G. (1999) Neural transplantation of hNT neurons for Huntington's disease. *Cell Transplant.* **8,** 143–151.

40. Baker, K. A., Hong, M., Sadi, D., and Mendez, I. (2000) Intrastriatal and intranigral grafting of hNT neurons in the 6-OHDA rat model of Parkinson's disease. *Exp. Neurol.* **162,** 350–360.

41. Zigova, T., Willing, A. E., Tedesco, E. M., Borlongan, C. V., Saporta, S., Snable, G. L., et al. (1999) Lithium chloride induces the expression of tyrosine hydroxylase in hNT neurons. *Exp. Neurol.* **157,** 251–258.

42. Hartley, R. S., Trojanowski, J. Q., and Lee, V. M. (1999) Differential effects of spinal cord gray and white matter on process outgrowth from grafted human NTERA2 neurons (NT2N, hNT). *J. Comp. Neurol.* **415,** 404–418.

43. Philips, M. F., Muir, J. K., Saatman, K. E., Raghupathi, R., Lee, V. M., Trojanowski, J. Q., et al. (1999) Survival and integration of transplanted postmitotic human neurons following experimental brain injury in immunocompetent rats. *J. Neurosurg.* **90,** 116–124.

44. Connor, B. and Dragunow, M. (1998) The role of neuronal growth factors in neurodegenerative disorders of the human brain. *Brain Res. Rev.* **27,** 1–39.

45. Hughes, P. E., Alexi, T., Walton, M., Williams, C. E., Dragunow, M., Clark, R. G., et al. (1999) Activity and injury-dependent expression of inducible transcription factors, growth factors, and apoptosis-related genes within the central nervous system. *Prog. Neurobiol.* **57,** 421–450.

46. Lin, S. Z., Chiang, Y. H., Wang, Y., Hayashi, T., Su, T. P., Hoffer, B. J., et al. (1999) Transplantation for chronic stroke: fetal cortical cells, fetal kidney cells and human derived cell line (hNT) neurons as graft source. *Soc. Neurosci. Abstr.* **29,** 1506.

47. Borlongan, C. V. and Freeman, T. B. (2000) Transplantation of human cultured neurons protects against 6-hydroxydopamine-induced parkinsonism in adult rats. *Soc. Neurosci. Abstr.* **30,** 209.

48. Tomac, A. C., Grinberg, A., Huang, S. P., Nosrat, C., Wang, Y., Borlongan, C., et al. (2000) Glial cell line-derived neurotrophic factor receptor alpha1 availability regulates glial cell line-derived neurotrophic factor signaling:evidence from mice carrying one or two mutated alleles. *Neuroscience* **95,** 1011–1023.

49. Chiang, Y. H., Lin, S. Z., Borlongan, C. V., Hoffer, B. J., Morales, M., and Wang, Y. (1999) Transplantation of fetal kidney tissue reduces cerebral infarction induced by middle cerebral artery ligation. *J. Cereb. Blood Flow Metab.* **19,**1329–1335.

50. Schinstine, M., Stull, N. D., and Iacovitti, L. (1996) Induction of tyrosine hydroxylase in hNT neurons. *Soc. Neurosci. Abstr.* **22,** 1959.

51. Iacovitti, L. and Stull, N. D. (1997) Expression of tyrosine hydroxylase in newly differentiated neurons from a human cell line (hNT). *NeuroReport* **8,** 1471–1474.

52. Borlongan, C. V., Saporta, S., Poulos, S. G., Othberg, A., and Sanberg, P. R. (1998) Viability and survival of hNT neurons determine degree of functional recovery in grafted ischemic rats. *NeuroReport* **9,** 2837–2842.

53. Tornatore, C., Baker-Cairns, B., Yadid, G., Hamilton, R., and Meyers, K. (1996) Expression of tyrosine hydroxylase in an immortalized human fetal astrocyte cell line; in vitro characterization and engraftment into the rodent striatum. *Cell Transplant.* **5,** 145–163.

54. Kondziolka, D., Wechsler, L., Goldstein, S., Meltzer, C., Thulborn, K. R., Gebel, J., et al. (2000) Transplantation of cultured human neuronal cells for patients with stroke. *Neurology* **55,** 565–569.

55. Zivin, J. A. (2000) Cell transplant therapy for stroke: hope or hype. *Neurology* **55,** 467.

Index